Echinoderm Aquaculture

Echinoderm Aquaculture

Editor:

Bruno Augusto Amato Borges

Delve Publishing

www.delvepublishing.com

Echinoderm Aquaculture

Editor: Bruno Augusto Amato Borges

708 3rd Avenue, 6th Floor

New York, NY 10017

United States of America

www.delvepublishing.com

ISBN: 978-1-68095-848-5

Library of Congress Control Number: 2016951601

About the Editor

Bruno Augusto Amato Borges

Bruno Augusto is Aquaculture Engineer from Federal University of Santa Catarina. His focus is on writing projects and development of new technologies in aquaculture. He is currently project service provider, consultant and develops research in the field of Biofloc Technology (BFT) for freshwater fish farming.

Contents

List of Contributors

Kyle A. Long
Whitney Laboratory for Marine Bioscience, University of Florida, 9505 Ocean Shore Blvd., St Augustine, FL 32080, USA

Carlos W. Nossa
Department of Ecology and Evolutionary Biology, Rice University, Houston, TX 77251-1892, USA

Mary A. Sewell
School of Biological Sciences, University of Auckland, Auckland, New Zealand

Nicholas H. Putnam
Department of Ecology and Evolutionary Biology, Rice University, Houston, TX 77251-1892, USA

Joseph F. Ryan
Whitney Laboratory for Marine Bioscience, University of Florida, 9505 Ocean Shore Blvd., St Augustine, FL 32080, USA
Department of Biology, University of Florida, Gainesville, FL 32611-8525, USA.

Vladimir S Mashanov,
Department of Biology, University of Puerto Rico, PR 00936-8377 San Juan, USA

Olga R Zueva
Department of Biology, University of Puerto Rico, PR 00936-8377 San Juan, USA

José E García-Arrarás
Department of Biology, University of Puerto Rico, PR 00936-8377 San Juan, USA

Kristen A Yankura,
Department of Biological Sciences, Carnegie Mellon University, Pittsburgh, PA 15213, USA

Megan L Martik,
Department of Biological Sciences, Carnegie Mellon University, Pittsburgh, PA 15213, USA

Charlotte K Jennings,
Department of Biological Sciences, Carnegie Mellon University, Pittsburgh, PA 15213, USA

Veronica F Hinman
Department of Biological Sciences, Carnegie Mellon University, Pittsburgh, PA 15213, USA

Aifu Yang,
Liaoning Key Lab of Marine Fishery Molecular Biology, Liaoning Ocean and Fisheries Science Research Institute, Dalian, Liaoning 116023, Peoples' Republic of China

Zunchun Zhou,
Liaoning Key Lab of Marine Fishery Molecular Biology, Liaoning Ocean and Fisheries Science Research Institute, Dalian, Liaoning 116023, Peoples' Republic of China

Yongjia Pan,
Liaoning Key Lab of Marine Fishery Molecular Biology, Liaoning Ocean and Fisheries Science Research Institute, Dalian, Liaoning 116023, Peoples' Republic of China

Jingwei Jiang,
Liaoning Key Lab of Marine Fishery Molecular Biology, Liaoning Ocean and Fisheries Science Research Institute, Dalian, Liaoning 116023, Peoples' Republic of China

Ying Dong,
Liaoning Key Lab of Marine Fishery Molecular Biology, Liaoning Ocean and Fisheries Science Research Institute, Dalian, Liaoning 116023, Peoples' Republic of China

Xiaoyan Guan,
Liaoning Key Lab of Marine Fishery Molecular Biology, Liaoning Ocean and Fisheries Science Research Institute, Dalian, Liaoning 116023, Peoples' Republic of China

Hongjuan Sun,
Liaoning Key Lab of Marine Fishery Molecular Biology, Liaoning Ocean and Fisheries Science Research Institute, Dalian, Liaoning 116023, Peoples' Republic of China

Shan Gao
Liaoning Key Lab of Marine Fishery Molecular Biology, Liaoning Ocean and Fisheries Science Research Institute, Dalian, Liaoning 116023, Peoples' Republic of China

Zhong Chen
Liaoning Key Lab of Marine Fishery Molecular Biology, Liaoning Ocean and Fisheries Science Research Institute, Dalian, Liaoning 116023, Peoples' Republic of China

Pablo A Ortiz-Pineda
University of Puerto Rico, Rio Piedras, Department of Biology, San Juan, PR, USA

Francisco Ramírez-Gómez
University of Puerto Rico, Rio Piedras, Department of Biology, San Juan, PR, USA

Judit Pérez-Ortiz
University of Puerto Rico, Rio Piedras, Department of Biology, San Juan, PR, USA

Sebastián González-Díaz
University of Puerto Rico, Rio Piedras, Department of Biology, San Juan, PR, USA

Francisco Santiago-De Jesús
University of Puerto Rico, Rio Piedras, Department of Biology, San Juan, PR, USA

Josue HernándezPasos
University of Puerto Rico, Rio Piedras, Department of Biology, San Juan, PR, USA

Cristina Del Valle-Avila
University of Puerto Rico, Rio Piedras, Department of Biology, San Juan, PR, USA

Carmencita Rojas-Cartagena
University of Puerto Rico, Rio Piedras, Department of Biology, San Juan, PR, USA

Edna C Suárez-Castillo
Yale University, Department of Molecular, Cellular and Developmental Biology, New Haven, CT, USA

Karen Tossas
University of Puerto Rico, Rio Piedras, Department of Biology, San Juan, PR, USA

Ana T Méndez-Merced
Universidad del Este- SUAGM, School of Science and Technology, Carolina, PR, USA

José L RoigLópez
University of Puerto Rico, Rio Piedras, Department of Biology, San Juan, PR, USA

Humberto Ortiz-Zuazaga
University of Puerto Rico, HPCf, High Performance Computing facility, San Juan, PR, USA

José E García-Arrarás
University of Puerto Rico, Rio Piedras, Department of Biology, San Juan, PR, USA

Bei Zhang
College of Food Science and Engineering, Ocean University of China, Qingdao, China

Changhu Xue
College of Food Science and Engineering, Ocean University of China, Qingdao, China

Xiaoqian Hu
College of Food Science and Engineering, Ocean University of China, Qingdao, China

Jie Xu
College of Food Science and Engineering, Ocean University of China, Qingdao, China

Zhaojie Li
College of Food Science and Engineering, Ocean University of China, Qingdao, China

Jingfeng Wang
College of Food Science and Engineering, Ocean University of China, Qingdao, China

Teruyoshi Yanagita
Department of Applied Biological Sciences, Saga University, Saga, Japan

Yong Xue
College of Food Science and Engineering, Ocean University of China, Qingdao, China

Yuming Wang
College of Food Science and Engineering, Ocean University of China, Qingdao, China

Domenico Schillaci
[1]Department STEBICEF, Università degli Studi di Palermo, Via Archirafi, 32, 90123, Palermo, Italy

Maria Grazia Cusimano
Department STEBICEF, Università degli Studi di Palermo, Via Archirafi, 32, 90123, Palermo, Italy

Vincenzo Cunsolo
Department Scienze chimiche, Università degli Studi di Catania, Viale A. Doria, 6, Catania, Italy

Rosaria Saletti
Department Scienze chimiche, Università degli Studi di Catania, Viale A. Doria, 6, Catania, Italy

Debora Russo
Department STEBICEF, Università degli Studi di Palermo, Via Archirafi, 32, 90123, Palermo, Italy

Mirella Vazzana
Department STEBICEF, Università degli Studi di Palermo, Via Archirafi, 32, 90123, Palermo, Italy

Maria Vitale
Istituto Zooprofilattico Sperimentale della Sicilia "A. Mirri", Via G. Marinuzzi, 3, 90129, Palermo, Italy

Vincenzo Arizza
Department STEBICEF, Università degli Studi di Palermo, Via Archirafi, 32, 90123, Palermo, Italy
Istituto EuroMediterraneo di Scienza e Tecnologia, Palermo, Italy

José E San Miguel-Ruiz
Department of Biology, University of Puerto Rico, Rio Piedras, Puerto Rico

Angel R Maldonado-Soto
Department of Biology, University of Puerto Rico, Rio Piedras, Puerto Rico

José E García-Arrarás
Department of Biology, University of Puerto Rico, Rio Piedras, Puerto Rico

Vladimir S Mashanov
Department of Biology, University of Puerto Rico, San Juan, PR 00936-8377, USA

Olga R Zueva,
Department of Biology, University of Puerto Rico, San Juan, PR 00936-8377, USA

Carmencita Rojas-Catagena
Department of Biology, University of Puerto Rico, San Juan, PR 00936-8377, USA

Jose E Garcia-Arraras
Department of Biology, University of Puerto Rico, San Juan, PR 00936-8377, USA

Ying Fan
Mariculture Institute of Shandong Province, Shandong Province Key Laboratory for Disease Control in Mariculture, No. 47 Guizhou Road, Qingdao 266002, Shandong, China

Xiaoqing Yu
Mariculture Institute of Shandong Province, Shandong Province Key Laboratory for Disease Control in Mariculture, No. 47 Guizhou Road, Qingdao 266002, Shandong, China

La Xu
Mariculture Institute of Shandong Province, Shandong Province Key Laboratory for Disease Control in Mariculture, No. 47 Guizhou Road, Qingdao 266002, Shandong, China

Shuxian Wang
Mariculture Institute of Shandong Province, Shandong Province Key Laboratory for Disease Control in Mariculture, No. 47 Guizhou Road, Qingdao 266002, Shandong, China

Haibin Ye
Mariculture Institute of Shandong Province, Shandong Province Key Laboratory for Disease Control in Mariculture, No. 47 Guizhou Road, Qingdao 266002, Shandong, China

Jing Diao
Mariculture Institute of Shandong Province, Shandong Province Key Laboratory for Disease Control in Mariculture, No. 47 Guizhou Road, Qingdao 266002, Shandong, China

Xiusheng Yang
Mariculture Institute of Shandong Province, Shandong Province Key Laboratory for Disease Control in Mariculture, No. 47 Guizhou Road, Qingdao 266002, Shandong, China

Tianbao Li
Mariculture Institute of Shandong Province, Shandong Province Key Laboratory for Disease Control in Mariculture, No. 47 Guizhou Road, Qingdao 266002, Shandong, China

Andreas Heyland
University of Guelph, 50 Stone Road East, Guelph, ON N1G 2W1, Canada

Jason Hodin
Hopkins Marine Station of Stanford University, Pacific Grove, CA 93950, USA

Margherita Perillo
Biology and Evolution of Marine Organisms, Stazione Zoologica Anton Dohrn, Napoli 80121, Italy
Present address: Department of Biology, Boston College, Chestnut Hill, MA, USA.

Yue Julia Wang
Department of Surgery and the McKusick Nathans Institute for Genetic Medicine, Johns Hopkins University, Baltimore, MD 21205, USA

Steven D. Leach
Department of Surgery and the McKusick Nathans Institute for Genetic Medicine, Johns Hopkins University, Baltimore, MD 21205, USA

Maria Ina Arnone
Biology and Evolution of Marine Organisms, Stazione Zoologica Anton Dohrn, Napoli 80121, Italy

Phoebe L Stewart
Department of Molecular Physiology and Biophysics, Vanderbilt University Medical Center, Nashville, TN USA
Department of Molecular and Medical Pharmacology, Crump Institute for Molecular Imaging, David Geffen School of Medicine at UCLA, Los Angeles, CA USA

Miriam Makabi
Department of Molecular and Medical Pharmacology, Crump Institute for Molecular Imaging, David Geffen School of Medicine at UCLA, Los Angeles, CA USA

Jennifer Lang
Department of Molecular Biosciences, University of Kansas, Lawrence, KS USA

Carrie Dickey-Sims
Stowers Institute for Medical Research, Kansas City, MO USA

Anthony J Robertson
Stowers Institute for Medical Research, Kansas City, MO USA

James A Coffman
Stowers Institute for Medical Research, Kansas City, MO USA

Kathy A Suprenant
Department of Molecular Biosciences, University of Kansas, Lawrence, KS USA

Syuto Hasegawa
Laboratory of Comparative Physiology, Graduate School of Fisheries Science, Hokkaido University, 3-1-1 Minato-cho, Hakodate 041-8611, Japan

Kazuhiro Ura
Graduate School of Fisheries Science, Hokkaido University, 3-1-1 Minato-cho, Hakodate 041-8611, Japan

Hiroyuki Tanaka
Laboratory of Marine Molecular Biology, Graduate School of Fisheries Science, Hokkaido University, 3-1-1 Minato-cho, Hakodate 041-8611, Japan

Takao Ojima
Laboratory of Marine Molecular Biology, Graduate School of Fisheries Science, Hokkaido University, 3-1-1 Minato-cho, Hakodate 041-8611, Japan

Yasuaki Takagi
Laboratory of Comparative Physiology, Graduate School of Fisheries Science, Hokkaido University, 3-1-1 Minato-cho, Hakodate 041-8611, Japan

Domenico Schillaci
Dip. STEBICEF, Università degli Studi di Palermo, Via Archirafi, 32-90123 Palermo, Italy

Maria Grazia Cusimano
Dip. STEBICEF, Università degli Studi di Palermo, Via Archirafi, 32-90123 Palermo, Italy

Angelo Spinello
Dip. STEBICEF, Università degli Studi di Palermo, Via Archirafi, 32-90123 Palermo, Italy
IEMEST, Istituto Euromediterraneo di Scienza e Tecnologia, Via Emerico Amari, 123-90139 Palermo, Italy

Giampaolo Barone
Dip. STEBICEF, Università degli Studi di Palermo, Via Archirafi, 32-90123 Palermo, Italy
IEMEST, Istituto Euromediterraneo di Scienza e Tecnologia, Via Emerico Amari, 123-90139 Palermo, Italy

Debora Russo
Dip. STEBICEF, Università degli Studi di Palermo, Via Archirafi, 32-90123 Palermo, Italy
IEMEST, Istituto Euromediterraneo di Scienza e Tecnologia, Via Emerico Amari, 123-90139 Palermo, Italy

Maria Vitale
Istituto Zooprofilattico Sperimentale della Sicilia "A. Mirri", Via G. Marinuzzi, 3-90129 Palermo, Italy.

Daniela Parrinello
Dip. STEBICEF, Università degli Studi di Palermo, Via Archirafi, 32-90123 Palermo, Italy

Vincenzo Arizza
Dip. STEBICEF, Università degli Studi di Palermo, Via Archirafi, 32-90123 Palermo, Italy
IEMEST, Istituto Euromediterraneo di Scienza e Tecnologia, Via Emerico Amari, 123-90139 Palermo, Italy

Camilla Björn
Pergamum AB, Arvid Wallgrens Backe 20, 413 46 Gothenburg, Sweden

Joakim Håkansson
Pergamum AB, Arvid Wallgrens Backe 20, 413 46 Gothenburg, Sweden
Present address: SP Technical Research Institute of Sweden, Medical Device Technology, Brinellgatan 4, 504 62 Borås, Sweden

Emma Myhrman
Pergamum AB, Arvid Wallgrens Backe 20, 413 46 Gothenburg, Sweden

Veronika Sjöstrand
Pergamum AB, Arvid Wallgrens Backe 20, 413 46 Gothenburg, Sweden

Tor Haug
Norwegian College of Fishery Science, Faculty of Biosciences, Fisheries and Economics, University of Tromsø, Tromsø N-9037, Norway

Kerstin Lindgren
Pergamum AB, Arvid Wallgrens Backe 20, 413 46 Gothenburg, Sweden

Hans-Matti Blencke
Norwegian College of Fishery Science, Faculty of Biosciences, Fisheries and Economics, University of Tromsø, Tromsø N-9037, Norway

Klara Stensvåg
Norwegian College of Fishery Science, Faculty of Biosciences, Fisheries and Economics, University of Tromsø, Tromsø N-9037, Norway

Margit Mahlapuu
Pergamum AB, Arvid Wallgrens Backe 20, 413 46 Gothenburg, Sweden

Marc Baeta
Institut de Ciències del Mar (CSIC), Passeig Marítim de la Barceloneta, 3749, 08003 Barcelona, Spain
Universitat Autònoma de Barcelona (UAB), Campus de Bellaterra, Edifici B, 08193 Cerdanyola del Vallès, Barcelona, Spain

Eve Galimany
Institut de Ciències del Mar (CSIC), Passeig Marítim de la Barceloneta, 3749, 08003 Barcelona, Spain
Instituto Español de Oceanografía, Centre Oceanogràfic de Balears, Moll de Ponent s/n, 07015 Palma de Mallorca, Spain

Montserrat Ramón
Institut de Ciències del Mar (CSIC), Passeig Marítim de la Barceloneta, 3749, 08003 Barcelona, Spain
Instituto Español de Oceanografía, Centre Oceanogràfic de Balears, Moll de Ponent s/n, 07015 Palma de Mallorca, Spain

Roy Vaughn
Department of Cell Biology, Microbiology and Molecular Biology, University of South Florida, 4202 E. Fowler Ave, Tampa, FL 33620, USA

Nancy Garnhart
Hubbard Center for Genome Studies, University of New Hampshire, 35 Colovos Rd, Durham, NH 03824, USA

James R Garey
Department of Cell Biology, Microbiology and Molecular Biology, University of South Florida, 4202 E. Fowler Ave, Tampa, FL 33620, USA

W Kelley Thomas
Hubbard Center for Genome Studies, University of New Hampshire, 35 Colovos Rd, Durham, NH 03824, USA

Brian T Livingston
Department of Biological, Sciences, California State University Long Beach, 1250 Bellflower Blvd, Long Beach, CA 90815, USA.

Anna Czarkwiani
Department of Genetics, Evolution and Environment, University College London, London, UK

Cinzia Ferrario
Department of Biosciences, University of Milan, Milan, Italy

David Viktor Dylus
Department of Genetics, Evolution and Environment, University College London, London, UK
Centre for Mathematics, Physics and Engineering in the Life Sciences and Experimental Biology, University College London, London, UK
Present address: Department of Ecology and Evolution & Center for Integrative Genomics, University of Lausanne, Lausanne, Switzerland

Michela Sugni
Department of Biosciences, University of Milan, Milan, Italy

Paola Oliveri
Department of Genetics, Evolution and Environment, University College London, London, UK
Research Department of Genetics, Evolution and Environment, University College London, Room 426, Darwin Building, Gower Street, London WC1E 6BT, UK

Preface

The cultivation of echinoderms is relatively little widespread as the cultivation of other aquatic organisms such as fish, shrimp and molluscs. According to FAO (2014) the world production of sea urchins and other echinoderms was 209,000 tons, accounting for 0.73% of world aquaculture production, but production of this group of animals was one of the more grew in the past nine years, with a growth of 231.9%.

Currently, the main producers of echinoderms are China, New Zealand, Norway, North America and Scotland. The sea urchin and sea cucumber are the most cultivated species. The marketing and export of echinoderms have increased a lot in the last 15 years in countries such as Ireland, the United States and Canada. The product is exported, especially to the countries of Asia, being its main consumer Japan. The echinoderms are used for a variety of applications from the food to the retardation of the development of diseases in humans. A food very exploited by market is the gonad of sea urchin, which reaches high values on the market and is consumed in several countries all over the world. According to scientists, studying the sea urchin can help in the search for treatments for diseases such as cancer, infertility, blindness and muscular dystrophy. Was discovered an extreme similarity between the human and the sea urchin genome. Another work, developed by British scientists has shown that the sea star can help in the discovery of new treatments for diseases such as asthma, fever and arthritis, besides having a strong anti-inflammatory feature. In addition, the consumption of sea cucumbers can contribute to the reduction of cardiovascular diseases, according to the Center of Marine Science at the University of Algarve. A lot of researches being done in search of drugs present in the oceans, sea animals produce an immense amount of compounds with different properties, so they must be encouraged research and improvement of knowledge in the area that presents immense potential for human health. For the responsible and sustainable development of this activity, must be evaluated and studied the populations and the management of the environment and of the natural stocks seeking prosperity in the use of this feature. This book gathers knowledge about the biology, ecology and human health applications, showing the great potential of the echinoderms in aquaculture.

Editor
Bruno Augusto Amato Borges

INTRODUCTION

Echinoderms are a very old phylum of marine animals whose name means "spiny skin" in Greek. This name is a misnomer, however, because not all echinoderms have spines. Their more fitting universal trait is common ancestry, including a unique water-based vascular system, and frequent five-fold symmetry. Though they do not always display five-fold symmetry (sea cucumbers are echinoderms, and they have bilateral symmetry), echinoderms are known for often playing games with the typical bilateral symmetry trend, as in sea urchins (radial symmetry), and the numerous starfish and sand dollars (five-fold symmetry). Echinoderms are one of several phyla that is exclusively marine.

The first known echinoderm is thought to be Arkarua, an ancient disc-like fossil about 1 cm in diameter with a five-fold pattern of dents that leads scientists to classify it as a likely echinoderm. This fossil dates to the Late Ediacaran, about 550 million years ago. Other than that, the first certain echinoderms appear in the Early Cambrian, about 530 million years ago. Containing 7,000 living species, echinoderms are the second-largest phylum of deuterostomes after the chordates (vertebrates), which are the dominant large-bodied phylum on land.

Very flexible, echinoderms are found at every ocean depth, from the intertidal zone to the abyssal zone, miles upon miles beneath the surface. There are two primary subphyla of echinoderms; the motile Eleutherozoa, which includes starfish, brittle stars, sea urchins, sand dollars, sea daisies, and sea cucumbers; and the sessile Pelmatozoa, which includes the crinoids (feather stars). The mobile subphyla crawls along the ocean floor using a muscular foot, and is specialized to consume bottom-dwellers that few other ocean animals can.

Echinoderms are significant because they are among the only large animals capable of surviving in the utter desert that characterizes the vast majority of the world's oceans. Their skeletons readily fossilize and provide important biogeographically information to paleontologists. Many limestone formations are made out of echinoderm skeletons, and some paleontologists believe that the evolutionary radiation of echinoderms was responsible for a sudden increase in the diversity of marine Mesozoic life.

Echinoderms such as sea urchins and sea cucumbers have been consumed all over the world for many years. They are highly valuable seafood which has led to the over exploitation of many wild stocks. In the past few decades, aquaculture of echinoderms has become a viable alternative to cope with the high international market demand and also to reduce pressure on wild populations. A reduction in the number of predators can cause sea urchins to overgraze vegetation and create 'urchin barrens'. This fundamentally changes an area's ecology and reduces local biodiversity. If, conversely, they are removed from an area by overfishing, their population can take a long time to recover. It is important to maintain a sustainable environment for sea urchins where they are taken from the wild. Aquaculture can be used to produce sea urchins for food and urchin barrens can be reseeded to help repopulate affected areas.

Sea urchins only live in marine waters and are widely distributed in the seas and oceans from the equator to the Polar Regions. There are about 850 extant species of sea urchin, and at least 17 species are commercially fished. Many species play a key role in the ecology of their environment.

Sea urchins are widely distributed in polar, temperate, and tropical oceans, where they are conspicuous members of most benthic marine communities. They play an important eco-logical role as herbivorous grazers, and their ability to alter algal community states has made them the subject of numerous ecological. Sea urchins are also used as a model organism in developmental studies and in schools to demonstrate cell division and early development: the purple urchin. Strongylocentroms purpuratus, was one of the first animal species to have its entire genome sequenced. Sea urchins and sea cucumbers are highly sought after delicacies growing in popularity globally. The demand for these species is rapidly outpacing natural stocks, and researchers and seafood industry personnel are now looking towards aquaculture as a means of providing a sustainable supply of these organism. Echinoderm Aquaculture is a practical reference on the basic biology and current culture practices for a wide range of geographically diverse echinoderm species.

There is an estimated total of 10 to 30 million species of organ-isms now living on earth. Of these, only about 770,000 plant and 1.5 million animal species are known to scientists. Most of the undescribed species are tropical insects and other small invertebrates. About 96% of the known plants and 85% of the animals live in terrestrial environments. Of these, only 3,000 plant species are cultivated and very few have great importance. Very few animals are cultivated. The bulk of our plant food comes from only seven species: wheat, rice, maize (corn), potato, barley, sweet potato, and cassava. Similarly, only seven categories of domesticated terrestrial animals provide most of our meat production: swine, cattle, poultry (chicken, duck, and turkey), sheep, goat, buffalo, and horse. There are about 20,000 species of fish, 200,000 species of aquatic invertebrates, and 14,000 species of aquatic plants. Very few of these are subjects of aquaculture, but relatively speaking, a greater diversity of aquatic species than of terrestrial species is cultivated. About 500 aquatic species have been cultivated commercially, although very few account for significant tonnages. A recent survey showed that at least 314 finfishes, 74 crustaceans, 69 mollusks, 43 algae, 13 angiosperms (including rice), 12 sponges, nine amphibians, four reptiles, three rotifers, two annelids, two mammals, and one echinoderm are subjects of aquaculture. Most of the important cultured spices are poikilothennic and carnivorous. Most species are virtually identical to wild stocks, and in many aquaculture species, seed stock is actually obtained from the wild.

Characteristics of Echinoderms

Echinoderms are characterized by radial symmetry, several arms (5 or more, mostly grouped 2 left - 1 middle - 2 right) radiating from a central body (= pentamerous). The body actually consists of five equal segments, each containing a duplicate set of various internal organs. They have no heart, brain, nor eyes, but some brittle stars seem to have light sensitive parts on their arms. Their mouth is situated on the underside and their anus on top (except feather stars, sea cucumbers and some urchins).

Echinoderms have tentacle-like structures called tube feet with suction pads situated at their extremities. These tube feet are hydraulically controlled by a remarkable vascular system. This system supplies water through canals of small muscular tubes to the tube feet (= ambulacral feet). As the tube feet press against a moving object, water is withdrawn from them, resulting in a suction effect. When water returns to the canals, suction is released. The resulting locomotion is generally very slow.

Ecology and range of Echinoderms

Echinoderms are exclusively marine. They occur in various habitats from the intertidal zone down to the bottom of the deep sea trenches and from sand to rubble to coral reefs and in cold and tropical seas.

Behavior of Echinoderms

Some echinoderms are carnivorous (for example starfish) others are detritus foragers (for example some sea cucumbers) or planktonic feeders (for example basket stars). Reproduction is carried out by the release of sperm and eggs into the water. Most species produce pelagic (= free floating) planktonic larvae which feed on plankton. These larvae are bilaterally symmetrical, unlike their parents (illustration of a larvae of a sea star below). When they settle to the bottom they change to the typical echinoderm features.

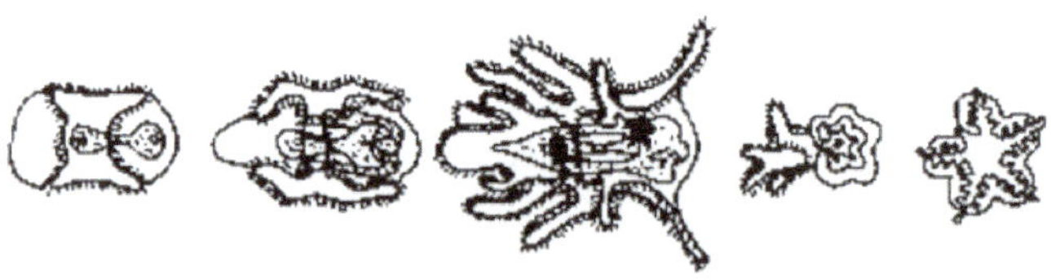

Echinoderms can regenerate missing limbs, arms, spines - even intestines (for example sea cucumbers). Some brittle stars and sea stars can reproduce asexually by breaking a ray or arm or by deliberately splitting the body in half. Each half then becomes a whole new animal.

Echinoderms are protected through their spiny skins and spines. But they are still preyed upon by shells (like the triton shell), some fish (like the trigger fish), crabs and shrimps and by other echinoderms like starfish which are carnivorous. Many echinoderms only show themselves at night (= nocturnal), therefore reducing the threat from the day time predators.

Echinoderms serve as hosts to a large variety of symbiotic organisms including shrimps, crabs, worms, snails and even fishes.

Characteristics of sea stars (or starfish)

Sea stars are characterized by radial symmetry, several arms (5 or multiplied by 5) radiating from a central body. Mouth and anus are close together on the underside, the anus is at the center of the disc together with the water intake (madreporite). The upper surface is often very colorful. Minute pincer-like structures called pedicellaria are present. These structures ensure that the surface of the arms stay free from algae. The underside is often a lighter color.

There are a few starfish that have 6 or 7 arms, for example Echinaster luzonicus or Protoreaster, some even more like the elven-armed sea star (Coscinasterias calamaria). Others normally have 5 arms but now have more arms, because after an injury an arm divided and grew into two arms.

Ecology and range or sea stars

The starfish lives everywhere in the coral reef and on sand or rocks.

Behavior of sea stars

The majority of sea stars are carnivorous and feed on sponges, bryozoans, ascidians and molluscs. Other starfishes are detritus feeders (detritus = organically enriched film that covers rocks) or scavengers. Some starfish are specialized feeders, for example the crown-of-thorns that feeds on life coral polyps.

Starfish have no hard mouth parts to help them capture prey. The stomach is extruded over the prey, thus surrounding the soft parts with the digestive organs. Digestive juices are secreted and the tissue of the prey liquefied. The digested food mass, together with the stomach is then sucked back in. This method can be observed, if you turn around a starfish, that sits on prey or sand - you will see the stomach retreating.

Starfish are well known for their powers of regeneration. A complete new animal can grow from a small fragment such as a arm. In some species (Linckia multifora and Echinaster luzonicus) one of the arms will virtually pull itself away, regenerates and forms a new animal. Autotomy (self-amputation) usually is a protective function, losing the body part to escape a predator rather than being eaten. But here it serves as a form of asexual reproduction. In other species of sea stars (Allostichaster polyplax and Coscinasterias calamaria) the body is broken into unequal parts (= fission) then the missing limbs regenerate.

The crown-of-thorns (Acanthaster planci) is one of the largest and the most venomous starfishes. It can reach 50 cm diameter and has numerous (10 to 20) spiny arms with formidable thorn like toxic spines. Don't touch them! A species of small cardinalfishes (Siphamia fuscolineata) and a commersal shrimp (Perliclimenes soror) live among those spines. The crown-of-thorns feed on live coral polyps. They "graze" the corals which are left behind white and dead. Their predators are the giant triton shell (Charonia tritonis) and some puffer fish. Scientist have also found out, that some crown of thorns are deterred from eating the coral polyps by the small crabs living among the coral branches (Trapezia sp). These crabs defend their coral host by breaking them off at the pedicellaria. Other small crabs (Tetralia sp) only pinch the tube feets of the starfish. Crown of thorns prefer corals that are not hosts to these crabs.

The cushion star (Culcita nouvaeguineae) doesn't look like a starfish at all, more like a large sea urchin without spines. Its pentagonal appearance gives only the slightest indication that this organism is related to other starfish.

Characteristics of sea urchins

Radial symmetrical body with an external chitinous skeleton and a centrally located jaw (called Aristotle's lantern) with horny teeth. The mouth consists of a complex arrangement of muscles and plates surrounding the circular opening. The anus is located on the upper surface. Some sea urchins have a spherical, bulb like cloaca (to store fecal material) that protrudes from the anal opening. It can be withdrawn into the shell. Depending on the species, movable spines of various sizes and forms are attached to the body. These spines often are sharp, pointed and in some cases even venomous. Pincer like pedicellaria for grabbing small prey. Some pedicellaria are also poisonous.

Ecology and range of sea urchins

Rubble and sand. An abundance of sea urchins can be a sign for bad water conditions.

Behavior of sea urchins

Locomotion by tube feet but also by movement of the spines on the underside of the body. Sea urchins are generally nocturnal, during the day they hide in crevices. However some sea urchins such as Diadema sometimes form large aggregations in open exposed areas. Despite their sharp spines sea urchins are easy game for some fishes, particularly triggerfishes and puffers. A triggerfish grabs the sea urchin with its hard beak like mouth by the spines or it blows some water towards the sea urchin and turns it on its back. The underside of a sea urchin has much shorter spines and those are easily crushed. During the breeding season the body cavity is crammed with eggs or sperms. This is one of the main reasons urchins are so attractive to fish predators (Japanese also like them for the same reason).

Some sea urchins are camouflaged. They hold on with their tube feet onto some bottom debris like rubble or pieces of seagrass and carry them on their back. Some even carry live soft corals or anemones. Most sea urchins are algal grazers but some feed on sponges, bryozonans and ascidians and others on detritus (detritus = organically enriched film that covers rocks). The sexes are separate and the young are formed indirectly by the fusion of sperm and eggs released into the water.

Many animals live in symbiotic relation with sea urchins. Even on the poisonous spines of the fire urchin (Asthenosoma varium) small shrimps (Periclimenes colemani) can be found. One shrimp (Stegopontonia commensalis) is striped black and white lengthwise and perfectly camouflaged and lives in spines of the long-spined sea urchin (Diadema setosum). Some cardinalfishes and juvenile shrimpfishes also like to take shelter in-between these spines, but even small cuttlefish hide there. It has been observed, that they change their coloring also to black and white. Some flatworms wrap around the thicker spines of the diadema sea urchin (Echinothrix calamaris).

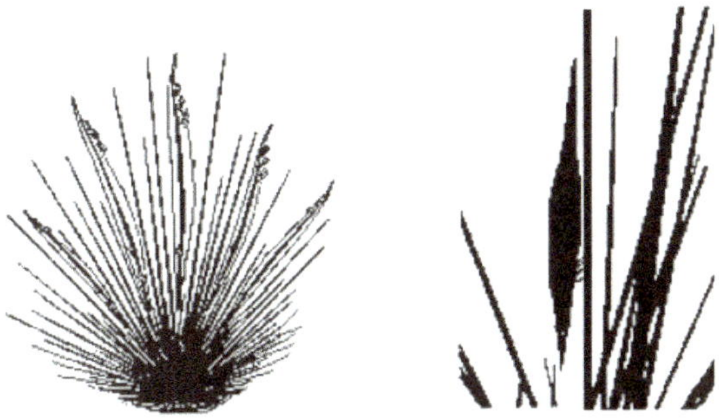

Characteristics of sea cucumbers

Unlike other echinoderms, holothurians don't have a distinct radial symmetry but are bilateral (distinct dorsal and ventral side). Holothurians are also called sea cucumbers. As their name suggests, they are cucumber shaped with an elongated, muscular, flexible body with a mouth at one end and the anus at the other. Around the mouth there is a number of tentacles (modified tube feet) used in food collecting. Sea cucumbers come in many sizes, from small species only a few centimeter in length to long snakelike animals which may stretch up to 2 meter!

Ecology and range of sea cucumbers

Rubble, rocks and sand. Also seen on some sponges in large aggregations.

Behavior of sea cucumbers

Most species feed on the rich organic film coating sandy surfaces. The crawl over the bottom ingesting sand. The edible particles (organic matter such as plankton, foraminifera and bacteria) are extracted when passing through their digestive tract and the processed sand is expelled from the anus (as worm-like excrements).

Sea cucumbers move by means of tube feet which extend in rows from the underside of the body. The tentacles surrounding the mouth are actually tube feet that have been modified for feeding.

Other holothurians feed on current-borne zooplankton. They bury in sand extruding their featherlike tentacles (Pseudocolochirus violaceus, Neothyondium magnum or Pentacta crassa). The tentacles have the same shape as soft corals or some anenemones. Large congregations of some small species are found on sponges. They apparently feed on substances secreted by the sponges as well as detritus from the surface.

Some species of holothurians have separate sexes others are hermaphrodites. The sea cucumbers hold on to exposed rocks or corals, raise their body to a upright position, rock back and forth and release the sperm and eggs into the sea.

Sea cucumbers have a remarkable capacity for regenerating their body parts. When attacked they shed a sticky thread like structure which is actually parts of their guts. The so called Cuverian threads are toxic (the poison is called holothurin) and can dissuade many potential predators. These structures quickly regenerate. (see photos below)

Holothurians host a variety of symbiotic organisms: crabs, shrimps, worms and even a very unusual fish. The pearlfish (Encheliophis homei and mourlani / Onuxodon margaritiferae) has a long slender, transparent body and lives in the gut cavity of the sea cucumber (Boshida argus, Thelanota ananas, Stichopus chloronotus). They also inhabit some starfish as well as pearl oyster shells. The fish leaves and enters (tail first) through the holothurian's anus. They probably feed on the gonads and other tissues of its host. It is said to leave at night to feed on small fishes and shrimps. Sea cucumbers are used in Asia as a base for soups.

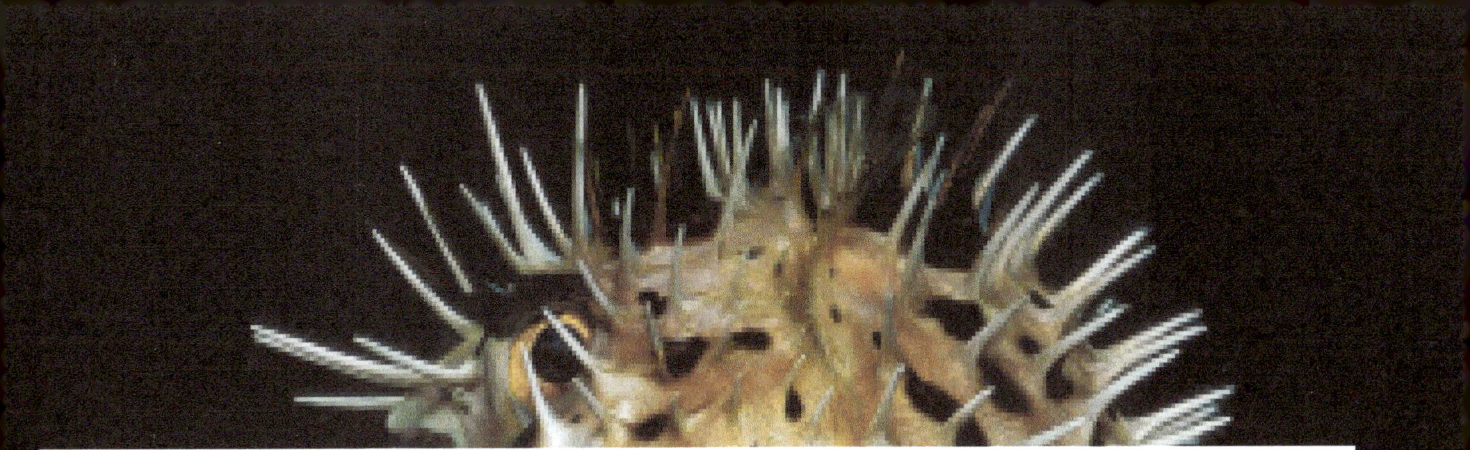

CHAPTER 1 ECHINODERM GENOMES

Kyle A. Long, Carlos W. Nossa, Mary A. Sewell, Nicholas H. Putnam and Joseph F. Ryan

Keywords: Echinoderms, Genome, Brittle star, Sea star, Sea cucumber

DATA DESCRIPTION

Echinodermata consists of five classes: Crinoidea (feather stars and sea lillies), Ophiuroidea (brittle stars and basket stars), Asteroidea (sea stars), Echinoidea (sea urchins, sea biscuits, and sand dollars), and Holothuroidea (sea cucumbers). These animals have a rich fossil record, unique biomechanical properties, experimentally tractable embryos, and as such have been a favorite subject of study for more than 150 years. Along with Hemichordata, Echinodermata form a clade called Ambulacraria that are the sister group of Chordata, which include vertebrates.

Echinoderms are easily recognized due to striking synapomorphies. The most obvious is their pentaradial (five-fold) body symmetry that is characteristic of adults (earlier stages exhibit bilateral symmetry). They have a unique water vascular system, which is characterized by canals connecting small tube feet on the lateral side of the animals. The water vascular system is used for essential functions such as feeding, locomotion, waste disposal, and respiration. The "spiny skin" from which these animals get their name is an endoskeleton made up of calcareous plates called ossicles that is composed of high magnesium calcite formed as solid test, plates or ossicles depending on the class of echinoderms. Lastly, the ossicles are connected by ligaments made of collagen that are normally rigid, but may become flexible upon various neuronal stimuli.

The vast majority of genomic work thus far has focused on the classic developmental model, the sea urchin *Stronglycentrotus purpuratus* [1]. The genome of this sea urchin has produced many important findings of great interest, including the discovery of a rearrangement event in evolution that led to an unusual Hox cluster organization [2], a well-characterized gene network for the specification of endoderm and mesoderm [3], and insight into the effect of ocean acidification on biomineralization [4]. More recently the genomes of the sea star *Patiria miniata*, the sea urchin *Lytechinus variegatus*, the sea cucumber *Parastichopus parvimensis*, and the brittle star *Ophiothrix spiculata* have been made available in Echinobase [5].

Here we provide genome sequences from species within three different echinoderm clades:*Australostichopus mollis*, commonly known as the brown sea cucumber (Fig. 1a), the brittle star *Ophionereis fasciata* (Fig. 1b), and *Patiriella regularis*, known as the New Zealand common cushion star (Fig. 1c). These species can be found in the shallow waters surrounding New Zealand.

Citation: Kyle A. Long, Carlos W. Nossa, Mary A. Sewell, Nicholas H. Putnam and Joseph F. Ryan, Echinoderm Genomes, DOI 10.1186/s13742-016-0125-6.

Figure 1: Photos of the echinoderm species whose genomes were sequenced in this study (a)*Australostichopus mollis*, commonly known as the brown sea cucumber; (b) the brittle star *Ophionereis fasciata* (b), and *Patiriella regularis*, known as the New Zealand common cushion star. Photo credits: (a) John A. Starmer, (b) Jennifer Howe, Victoria University of Wellington, and (c) username kiwi_kid on flickr (http://tinyurl.com/pregularis).

These data can be used for the gene family phylogenetic analyses, domain/gene losses, and presence of small non-coding RNAs among other applications (e.g., [6]). They will be particularly useful in a comparative framework with existing Echinoderm genomes, for example to identify highly conserved non-coding regions. Finally, these data will be a key resource for labs working on these animals in the lab or in the field (e.g., designing markers, probes, and genome-editing constructs).

SAMPLE PREPARATION AND SEQUENCING

Raw sequence data for all three species were produced from a single male, a single female, and the offspring of these two adults. For each individual sequenced, reads were barcoded and raw reads were submitted to the European Nucleotide Archive (ENA) separately. This sequencing strategy was originally used in order to employ a genetic mapping approach to genome assembly [7], and is not ideal for the standard*de novo* assembly strategy described herein. Nevertheless, when it became apparent that genetic mapping approach would not be feasible due to personnel, we applied a more traditional approach, and generated a useful set of data.

Parental tissue and larvae from each species were stored in ethanol prior to DNA extraction. DNA was extracted using Zymo DNA clean kit and was used to prepare Nextera libraries for 2 × 100 PE (paired end) sequencing on Illumina HiSeq2000 at BGI Shenzhen.

ASSEMBLY

For all steps of the assembly, all reads from each species were pooled (i.e., 2 parents and offspring). We used ALLPATHS-LG v44837 [8] to correct errors in the raw reads. We used Cutadapt v1.4.2 [9] to remove adapter sequences. We assembled these data using three assemblers: SOAPdenovo2 r2.04 [10], ABySS 3.81 [11], and Platanus 1.1.2 [12]. For each program we used a range of k-mer values: 31, 39, 45, 55, 63 for SOAP and ABySS, and 32, 39, 45, 55, 64 for Platanus. This provided a total of 15 assemblies for each organism. Each assembly/k-mer combination was evaluated using N50 values and number of conserved eukaryotic genes recovered by CEGMA 2.4 [13]. In all cases, SOAPdenovo assemblies were deemed superior to those produced by ABySS and Platanus (Table 1). The sequences represent coverages of 12.5x, 22.5x, and 21.4x for the brittle star, sea star, and sea cucumber, respectively.

Table 1: N50 values and CEGMA scores for each Assembly

Species	Assembler	K-mer Value	N50 (bp)	# Contigs	CEGMA Score (Partial/Complete)
O. fasciata	**SOAPdenovo**	31	484	3,968,282	69/16
		39	451	4,740,140	69/17
		45	**501**	**4,814,066**	**71/20**
		55	449	5,961,782	63/9
		63	528	4,184,863	57/5
	ABySS	31	70	32,912,859	34/6
		39	99	24,109,265	32/1
		45	116	19,642,981	26/1
		55	146	18,334,263	0/0
		63	166	13,153,732	0/0
	Platanus	32	103	126	0/0
		39	104	129	0/0
		45	108	119	0/0
		55	105	130	0/0
		64	111	119	0/0
P. regularis	**SOAPdenovo**	31	1383	19,728	1/0
		39	469	2,348,237	17/1
		45	488	2,596,707	15/0
		55	470	3,424,052	22/1
		63	**557**	**3,006,458**	**50/2**
	ABySS	31	70	14,653,947	12/1
		39	96	11,632,456	14/2
		45	116	9,818,773	15/3
		55	158	6,986,935	22/3
		63	191	5,068,075	19/4
	Platanus	32	155	1,923,405	0/0
		39	157	2,475,110	0/0
		45	161	2,814,106	0/0
		55	158	3,044,990	0/0
		64	143	3,092,862	0/0
A. mollis	**SOAPdenovo**	31	847	2,132,880	77/16
		39	564	3,162,372	62/12
		45	577	3,457,710	66/13
		55	475	4,839,379	66/17
		63	**626**	**3,712,641**	**87/18**
	ABySS	31	76	19,911,844	44/5
		39	104	15,084,417	48/5
		45	127	12,518,443	52/7
		55	171	8,798,225	42/5
		63	204	6,232,955	32/0
	Platanus	32	214	2,653,880	0/0
		39	182	3,763,387	0/0
		45	164	4,653,306	0/0
		55	104	102	0/0
		64	106	99	0/0

Assemblies deemed to be the best for each species based on N50 and CEGMA scores are in bold and were used for all downstream analyses

GENE PREDICTION

We used Augustus v3.0.3 [14] to generate *ab initio* gene predictions for the best assemblies of each of the echinoderm species. We created a training set with the *Strongylocentrotus purpuratus* v3.1 scaffolds and corresponding predicted gene models from Echinobase [5]. We used "generic," "human," and the custom model "strongylocentrotus_purpuratus" as values for the -species parameter. To compare the three different sets of gene predictions we BLASTed both human and *S. purpuratus* protein models against the protein predictions from Augustus using BLASTP v2.2.31+ [15] with an E-value cutoff of 1e-6 and limiting to a single target sequence (blastp -query query.fa -db aug.fa -evalue 1e-6 -outfmt 6 -max_target_seqs 1). For all species BLAST searches against the Augustus protein sequences generated with the custom *S. purpuratus* model resulted in the most hits. We therefore chose these predictions as our final sets.

We generated 49,301 *A. mollis*, 102,838 *O. fasciata*, and 1135 *P. regularis* gene predictions. In the case of *A. mollis* and *O. casciata* these numbers are substantially higher than the 22,709 published *S. purpuratus* gene models. The high number in*A. mollis* and *O. casciata* might be due to multiple fragmented predictions representing single genes. *P. regularis* has substantially fewer predictions, which might suggest that in most cases, the scaffolds were too short even to predict partial genes.

DATA AVAILABILITY

All data including sequencing reads and assemblies have been submitted to the ENA under the following project accessions: *Australostichopus mollis* = PRJEB10682; *Patiriella regularis* = PRJEB10600; *Ophionereis fasciata* = PRJEB10339. Supporting data is also archived in the*GigaScience* GigaDB database [16], and additional resources are available fromhttp://ryanlab.whitney.ufl.edu/genomes/.

ACKNOWLEDGEMENTS

The authors would like to acknowledge Mansi Srivastava who was the catalyst for the collaboration between these authors. Josefina Peters-Didier assisted MS with collections and spawning of A. mollis. KL received support from the National Science Foundation Research Experience for Undergraduates (REU) Program (DBI-1156528).

REFERENCES

1. Kondo M, Akasaka K. Current status of echinoderm genome analysis-what do we know? Curr Genomics. 2012;13(2):134.CrossRefPubMedPubMedCentral
2. Cameron RA, Rowen L, Nesbitt R, Bloom S, Rast JP, Berney K, et al. Unusual gene order and organization of the sea urchin hox cluster. J Exp Zool B Mol Dev Evol. 2006;306(1):45.CrossRefPubMed
3. Davidson EH, Rast JP, Oliveri P, Ransick A, Calestani C, Yuh C-H, et al. A genomic regulatory network for development. Science. 2002;295(5560):1669–78. CrossRefPubMed
4. Kelly MW, Padilla-Gamiño JL, Hofmann GE. Natural variation and the capacity to adapt to ocean acidification in the keystone sea urchin Strongylocentrotus purpuratus. Glob Chang Biol. 2013;19(8):2536–46. CrossRefPubMed
5. Cameron RA, Samanta M, Yuan A, He D, Davidson E. SpBase: the sea urchin genome database and web site. Nucleic Acids Res. 2009;37 suppl 1:D750–4. CrossRefPubMedPubMedCentral
6. ZwaryczAS, Nossa CW, Putnam NH, Ryan JF. Timing and scope of genomic expansion within Annelida: evidence from homeoboxes in the genome of the earthworm Eisenia fetida. Genome Biol Evol. 2015:evv243.
7. Nossa CW, Havlak P, Yue J-X, Lv J, Vincent KY, Brockmann HJ, et al. Joint assembly and genetic mapping of the Atlantic horseshoe crab genome reveals ancient whole genome duplication. GigaScience. 2014;3(1):1–21.CrossRef
8. Gnerre S, MacCallum I, Przybylski D, Ribeiro FJ, Burton JN, Walker BJ, et al. High-quality draft assemblies of mammalian genomes from massively parallel sequence data. Proc Natl Acad Sci. 2011;108(4):1513–8. CrossRefPubMedPubMedCentral
9. Martin M. Cutadapt removes adapter sequences from high-throughput sequencing reads. EMBnet J. 2011;17(1):10–2.CrossRef
10. Luo R, Liu B, Xie Y, Li Z, Huang W, Yuan J, et al. SOAPdenovo2: an empirically improved memory-efficient short-read de novo assembler. GigaScience. 2012;1(1):18.CrossRefPubMedPubMedCentral
11. Simpson JT, Wong K, Jackman SD, Schein JE, Jones SJ, Birol I. ABySS: a parallel assembler for short read sequence data. Genome Res. 2009;19(6):1117–23. CrossRefPubMedPubMedCentral
12. Kajitani R, Toshimoto K, Noguchi H, Toyoda A, Ogura Y, Okuno M, et al. Efficient de novo assembly of highly heterozygous genomes from whole-genome shotgun short reads. Genome Res. 2014;24(8):1384–95. CrossRefPubMedPubMedCentral
13. Parra G, Bradnam K, Korf I. CEGMA: a pipeline to accurately annotate core genes in eukaryotic genomes. Bioinformatics. 2007;23(9):1061–7.CrossRefPubMed
14. Stanke M, Morgenstern B. AUGUSTUS: a web server for gene prediction in eukaryotes that allows user-defined constraints. Nucleic Acids Res. 2005;33 suppl 2:W465–7.CrossRefPubMedPubMedCentral
15. Camacho C, Coulouris G, Avagyan V, Ma N, Papadopoulos J, Bealer K, et al. BLAST+: architecture and applications. BMC Bioinformatics. 2009;10(1):421. CrossRefPubMedPubMedCentral
16. Long KA, Nossa CW, Sewell MA, Putnam NH, Ryan JF. Supporting data for "Low coverage sequencing of three echinoderm genomes: the brittle star Ophionereis fasciata, the sea star Patiriella regularis, and the sea cucumber Australostichopus mollis". GigaScience Database. 2016. http://dx.doi.org/10.5524/100194.

CHAPTER 2

TRANSCRIPTOMIC CHANGES DURING REGENERATION OF THE CENTRAL NERVOUS SYSTEM IN AN ECHINODERM

Vladimir S Mashanov, Olga R Zueva and José E García-Arrarás

Keywords: Transcriptome, RNA-seq, Gene expression, Regeneration, Echinoderm, Nervous system, Transcription factors, Injury, Extracellular matrix

BACKGROUND

Echinoderms constitute a phylum of marine invertebrates closely related to chordates. Their phylogenetic position as a non-chordate deuterostome phylum combined with the ability to regenerate various body parts makes them a valuable group, which can give unique insights into fundamental issues of regenerative biology and provide new clues to finding better treatment of human conditions. Recent analyses of cellular events underlying post-traumatic regeneration in echinoderms (reviewed in [1, 2]) identified interesting parallels with corresponding processes in regeneration-competent vertebrates, pointing to reparative mechanisms that might have been evolutionary preserved throughout Deuterostomia and could potentially be re-activated in poorly regenerating vertebrates. However, the paucity of genomic and transcriptomic information has precluded major progress in understanding key regulators of regeneration at the molecular level [3, 4].

The body of an echinoderm has radial organization and is composed of multiple (usually five) units, which are arranged as sectors or rays around the central axis connecting the mouth and the anus. Each of these radial units is supplied with a set of major organs, which in sea cucumbers (Figure 1) include a radial nerve cord (a homolog of the chordate neural tube [5]), a canal of the water-vascular system, and a longitudinal muscle band (Figure 2A, B).

Figure 1: The model organism used in this study: Holothuria glaberrima Selenka, 1867 (Echinodermata: Holothuroidea).

Citation: Mashanov et al.: Transcriptomic changes during regeneration of the central nervous system in an echinoderm. BMC Genomics 2014 15:357. doi:10.1186/1471-2164-15-357.

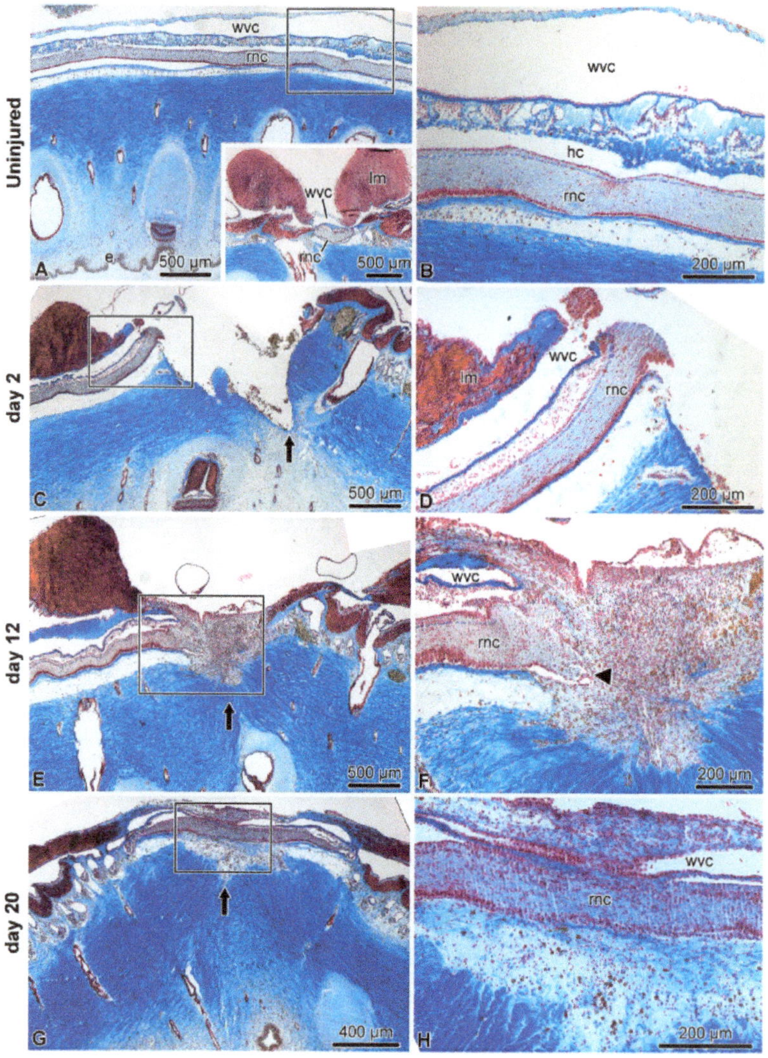

Figure 2: Histological organization of the radial organ complex in non-injured (A, B) and regenerating animals on day 2 (C, D), 12 (E, F), and 20 (G, H) post injury. All micrographs are longitudinal paraffin sections, except for A insert, which is a cross section. All sections were stained with Heindenhein's azan. The right column (B, D, F, H) shows high magnification views of the boxed areas in the corresponding images of the left column (A, C, E, G, respectively). e, epidermis; hc, hyponeural canal; lm, longitudinal muscle; rnc, radial nerve cord; wvc, water-vascular canal. Arrows show the location of the plane of injury. The arrowhead in **F** shows the growing tip of the regenerating radial nerve cord.

Our previous research [1, 6–8] showed that following a transverse cut, the injured organs of the radial organ complex on either side of the wound start growing across the wound gap and eventually reconnect to restore the anatomical continuity (Figure 2). The newly regenerated structures then completely re-acquire their normal tissue architecture and resume their functions. Radial nerve cord regeneration involves extensive dedifferentiation of radial glial cells in the vicinity of the injury. These dedifferentiated glial cells play the key role in subsequent regeneration through extensive proliferation, ECM invasion, and differentiation into new neurons and glial cells. The newly produced neurons are thought to be functionally integrated into the CNS circuitry, as they survive for extended periods of time and form typical synaptic connections [1, 7].

The main goal of the present study is to help understand the molecular basis underlying the extensive regenerative capacity of the central nervous system in the sea cucumber Holothuria glaberrima by providing an outline of the transcriptomic landscape and thus identifying possible directions of future research. To this end, we used deep RNA sequencing on both the 454 and Illumina platforms to analyze changes in the transcriptome that occurred on day 2, day 12, and day 20 after injury. These time points were chosen based on our previous studies of cellular events in the regenerating radial nerve cord of H. glaberrima[1]. Day 2 post-injury (Figure 2C, D) is the early post-injury phase of extensive dedifferentiation of radial glia, axonal degeneration, and programmed cell death in the injured radial nerve. Day 12 post injury (Figure 2E, F) corresponds to a period of active growth across the wound gap; dividing cells are most abundant at this stage. Day 20 (Figure 2G, H) post injury is a late regeneration phase, when the two growing regenerates have restored their anatomical continuity and started to resume their typical histological architecture [1, 7]. The present study provides first insights into gene expression changes that underlie these previously described cellular events.

RESULTS AND DISCUSSION

Sequencing and Assembly: Technical Information about the Assembly

The present study was performed on the brown rock sea cucumber Holothuria glaberrimaSelenka, 1867 (Echinodermata: Holothuroidea), an established model organism in echinoderm regenerative biology. We injured the radial nerve cord and the surrounding tissues, including the body wall connective tissue, water-vascular canal, and the longitudinal muscle band (Figure 2C), by performing a single transverse cut through these organs in the mid-ventral ambulacrum at about the mid-body level. We then harvested the regenerating tissues on day 2, day 6, day 12, and day 20 post-injury, as well as non-injured radial nerve cord and used the samples for sequencing library preparation. The major steps involved in our sequencing and data analysis pipeline are outlined in Figure 3.

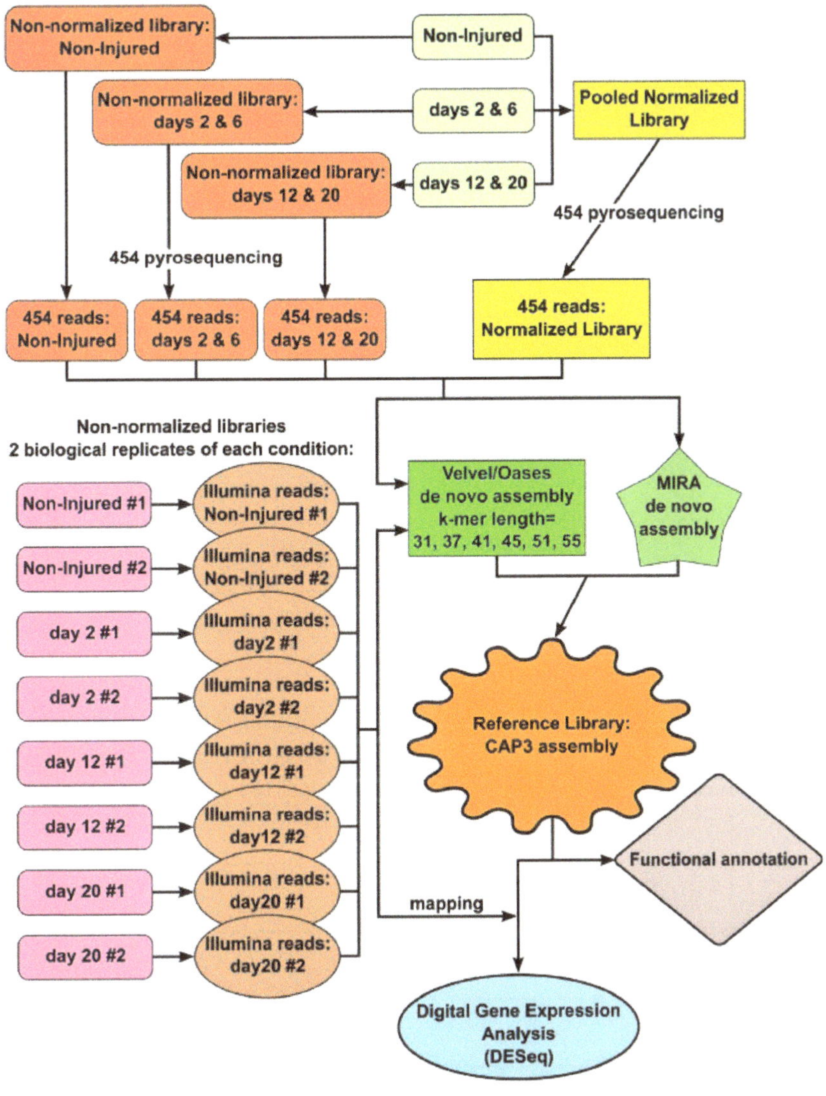

Figure 3: Summary of major steps involved in library preparation, sequencing, assembly, and annotation workflow.

The high-throughput transcriptome sequencing yielded a total of 2,428,740 Roche 454 reads with the average/modal length of 595/546 bases and 331,931,211 single-end Illumina reads (75 bp) (Table 1). We first combined the reads from all libraries to assemble a reference transcriptome for annotation purposes. Prior to the de-novo assembly step, we subjected the raw reads from both sequencing platforms to a rigorous cleaning/clipping procedure (see Methods for details) leaving 1,620,029 Roche 454 reads (with the average/modal length of 534/523 bases, respectively) and 246,993,807 Illumina reads (with the average/modal length of 57/61 bases, respectively) for further analysis (Table 1). All cleaned 454 reads were assembled with MIRA (Figure 3) into 125,828 contigs with the mean length of 676 bp and N50 (i.e., the contig length such that 50% of the total assembled sequence is contained in contigs of this size or longer) of 720 bp (Table 2). Illumina reads were assembled with Velvet and Oases (Figure 3). De Bruijn graph assemblers, such as Velvet/Oases, are known to be sensitive to the value of parameter k, which is under the control of the user. It has been shown that the use of a single k-mer length may result in suboptimal de novo transcriptome assemblies [9]. For a given assembly, the best k-mer value depends on the coverage depth (which is distributed over wide ranges in non-normalized libraries), the sequencing error rate, and the complexity of the transcriptome. Changing the k-mer value in either direction (i.e., increasing or decreasing) is associated with both pros and cons. Using higher k-mer values increases specificity and leads to a better assembly of highly expressed transcripts, whereas lower k-mer values improves sensitivity and allows better reconstruction of weakly expressed transcripts [9, 10]. Therefore, seven separate assembly runs were performed with different k-mer length (ranging from 31 to 55). They yielded between 50,903 and 237,177 contigs (assemblies with the higher k-mer value producing the lower number of contigs), with the mean length of 674 – 800 bp, the maximum length of 16,096 – 27,112, and N50 ranging between 1,377 and 1,555 (Table 2). Then all MIRA and Velvet contigs were pooled together and used as an input for the final assembly run with the CAP3 program (Figure 3), which produced 70,173 contigs with dramatically improved mean length (1,558 bp) and N50 (2,767 bp) and the maximum length of 27,089 bp (Table 2). We refer to this final set of 70,173 contigs (available in the LabArchives notebook [11]) as a reference library, which represents the transcriptional diversity in the normal and regenerating radial organ complex of the sea cucumber H. glaberrima, and which we used for functional annotation and read mapping (Figure 3). Previously, 53 of these contigs were found to correspond to 36 long terminal repeat (LTR) transposons. Homology analysis and expression pattern of these mobile genetic elements were characterized in detail elsewhere [12, 13] and, therefore, will not be covered in this study.

Table 1: Summary statistics of sequencing runs and read processing

	Total number	**Total number**	**Mean read**	**Modal (most frequent)**	**Read length**
	of reads	**of bases**	**length**	**read length**	**range**
Total Raw 454 reads	2,428,740	1,444,989,971	595	546	49 – 1,201
Total Cleaned 454 reads	1,620,029	864,494,207	534	523	60 – 994
Raw Illimina Norm #1	41,062,198	3,079,664,850	75	75	75
Raw Illimina Norm #2	41,396,915	3,104,768,625	75	75	75
Raw Illimina d2 #1	40,461,887	3,034,641,525	75	75	75
Raw Illimina d2 #2	42,713,246	3,203,493,450	75	75	75
Raw Illimina d12 #1	42,485,405	3,186,405,375	75	75	75
Raw Illimina d12 #2	42,718,652	3,203,898,900	75	75	75
Raw Illimina d20 #1	38,333,067	2,874,980,025	75	75	75
Raw Illimina d20 #2	42,759,841	3,206,988,075	75	75	75
Total Raw Illumina	**331,931,211**	**24,894,840,825**	**75**	**75**	**75**
Cleaned Illimina Norm #1	30,668,029	1,760,422,119	57	61	32 – 61
Cleaned Illimina Norm #2	29,685,397	1,717,566,430	58	61	32 – 61
Cleaned Illimina d2 #1	28,982,010	1,648,498,857	57	61	32 – 61
Cleaned Illimina d2 #2	30,382,668	1,721,162,425	57	61	32 – 61
Cleaned Illimina d12 #1	33,755,449	1,947,272,365	58	61	32 – 61
Cleaned Illimina d12 #2	33,893,212	1,956,101,048	58	61	32 – 61
Cleaned Illimina d20 #1	29,670,665	1,697,625,162	57	61	32 – 61
Cleaned Illimina d20 #2	29,956,377	1,693,992,519	57	61	32 – 61
Total Cleaned Illumina	**246,993,807**	**14,142,640,925**	**57**	**61**	32–61

Table 2: Summary statistics for the intermediate and final assembly steps

	Total number of contigs	Total number of bases	Mean length, nt	Length range, nt	N50
Normalized and non-normalized 454 reads combined	125,828	85,117,028	676	100 – 20,566	720
Velvet (k-mer length=31)	237,177	162,204,944	684	100 – 16,096	1,555
Velvet (k-mer length=37)	188,901	127,287,870	674	100 – 27,112	1,496
Velvet (k-mer length=41)	158,033	111,038,497	703	100 – 27,083	1,554
Velvet (k-mer length=45)	133,348	93,061,666	698	100 – 25,388	1,464
Velvet (k-mer length=51)	87,923	63,748,786	725	100 – 23,585	1,377
Velvet (k-mer length=55)	50,903	40,710,596	800	100 – 23,018	1,386
Reference Assembly (CAP3)	**70,173**	**109,360,184**	**1,558**	**100 – 27,089**	**2,767**

To assess the accuracy of our sequencing and assembly pipeline, we re-sequenced 16 randomly selected contigs from the reference library using Sanger technology (Table 3). In total, 34,708 nt of 44,096 nt (~79%) were analyzed by re-sequencing. The overall error rate (determined as a percentage of mismatch between Sanger sequencing and next-generation RNA-seq) was 0.19%, with insertions, deletions, and substitutions affecting 29 nt, 16 nt, and 20 nt, respectively. Eleven of the resequenced transcripts had a complete open reading frame (ORF), in four contigs both the 5' and 3' ends of the ORF were incomplete, in one contig, only the 5' end was missing, and one contig was likely a non-coding RNA. One of the 14 ORF-containing contigs had a frameshift error.

Table 3: Validation of assembled contigs by re-sequencing using Sanger technology

ContigID	Best blastx match	ORF complete?	Frame-shifts in ORF	Contig length, bp	Checked by Sanger sequencing, bp	Insertions	Deletions	Substitutions	% Discrepancy
Contig3756	CENPS	yes	no	665	564	8	0	1	1.60%
Contig7869	Myc	yes	no	3,268	2,922	0	0	0	0.00%
Contig54848	HuD/ Elav	yes	1	4,404	3,431	9	2	3	0.41%
Contig19303	Lgr5	both 5' and 3' ends are missing	no	1,510	1,510	0	0	0	0.00%
Contig40997	NeuroD	both 5' and 3' ends are missing	no	745	470	0	1	2	0.64%
Contig1064	NF1A	yes	no	5,279	3,825	3	9	1	0.34%
Contig32193	NvProtien1	5' end is missing	no	2,411	1,986	0	1	0	0.05%
Contig51632	Oct	both 5' and 3' ends are missing	no	1,947	1947	0	0	0	0.00%
Contig46302	Piwi	yes	no	3,430	2,929	0	0	3	0.10%
Contig67140	Sox	yes	no	3,096	2,889	2	0	0	0.07%
Contig20197	unknown	NA	NA	1,339	1,252	2	0	0	0.16%
Contig68289	Wnt3	yes	no	2,561	990	2	0	2	0.40%
Contig50192	DCLK	yes	no	3,152	2,474	0	0	3	0.12%
Contig8186	Lhx	yes	no	5,430	3,041	0	0	5	0.16%
Contig3071	FoxJ1	yes	no	2,523	2,404	2	3	0	0.21%
Contig10179	Klf	yes	no	2,336	2,074	1	0	0	0.05%
			TOTAL:	**44,096**	**34,708**	**29**	**16**	**20**	**0.19%**

In order to determine the proportion of the contigs in the reference library, which corresponded to known proteins, we performed a BLASTX search against publicly available protein databases. In the first round of search, the sequences were matched against the Swissprot database. The contigs, which did not produce hits passing the significance threshold corresponding to anE-value < 1e-6 were subsequently subjected to a second round of BLASTX search against the non-redundant (nr) NCBI protein database with the same threshold. Overall, 24,324 (or 33.66%) contigs had significant BLAST hits. The results of the BLAST analysis are listed.

We also annotated the sea cucumber transcriptome by performing reciprocal best BLAST hit analysis (with a threshold e-value < 1e-6) versus the NCBI's collection of the sea urchinStrongylocentrotus purpuratus predicted protein sequences [14], the echinoderm species whose genome has been most thoroughly characterized so far. There were 23,637 one-way BLASTX matches between the H. glaberrima contigs and the S. purpuratus proteins and 19,869 one-way TBLASTN matches between the sea urchin proteins and the contigs of our transcriptome. The number of reciprocal blast matches (putative orthologous sequences) was 8,577.

We, obviously, do not expect all our 70,173 contigs to represent individual sea cucumber genes. It is impossible to give an exact answer on the number of unique genes represented by de novo assembled contigs without having complete genomic information. Thus, we can only provide an educated guess here. The genome of the sea urchin S. purpuratus[15] encodes ~23,300 genes, of which ~7,000 are presumed orthologs of mammalian genes. In the course of our analysis, a similar number (8,522) of unique mouse proteins showed significant similarity to sequences contained in the assembled transcriptome of H. glaberrima. If we assume the same level of genomic similarity between mammals and each of the two echinoderm species, the similar number of "mammalian genes" observed in the sea urchin genome and the sea cucumber transcriptome would imply that (i) sea cucumbers have roughly the same number of genes as sea urchins (~23,000), and (ii) most of these genes are represented in our reference library constructed from mRNA samples from non-injured and regenerating animals. If the above reasoning is correct, there is a ~3× redundancy in our assembly.

A similar redundancy ratio (~2.4×) can be obtained, if we consider that, when blasted against the reference mouse proteome, 20,619 of 70,173 contigs of the sea cucumber reference library matched 8,522 unique mouse proteins. Some of this redundancy is part of the "natural" variation that is expected from differences in mRNA processing (such as splicing). In fact, when manually inspecting assembled contigs, we saw polymorphic transcripts, which were especially common among retroelements. Additional variation can be due to the limitations of the sequencing techniques and/or de novo assembly programs.

Differential Gene Expression in Radial Organ Complex Regeneration

To characterize differentially regulated genes involved in regeneration, we mapped the reads from each of the eight Illumina libraries (two biological replicates for each of the four conditions — non-injured animals and regenerating organs on days 2, 12 and 20 post-injury) to the contigs of the reference library (Figure 3). Gene expression values in each pair of replicates showed high correlation (Pearson's product-moment correlation coefficient $r = 0.94 - 0.98$, $p < 2.2e\text{-}16$). We then used the DESeq package [16] to identify sets of significantly up- and down-regulated genes in the regenerating tissues as compared with the intact animals (Figure 4). Upon imposing a threshold adjusted P value (P_{adj}) of less than 0.001 and a threshold fold change of 2, we identified a total of 4,023 and 3,257 transcripts, which were differentially up-regulated and down-regulated, respectively. The distribution of unique and overlapping sets of the differentially expressed genes at different time points is summarized in Figure 4B. Note that the early post-injury phase (day 2) is characterized by the highest values of both the total number of differentially expressed transcripts and the number of unique up-regulated and down-regulated genes. On the other hand, in line with the general histological similarity between the late regenerates and uninjured radial organs (Figure 2A, B, G, H) [1, 7], the samples analyzed on day 20 post-injury showed the lowest number of differentially regulated transcripts.

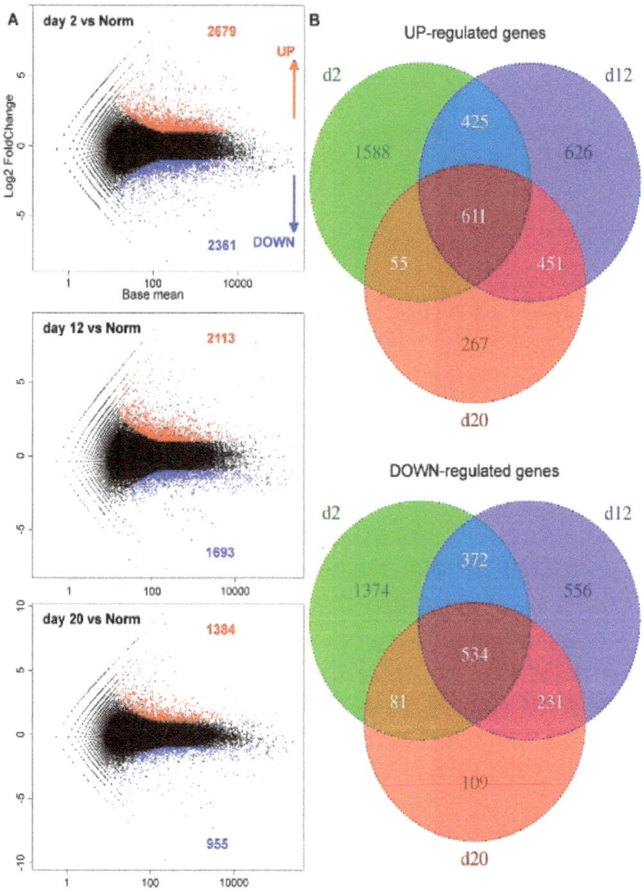

Figure 4: Differential gene expression analysis of the regenerating radial organ complex. (A) DESeq scatter plots showing comparisons of gene expression levels between regenerating and uninjured animals. Each dot represents the mean expression level for a given contig. Differentially up-regulated contigs (with false discovery rate < 0.001 and a threshold log2 fold change of ±1) are shown in red and differentially down-regulated contigs are shown in blue. Numbers above or below the plots refer to the total number or up- or down-regulated contigs, respectively, at a given time point of regeneration. (B) Venn diagrams showing the numbers of common and distinct differentially up-regulated (top) and down-regulated (bottom) genes at different time points of regeneration.

At all three time points after the injury, the majority (~63-85%) of the significant differentially expressed contigs were up- or down-regulated between 2- and 4-fold relative to the normal animals, suggesting that relatively small changes in transcript abundance of most genes are sufficient for regeneration to occur. Among the most extreme outliers in our differential expression analyses were sequences that were identified as retrotransposon-derived transcripts. Some of them showed an over 50-fold change in expression during regeneration. This unexpected finding prompted us to undertake a separate study of these mobile genetic elements that has already been published elsewhere [12].

To verify the validity of our approach and the accuracy of the large-scale digital gene expression assay, we selected 21 genes for analysis by quantitative real-time RT-PCR (qRT-PCR). This dataset also includes previously determined expression values for three LTR retroelements (Gypsy-1_Hg, Gypsy-2_Hg, and Gypsy-20_Hg) characterized earlier elsewhere [12]. Our goal was to validate the sensitivity of the assay at different levels of transcript abundances. Therefore, we chose not only highly over-expressed or down-regulated genes, but also genes with moderate fold change and genes, whose transcript abundance remained stable throughout the experiment. We found highly significant strong positive correlation between the RNA-seq and qRT-PCR data (Pearson's r=0.954, p <2.2e-16) (Figure 5). The corresponding fold change ratios (relative to the uninjured animals) for both quantification techniques are listed. This correlation analysis suggests that our RNA-seq data reliably represent relative changes of mRNA levels in the regenerating tissues.

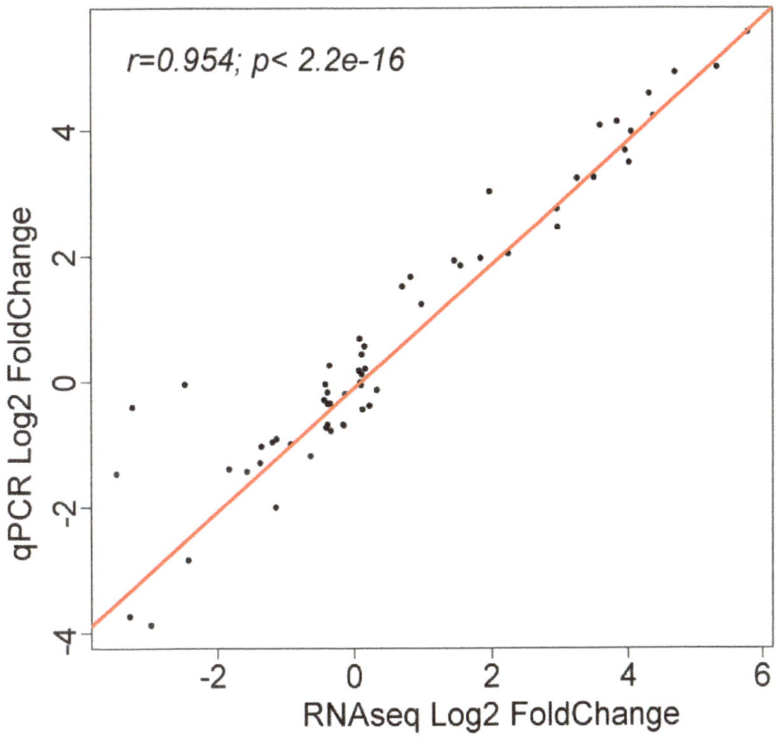

Figure 5: Comparison of mRNA expression levels as determined by RNA-seq and qRT-PCR. Comparisons were performed for 21 genes at three time points of regeneration relative to uninjured animals (for the list of genes and numerical values). The red line is the linear regression line.

Functional Annotation of Differentially Expressed Genes at Different Time Points of Regeneration

From the biological perspective, it is important to know which processes predominate or become suppressed at different time points during regeneration. For functional annotation, we employed an approach, which was conceptually similar to that described by Stewart et al. [17] for the analysis of the limb blastema transcriptome in axolotl. In order to gain insight into the biological meaning behind the large numbers of differentially regulated transcripts, we analyzed the lists of genes that showed significant changes in expression levels (with the cut-off set at ±1 log2 fold change, and $P_{adj} < 0.001$) with DAVID v6.7 [18, 19] to determine enriched functional categories. To be able to use this tool, we had to match the contigs resulting from our de novo assembly of the sea cucumber transciptome to genes of some well-annotated model organism. We thus annotated the assembled transcriptome of H. glaberrima against the non-redundant reference proteome of the mouse. Our assembled contigs matched to 8,522 unique mouse genes at the BLASTX cut-off value of 1e-6. This entire list of annotated contigs representing both uninjured and regenerating animals was submitted to DAVID along with the lists of differentially expressed genes and used as a custom reference background. Functional annotation clustering was performed using Gene Ontology (GO_TERM_BP_FAT, GO_TERM_CC_FAT, GO_TERM_MF_FAT) and pathway (KEGG, BIOCARTA, PANTHER) annotations. The functional clustering algorithm in DAVID groups together statistically enriched annotation categories if they share significant degrees of similarity in terms of associated gene members. The EASE score (modified Fisher exact p-value) threshold was set at 0.05. Functional annotation clusters with enrichment scores (geometric mean of EASE scores, in negative log scale) > 1.3 (which is equivalent to a non-log scale value of 0.05) were considered significant. The following major findings emerged from this analysis (Figure 6).

1. Among the most enriched functional categories associated with both up-regulated and down-regulated genes at all time points were those related to synthesis and organization of the extracellular matrix (ECM) components, ECM remodeling, and interaction between cells and the extracellular matrix (see below for a detailed discussion).

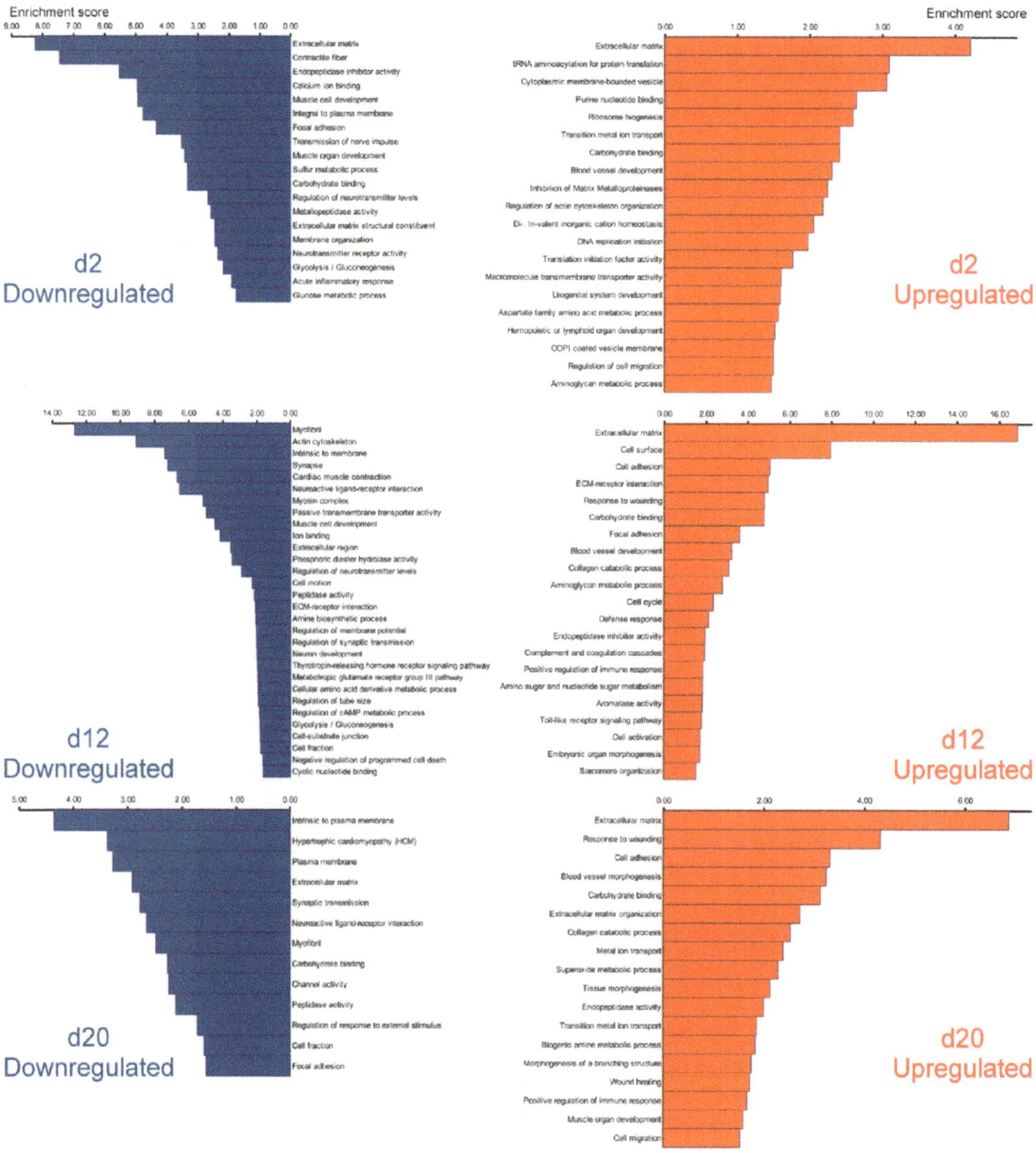

Figure 6: Diagram showing clustering of functional annotation terms associated with differentially expressed genes during the radial organ complex regeneration. Differentially expressed genes that are up-regulated (right column) or down-regulated (left column) at a given time point of regeneration were analyzed for significantly enriched functional annotation terms, which were clustered using DAVID. Functional annotation clusters with enrichment scores (in negative log scale) > 1.3 are shown. The corresponding detailed data resulting from DAVID output are listed.

2. Functional annotation terms associated with normal physiology, differentiation, and development of the nervous tissue were over-represented among the down-regulated genes on days 2 and 12 post injury, and some of those annotation categories were also enriched among negatively regulated genes even as late as after 20 days post-injury. This observation correlates well with previously published morphological studies showing extensive dedifferentiation in the regenerating radial nerve at these time points [1, 7]. Interestingly, annotation categories related to glycolysis were also enriched among the down-regulated genes on day 2 and 12.

Since our tissue samples, in addition to the radial nerve cord per se, also contained some adjacent tissues, including the contractile epithelium of the water-vascular canal and the coelomic myoepithelium of the body wall (see Methods), annotation terms associated with structural components of muscular tissue, as well as with muscle development and physiology, also appeared in the analysis. These terms were enriched in sets of down-regulated genes at all three time points of regeneration corroborating earlier observations of muscular dedifferentiation during body wall regeneration in sea cucumbers [6].

3. On day 2 post-injury, the up-regulated genes were associated with over-represented terms related to initiation of DNA replication and protein translation. On day 12, positively regulated genes were characterized by continued over-representation of terms associated with DNA synthesis and cell cycle. These data corroborate our cell proliferation assay, which suggested that the peak of cell division in the regenerating radial nerve cord occurs on days 8 thru 12 after injury [1]. On day 12, up-regulated genes are also enriched in annotation categories associated with activation of the innate immune response. The latter group of functional terms remains over-represented among up-regulated genes on day 20.
4. All three regeneration time points are characterized by over-representation of annotation categories associated with developmental morphogenesis, indicating involvement of ontogenic processes in regeneration.

In order to get further insight into mechanisms controlling gene expression in post-traumatic regeneration of the radial nerve cord, we performed clustering of differentially expressed genes using the unsupervised clustering algorithm implemented in the AutoSOME program [20] to identify sets of co-regulated genes with similar patterns of expression throughout the time course of regeneration. This gene co-expression analysis yielded eight distinct clusters with more than 100 genes in each (Figure 7). In two of these clusters (clusters 1 and 3), the median gene expression in regenerating animals was consistently higher at all three regeneration stages when compared with uninjured animals. These genes were enriched in functional annotation terms related to the extracellular matrix, initiation of DNA replication and translation, positive regulation of immune response, wound healing, and cell migration. Three other clusters (clusters 2, 4, 5) contained genes, whose median expression at all three time points was lower that in normal animals. Associated enriched functional annotation terms were related to extracellular matrix, transmission of nerve impulse, regulation of neurotransmitter activity, muscle cell and neuron development, cytoskeleton, MAPK signaling pathway, negative regulation of cell death, gated channel activity, and carbohydrate binding. Clusters 6 and 7 contained genes, whose expression initially either remained unchanged or somewhat decreased at the early post-injury stage (day 2), but then was elevated during the growth (day 12) and late regeneration (day 20) phases. These genes were associated with positive regulation of developmental processes, extracellular matrix, cell adhesion, serine peptidase activity, endocytosis. Cluster 8 contained genes, which were markedly down-regulated at the early post-injury stage (day 2), but whose expression than returned to the normal levels as regeneration progressed. Enriched functional annotation terms associated with this cluster included extracellular matrix, neuromuscular junction development, and synaptic transmission.

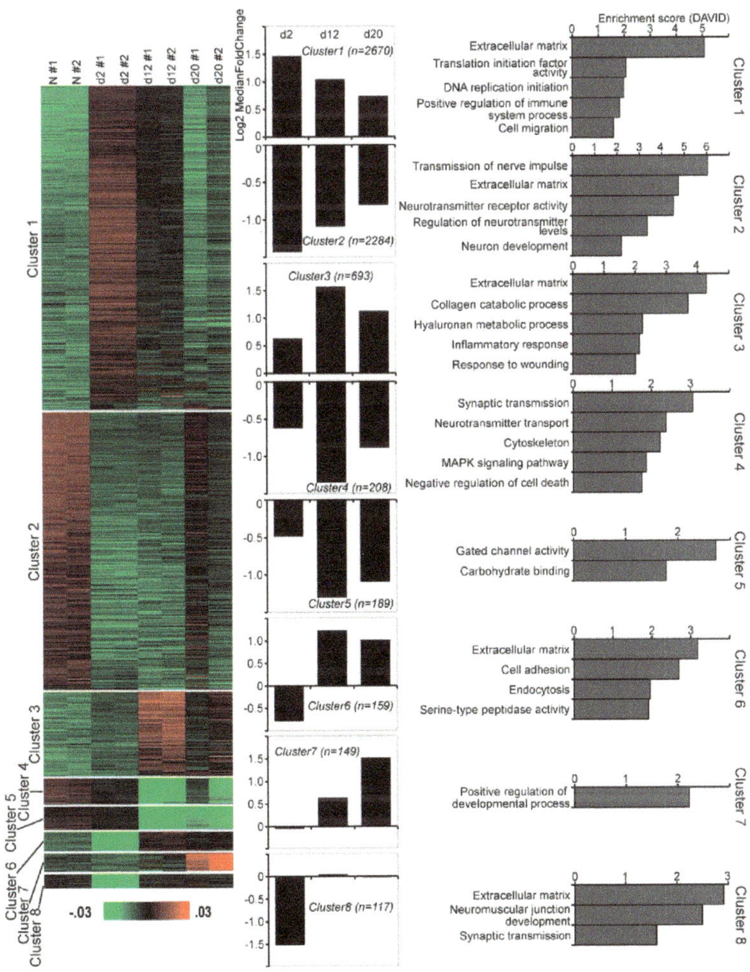

Figure 7: Clustering of co-expressed genes. The left row shows a heatmap produced by AutoSOME with clusters of co-expressed genes. Only eight clusters containing more than 100 genes are shown. The middle column shows median log2 fold-change in expression (vs uninjured animals) of the genes contained in each cluster. The clustered transcripts were analyzed with DAVID for significantly enriched functional annotation terms. The right column shows representative functional categories and their enrichment score.

Putative Regulation of Groups of Co-Expressed Genes by Transcription Factors (TFs)

We then sought to get insight into which upstream transcription factors (TFs) may regulate coordinated expression of genes within each of the eight clusters identified above (Figure 7). Over-representation of transcription factor binding sites associated with co-expressed genes of each cluster was predicted using oPOSSUM software [21]. We considered only those of the predicted TFs whose significant ($p < 0.05$) changes in expression level correlated in time (at least at one time point) with changes in the median gene expression within the respective cluster. This analysis yielded 11 TFs, which putatively regulate regeneration-related genes (Figure 8) and therefore constitute good candidates for further functional analysis. It is worth noting that there was some overlap between the clusters. Thus, CTCF and NfkB1 were associated with four and three clusters, respectively, and SRF, Fli1, and PLAG1 were associated with two different clusters each. Some of the TFs (for example, SFR, NfkB1) were identified as functioning as both transcriptional activators and transcriptional repressors in different clusters.

Cluster	Transcription Factor	Possible role	Contig #	Blast Hit, Accession	Blast Hit, E-value	Expression Profile (log2 fold change vs Norm)			Z-score	Fisher score
						Day 2	Day 12	Day 20		
Cluster1	Liver X receptor alpha (NR1H-like)	Transcriptional activator	Contig44610	Q5E9B6	9.00E-114	1.15	NS	NS	6.08	2.00
	Serum response factor (SRF)	Transcriptional activator	Contig21621	Q90718	2.00E-069	NS	0.66	1.24	7.16	1.69
	CCAAT/enhancer-binding protein (CEBP-like)	Transcriptional activator	Contig39	Q05826	2.00E-018	0.63	NS	NS	5.58	1.46
	Transcriptional repressor CTCF	Transcriptional repressor	Contig31852	Q08705	8.00E-074	NS	-0.54	-0.51	8.06	1.39
Cluster2	Serum response factor (SRF)	Transcriptional repressor	Contig21621	Q90718	2.00E-069	NS	0.66	1.24	7.16	1.69
	Fli1	Transcriptional repressor	Contig30142	Q01543	2.00E-123	NS	0.96	0.60	11.68	4.86
	Zinc finger protein PLAG1	Transcriptional activator	Contig4325	Q58NQ5	2.00E-020	NS	-2.29	NS	7.30	4.02
	Transcription factor COE3 (Ebf3)	Transcriptional activator	Contig64380	O08791	0.00E+000	-0.47	-0.46	NS	15.72	2.12
Cluster3	Nuclear factor NF-kappa-B p105 subunit (NfkB1)	Transcriptional activator	Contig6439	Q6F3J0	3.00E-110	0.48	NS	NS	14.41	2.03
	Estrogen-related receptor beta (Esrrb)	Transcriptional activator	Contig65136	P11475	6.00E-082	NS	0.73	0.63	7.04	1.55
	Transcriptional repressor CTCF	Transcriptional repressor	Contig31852	Q08705	8.00E-074	NS	-0.54	-0.51	6.51	0.98
Cluster4	Pre-B-cell leukemia transcription factor 3 (PBX3)	Transcriptional activator	Contig30212	P40426	0.00E+000	-0.53	-0.69	NS	6.18	7.57
Cluster5	Fli1	Transcriptional repressor	Contig30142	Q01543	2.00E-123	NS	0.96	0.60	8.08	2.56
	Interferon regulatory factor 2 (IRF2)	Transcriptional activator	Contig13671	Q8R4E0	2.00E-043	-0.78	NS	NS	5.00	2.30
Cluster6	Zinc finger protein PLAG1	Transcriptional repressor	Contig4325	Q58NQ5	2.00E-020	NS	-2.29	NS	9.37	2.48
	Nuclear factor NF-kappa-B p105 subunit (NfkB1)	Transcriptional repressor	Contig6439	Q6F3J0	3.00E-110	0.48	NS	NS	6.20	2.42
	Transcriptional repressor CTCF	Transcriptional repressor	Contig31852	Q08705	8.00E-074	NS	-0.54	-0.51	4.76	2.27
Cluster7	Transcriptional repressor CTCF	Transcriptional repressor	Contig31852	Q08705	8.00E-074	NS	-0.54	-0.51	5.74	2.68
Cluster8	Nuclear factor NF-kappa-B p105 subunit (NfkB1)	Transcriptional repressor	Contig6439	Q6F3J0	3.00E-110	0.48	NS	NS	6.27	2.11

- significant UP-regulation (padj < 0.05)
- significant DOWN-regulation (padj < 0.05)
NS -non-significant

Figure 8: Putative regulation of co-expressed transcripts by transcription factors (TFs). Co-expressed genes contained in clusters identified with AutoSOME (Figure 7) were analyzed for over-representation of putative TF binding sites using oPOSSUM.

Among these predicted 11 TFs, there are genes that have been previously implicated in post-traumatic processes and developmental processes in various organisms. For example, NfkB1 is known to affect expression of broad range of downstream genes involved in various biological processes including immunity, differentiation, and programmed cell death. For example, elevated NfkB signaling has been previously shown to activate the Wnt signaling pathway and thus induce dedifferentiation of nonstem cells [22]. Another TF, serum response factor (SRF), was shown to be one of the key genes involved in post-traumatic regeneration initiation in planarians [23]. In gastric ulcer healing, SRF promotes re-epithelialization and muscle regeneration through activation of cell migration and proliferation [24]. Moreover, SRF is also implicated in neuronal cell migration and axonal guidance through regulation of components of the actin cytoskeleton [25]. Still another gene, CCAAT/enhancer binding protein (CEBP) was suggested to be involved in regulation of the innate immune response during tissue injury repair [26].

Interestingly, the candidate TFs identified in this study included not only previously known regeneration-related genes, but also factors not previously known to act as regulators of post-traumatic tissue regrowth, such as, for example, liver X receptor, Fli1, PLAG1, Ebf3, Esrrb. The potential role of these genes in regeneration deserves further attention in future research.

Discussion of Selected Functional Gene Groups

The preceding section provided an overview of unbiased functional characterization of the regeneration-associated genes at the global, transcriptome-wide level. Below, we further zoom in on certain groups of differentially expressed genes, which we picked from our database based on the results of the above analysis, prior biological knowledge, or both.

Cancer-Related Pathways

As revealed by DAVID analysis, the set of differentially expressed genes at the early post-injury stage (day 2) includes many known cancer-related genes. For example, pathway mapping revealed differential regulation of Wnt receptors (Fzd3 and Fzd4) and ligands (Wnt9, Wnt2, and Wnt6). Another notable observation is down-regulation of survivin (Birc5). This gene codes for a multifunctional protein, highly expressed in most human cancers, and implicated in both suppression of programmed cell death [27] and regulation of cell division [28–30]. We have previously shown that elevated expression ofsurvivin in regenerating sea cucumber intestinal tissues correlated with low levels of apoptosis [31]. Reduced expression of this gene in the regenerating radial nerve cord may thus be at least in part explained as being associated with extensive programmed cells death in the vicinity of the injury [1].

Differential expression of oncogenes early in regeneration has been also reported for other model systems, including early limb blastema in axolotl [17] and can be explained by the fact that both regeneration and cancer progression are developmental processes that share a number of key mechanisms including cell division, programmed cell death, and differentiation.

ECM-Related Pathways

As mentioned above, among the most significantly over-represented pathways associated with differentially expressed genes at all three time points of the radial nerve cord regeneration were those related to the extracellular matrix (ECM) components and ECM-cell interactions. One example is significantly reduced expression of genes coding for basal lamina proteins on day 2 (Col4a1, Lama4), but up-regulation of the fibrillar collagens Col5a1 andCol11a2. These data correlate well with previous observations of breakdown of the basal lamina during the early post-injury phase [7].

Of particular interest is concurrent up-regulation of both matrix metalloproteases (MMPs) and their inhibitors, TIMPs, in the regenerating radial organ complex. By breaking down components of the connective tissue matrix at the injury site and thus affecting cell migration and proliferation, epithelialization, differentiation, and apoptosis, MMPs are known to facilitate wound healing and regeneration [32–34]. However, up-regulation of TIMPs is as important for precise regulation of the regenerative processes, as they protect newly synthesized ECM matrix from degradation [32].

Proper interactions between cells and the ECM are essential both for correct tissue organization and signal transduction. Among the main cell adhesion molecules mediating these interactions are integrins [35–37]. In regeneration of the radial nerve in the sea cucumber, integrin betaand integrin alpha-8 are significantly up-regulated. Integrins affect cytoskeletal organization through Cdc42, a member of the Rho family of small GTPases [38,39]. Expression levels of the Cdc42 transcript were consistently highly elevated in the sea cucumber radial nerve regeneration, suggesting a possible involvement of the integrin-Cdc42 pathway in control of ECM-cell interactions in this animal model.

Besides integrins, other cell adhesion molecules are also up-regulated during the early post-injury phase. These include selectins. Interestingly, expression of selectins is known to be induced by NF-kB [40], which was putatively identified as one of the putative key transcription factors driving differential gene expression during the early post-injury stage of neural regeneration in our model (see above).

Pluripotency Factors

As has been documented by morphological studies [1, 6, 7], regeneration of the radial organs in sea cucumbers involves extensive dedifferentiation of specialized cells of the adult tissues in the vicinity of the injury. The dedifferentiation step endows these cells with increased proliferative potential, as well as with the ability to migrate and subsequently re-acquire their initial morphology or, more interestingly, transdifferentiate to other cell types. As this dedifferentiation phase is an indispensable component of the regenerative response, we asked if it requires elevated expression of factors known to be involved in acquisition and/or control of pluripotency. We, therefore, searched our assembled transcriptome for homologs of those genes and analyzed their expression values based on our RNA-seq read count data (Figure 9). Out of 15 factors analyzed, 11 were found to be expressed in both normal and regenerating tissues. Of those 11 genes, only the Myc homolog was expressed at significantly higher levels in regeneration. Myc genes have a wide variety of functions affecting the developmental potential of cells, including driving cells into the cell cycle, promoting the open state of chromatin, and DNA replication. C-Myc is one of the Yamanaka factors involved in the production of induced pluripotent cells [41] from specialized mammalian cells. During this reprogramming procedure, Myc represses differentiation-associated genes [42, 43]. The functional role of Myc in echinoderm regeneration is a subject of ongoing research in our group.

Gene	Contig ID	Blast Hit, Accession #	Blast Hit, E-value	Expression Profile (log2 fold change vs Norm)		
				Day 2	Day 12	Day 20
Myc	Contig7869	Q17103	4.01E-030	2.91	1.78	1.36
Klf2	Contig10179	Q9Y5W3	2.40E-042	NS	NS	NS
Sox2	Contig67140	P48430	1.57E-060	NS	NS	NS
Oct1*	Contig51632	P16143	4.90E-070	NS	NS	NS
Lgr5	Contig13383	O75473	2.81E-041	NS	NS	NS
Bmi-1	Contig31759	Q5SDR3	5.00E-086	-1.16	-1.07	NS
Ephb2	Contig44803	P54763	1.00E-084	NS	NS	NS
Msi2	Contig55987	Q96DH6	1.00E-091	NS	NS	NS
Sall4	Contig39977	Q9UJQ4	1.00E-041	NS	NS	NS
Zic3	Contig4371	Q62521	6.00E-084	NS	NS	NS
Lin28	Contig3877	Q8JHC4	3.00E-035	NS	NS	NS
Nanog			not detected in the transcriptome			
TERT			not detected in the transcriptome			
Dppa2			not detected in the transcriptome			
Gdf3			not detected in the transcriptome			

- significant up-regulation (p < 0.001; Log2FoldChange > 1) - significant down-regulation (p < 0.001; Log2FoldChange < -1) NS - non-significant

Figure 9: Expression of pluripotency factors in regeneration of the radial organ complex in the sea cucumber H. glaberrima . Summary of the digital expression assay and homology search against the UniProt database. Red and blue colors indicate significant (adjusted p < 0.001, more than two fold change in expression level) up-regulation or down-regulation, respectively, in the regenerating tissues as opposed to uninjured organs. NS, no significant changes in expression was observed. *Note that the gene designated as Oct1 is included in this list because it is the only Oct homolog found in the H. glaberrima transcriptome. Likewise, Oct1/2 is the only Oct in the sea urchin genome [44]. In mammals, there are a number of other Oct genes that perform various essential roles. One of them, Oct4 was used to induce pluripotency in mammalian somatic cells [41, 45].

To our surprise, we also found that expression of the Bmi-1 homolog was significantly reduced in the regenerating tissues on days 2 and 12 post-injury. The remaining 9 genes were expressed at the same level both in the non-injured and regenerating tissues.

A possible explanation that can be proposed to account for the observed data is that although the analyzed pluripotency factors do not show any large scale over-expression in regeneration, most of them are still expressed at a certain level both in the normal and regenerating tissues. Therefore, it may be hypothesized that this basal level constitutes sufficient environment for dedifferentiation to be triggered by Myc upregulation alone. Interestingly, a similar situation has been reported in non-injured and regenerating tissues of lower vertebrates, such as Danio rerioand Xenopus, where pluripotency markers are never shut off completely under normal conditions. Even though the expression level of these genes remained largely at the same level after the injury, it was hypothesized that this basal expression is neverhteless sufficient to facilitate regeneration upon damage [46].

Genes Associated with Neurogenesis

Post-traumatic neurogenesis in the lesioned radial nerve cord involves extensive proliferation of radial glial cells, which then act as progenitors generating both new glial cells and neurons [1,7]. Surprisingly, our functional analysis of differentially expressed genes did not yield statistically significant enrichment of neurogenesis-associated functional categories. The only exception was the enrichment of the category "neuron development" in the set of down-regulated genes during the growth phase on day 12 post-injury. We then manually searched our databases for homologs of genes known to be implicated in maintenance of neural stem cells and in specification of the neuronal and glial lineages in other animal models (25 genes in total) (see e.g., [47–56]). The results of this analysis are summarized in Figure 10. The most probable explanation why neurogenesis was not a dominant theme in the functional analysis of differentially expressed genes is that most of the neurogenesis-related genes, whose homologs were identified in the transcriptome of H. glaberrima, were already expressed in the non-injured radial nerve cord. Their expression level showed some fluctuations in the regenerating tissues, but not above the cut-off threshold (twofold change in expression, adjusted p-value < 0.001), with the exception of up-regulated Sox11, Bmp7 and down-regulated Prom1, Isl1, and Lhx3.

Gene	Contig ID	Blast Hit, Accession #	Blast Hit, E-value	Expression Profile (log2 fold change vs Norm) Day 2	Day 12	Day 20	Known function	Reference
Ascl1 (Mash1)			Not present in the transcriptome				Essential for embryonic and adult neurogenesis	Kim et al.,2011
Neurog2			Not present in the transcriptome				Maintaining neuronal progenitors in an undifferentiated state	Roybon et al., 2009
Nestin			Not present in the transcriptome				Required for survival and renewal of neural stem cells	Morrens et al., 2012
Vimentin			Not present in the transcriptome				Marker of neural stem cells	Li et al., 2011
REST			Not present in the transcriptome				Represses neuronal genes in non-neural tissues and in undifferentiated neuronal progenitors	Covey et al., 2012
Ifitm3			Not present in the transcriptome				Enriched in adult neuronal stem cells	Beckervordersandforth et al., 2010
Rarres2			Not present in the transcriptome				Enriched in adult neuronal stem cells	Beckervordersandforth et al., 2010
111001D15RIK			Not present in the transcriptome				Enriched in adult neuronal stem cells	Beckervordersandforth et al., 2010
Sox11	Contig66300	P40650	6.00E-041	1.25	1.26	NS	Enriched in adult neuronal stem cells	Beckervordersandforth et al., 2010
Thbs4	Contig55285	Q06441	0.00E+000	NS	0.63	0.62	Enriched in adult neuronal stem cells	Beckervordersandforth et al., 2010
Musashi	Contig55987	Q96DH6	1.00E-091	NS	NS	NS	Proliferation and maintenance of neural stem cells	Beckervordersandforth et al., 2010
Prom1	Contig46899	Q9W735	5.00E-081	1.25	-0.84	NS	Control of symmetric vs asymmetric divisions in neural stem cells	Morrens et al., 2012
DCLK1	Contig50192	O15075	0.00E+000	0.60	NS	NS	Controls mitolitc divisions of neuronal progenitors	Shu et al., 2006
Isl1	Contig7476	P53405	9.00E-158	-0.93	-1.00	1.05	Motor neuron fate specification (late stages)	Bréjot et al., 2006
Lhx3	Contig56124	Q90421	9.00E-122	-0.99	1.00	-0.84	Motor neuron fate specification	Bréjot et al., 2006
Olig2			Not present in the transcriptome				Motor neuron fate specification	Bréjot et al., 2006
Meis2	Contig10671	O14770	4.00E-049	NS	NS	NS	Regulation of GABAergic neurogenesis	Beckervordersandforth et al., 2010
Nr4a2 (Nurr1)	Contig55201	Q08E53	6.00E-136	-0.89	-0.79	NS	Dopamine neuron development	Park et al., 2006
NeuroD1	Contig40997	Q60867	3.00E-054	NS	NS	NS	Neuronal differentiation	Roybon et al., 2009
Reln	Contig58365	P78509	0.00E+000	-0.67	NS	-0.53	Induction of radial glia phenotype	Keilani and Sugaya, 2008
FoxJ1	Contig3071	Q92949	2.00E-056	0.49	0.77	0.62	Required for differentiation of ependyma and specification of adult neural stem cells	Jacquet et al., 2009
Prox1	Contig34290	Q91018	1.00E-086	0.67	NS	0.51	Required for interneuron differentiation	Mishra et al., 2008
Wnt3	Contig68289	Q2LMP1	8.00E-131	NS	-0.49	NS	Precursor cell survival, proliferation, neuronal differentiation, dendrite arborization, synapse formation	Morrens et al., 2012
Bmp7	Contig55448	P18075	8.00E-089	1.16	0.47	NS	Regulates the balance between precursor cell quiescence and proliferation.	Morrens et al., 2012
Rbfox3 (NeuN)	Contig68376	A6NFN3	5.00E-018	NS	-0.43	NS	Marker of mature neurons	

- significant up-regulation (p < 0.001; Log2FoldChange > 1) - significant down-regulation (p < 0.001; Log2FoldChange < -1) - trend (p < 0.05) NS - non-significant

Figure 10: Expression of neurogenesis-associated genes in regeneration of the radial organ complex in the sea cucumber H. glaberrima . Summary of the digital expression assay and homology search against the UniProt database. Red and blue colors indicate significant (adjusted p < 0.001, more than two fold change in expression level) up-regulation or down-regulation, respectively, in the regenerating tissues as opposed to uninjured organs. NS, no significant changes in expression were observed.

To our surprise, homologs of some of the key markers of vertebrate neural stem cells, including nestin and vimentin, were absent from the sea cucumber transcriptome. Likewise, transcripts of some of the important pro-neural factors [55], such as Mash1 and Neurog2 were also not detected in either the normal or regenerating animals. This may be due to the fact that the program of post-traumatic neurogenesis in sea cucumbers is not known and, obviously, may not entirely consist of the mechanisms that regulate neurogenesis in vertebrates, whose genes were used as the reference.

Prevention of Excitotoxic Neuronal Cell Death

One of the adaptations that may facilitate neural regeneration in the sea cucumber H. glaberrima is inhibition of excitotoxicity (Figure 11). In vertebrates, nervous system trauma results in two types of degenerative processes: the primary damage as the direct result of the injury, which affects the cells in the immediate vicinity of the trauma, and the secondary degeneration, which significantly increases the amount of the affected tissue and is mediated by endogenous mechanisms triggered by the initial injury [57, 58]. Excitotoxicity is one of these secondary pathological processes. It involves neuronal death via excessive stimulation by glutamate, whose extracellular concentration increases after CNS injury. Our results suggest that transection of the radial organ complex in the sea cucumber H. glaberrima results in significant down-regulation of glutamate carboxypeptidase II (GCPII) in the vicinity of the injury (Figure 11). This enzyme hydrolyzes NAAG (N-acetyl-aspartyl-glutamate) to glutamate and N-acetylaspartate. It can be hypothesized that down-regulation of GCPII will eventually result in inhibition of NAAG hydrolysis and accumulation of intact NAAG, which is known to further decrease the amount of glutamate released from presynaptic terminals and therefore prevent excitotoxic neuronal cell death [59]. Moreover, NAAG has been reported to stimulate production of important regulators of neuron survival, such as TGF- β[59, 60]. A sequence strongly related to TGF- β is present in the sea cucumber transcriptome. It shows a trend toward up-regulation on day 2 and is significantly up-regulated above the cut-off threshold on days 12 and 20 (Figure 11). On the other hand, the transcripts coding for another enzyme involved in glutamate production (Abat) and for the glutamate receptors (Glur1, Glur4) are also significantly down-regulated. Suppression of pathways leading to induction of excitotoxic neuronal death and concomitant induction of factors involved in neuronal survival can represent one of the adaptations aimed at minimizing the traumatic impact of the neuronal injury and sparing the maximum possible number of neurons, so that less of them will need to be replenished in the course of subsequent regeneration. A promising further line of research would be to experimentally test if changes in the severity of excitotoxicity would actually affect the extent of tissue damage and the rate of re-connection in the injured radial nerve cord.

A

Gene	Contig #	Best Blast Hit, Accession #	Best Blast Hit, E-value	Expression Profile (log2 fold change vs Norm)		
				Day 2	Day 12	Day 20
Glutamate carboxypeptidase II (GCPII)	Contig33698	O77564	1.96E-157	-2.13	-1.51	-1.54
Glutamate receptor 1 (Glur1)	Contig69406	P19490	1.04E-059	-1.92	-1.73	-1.61
GABA aminotransferase (Abat)	Contig30958	P80404	4.65E-138	-1.71	-0.79	-0.85
Transforming growth factor beta-2 (Tgfb2)	Contig851	P21214	6.05E-047	0.63	1.50	1.11

- significantly up-regulated - significantly down-regulated

B

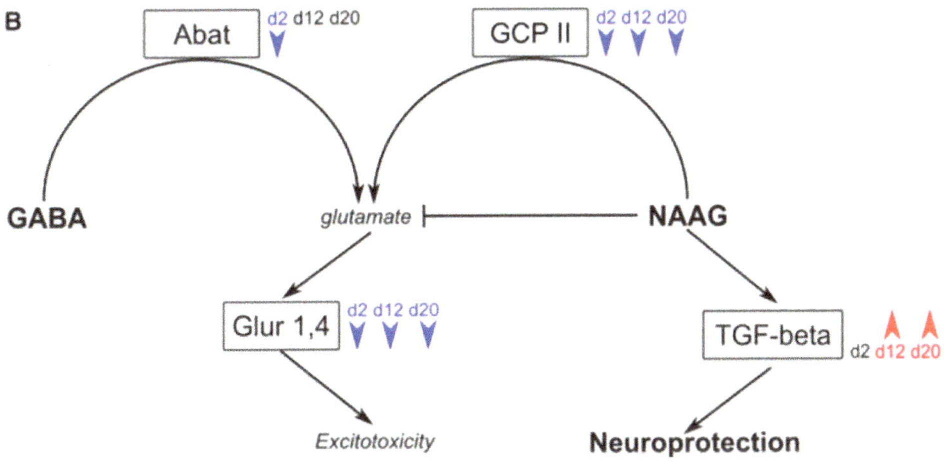

Figure 11: Suppression of excitotoxicity as a possible adaptation contributing to efficient neural regeneration in echinoderms. (A) The relevant genes in the transcriptome of the sea cucumber H. glaberrima. Their expression levels (relative to the uninjured tissues) and results of homology search against the UniProt database. Red and blue colors indicate significant (adjusted $p < 0.001$, more than two fold change in expression level) up-regulation or down-regulation, respectively. (B) A diagram illustrating a hypothetical mechanism of excitotoxicity suppression in the injured radial nerve cord of H. glaberrima. Down-regulation of genes coding for two enzymes, GCPIIand Abat, results in decreased glutamate production. Decreased activity of GCPII also leads to accumulation of its substrate, NAAG, which further inhibits release of glutamate from presynaptic terminals, but stimulates production of TGF-beta, which promotes neuronal survival. Down-regulation of ionotropic glutamate receptors (Glur1, Glur4) reduces the overstimulation of the post-synaptic membrane by glutamate and thus prevents downstream neurotoxic cascades from being triggered.

CONCLUSIONS

This study was conceived as a first stage in exploring molecular mechanisms behind the observed cellular processes in echinoderm CNS regeneration. In general, functional annotations of the differentially expressed genes corroborate well our previous morphological data, but also open up potential new avenues for future research. As mentioned above, our results point out to the important role of the ECM remodeling in regeneration of the radial complex in the sea cucumber. So far, reorganization of the connective tissue in regenerating echinoderms has received little attention. There have been just two experimental studies addressing this issue directly and in both cases they were focused on regeneration of the digestive tube only [33, 34].

Another promising line of future research would be to experimentally test the predictions of this study suggesting the existence of mechanisms suppressing excitotoxicity in the injured CNS. If these mechanisms actually exist, their understanding will be valuable for devising new therapies to prevent excitocytotic neuronal death following, for example, brain and spinal cord injury.

One of the most important outcomes from this study is a predicted list of putative transcription factors, which presumably control differential expression of large groups of downstream genes and thus occupy key positions in regulatory networks controlling regeneration. These genes represent promising candidates for future functional analysis.

Among other interesting findings is that post-traumatic regeneration of the radial nerve cord did not involve large-scale over-expression of pluripotency factors. Many of these genes were already expressed in intact tissues, and only Myc showed up-regulation after injury.

We are well aware of the limitations in our study. For example, by design, our functional annotation was dependent on matching the sea cucumber contigs against a well studied proteome from another organism (in this case, mouse). This approach could have led to many potentially relevant sea cucumber-specific sequences being excluded from the analysis. This issue cannot be resolved without carrying out a separate study aimed at characterization of the 'new' or 'unknown' sequences, which do not have significant homologs in current databases.

The present paper, like most of other high-throughput studies of gene expression, uses mRNA abundance levels to get insight into how changes in gene expression might affect the phenotype of tissues and cells, although it is largely the quantity of the protein that directly determines the phenotype. However, for the time being, transcriptomic approaches are justified by the fact that the current methodologies for direct quantification of protein expression are either less reliable or more laborious and expensive than mRNA-based studies.

Notwithstanding the limitations, the results and predictions reported above are valuable, because they provide a number of clearly defined testable hypotheses, whereas the associated pitfalls and limitations are well known to many researchers working with 'non-model' organisms and are not unresolvable in the future.

METHODS

Sea Cucumber Collection, Maintenance, and Radial Nerve Cord Injury

Adult individuals of the brown rock sea cucumber Holothuria glaberrima Selenka, 1867 were collected from the intertidal zone of the Atlantic coast of Puerto Rico. The reader is referred to our previous publications for detailed description of the injury paradigm and surgical procedures [1, 8, 12]. Briefly, the animals were brought to the laboratory and induced to eviscerate (autotomize their viscera) by injecting 0.35 M KCl into the coelomic cavity. The sea cucumbers needed to be eviscerated, because our surgery involved cutting the radial nerve cord from the inside of the body. In order to perform the transection we needed to get access to the inner side of the body wall. We did so by anesthetizing the animals in 0.2% chlorobutanol (Sigma) for 10–30 min and then exposing the coelomic surface of the body wall through the anus by pushing a glass rod against a radial region of the epidermis at the mid-body level. It was only possible after the animals have been induced to autotomize the viscera. H. glaberrimadoes not survive penetrating injuries to the body wall (when there is a direct communication between the coelom and the environment). Nevertheless, it readily regenerates if the injury is made from the coelomic side of the body wall without disrupting the epidermis. We thus cut the radial organs of the mid-ventral radius (including the longitudinal muscle band, water-vascular canal, the radial nerve cord, and the underlying connective tissue), but not the epidermis, with a sharp razor blade. The operated animals were returned to the aquaria and kept at room temperature in well-aerated seawater, which was changed regularly. All experiments were conducted in accordance with the NIH and University of Puerto Rico guidelines for the care and use of laboratory animals.

RNA Extraction and Library Preparation

Samples for high-throughput sequencing were prepared as previously described [12]. From each regenerating animal, we excised the region of the injury gap (~3–4 mm wide) plus ~3 mm of stump ('old') tissue on either side of the injury plane. The wet weight of an individual tissue sample was around 10–15 mg. Tissue samples of comparable size and weight were also excised from uninjured animals. During tissue sampling, every effort was made to separate the radial nerve cord from surrounding tissues. However, isolation of the pure nerve cord by surgical means turned out to be practically impossible. Therefore, our tissue samples also consistently contained small amounts of the surrounding connective tissue, an accompanying segment of the water-vascular canal and a stretch of the contractile coelomic epithelium of the body wall because of close anatomical proximity of these structures to the radial nerve cord (Figure 2). For the 454 platform, we generated three non-normalized libraries representing uninjured animals (38 individuals), days 2 and 6 post-injury (63 and 71 animals, respectively), and days 12 and 20 post-injury (62 and 66 animals, respectively). In addition, equal quantities of the above samples were combined to prepare a normalized library. The samples extracted from the regenerating animals on day 6 post-injury were only used for 454 sequencing to increase transcript diversity in the final assembled transcriptome, and were not subjected to sequencing on the Illumina platform (see below). Total RNA was extracted using TRI reagent (Sigma), assessed for quality on an Agilent 2100 Bioanalyzer with the RNA 6000 Nano chips, and subjected to two rounds of poly(A) selection using Poly(A)Purist technology (Ambion). Normalization procedure was performed with a TRIMMER kit (Evrogen) following the manufacturer's protocol. The normalized cDNA was amplified using Advantage 2 Polymerase Mix (Clontech).

For Illumina sequencing two non-normalized libraries were prepared for each of the four conditions: (i) uninjured radial organ complex (total RNA samples were pooled from 4 and 3 animals for the first and second libraries, respectively); (ii) day 2 post-injury (20 and 19 animals were used); (iii) day 12 post-injury (20 animals were used for each of the libraries); and (iv) day 20 post-injury (15 animals were used for each of the libraries). The final stages in library preparation and sequencing were performed by sequencing service providers at the DNA Facility of the University of Iowa (Genome Sequencer FLX System, Roche) and the Genome Sequencing and Analysis Core Facility of the Duke Institute for Genome Sciences and Policy (Illumina Genome Analyzer IIx, Illumina). Raw sequencing reads from both the 454 and Illumina platforms were deposited at NCBI Sequence Read Archive (SRA) under accession number NCBI:SRA051990 [61].

De novo Assembly Pipeline

The first round of filtering/cleaning of 454 reads was performed with SeqClean [62] and included removal of synthetic adaptor/primer sequences used in library preparation and screening for E. coli contamination (GenBank:U00096). The reads were then processed with the standalone version of PRINSEQ tool (v 0.14.2) [63] with the following filtering parameters: minimum length — 60 bases, maximum length — 1100 bases (twice the mean length), maximum allowed percentage of N's – 1%, minimum mean quality — 17. Also, exact duplicates and reverse complement exact duplicates were removed.

Raw Illumina sequencing reads were quality checked using the FastQC tool (v 0.9.1) [64] and then trimmed using FASTX-Toolkit (v 0.0.13) [65] to remove Illumina adapters, discard bases with quality < 20, and reads shorter than 32 bases.

We applied a complex pipeline to assemble the pooled reads from all libraries into a single reference contig set (Figure 3). First, we used MIRA 3.2.1 [66] to assemble all normalized and non-normalized 454 reads. The Illumina reads were assembled with Velvet (v.1.1.03) [10] and Oases (v.0.1.21). Separate assembly runs were performed at different kmer lengths (31 to 55). Cleaned 454 reads were introduced into each of these runs to improve the assembly. The contigs longer than 100 bp resulting from the Velvet/Oases assemblies were combined together with 454 contigs for the final assembly with CAP3, to produce what we refer to as the reference library. The files containing the contigs, their annotations, and expression values were parsed and analyzed using custom written scripts, which are available from the authors by request.

Differential Gene Expression in Regeneration

For digital expression assay based on RNA-seq data, we used Bowtie 2 [67] to map Illumina reads to the contigs of the reference library. The raw read counts for each of the eight Illumina libraries (four conditions, two biological replicates per condition, see above) were used as input to the DESeq R package [16] to perform pairwise differential expression analysis between the intact and regenerating animals. The estimateSizeFactors function of the DESeq package was used to normalize gene counts. The resulting P values were adjusted for multiple testing with Benjamini-Hochberg procedure. Genes with an adjusted P value < 0.001 and a fold change greater than 2 were considered differentially expressed.

In order to validate the changes in gene expression determined by RNA-seq, we selected 21 genes with different levels of transcript abundances for real-time PCR analysis. Poly(A) RNA was extracted as described above. PCR primers were designed using Primer Premier 5.0 software (PREMIER Biosoft International). RNA was reverse transcribed with random hexamer primers and SuperScript II reverse transcriptase (Invitrogen). Template cDNA was diluted 10-fold or 100-fold and used at 2 μl per 25 μl of PCR reaction with Brilliant SYBR Green Master Mix or Brilliant II SYBR Green Master Mix (Agilent) following the manufacturer's protocol. Real-time PCR reactions were run on Mx3005P qPCR System (Stratagene). The reactions were performed on three independent samples per condition (biological replicates). Each sample was analyzed at least twice, making sure that the difference between technical replicates was less than 0.5 Ct [68]. PCR efficiencies were evaluated by running five 10-fold dilutions of the cDNA template in a PCR reaction and were considered acceptable if the corresponding slope values determined by the MxPro QPCR software (Strategene) lay between -3.2 and -3.5 and the R^2 was above 0.98. All expression values were normalized relative to the 'normalization factor' calculated with the geNorm Visual Basic Application for Microsoft Excel [69] from the expression of values of four genes (Rpl18a, Atp6l, Eef2, and Sod), which were identified among the least changing transcripts across the experimental conditions by RNA-seq.

Annotation

The assembled contigs of the reference library were used as input for the BLASTX homology search. Initially, the sea cucumber sequences were compared with the Swiss-Prot database with a cut-off significance threshold set at 1e-6. Those contigs that lacked matches were then subjected to a second round of BLASTX search against the larger NCBI non-redundant database. In addition, the sea cucumber transcriptome was also annotated versus the NCBI's collection of the sea urchin predicted protein sequences [14] by performing reciprocal best BLAST hit analysis (with a threshold e-value < 1e-6).

Functional annotation of differentially expressed genes was performed with DAVID Gene Ontology web server [18]. In order to be able to use this tool, we matched all 70,173 contigs of our assembled transcriptome to the non-redundant reference proteome of the mouse [70], release 2012_05 using BLASTX with the cut-off e-value of 1e-6. Overall, our assembled contigs showed significant homology to 8,522 mouse genes. We then submitted the annotated lists of differentially expressed genes as an input to DAVID and analyzed them against the background of all annotated genes of our reference library. For pathways of interest, KEGGanim [71] was used to generate diagrams showing changes in expression level of individual genes. This approach allows to observe expression dynamics in the context of specific pathway interactions.

Expression Profile Clustering

Unsupervised gene expression profile clustering of differentially expressed genes (i.e., the genes showing more than two-fold change in expression with adjusted P < 0.01) was performed using AutoSOME 2.0 software [20]. Prior to clustering, the original count data were subjected to variance stabilization transformation in the DESeq package. The parameters were set as suggested by the authors of the program (running mode: precision, number of ensemble runs: 500, P-value threshold: 0.05). Unit variance, median centering (rows), and sum squares (both rows and columns) normalization procedures were applied. Eight clusters containing more than 100 contigs were visualized on a heatmap (Figure 7) and considered for further analysis.

Identification of Putative Regeneration-Associated Transcription Factors

Over-represented transcription factors associated with co-expressed genes were predicted with oPOSSUM v3.0 software [21]. Lists of gene identifiers corresponding to each of the cluster identified by AutoSOME were used as input. The list of gene identifiers corresponding to the entire reference library was used as the background set. The JASPAR CORE collection was used as a set of transcription binding site matrices. The conservation cutoff and the matrix score threshold were left at their default values of 0.4 and 85%, respectively.

In order to select potentially relevant transcription factors, we followed the suggestions of the software's authors and plotted the Z-score against Fisher score for the transcription factors associated with the genes in each of the cluster. The genes, which showed clear segregation of the scores, were considered for further analysis, if changes in their expression were significant (adjusted $P < 0.05$) at least at one of the three analyzed time points. If the expression level of the putative transcription factor changed in the same direction as the median gene expression in the gene cluster, this transcription factor was considered a transcriptional activator. If the change in expression was in the reverse direction, the transcription factor was considered a transcriptional repressor.

AVAILABILITY OF SUPPORTING DATA

Raw sequencing reads supporting the results of this article are available in the NCBI SRA repository [61]. The contigs of the reference library, representing the transcriptomic diversity in the normal and regenerating radial organ complex of H. glaberrima are available in the LabArchives notebook [11].

ACKNOWLEDGEMENTS

The authors acknowledge the assitance of Rey Rosa in animal collection and maintenance. We also thank Prof. B. Galliot, Prof. L. Moroz, and Prof. H. Ortiz-Zuazaga for stimulating discussions and valuable suggestions. The study was supported by the NIH (Grants 1SC1GM084770-01, 1R03NS065275-01), the NSF (Grants IOS-0842870, IOS-1252679), several NSF and NIH equipment funds for the Sequencing Genomic Facility (SGF UPRRP) and the University of Puerto Rico.

AUTHORS' CONTRIBUTIONS

VSM, ORZ, and JEGA conceived the study, analyzed the data, and interpreted the results. VSM and ORZ carried out experimental procedures. VSM drafted the manuscript. VSM, ORZ, and JEGA finalized the manuscript. All authors read and approved the final version of the manuscript.

REFERENCES

1. Mashanov VS, Zueva OR, García-Arrarás JE: Radial glial cells play a key role in echinoderm neural regeneration. BMC Biol. 2013, 11: 49-[http://www.biomedcentral.com/1741-7007/11/49],
2. Mashanov VS, Zueva O, García-Arrarás JE: Chapter seven - postembryonic organogenesis of the digestive tube: why does it occur in worms and sea cucumbers but fail in humans?. Mechanisms of Regeneration, Volume 108 of Current Topics in Developmental Biology. 2014, San Diego: Academic Press, 185-216. [http://www.sciencedirect.com/science/article/pii/B9780123914989000061],
3. Rojas-Cartagena C, Ortíz-Pineda P, Ramírez-Gómez F, Suárez-Castillo EC, Matos-Cruz V, Rodríguez C, Ortíz-Zuazaga H, García-Arrarás JE: Distinct profiles of expressed sequence tags during intestinal regeneration in the sea cucumber Holothuria glaberrima. Physiol Genomics. 2007, 31 (2): 203-215.
4. Sun L, Chen M, Yang H, Wang T, Liu B, Shu C, Gardiner DM: Large scale gene expression profiling during intestine and body wall regeneration in the sea cucumber Apostichopus japonicus. Comp Biochem Physiol Part D Genomics Proteomics. 2011, 6 (2): 195-205.
5. Heinzeller T, Welsch U: The echinoderm nervous system and its phylogenetic interpretation. Brain evolution and cognition. Edited by: Roth G, Wullimann M. 2001, New York: Wiley-Spektrum, 41-75.
6. García-Arrarás JE, Dolmatov IY: Echinoderms: potential model systems for studies on muscle regeneration. Curr Pharm Des. 2010, 16 (8): 942-955.
7. Mashanov VS, Zueva OR, Heinzeller T: Regeneration of the radial nerve cord in a holothurian: a promising new model system for studying post-traumatic recovery in the adult nervous system. Tissue Cell. 2008, 40 (5): 351-372.
8. Miguel-San Ruiz JE, García-Arrarás JE: Common cellular events occur during wound healing and organ regeneration in the sea cucumber Holothuria glaberrima. BMC Dev Biol. 2007, 7: 115-
9. Surget-Groba Y, Montoya-Burgos JI: Optimization of de novo transcriptome assembly from next-generation sequencing data. Genome Res. 2010, 20 (10): 1432-1440.
10. Zerbino DR, Birney E: Velvet: algorithms for de novo short read assembly using de Bruijn graphs. Genome Res. 2008, 18 (5): 821-829.
11. Contigs of the reference library available in the LabArchives notebook. [doi:10.6070/H4PN93J1],
12. Mashanov VS, Zueva OR, García-Arrarás JE: Posttraumatic regeneration involves differential expression of long terminal repeat (LTR) retrotransposons. Dev Dyn. 2012, 241 (10): 1625-1636.
13. Mashanov VS, Zueva OR, García-Arrarás JE: Retrotransposons in animal regeneration: overlooked components of the regenerative machinery?. Mob Genet Elem. 2012, 2 (5): 244-247.
14. NCBI collection of predicted proteins of the sea urchin Strongylocentrotus purpuratus. [ftp://ftp.ncbi.nih.gov/genomes/Strongylocentrotus_purpuratus/protein/],
15. Sodergren E, Weinstock GM, Davidson EH, Cameron RA, Gibbs RA, Angerer RC, Angerer LM, Arnone MI, Burgess DR, Burke RD, Coffman JA, Dean M, Elphick MR, Ettensohn CA, Foltz KR, Hamdoun A, Hynes RO, Klein WH, Marzluff W, McClay DR, Morris RL, Mushegian A, Rast JP, Smith LC, Thorndyke MC, Vacquier VD, Wessel GM, Wray G, Consortium SUGS, et al: The genome of the sea urchin Strongylocentrotus purpuratus. Science. 2006, 314 (5801): 941-952.
16. Anders S, Huber W: Differential expression analysis for sequence count data. Genome Biol. 2010, 11 (10): R106-
17. Stewart R, Rascón CA, Tian S, Nie J, Barry C, Chu LF, Ardalani H, Wagner RJ, Probasco MD, Bolin JM, Leng N, Sengupta S, Volkmer M, Habermann B, Tanaka EM, Thomson JA, Dewey CN: Comparative RNA-seq analysis in the unsequenced axolotl: the oncogene burst highlights early gene expression in the blastema. PLoS

Comput Biol. 2013, 9 (3): e1002936-[doi:10.1371/journal.pcbi.1002936],

18. The Database for Annotation, Visualization and Integrated Discovery (DAVID) v6.7. [http://david.abcc.ncifcrf.gov/],
19. Huang DW, Sherman BT, Lempicki RA: Systematic and integrative analysis of large gene lists using DAVID bioinformatics resources. Nat Protoc. 2009, 4: 44-57. [doi:10.1038/nprot.2008.211],
20. Newman AM, Cooper JB: AutoSOME: a clustering method for identifying gene expression modules without prior knowledge of cluster number. BMC Bioinformatics. 2010, 11: 117-[doi:10.1186/1471-2105-11-117],
21. Sui SJH, Fulton DL, Arenillas DJ, Kwon AT, Wasserman WW: oPOSSUM: integrated tools for analysis of regulatory motif over-representation. Nucleic Acids Res. 2007, 35 (Web Server issue): W245-W252. [doi:10.1093/nar/gkm427],
22. Schwitalla S, Fingerle AA, Cammareri P, Nebelsiek T, Göktuna SI, Ziegler PK, Canli O, Heijmans J, Huels DJ, Moreaux G, Rupec RA, Gerhard M, Schmid R, Barker N, Clevers H, Lang R, Neumann J, Kirchner T, Taketo MM, van den Brink, Sansom OJ, Arkan MC, Greten FR: Intestinal tumorigenesis initiated by dedifferentiation and acquisition of stem-cell-like properties. Cell. 2013, 152 (1-2): 25-38. [doi:10.1016/j.cell.2012.12.012],
23. Wenemoser D, Lapan SW, Wilkinson AW, Bell GW, Reddien PW: A molecular wound response program associated with regeneration initiation in planarians. Genes Dev. 2012, 26 (9): 988-1002. [doi:10.1101/gad.187377.112],
24. Tarnawski AS, Ahluwalia A: Molecular mechanisms of epithelial regeneration and neovascularization during healing of gastric and esophageal ulcers. Curr Med Chem. 2012, 19: 16-27.
25. Knöll B: Serum response factor mediated gene activity in physiological and pathological processes of neuronal motility. Front Mol Neurosci. 2011, 4: 49-[doi:10.3389/fnmol.2011.00049],
26. Huang GN, Thatcher JE, McAnally J, Kong Y, Qi X, Tan W, DiMaio JM, Amatruda JF, Gerard RD, Hill JA, Bassel-Duby R, Olson EN: C/EBP transcription factors mediate epicardial activation during heart development and injury. Science. 2012, 338 (6114): 1599-1603. [doi:10.1126/science.1229765],
27. Marusawa H, Matsuzawa SI, Welsh K, Zou H, Armstrong R, Tamm I, Reed JC: HBXIP functions as a cofactor of survivin in apoptosis suppression. EMBO J. 2003, 22 (11): 2729-2740. [doi:10.1093/emboj/cdg263],
28. Xia F, Canovas PM, Guadagno TM, Altieri DC: A survivin-ran complex regulates spindle formation in tumor cells. Mol Cell Biol. 2008, 28 (17): 5299-5311. [doi:10.1128/MCB.02039-07],
29. Ruchaud S, Carmena M, Earnshaw WC: The chromosomal passenger complex: one for all and all for one. Cell. 2007, 131 (2): 230-231. [doi:10.1016/j.cell.2007.10.002],
30. Li F, Cheng Q, Ling X, Stablewski A, Tang L, Foster BA, Johnson CS, Rustum YM, Porter CW: Generation of a novel transgenic mouse model for bioluminescent monitoring of survivin gene activity in vivo at various pathophysiological processes: survivin expression overlaps with stem cell markers. Am J Pathol. 2010, 176 (4): 1629-1638. [doi:10.2353/ajpath.2010.090414],
31. Mashanov VS, Zueva OR, Rojas-Catagena C, García-Arrarás JE: Visceral regeneration in a sea cucumber involves extensive expression of survivin and mortalin homologs in the mesothelium. BMC Dev Biol. 2010, 10: 117-[doi:10.1186/1471-213X-10-117],
32. Bellayr IH, Mu X, Li Y: Biochemical insights into the role of matrix metalloproteinases in regeneration: challenges and recent developments. Future Med Chem. 2009, 1 (6): 1095-1111. [doi:10.4155/fmc.09.83],
33. Lamash NE, Dolmatov IY: Proteases from the regenerating gut of the holothurian Eupentacta fraudatrix. PLoS One. 2013, 8 (3): e58433-[doi:10.1371/journal.pone.0058433],
34. Quiñones JL, Rosa R, Ruiz DL, García-Arrarás JE: Extracellular matrix remodeling and metalloproteinase involvement during intestine regeneration in the sea cucumber Holothuria glaberrima. Dev Biol. 2002, 250: 181-197.
35. Zargham R, Thibault G: Alpha 8 integrin expression is required for maintenance of the smooth muscle cell differentiated phenotype. Cardiovasc Res. 2006, 71: 170-178. [doi:10.1016/j.cardiores.2006.03.003],
36. Benoit YD, Lussier C, Ducharme PA, Sivret S, Schnapp LM, Basora N, Beaulieu JF: Integrin alpha8beta1 regulates adhesion, migration and proliferation of human intestinal crypt cells via a predominant RhoA/ROCK-dependent mechanism. Biol Cell. 2009, 101 (12): 695-708. [doi:10.1042/BC20090060],
37. Lahlou H, Muller WJ: β1-integrins signaling and mammary tumor progression in transgenic mouse models: implications for human breast cancer. Breast Cancer Res. 2011, 13 (6): 229-[doi:10.1186/bcr2905],
38. Stengel K, Zheng Y: Cdc42 in oncogenic transformation, invasion, and tumorigenesis. Cell Signal. 2011, 23 (9): 1415-1423. [doi:10.1016/j.cellsig.2011.04.001],
39. Chen C, Wirth A, Ponimaskin E: Cdc42: an important regulator of neuronal morphology. Int J Biochem Cell Biol. 2012, 44 (3): 447-451. [doi:10.1016/j.biocel.2011.11.022],
40. Newton K, Dixit VM: Signaling in innate immunity and inflammation. Cold Spring Harb Perspect Biol. 2012, 4 (3): [doi:10.1101/cshperspect.a006049],
41. Takahashi K, Yamanaka S: Induction of pluripotent stem cells from mouse embryonic and adult fibroblast cultures by defined factors. Cell. 2006, 126 (4): 663-676. [doi:10.1016/j.cell.2006.07.024],
42. Sridharan R, Tchieu J, Mason MJ, Yachechko R, Kuoy E, Horvath S, Zhou Q, Plath K: Role of the murine reprogramming factors in the induction of pluripotency. Cell. 2009, 136 (2): 364-377. [doi:10.1016/j.cell.2009.01.001],

43. Wey A, Knoepfler PS: c-myc and N-myc promote active stem cell metabolism and cycling as architects of the developing brain. Oncotarget. 2010, 1 (2): 120-130.

44. Range R, Lepage T: Maternal Oct1/2 is required for Nodal and Vg1/Univin expression during dorsal-ventral axis specification in the sea urchin embryo. Dev Biol. 2011, 357 (2): 440-449. [doi:10.1016/j.ydbio.2011.07.005],

45. Tantin D: Oct transcription factors in development and stem cells: insights and mechanisms. Development. 2013, 140 (14): 2857-2866. [doi:10.1242/dev.095927],

46. Christen B, Robles V, Raya M, Paramonov I, Belmonte JCI: Regeneration and reprogramming compared. BMC Biol. 2010, 8: 5-[doi:10.1186/1741-7007-8-5],

47. Beckervordersandforth R, Tripathi P, Ninkovic J, Bayam E, Lepier A, Stempfhuber B, Kirchhoff F, Hirrlinger J, Haslinger A, Lie DC, Beckers J, Yoder B, Irmler M, Götz M: In vivo fate mapping and expression analysis reveals molecular hallmarks of prospectively isolated adult neural stem cells. Cell Stem Cell. 2010, 7 (6): 744-758. [doi:10.1016/j.stem.2010.11.017],

48. Bréjot T, Blanchard S, Hocquemiller M, Haase G, Liu S, Nosjean A, Heard JM, Bohl D: Forced expression of the motor neuron determinant HB9 in neural stem cells affects neurogenesis. Exp Neurol. 2006, 198: 167-182. [doi:10.1016/j.expneurol.2005.11.026],

49. Covey MV, Streb JW, Spektor R, Ballas N: REST regulates the pool size of the different neural lineages by restricting the generation of neurons and oligodendrocytes from neural stem/progenitor cells. Development. 2012, 139 (16): 2878-2890. [doi:10.1242/dev.074765],

50. Kim EJ, Ables JL, Dickel LK, Eisch AJ, Johnson JE: Ascl1 (Mash1) defines cells with long-term neurogenic potential in subgranular and subventricular zones in adult mouse brain. PLoS One. 2011, 6 (3): e18472-[doi:10.1371/journal.pone.0018472],

51. Li H, Jin G, Qin J, Yang W, Tian M, Tan X, Zhang X, Shi J, Zou L: Identification of neonatal rat hippocampal radial glia cells in vitro. Neurosci Lett. 2011, 490 (3): 209-214. [doi:10.1016/j.neulet.2010.12.054],

52. Misra K, Mishra K, Gui H, Matise MP: Prox1 regulates a transitory state for interneuron neurogenesis in the spinal cord. Dev Dyn. 2008, 237 (2): 393-402. [doi:10.1002/dvdy.21422],

53. Morrens J, Broeck WVD, Kempermann G: Glial cells in adult neurogenesis. Glia. 2012, 60 (2): 159-174. [doi:10.1002/glia.21247],

54. Park CH, Kang JS, Kim JS, Chung S, Koh JY, Yoon EH, Jo AY, Chang MY, Koh HC, Hwang S, Suh-Kim H, Lee YS, Kim KS, Lee SH: Differential actions of the proneural genes encoding Mash1 and neurogenins in Nurr1-induced dopamine neuron differentiation. J Cell Sci. 2006, 119 (Pt 11): 2310-2320. [doi:10.1242/jcs.02955],

55. Roybon L, Hjalt T, Stott S, Guillemot F, Li JY, Brundin P: Neurogenin2 directs granule neuroblast production and amplification while NeuroD1 specifies neuronal fate during hippocampal neurogenesis. PLoS One. 2009, 4 (3): e4779-[doi:10.1371/journal.pone.0004779],

56. Shu T, Tseng HC, Sapir T, Stern P, Zhou Y, Sanada K, Fischer A, Coquelle FM, Reiner O, Tsai LH: Doublecortin-like kinase controls neurogenesis by regulating mitotic spindles and M phase progression. Neuron. 2006, 49: 25-39. [doi:10.1016/j.neuron.2005.10.039],

57. Mark LP, Prost RW, Ulmer JL, Smith MM, Daniels DL, Strottmann JM, Brown WD, Hacein-Bey L: Pictorial review of glutamate excitotoxicity: fundamental concepts for neuroimaging. AJNR Am J Neuroradiol. 2001, 22 (10): 1813-1824.

58. Weil ZM, Norman GJ, DeVries AC, Nelson RJ: The injured nervous system: a Darwinian perspective. Prog Neurobiol. 2008, 86: 48-59. [doi:10.1016/j.pneurobio.2008.06.001],

59. Bar□inka C, Rojas C, Slusher B, Pomper M: Glutamate carboxypeptidase II in diagnosis and treatment of neurologic disorders and prostate cancer. Curr Med Chem. 2012, 19 (6): 856-870.

60. Krieglstein K, Strelau J, Schober A, Sullivan A, Unsicker K: TGF-beta and the regulation of neuron survival and death. J Physiol Paris. 2002, 96 (1-2): 25-30.

61. Raw sequencing reads supporting the results of this article deposited with the NCBI SRA repository. [https://www.ncbi.nlm.nih.gov/sra/?term=SRA051990],

62. SeqClean tool. [http://compbio.dfci.harvard.edu/tgi/software/],

63. Schmieder R, Edwards R: Quality control and preprocessing of metagenomic datasets. Bioinformatics. 2011, 27 (6): 863-864. [doi:10.1093/bioinformatics/btr026],

64. FastQC Tool: [http://www.bioinformatics.babraham.ac.uk/projects/fastqc/]

65. FASTX-Toolkit: [http://hannonlab.cshl.edu/fastx_toolkit/index.html]

66. Chevreux B, Pfisterer T, Drescher B, Driesel AJ, Müller WEG, Wetter T, Suhai S: Using the miraEST assembler for reliable and automated mRNA transcript assembly and SNP detection in sequenced ESTs. Genome Res. 2004, 14 (6): 1147-1159. [doi:10.1101/gr.1917404],

67. Langmead B, Trapnell C, Pop M, Salzberg SL: Ultrafast and memory-efficient alignment of short DNA sequences to the human genome. Genome Biol. 2009, 10 (3): R25-[doi:10.1186/gb-2009-10-3-r25],

68. Nolan T, Hands RE, Bustin SA: Quantification of mRNA using real-time RT-PCR. Nat Protoc. 2006, 1 (3): 1559-1582. [doi:10.1038/nprot.2006.236],

69. Vandesompele J, Preter KD, Pattyn F, Poppe B, Roy NV, Paepe AD, Speleman F: Accurate normalization of real-time quantitative RT-PCR data by geometric averaging of multiple internal control genes. Genome Biol. 2002, 3 (7): RESEARCH0034-

70. Mouse reference proteome. [https://www.ebi.ac.uk/reference_proteomes/],

71. Adler P, Reimand J, JÄnes J, Kolde R, Peterson H, Vilo J: KEGGanim: pathway animations for high-throughput data. Bioinformatics. 2008, 24 (4): 588-590. [doi:10.1093/bioinformatics/btm581],

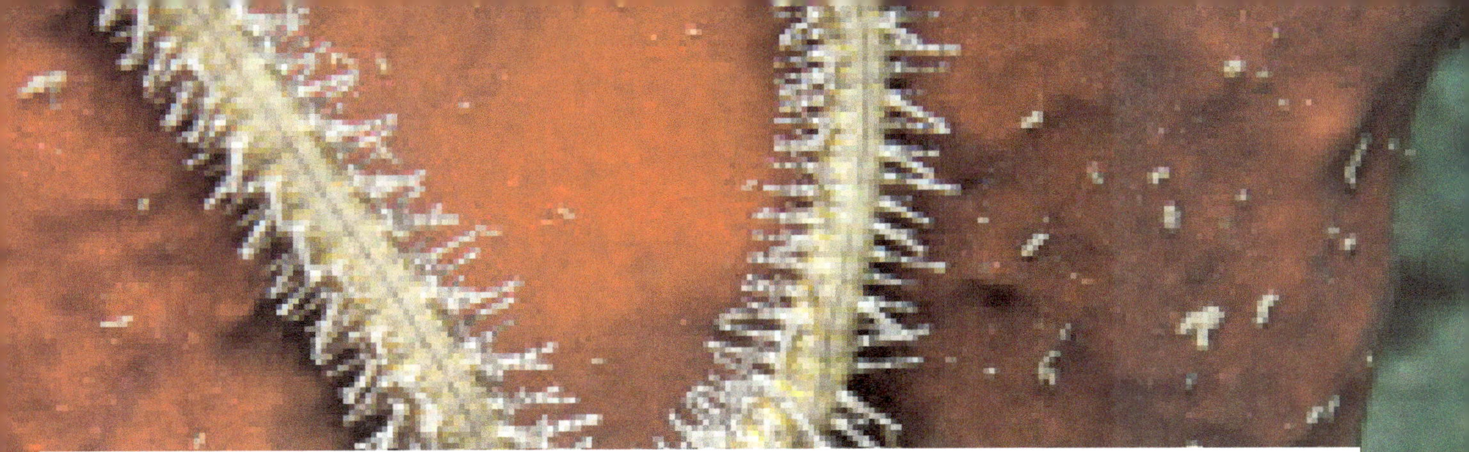

CHAPTER 3 EVOLUTION OF LARVAL DEVELOPMENT IN ECHINODERMS

Kristen A Yankura, Megan L Martik, Charlotte K Jennings, Veronica F Hinman

BACKGROUND

The astonishing diversity of animal forms, coupled with the complex life histories typical of many marine invertebrates, presents numerous challenges in inferring the ancestral character of members of the closely related phyla collectively known as the deuterostomes. Modern molecular phylogenies place four phyla within the monophyletic deuterostomes: Echinodermata and Hemichordata comprise a distinct clade called the Ambulacraria [1–3] that is a sister group to Chordata [4]. Xenoturbella is a recent out-group addition to the Ambulacraria [5].

Within the Ambulacraria, the free-swimming, bilaterally symmetric larvae of echinoderms, especially the bipinnaria larva of sea stars and the auricularia larva of sea cucumbers, share many similarities with the tornaria larva of indirectly developing hemichordates. These microscopic larvae have an apical concentration of serotonergic neurons [6] and one or two concentrations, or bands, of cilia used to feed and swim in the plankton [7, 8]. Neurons lie beneath this ciliated epithelium and innervate the bands [9]. Similarities in larval form initially provided the basis for many of the hypotheses surrounding the evolutionary origins of the chordates and, in particular, the centralized nervous system. These hypotheses, in which a microscopic larval stage is assumed ancestral to the entire deuterostome clade, propose that a centralized nervous system evolved from an infolding of the larval ciliary bands [10–12].

Not all Ambulacrarians develop through a larval stage, however, and recent comparisons of regulatory gene expression have revealed that orthologs of many genes and signaling molecules involved in vertebrate neural patterning are expressed in spatially restricted domains along the anterior-posterior (AP) axis of the direct developing vermiform hemichordate juvenile*Saccoglossus kowalevskii* [13, 14], a species that does not develop via a tornaria larva. This general correspondence in AP position of orthologs between the vertebrate and hemichordate nervous systems implies some homology in axial patterning. In addition, the broad ectodermal expression of these genes in hemichordates suggests that a diffuse panectodermal neural domain was the ancestral state of the deuterostome nervous system and that the centralization event occurred later within the lineage leading to the chordates. These findings and those of other

Citation: Yankura et al.: evolution of larval development in echinoderms. BMC Biology 2010 8:143. doi:10.1186/1741-7007-8-143.

researchers (reviewed in [15, 16]) therefore negate the need to invoke a ciliated ancestor for deuterostomes as suggested by Garstang [10]. In addition, general homologies in axial expression patterns between vertebrates and direct developing protostomes have been observed as well [17], suggesting that common patterning mechanisms have been employed since the radiation of Bilaterians. Indirect developing larval forms have less obviously distinguished body axes and no strong homology of axial patterning and, as a result, appear derived and possibly secondarily simplified in comparison.

Here we examine the expression of regulatory gene orthologs that have known or suspected roles in patterning the axial neuroectoderm of many protostome and deuterostome embryos within the indirectly developing sea star, *Patiria miniata* (previously *Asterina miniata*), which forms a typical bipinnaria larva. We show that these genes are expressed in diffuse concentric ectodermal domains that pattern the early embryonic axis. Furthermore, we observe in the sea star a general correspondence of domains of orthologous gene expression to those found along the AP and dorsal-ventral (DV) axis of direct developing deuterostomes. In addition, we detect expression of retinal determining gene orthologs in the mesoderm of echinoderm larvae, but not within the ectoderm of gastrulating embryos. We discuss the role of this implied modularity of regulatory patterning during evolution.

RESULTS AND DISCUSSION

Isolation of Sea Star Transcription Factors

P. miniata orthologs of regulatory genes that have known or suspected roles in patterning the axial neuroectoderm of many protostome and deuterostome embryos were isolated using a candidate gene approach. Recombinants for the following seven genes that encode homeodomain proteins were obtained: *retinal homeobox* (*rx*), *optix-like homeobox 3* (*six3*),*gastrulation and brain-specific homeobox* (*gbx*), *lim domain homeobox 2* (*lhx2*) and *paired box homeobox 6* (*pax6*), as well as members of the Nkx gene family, *nk2.1* and *nk1*. We also identified partial sequences of *eyes absent* (*eya*), the ets family gene *pea3*, and two C2H2 zinc-finger genes, *zic* and *krupple-like factor 13* (*klf13*). The following four winged-helix forkhead box genes were isolated: *foxq2*, *foxj1*, *foxd* and *foxg*. A complete list of orthologs, sequence lengths, and orthology of gene sequences is provided.

Animal-Vegetal Patterning of the Sea Star Blastula Ectoderm

Sea star late blastulae have a morphologically distinct animal-vegetal (AV) axis that is first readily observed when elongation of cells at the vegetal pole results in a noticeable thickening of epithelium termed the *vegetal plate* [18]. During gastrulation, the vegetal plate invaginates to produce the mesoderm and endoderm of the larva, leaving the remaining animal epithelium as ciliated ectoderm [18]. At this stage, no obvious morphological differences in ectodermal cell type have been observed, but we nonetheless reveal here a remarkable complexity of regulatory states within the ectoderm (summarized in Figure 1).

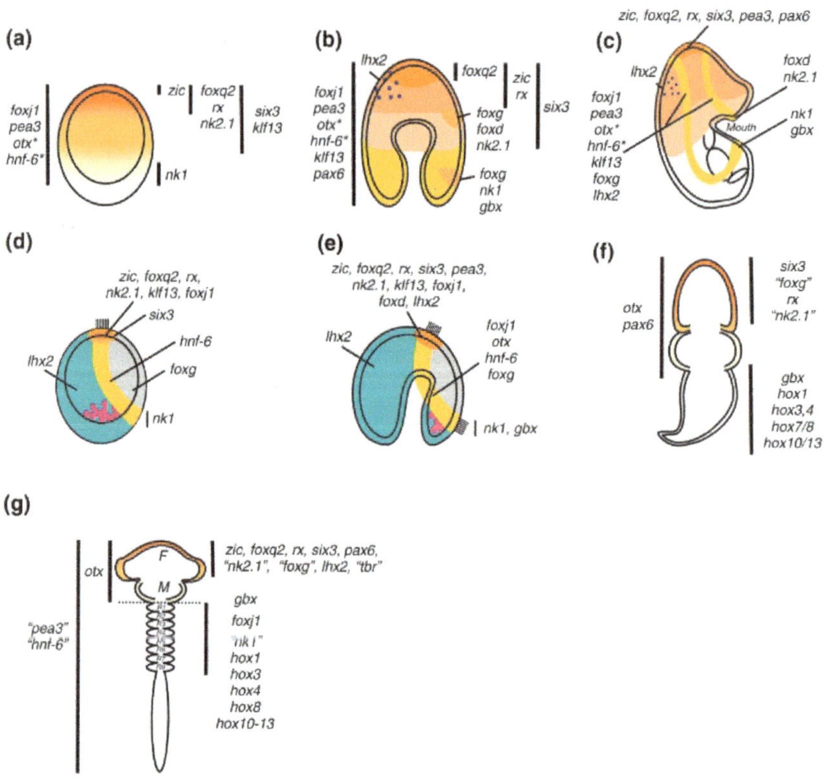

Figure 1: Comparison of orthologous neuroectodermal gene expression domains among the deuterostomes. (A-E) Indirectly developing echinoderms. (F) Directly developing hemichordate. (G) Generalized vertebrate. Sea stars (Figures 1A-1C) and sea urchins (Figures 1D and 1E) are viewed laterally; animal pole is up and oral side is right. Figures 1F and 1G are dorsal views; anterior is up. Genes are listed beside their cognate expression domains. Vertical bars in Figures 1A, 1B, 1F and 1G approximate domain boundaries. The orange to yel-

low gradient in Figures 1A, 1F and 1G reflects a general conservation of anterior (animal)-most axial patterning among the three phyla. (A)Nested, concentric expression domains pattern the animal-vegetal (AV) axis of blastulae; asterisks denote previously reported expression [18, 19]. (B) Concentric domains of *zic,foxq2*, *rx* and *si x3* persist in gastrulae (orange to peach gradient); additional oral (for example, *foxg*, *foxd* and *gbx*; light orange) and aboral (for example, *lhx2*; purple) domains are evident. Genes (left) are broadly expressed. (C) Expression in larval animal pole domain (orange to peach) and/or ciliary bands (gold). (D) Sea urchin animal pole (orange and light orange), ciliary band (gold), aboral ectoderm (turquoise) and oral ectoderm (*foxg*; gray) are molecularly distinct territories in blastulae. (E) Expression is maintained in gastrulae animal pole (orange) and ciliary band (gold). (D and E) Pink circles represent skeletogenic mesoderm. See references [20, 22, 23, 41–49]. (F) Orthologs expressed in hemichordate anterior, middle and posterior body segments show corresponding expression in the vertebrate forebrain, midbrain and hindbrain, respectively; data are summarized from Lowe *et al.* [13]. (G) Expression in generalized vertebrate centralized nervous system. F, forebrain; M, midbrain; R1-R8, rhombomeres of hindbrain. *zic* [50];*pea3* [51]; *hnf-6/onecut* [52]; and *tbr* [53]; *foxj1* [54]; *hox* genes [55]. See references [24] and [26–32]. Echinoderm gene names (quotations) are substituted for simplicity in Figures 1F and 1G.

Transcripts of sea star regulatory genes are localized throughout the animal ectoderm in overlapping concentric domains along the AV axis. Some of these transcripts, such as those of*zic*, *foxq2*, *rx* and *nk2.1*, are found only in the animal-most ectoderm (Figures 2A-2D). Of these genes, *zic* appears to be most closely localized to the animal pole, while expression of *foxq2* and*rx* overlaps with *zic*, yet extends further. Transcripts of *six3* and *klf13* (Figures 2E and 2F) also are detected in the animal-most ectoderm of blastulae; however, they show a still broader distribution. Although there is no clear cell morphology that demarcates the boundary between the animal ectoderm and the vegetal plate endomesoderm, we observe a ring of *nk1* expression above the vegetal plate (Figure 2G) that partially overlaps with endodermally localized *gatae*transcripts (Figure 2H). The *nk1* expression domain therefore likely marks the vegetal-most ectoderm of the blastula. *foxj1* and *pea3* (Figures 2I and 2J) are expressed throughout the ectoderm.

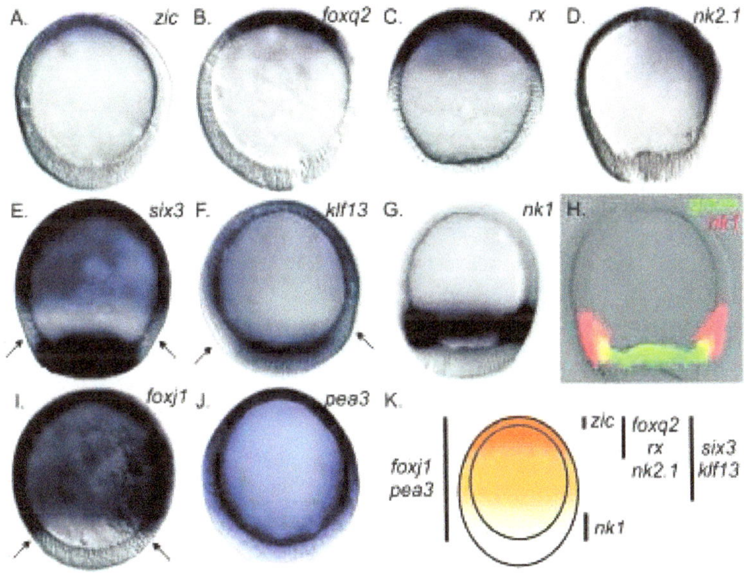

Figure 2: Nested concentric expression domains pattern the axial ectoderm of sea star, *P. miniata* , blastulae. Embryos are oriented with the animal pole up. (A-G)Whole mount *in situ* hybridization (*WMISH*). (A) *zic*, (B) *foxq2*, (C) *rx*, and (D) *nk2.1*expression is restricted to the animal-most ectoderm. Transcripts of (E) *six3* and (F) *klf13*are detected in the ectoderm and in the vegetal plate endomesoderm. Arrows in (E) and(F) point to a clearing above the vegetal pole where no or few transcripts are detected. (G)*nk1* transcripts are localized to a ring above the vegetal pole. (H) The boundary between the vegetal-most ectoderm (*nk1*, red) and the endoderm (*gatae*, green) as visualized by fluorescence *in situ* hybridization (*FISH*). Colocalization is in yellow. (I and J) *WMISH*. Transcripts of (I) *foxj1* and (J) *pea3* are detected throughout the entire ectoderm. *pea3* is weakly detected in the vegetal plate endomesoderm. Arrows in Figure 1J point to the limits of *foxj1* expression . (K) Schematic shows the patterns described above as five nested domains of expression along the AV axis. For simplicity, *nk2.1* is grouped here with the concentric domains of *foxq2* and *rx* expression. Gene names are listed next to their cognate expression domains. Vertical bars approximate the expression boundaries of associated genes. The color gradient spans the animal (orange) to vegetal (yellow) limits of the ectoderm.

Taken together, the spatial expression of these regulatory genes demonstrates that the ectoderm of the sea star blastula is patterned along the AV axis in at least five nested concentric domains (summarized in Figure 2K).

Regulatory Gene Expression within the Ectoderm of the Ciliary Bands and Animal Pole Domain

During gastrulation, the ectoderm appears to undergo very little morphological change other than the coalescence of cilia within two bands: a preoral ciliary band that loops above the opening of the mouth and a postoral ciliary band that loops below it and around the aboral surface at the "back" of the embryo (Figure 3A) [7, 18]. Transcripts of several genes that were distributed broadly throughout the ectoderm prior to gastrulation are later expressed within the ectoderm of the ciliary bands of the larva following gastrulation (for example, *foxj1* and *klf13* in Figures 3B-3E, *pea3* as summarized in Figure 1 and as previously reported for *otx* and *hnf-6/onecut* expression [18, 19]). At present, it is unclear if these patterns of expression reflect a migration of ectodermal cells to the sites of the future ciliary bands or if there is another patterning mechanism that restricts the earlier broad expression. Other transcripts are first detected at this stage within the ectoderm on the oral side of the gastrula and then later within the ciliary bands (for example, transcripts of *foxg*, *foxd* and *gbx*; Figures 3F and 3G, 3H and 3I, and 3J and 3K, respectively). A two-probe whole mount *in situ* hybridization (*WMISH*) of *foxg* and *lhx2*, a gene localized to the aboral ectoderm, further highlights the oral side restriction of *foxg* transcripts in the gastrula (Figure 3).

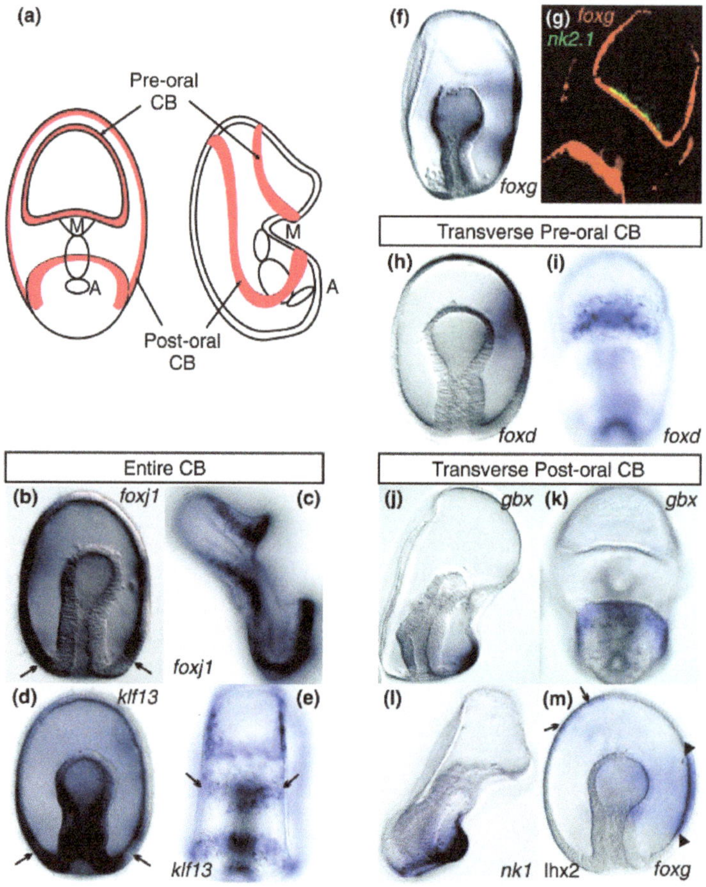

Figure 3: Heterogeneous regulatory patterning of the larval ciliary bands as visualized by *WMISH*. (A) Schematic describes the position of the two larval ciliary bands (red) from oral (left) and lateral (right) views. A, anus; CB, ciliary band; M, mouth. (B-F) *WMISH*. Expression of (B and C) *foxj1* and (D and E) *klf13* is initially broad throughout(B and D) the ectoderm of gastrulae, then later is restricted to (C and E) the larval ciliary bands. Arrows in Figure 3B show the vegetal limits of *foxj1* expression. Arrows in Figure 3D point to a clearing above the vegetal pole where transcripts of *klf13* were detected. *klf13* transcripts are additionally detected in an ectodermal territory near the mouth (arrows in Figure 3E). (F) *foxg* is first expressed within two ectodermal domains on the oral side of gastrulae. (G) *FISH* of *nk2.1* (green) and ciliary band marker *foxg* (red) highlights *nk2.1* expression in only the transverse preoral ciliary band. Colocalization is shown in yellow. (H-M) *WMISH*. *foxd* is expressed within a single domain in (H) the oral side ectoderm of gastrulae and (I) in the transverse, preoral larval ciliary band. *gbx* is expressed in one domain in (J) the oral side ectoderm in gastrulae and in (K) the transverse postoral larval ciliary band. (L) *nk1* is expressed in the transverse postoral ciliary band in the larva. (M) A two-probe *WMISH* shows *lhx2* expression in a spotted pattern in the aboral ectoderm (arrows, left) opposite of *foxg* expression (arrowheads, right). Embryos are oriented with the animal pole up and laterally, except in Figures 3E, 3G, 3I and 3K, which are oral views. In lateral views, the oral side is to the right.

The expression patterns at this later stage also show that the regulatory state of the early larval ciliary bands is heterogeneous, for example, *nk2.1* and *foxd* are expressed in part of the preoral ciliary band directly above the mouth (Figures 3G and 3I, respectively), while *gbx* and *nk1* are localized to part of the postoral ciliary band below the mouth (Figures 3K and 3L, respectively). Therefore, while the regulatory state of the ciliary band ectoderm can be defined by a suite of transcription factors (that is, *klf13*, *foxj1*, *pea3*, *foxg*, *otx* and *hnf-6/onecut*), they are further subdivided into pre- and postoral regions on the basis of the localization of *foxd*, *nk2.1*, *nk1* and *gbx*.

Other transcripts that we detected within the ectoderm of the blastula remained within the animal ectoderm as gastrulation proceeded in what we define here as the animal pole domain. Unlike sea urchins, the sea star, *P. miniata*, does not appear to have a morphologically distinct animal pole domain at this stage. Transcripts of *foxq2*, *pax6* and *pea3* (Figures 4A-

4C) tightly localize to the animal pole ectoderm, although their vegetal boundaries do not exactly coincide. Transcripts of *zic*, *rx* and *six3* are expressed within the animal pole domain as well, but even more vegetally throughout the animal ectoderm (Figures 4D-4F). The vegetal boundary of the animal pole domain therefore is not clearly defined by regulatory gene expression. The preoral and postoral ciliary bands run through the sea star animal pole domain as demonstrated by a two-probe fluorescence *in situ* hybridization (*FISH*) using the ciliary band marker, *foxg*, and the animal pole domain gene, *pax6* (Figure 4G). Thus, despite its lack of morphological regionalization, the animal pole has a distinct regulatory state, as defined by *foxq2*, *pax6*, *pea3*,*zic*, *rx* and *six3* expression, suggesting that it is a unique territory within the sea star. It is not yet clear whether these genes remain expressed in all cells of the sea star animal pole domain during later stages of larval development or if expression becomes refined to only subsets of cells within this domain.

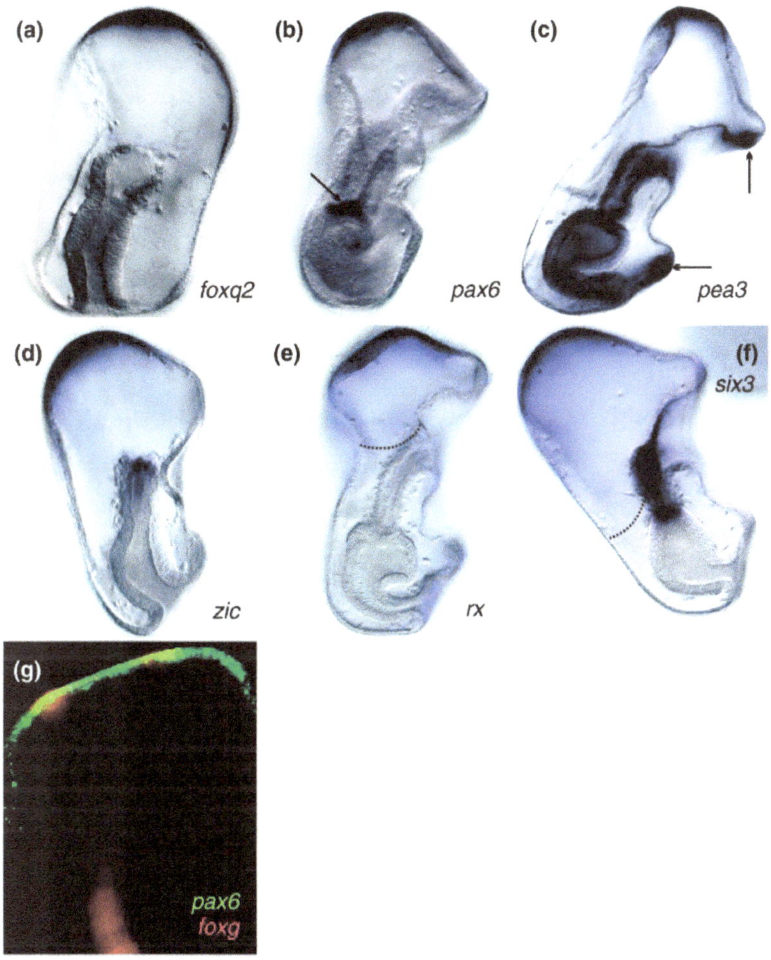

Figure 4: Gene expression molecularly defines the animal pole domain in the sea star. Embryos are shown laterally, with the animal pole up and oral side to the right. (A-F)*WMISH*. Expression of (A) *foxq2*, (B) *pax6*, (C) *pea3*, (D) *zic*, (E) *rx* and (F) *six3* within the apical-most ectoderm defines the animal pole domain within late gastrulae (Figures 4A, 4D and 4F) and early larvae (Figures 4B, 4C and 4E). The vegetal limits of this domain are variable (see dotted lines in Figures 4E and 4F). Transcripts of *pea3* additionally localize within the ectoderm of the larval ciliary bands (arrows in Figure 4C). *pax6*expression in mesodermally derived coelom (arrow in Figure 4B). (G) *FISH* demonstrates that the ciliary bands, as marked by *foxg* (red), run through the ectoderm of the animal pole domain, as marked by *pax6* (green). Colocalization is shown in yellow.

Comparisons of Ectodermal Patterning between Sea Urchin and Sea Star Embryos

At first inspection, the expression patterns of many genes appear markedly different in the earlier blastula stages of sea urchin and sea star embryos. The later restrictions within the animal pole or ciliary bands are, with some exceptions, more similar (Figure 1). We suggest that the sea urchin embryo may simply undergo a relatively more rapid specification of these territories, with an associated loss of intermediate domains that we observe in the sea star. Indeed, a careful examination of expression patterns in sea urchin has recently shown that the apical plate in sea urchin consists of at least two regulatory domains: an inner animal pole domain flanked by a ring of *six3* expression [20]. These two domains in the sea urchin hatched blastula may therefore represent a more apically compressed version of the nested, concentric regulatory domains found in the sea star blastula.

Some of the patterning differences between sea urchin and sea star ectoderm also seem to account for the differences observed in the localization of the pan-neuronal marker, synaptotagmin-B [21]. For example, similar to the patterns of gene expression that we describe here, synaptotagmin-B is detected broadly throughout the ectoderm of the sea star gastrula, but in the larva it is found primarily in neurons associated with the ciliary bands and animal pole [9]. In the sea urchin, however, synaptotagmin-B is already localized to the animal pole domain and the presumptive ciliary band by the gastrula stage [9].

Although expression of many of the genes within the ciliary bands of the sea star appears conserved in the sea urchin, *nk2.1* and *foxd* show clear differences in expression that may be associated with the evolutionary transition from a double looping of the ciliary band around the body of the sea star bipinnaria and hemichordate tornaria to a single looping of the ciliary band observed in sea urchins. This single ciliary band in the sea urchin develops at the junction between oral and aboral ectoderm. *nk2.1* and *foxd* are expressed in part of the preoral ciliary band of the oral hood of the sea star, while sea urchin orthologs of these are found in the animal plate ectoderm (compare Figures 1C-1E). Interestingly, both of these genes in sea urchin appear enriched on the oral side of the embryos [22, 23]. Thus, we speculate that the preoral ciliary band may have been compressed into the oral-side animal plate territory in sea urchins and that this region within the sea urchin may therefore constitute a different territory than the remaining animal plate.

Conservation of Anterior (animal)-Most Regulatory Patterning with other Deuterostomes

Comparisons of the regulatory gene expression patterns that we observed in these indirectly developing sea star embryos with those known in directly developing bilaterians illuminate additional surprising patterns of conservation. We observe a general mapping of gene expression patterns along body axes (compare Figures 1A-1C with Figures 1F-1G). For instance, in the sea star, *foxq2*, *rx*, *pax6* and *six3* orthologs are apically expressed within the ectoderm.*foxq2* expression in the amphioxus, a basal chordate, is restricted to the anterior-most end of the embryo [24]. Orthologs of *rx* and *pax6* are expressed in the anterior-most neuroectoderm in the hemichordate *Saccoglossus* [13], and they also pattern the anteriorly localized eye primordium in vertebrates [25, 26]. The *Drosophila rx* ortholog is required for brain development [27]. Orthologs of *six3* and *otx* are expressed in anterior neuroectoderm in members of all three deuterostome phyla [13, 28, 29]. The most vegetal ectoderm in sea stars is characterized by the presence of *nkl* and *gbx* transcripts. In vertebrates, a *gbx* ortholog establishes the midbrain-hindbrain boundary [30]. The zebrafish ortholog of *nkl*, *sax2*, is expressed within the midbrain-hindbrain boundary as well, although its expression is not exclusive to this territory [31]. Expression of *nkl* and *gbx* in sea stars, and possibly sea urchins, marks the vegetal (posterior)-most ectoderm.

There is some evidence of additional conservation between the DV and oral-aboral axes as well. The mouse ortholog of *nk2.1* (*nkx2.1*) is involved in the formation of motor neurons in the ventral telencephalon [32]. *Saccoglossus nk2.1* orthologs also show a ventral bias in expression [13]. Furthermore, *foxg* plays a role in ventral forebrain development, while *lhx2* specifies dorsal telencephalic fates [32]. We similarly show that expression of sea star orthologs of *foxg*and *nk2.1* is restricted to the oral (ventral) ectoderm, while *lhx2* orthologs are expressed within the aboral (dorsal) ectoderm (Figure 1B).

While in these comparisons we do not intend to convey a tight homology in gene expression patterns across deuterostome phyla, we predict that similarities in the overall patterning are an ancestral innovation and perhaps evidence of maintenance of some elements of a developmental gene regulatory network (GRN) inherited from a common ancestor. Conservation, however, is not maintained for orthologs of genes expressed within regions posterior to the midbrain-hindbrain boundary in chordates and hemichordates as *nkl* marks the vegetal-most ectoderm. Also, the overlapping expression of *hox* gene orthologs needed to pattern the posterior of many embryos are found only later in echinoderm development within the mesoderm of the rudiment [33].

Separation of "Retinal" from "Anterior Neural" Regulatory Patterning

Vertebrate orthologs of transcription factors such as*pax6*, *six3* and *rx* play known roles in pattering and specifying anterior vertebrate sensory systems, most notably the eyes [25, 26,34]. Furthermore, orthologs of *pax6*, the *six* gene family members and *eya* operate in a similar gene network for retinal determination in both vertebrates and *Drosophila* (as reviewed in [35]).

Having established that orthologs of many regulatory genes involved in anterior neural specification are also expressed within the anterior ectoderm of sea star embryos, we sought to determine if orthologs of transcription factors involved within the retinal determination network are also expressed within echinoderm embryos.

We have already shown that the sea star *pax6* ortholog is expressed within the animal pole domain (Figure 4B), although it is not expressed within the ectoderm of the sea urchin embryo (Figure 5). Transcripts of both *pax6* and *eya* in both sea urchins and sea stars, however, are detected in the mesoderm of midgastrulae and then more prominently in one mesodermally derived coelom in late gastrulae (Figures 5A-5D and Figures 5G-5L). While we were unable to obtain a *six1.2* ortholog from the sea star, this gene is expressed also within the mesodermal coelom in the sea urchin (Figures 5E and 5F).

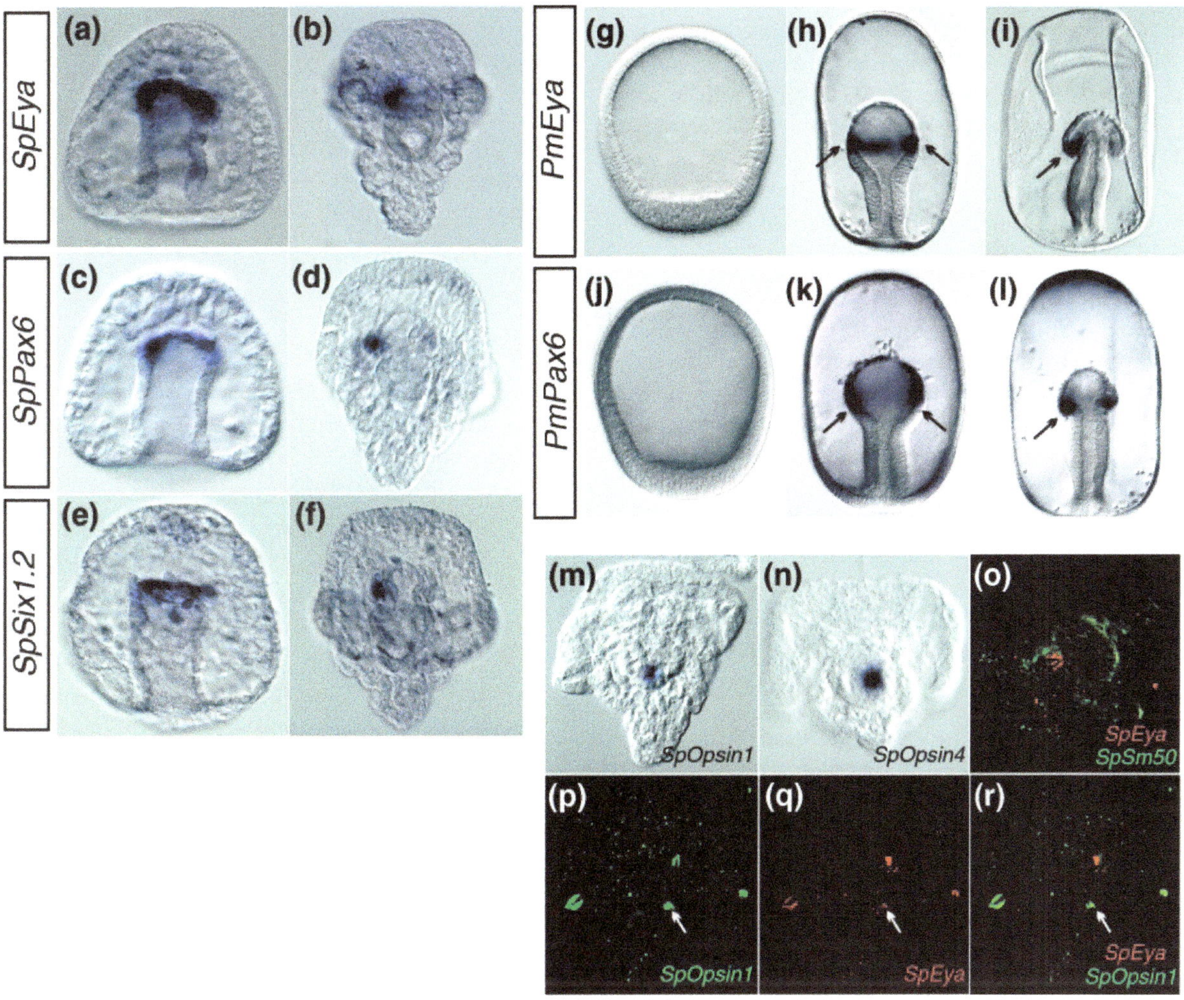

Figure 5: Retinal determination orthologs are expressed within sea urchin, (*Strongylocentrotus purpuratus, Sp*), and sea star mesoderm. (A-N) *WMISH*.(A and B) *SpEya*, (C and D) *SpPax6* and (E and F) *SpSix1.2* are expressed at the tip of the archenteron in gastrulae (Figures 5A, 5C and 5E) and in a mesodermally derived coelom in early pluteus larvae (Figures 5B, 5D and 5F). *PmEya* (Figures 5G-5I) and *PmPax6* (Figures 5J-5L) expression is first detected in the mesoderm of the archenteron bulb in midgastrulae (Figures 5H and 5K; arrows) and then more prominently in a mesodermally derived coelom in late gastrulae (Figures 5I and 5L; arrow). *PmPax6* transcripts are also found broadly throughout the ectoderm (Figure 5K), with more pronounced expression in the apical ectoderm (Figure 5L). *SpOpsin1* (Figure 5M) and *SpOpsin4* (Figure 5N) are expressed in 1-week-old larvae. (O-R) *FISH* in 1-week-old larvae. *SpEya* (Figure 5O) is shown relative to the skeletal marker *SpSm50*, which was used to orient the embryo. Figures 5P-5R show transcripts of *SpOpsin1* (green) colocalizing with those of *SpEya* (red). Colocalization is shown in yellow. Arrows in Figures 5P, 5Q and 5R point to expression. Additional areas of expression within the 1-week-old larvae may be a result of nonspecific staining.

In the sea urchin, expression of two members of the light-sensing rhodopsin family of G-coupled protein receptors, *opsin1* and *opsin4*, has been shown as early as 1 week [36]. We were unable to obtain sea star opsin sequences; however, we confirmed the expression of *opsin1* and *opsin4*1-week-old sea urchin larvae (Figures 5M and 5N). The morphology of the late larval sea urchin embryos makes it difficult to decipher the precise location of these transcripts within the embryo. We therefore sought to determine if *opsins* collocalize with *eya*, which we show is expressed likely within one or both coeloms (depending on developmental timing) in 1-week-old larvae (Figure 5O). Using a two-probe *FISH*, we observe that transcripts of *opsin1* colocalized with those of *eya* in 1-week-old sea urchin larvae (Figures 5P-5R). Expression of retinal determination orthologs within the mesoderm of gastrulae and larvae allow for the possibility that these genes operate within a common GRN.

The tightly coupled GRNs for anterior neural and visual sensory structures that are found in vertebrates and also in invertebrates, such as *Drosophila*, therefore are spatially separated in echinoderms. The presence of gene transcripts of *pax6*, *rx* and *six3*, but not, for example, *eya*, within the animal ectoderm of sea star bipinnaria larva may indicate a partial retention of an ancestral retinal determination network that once operated within this embryonic territory. This might also explain the absence of apically localized rhabdomeric eyespots, which are characteristic of the indirectly developing tornaria larvae of some hemichordates but were likely lost in the echinoderm lineage [16].

CONCLUSIONS

Inferring Ancestral States

The detailed expression analyses reported here support the hypothesis that indirectly developing planktonic echinoderm embryos likely utilize ancient regulatory mechanisms for various anterior neuroectodermal and/or sensory developmental processes that are potentially conserved throughout the Bilateria. Compared to vertebrates and well-studied protostome model organisms such as *Drosophila*, however, echinoderm embryos separate the deployment of these subcircuits in space and/or time. Thus, echinoderm embryos may have conserved sets of genetic regulatory relationships for "head/anterior brain" within the ectoderm of the early blastula and others for "retinal determination" within the mesodermal coelom of the gastrula.

Much of the difficulty in inferring the ancestral state of the deuterostomes and the mysteries of the origin of the phylum to which we belong arises from the complex life histories found within extant lineages [37]. Given the conservation of complex sensory and AP patterning between protostomes such as *Drosophila* and vertebrates, the parsimonious explanation is that ancestral developmental regulatory interactions, perhaps even entire GRN subcircuits, have been uncoupled along the lineage of echinoderms, possibly coincident with a simplification in early development. However, until a greater breath of taxa have resolved GRNs, we cannot know the flexibility with which modular subcircuits can be deployed during evolution of alternative body plans or if intercalation of GRN subcircuits occurs independently or coincident with an increase in complexity.

METHODS

Sea star and Sea Urchin Embryo Culture and Characterization of Gene Expression

P. miniata embryos, previously named *Asterina miniata*, and *Strongylocentrotus purpuratus*embryos were cultured as described previously [18, 38]. Partial gene sequences were obtained via screening a 3-day-old (late-gastrula stage) *P. miniata* arrayed cDNA library using *S. purpuratus* sequence-specific probes and low stringency conditions as previously described [39]. Whole mount *in situ* hybridization (*WMISH*) was performed as described previously [18].

Two-color FISH

WMISH was performed essentially as described by Hinman *et al.* [18], with modifications to detect riboprobes using fluorescence as described by Denkers *et al.* [40]. In brief, both digoxygenin (DIG) and 2,4-dinitrophenol (DNP) labeled riboprobes were used. Hybridized probes were detected using anti-DIG antibody (1:2,000; Roche: Indianapolis, IN, USA) and anti-DNP antibody (1:1,000; PerkinElmer: Chicago, IL, USA), both conjugated to horseradish peroxidase, and the Tyramide Signal Amplification (TSA) Plus Fluorescence Systems Kit (PerkinElmer). A CyIII- or fluorescein-labeled tyramide was deposited near the *in situ* riboprobe in a reaction catalyzed by horseradish peroxidase, allowing for fluorescence detection of DIG- and DNP-labeled riboprobes.

ACKNOWLEDGEMENTS

The authors thank Dr Daniel Brown, Brenna McCauley, Alys Cheatle and three anonymous reviewers for helpful comments on the manuscript. This work was partially supported by National Science Foundation grant 0844948 to VFH, and CKJ and MLM were funded by a Howard Hughes Medical Institute Undergraduate Education Research Grant.

AUTHORS' CONTRIBUTIONS

VFH and KAY conceived of the study, participated in its design and drafted the manuscript. KAY cloned sea star orthologs and performed *WMISH* and *FISH*. CKJ cloned sea urchin orthologs and performed WMISH. MLM cloned sea star orthologs and performed *WMISH*and *FISH*. All authors read and approved the final manuscript.

REFERENCES

1. Bromham LD, Degnan BM: Hemichordates and deuterostome evolution: robust molecular phylogenetic support for a hemichordate + echinoderm clade. Evol Dev. 1999, 1: 166-171.
2. Castresana J, Feldmaier-Fuchs G, Yokobori S, Satoh N, Paabo S: The mitochondrial genome of the hemichordate *Balanoglossus carnosus* and the evolution of deuterostome mitochondria. Genetics. 1998, 150: 1115-1123.
3. Cameron CB, Garey JR, Swalla BJ: Evolution of the chordate body plan: new insights from phylogenetic analyses of deuterostome phyla. Proc Natl Acad Sci USA. 2000, 97: 4469-4474.
4. Turbeville JM, Schulz JR, Raff RA: Deuterostome phylogeny and the sister group of the chordates: evidence from molecules and morphology. Mol Biol Evol. 1994, 11: 648-655.
5. Bourlat SJ, Juliusdottir T, Lowe CJ, Freeman R, Aronowicz J, Kirschner M, Lander ES, Thorndyke M, Nakano H, Kohn AB, Heyland A, Moroz LL, Copley RR, Telford MJ: Deuterostome phylogeny reveals monophyletic chordates and the new phylum Xenoturbellida. Nature. 2006, 444: 85-88.
6. Byrne M, Nakajima Y, Chee FC, Burke RD: Apical organs in echinoderm larvae: insights into larval evolution in the Ambulacraria. Evol Dev. 2007, 9: 432-445.
7. Hyman LH: The Invertebrates: Echinodermata. 1955, New York: McGraw-Hill
8. Hyman LH: The Invertebrates: Smaller Coelomate Groups. 1959, New York: McGraw-Hill
9. Nakajima Y, Kaneko H, Murray G, Burke RD: Divergent patterns of neural development in larval echinoids and asteroids. Evol Dev. 2004, 6: 95-104.
10. Garstang W: Preliminary note on a new theory of the phylogeny of the Chordata. Zool Anz. 1894, 17: 122-125.
11. Lacalli TC: Apical organs, epithelial domains, and the origin of the chordate central nervous system. Am Zool. 1994, 34: 533-541.
12. Nielsen C: Origin of the chordate central nervous system - and the origin of chordates. Dev Genes Evol. 1999, 209: 198-205.
13. Lowe CJ, Wu M, Salic A, Evans L, Lander E, Stange-Thomann N, Gruber CE, Gerhart J, Kirschner M: Anteroposterior patterning in hemichordates and the origins of the chordate nervous system. Cell. 2003, 113: 853-865.

14. Lowe CJ, Terasaki M, Wu M, Freeman RM, Runft L, Kwan K, Haigo S, Aronowicz J, Lander E, Gruber C, Smith M, Kirschner M, Gerhart J: Dorsoventral patterning in hemichordates: insights into early chordate evolution. PLoS Biol. 2006, 4: e291-Central
15. Lacalli TC: Protochordate body plan and the evolutionary role of larvae: old controversies resolved?. Can J Zool. 2005, 83: 216-224.
16. Brown FD, Prendergast A, Swalla BJ: Man is but a worm: chordate origins. Genesis. 2008, 46: 605-613.
17. Arendt D, Denes AS, Jekely G, Tessmar-Raible K: The evolution of nervous system centralization. Philos Trans R Soc Lond B Biol Sci. 2008, 363: 1523-1528.Central
18. Hinman VF, Nguyen AT, Davidson EH: Expression and function of a starfish Otx ortholog, AmOtx: a conserved role for Otx proteins in endoderm development that predates divergence of the eleutherozoa. Mech Dev. 2003, 120: 1165-1176.
19. Otim O, Hinman VF, Davidson EH: Expression of AmHNF6, a sea star orthologue of a transcription factor with multiple distinct roles in sea urchin development. Gene Expr Patterns. 2005, 5: 381-386.
20. Wei Z, Yaguchi J, Yaguchi S, Angerer RC, Angerer LM: The sea urchin animal pole domain is a Six3-dependent neurogenic patterning center. Development. 2009, 136: 1179-1189.Central
21. Burke RD, Osborne L, Wang D, Murabe N, Yaguchi S, Nakajima Y: Neuron-specific expression of a synaptatogmin gene in the sea urchin *Strongylocentrotus purpuratus*. J Comp Neurol. 2006, 496: 244-251.
22. Tu Q, Brown CT, Davidson EH, Oliveri P: Sea urchin Forkhead gene family: phylogeny and embryonic expression. Dev Biol. 2006, 300: 49-62.
23. Takacs CM, Amore G, Oliveri P, Poustka AJ, Wang D, Burke RD, Peterson KJ: Expression of an NK2 homeodomain gene in the apical ectoderm defines a new territory in the early sea urchin embryo. Dev Biol. 2004, 269: 152-164.
24. Yu JK, Holland ND, Holland LZ: AmphiFoxQ2, a novel winged helix/forkhead gene, exclusively marks the anterior end of the amphioxus embryo. Dev Genes Evol. 2003, 213: 102-105.
25. Bailey TJ, El-Hodiri H, Zhang L, Shah R, Mathers PH, Jamrich M: Regulation of vertebrate eye development by Rx genes. Int J Dev Biol. 2004, 48: 761-770.
26. Pichaud F, Desplan C: Pax genes and eye organogenesis. Curr Opin Genet Dev. 2002, 12: 430-434.
27. Davis RJ, Tavsanli BC, Dittrich C, Walldorf U, Mardon G: *Drosophila* retinal homeobox (drx) is not required for establishment of the visual system, but is required for brain and clypeus development. Dev Biol. 2003, 259: 272-287.
28. Simeone A, Acampora D, Mallamaci A, Stornaiuolo A, D›Apice MR, Nigro V, Boncinelli E: A vertebrate gene related to orthodenticle contains a homeodomain of the bicoid class and demarcates anterior neuroectoderm in the gastrulating mouse embryo. EMBO J. 1993, 12: 2735-2747.
29. Oliver G, Mailhos A, Wehr R, Copeland NG, Jenkins NA, Gruss P: Six3, a murine homologue of the sine oculis gene, demarcates the most anterior border of the developing neural plate and is expressed during eye development. Development. 1995, 121: 4045-4055.
30. Rhinn M, Brand M: The midbrain-hindbrain boundary organizer. Curr Opin Neurobiol. 2001, 11: 34-42.
31. Bae YK, Shimizu T, Muraoka O, Yabe T, Hirata T, Nojima H, Hirano T, Hibi M: Expression of *sax1/nkx1.2* and *sax2/nkx1.1* in zebrafish. Gene Expr Patterns. 2004, 4: 481-486.
32. Hebert JM, Fishell G: The genetics of early telencephalon patterning: some assembly required. Nat Rev Neurosci. 2008, 9: 678-685.Central
33. Arenas-Mena C, Cameron AR, Davidson EH: Spatial expression of Hox cluster genes in the ontogeny of a sea urchin. Development. 2000, 127: 4631-4643.
34. Carl M, Loosli F, Wittbrodt J: Six3 inactivation reveals its essential role for the formation and patterning of the vertebrate eye. Development. 2002, 129: 4057-4063.
35. Wawersik S, Maas RL: Vertebrate eye development as modeled in *Drosophila*. Hum Mol Genet. 2000, 9: 917-925.
36. Raible F, Tessmar-Raible K, Arboleda E, Kaller T, Bork P, Arendt D, Arnone MI: Opsins and clusters of sensory G-protein-coupled receptors in the sea urchin genome. Dev Biol. 2006, 300: 461-475.
37. Swalla BJ: Building divergent body plans with similar genetic pathways. Heredity. 2006, 97: 235-243.
38. Ettensohn CA, Wessel GM, Wray GA: Development of Sea Urchins, Ascidians, and Other Invertebrate Deuterostomes: Experimental Approaches. 2004, San Diego: Elsevier Academic Press
39. Hinman VF, Nguyen AT, Cameron RA, Davidson EH: Developmental gene regulatory network architecture across 500 million years of echinoderm evolution. Proc Natl Acad Sci USA. 2003, 100: 13356-13361.Central
40. Denkers N, Garcia-Villalba P, Rodesch CK, Nielson KR, Mauch TJ: FISHing for chick genes: triple-label whole-mount fluorescence in situ hybridization detects simultaneous and overlapping gene expression in avian embryos. Dev Dyn. 2004, 229: 651-657.
41. Materna SC, Howard-Ashby M, Gray RF, Davidson EH: The C_2H_2 zinc finger genes of*Strongylocentrotus purpuratus* and their expression in embryonic development. Dev Biol. 2006, 300: 108-120.
42. Burke RD, Angerer LM, Elphick MR, Humphrey GW, Yaguchi S, Kiyama T, Liang S, Mu X, Agca C, Klein WH, Brandhorst BP, Rowe M, Wilson K, Churcher AM, Taylor JS, Chen N, Murray G, Wang D, Mellott D, Olinski R, Hallbook F, Thorndyke MC: A genomic view of the sea urchin nervous system. Dev Biol. 2006, 300: 434-460.Central
43. Rizzo F, Fernandez-Serra M, Squarzoni P, Archimandritis A, Arnone MI: Identification and developmental expression of the *ets* gene family in the sea urchin (*Strongylocentrotus purpuratus*). Dev Biol. 2006, 300: 35-48.

44. Howard-Ashby M, Materna SC, Brown CT, Chen L, Cameron RA, Davidson EH: Identification and characterization of homeobox transcription factors genes in*Strongylocentrotus purpuratus*, and their expression in embryonic development. Dev Biol. 2006, 300: 74-89.

45. Su YH, Li E, Geiss GK, Longabaugh WJ, Kramer A, Davidson EH: A perturbation model of the gene regulatory network for oral and aboral ectoderm specification in the sea urchin embryo. Dev Biol. 2010, 329: 410-21.

46. Otim O, Amore G, Minokawa T, McClay DR, Davidson EH: *SpHnf6*, a transcription factor that executes multiple functions in sea urchin embryogenesis. Dev Biol. 2004, 273: 226-243.

47. Poustka AJ, Kuhn A, Radosavljevic V, Wellenreuther R, Lehrach H, Panopoulou G: On the origin of the chordate central nervous system: expression of *onecut* in the sea urchin embryo. Evol Dev. 2004, 6: 227-236.

48. Minokawa T, Rast JP, Arenas-Mena C, Franco CB, Davidson EH: Expression patterns of four different regulatory genes that function during sea urchin development. Gene Expr Patterns. 2004, 4: 449-456.

49. Yuh CH, Brown CT, Livi CB, Rowen L, Clarke PJC, Davidson EH: Patchy interspecific sequence similarities efficiently identify positive *cis*-regulatory elements in the sea urchin. Dev Biol. 2002, 246: 148-161.

50. Aruga J: The role of *Zic* genes in neural development. Mol Cell Neurosci. 2004, 26: 205-221.

51. Munchberg SR, Ober EA, Steinbeisser H: Expression of the Ets transcription factors erm and pea3 in early zebrafish development. Mech Dev. 1999, 88: 233-236.

52. Hong SK, Kim CH, Yoo KW, Kim HS, Kudoh T, Dawid IB, Huh TL: Isolation and expression of a novel neuron-specific onecut homeobox gene in zebrafish. Mech Dev. 2002, 112: 199-202.

53. Bulfone A, Smiga SM, Shimamura K, Peterson A, Puelles L, Rubenstein JL: T-brain-1: a homolog of Brachyury whose expression defines molecularly distinct domains within the cerebral cortex. Neuron. 1995, 15: 63-78.

54. Aamar E, Dawid IB: Isolation and expression analysis of *foxj1* and *foxj1.2* in zebrafish embryos. Int J Dev Biol. 2008, 52: 985-991.Central

55. Tümpel S, Wiedemann LM, Krumlauf R: *Hox* genes and segmentation of the vertebrate hindbrain. Curr Top Dev Biol. 2009, 88: 103-137.

CHAPTER 4

RNA SEQUENCING ANALYSIS TO CAPTURE THE TRANSCRIPTOME LANDSCAPE DURING SKIN ULCERATION SYNDROME PROGRESSION IN SEA CUCUMBER APOSTICHOPUS JAPONICUS

Aifu Yang, Zunchun Zhou, Yongjia Pan, Jingwei Jiang, Ying Dong, Xiaoyan Guan, Hongjuan Sun, Shan Gao and Zhong Chen

Keywords: Sea cucumber (Apostichopus japonicus), Skin ulceration syndrome, RNA sequencing, Dynamic expression profiles

BACKGROUND

The transcriptome is a set of all RNA transcripts including rRNA, tRNA, mRNA and non-coding RNA produced in one type of cell or a population of certain types of cells at a particular stage in an organism [1]. Unlike the genome, which is roughly fixed for a certain cell type, the transcriptome is considered to be highly dynamic [2–7]. These transcriptomic changes are the prelude to the impact of protein translation on the phenotype of the organism. Transcriptome analysis is thus essential for elucidating the underlying molecular constituents of cells and tissues in various biological progresses. With the development of RNA-sequencing (RNA-seq) technology, it provides a far more precise measurement of the levels of transcripts and their isoforms than other methods [8]. Many biologically related issues, such as the expression levels of specific genes, the presence of novel transcripts and fusion transcripts and the events of differential splicing, allele-specific expression and RNA editing can be determined accurately by RNA-seq technology [1, 9–11]. Among several RNA-seq technologies, the Illumina sequencing platform is a particularly attractive approach to transcriptomic studies, because it is relatively low cost, and its coverage and depth are far greater than other currently available sequencing technologies [12–15]. Extensive studies have taken the advantage of the RNA-seq strategy to interpret the dynamic transcriptome in many species, such as maize [16–18], Drosophila [2], locust [4], Xenopus [5], and sea cucumber [6]. Other studies have also focused on whole transcriptome analysis to provide an improved understanding of the processes of tumorigenesis and cancer metastasis in human [19, 20].

The sea cucumber *A. japonicus* is one of the most valuable sea foods in East Asian countries because it has a high nutritional value. With increasing market demands for related products, sea cucumber aquaculture has developed rapidly in recent years, and has now become a vigorous industry along the northern coasts of China. However, various diseases in sea cucumber caused by bacteria, viruses or protozoa have severely limited the stable development of this important industry.

Citation: Aifu Yang, Zunchun Zhou, Yongjia Pan, Jingwei Jiang, Ying Dong, Xiaoyan Guan, Hongjuan Sun, Shan Gao and Zhong Chen, RNA sequencing analysis to capture the transcriptome landscape during skin ulceration syndrome progression in sea cucumber Apostichopus japonicas, DOI 10.1186/s12864-016-2810-3.

Among these diseases, SUS is the most common and serious, usually resulting in high mortality. The SUS is characterized by skin ulceration, evisceration, general atrophy, swollen mouth and anorexia [15, 21, 22]. Some studies have focused mainly on the investigation, isolation and identification of the pathogens responsible for SUS epidemics [22–27]. Recently, investigations of the microRNAs (miRNAs) [15, 28] and genes involved in immune pathways [29] associated with SUS in *A. japonicus* have been reported. However, these molecular studies do not provide a comprehensive understanding of the immune response against SUS in *A. japonicus* and further studies of the molecular mechanisms involved in the SUS epidemics of *A. japonicus* are required.

SUS infection is a dynamic progress. After infection, skin ulceration begins with the appearance of one or more small white ulcerative specks, followed by deep and enlarged ulcerative lesions, leading to exposure of the underlying muscle and spicules. Finally, the ulcerative specks develop into extensive lesions and in severe case, many of the infected sea cucumbers lose the ability to attach to the tanks [21, 22, 26]. During the onset and progression of SUS, besides the appearance of obvious white ulceration on the skin, the molecular changes triggered by the disease comprise a process that is both dynamic and complicated. Systematic investigations of the multiple pathways involved in the multi-stage SUS progression and its polygenic regulation are required to clarify the origin and development of the disease. In the present study, we screened 21 cDNA libraries from populations of SUS-affected and healthy *A. japonicus* by high-throughput sequencing using the Illumina HiSeq™ 2000 platform. We compared the differentially expressed genes (DEGs) in four different tissue types (body walls, BW; intestines, Int; respiratory trees, RT; and coelomocytes, C) in SUS-affected and healthy samples, and also investigated the dynamic changes in the transcriptome of BW during the three-stage SUS progression. Furthermore, GO and KEGG enrichment analyses were conducted to elucidate the main functions of DEGs and to provide an overview of the regulation of genes during SUS progression. These results expand our understanding of the complex molecular mechanism of SUS, provide a framework for further investigations of SUS progression, and will be useful in determining strategies aimed at the prevention of SUS caused by bacterial pathogens in sea cucumber.

RESULTS

Transcriptome Sequencing and Analysis

To obtain a comprehensive gene expression profile for SUS progression in *A. japonicus,* we first constructed 13 cDNA libraries (from D1 to D13) for RNA-seq that represented the ulcerative body wall and distal normal tissue at all three SUS stages (Fig. 1), as well as four tissues (BW, Int, RT and C) obtained from SUS-affected and healthy individuals. Based on the analysis of the gene expression profiles in the first 13 cDNA libraries, we constructed eight cDNA libraries (from T1 to T8) for RNA-seq that represented the ulcerative BW of the three SUS stages and the BW of healthy individuals using the Illumina HiSeq™ 2000 platform. A total of 553 million (553,212,906) raw reads were generated from 21 libraries; these were deposited in the NCBI SRA database (accession number: SRP050068). After trimming the raw reads, a total of 542 million (541,944,042) high-quality clean reads were obtained (Table 1). The map reference transcriptome containing 161,174,656 Illumina reads of the first RNA-seq (Table 1), 104,067,712 Illumina reads from our previous studies [30], 1,076,411 pre-processed 454 sequences [31] and 4,786 expressed sequence tags (ESTs) from the public database were included in the *A. japonicus* transcriptome analysis (Table 2). According to the method described by Zhou et al., the Illumina reads were combined with the 454 sequencing data and Sanger sequences to improve the quality of the assembled transcriptome [30]. Assembly of these sequences generated 93,163 unigenes with an average length of 1,052 bp (N50 = 1,575 bp). The total assembled unigenes included 39,744 transcripts attributed to the different sequence splicing of 17,252 genes. Each of these unigenes contained at least two sequences with pairwise sequence similarity greater than 70 %. Of all the unigenes obtained, 30,315 showed significant matches to Nr, 11,771 to Nt databases, 26,412 to SwissProt, 21,316 to KEGG and 13,976 to GO. Altogether, 33,860 unigenes had significant matches, with at least one match to these databases for each of the unigenes (Table 2). Approximately 74.19 % to 78.84 % (Table 1) of the total clean reads from all the libraries were mapped to the 93,163 unigenes obtained from the reference transcriptome data.

Figure 1: Features associated with each stage of SUS and indicators of the sampling positions. aHealthy *A. japonicus*. b and c SUS and its distal normal BW at stage I of SUS progression.d and e SUS and its distal normal BW at stage II of SUS progression. f and g SUS and its distal normal BW at stage III of SUS progression.

Table 1: Summary of the alignment statistics for cDNA libraries in *A. japonicus*

Sample ID	Libraries name	Total Reads	Total Clean Reads (Percentage)	Total Mapped Reads (Percentage)
1	H-BW (D1)	12,378,414	12,098,662 (97.74 %)	9,410,905 (77.78 %)
2	H-Int (D2)	12,602,105	12,430,716 (98.64 %)	9,576,807 (77.04 %)
3	H-RT (D3)	12,674,851	12,498,671 (98.61 %)	9,454,890 (75.65 %)
4	H-C (D4)	13,579,708	12,972,695 (95.53 %)	9,914,340 (76.42 %)
5	I-SUS-BW (D5)	12,791,397	12,451,146 (97.34 %)	9,473,854 (76.09 %)
6	I-N-BW (D6)	11,709,219	11,400,096 (97.36 %)	8,611,908 (75.54 %)
7	II-SUS-BW (D7)	12,426,665	12,262,633 (98.68 %)	9,408,950 (76.73 %)
8	II-N-BW (D8)	11,529,198	11,225,980 (97.37 %)	8,409,997 (74.92 %)
9	II-SUS-Int (D9)	13,183,442	13,008,102 (98.67 %)	10,255,121 (78.84 %)
10	II-SUS-RT (D10)	13,080,922	12,887,324 (98.52 %)	10,057,462 (78.04 %)
11	II-SUS-C (D11)	13,220,635	13,044,801 (98.67 %)	10,064,650 (77.15 %)
12	III-SUS-BW (D12)	12,239,111	12,073,883 (98.65 %)	9,327,113 (77.25 %)
13	III-N-BW (D13)	13,050,949	12,819,947 (98.23 %)	9,831,997 (76.69 %)
Total reads for the first RNA-seq		164,466,616	161,174,656 (98.00 %)	123,797,994 (76.81 %)
14	H-BW (T1)	51,200,576	50,022,670 (97.7 %)	38,946,697 (77.86 %)
15	H-BW (T2)	44,371,450	43,357,830 (97.72 %)	32,165,535 (74.19 %)
16	I-SUS-BW (T3)	51,818,216	50,866,582 (98.16 %)	39,225,086 (77.11 %)
17	I-SUS-BW (T4)	49,027,132	47,952,352 (97.81 %)	37,436,180 (78.07 %)
18	II-SUS-BW (T5)	40,931,866	40,071,484 (97.9 %)	30,984,132 (77.32 %)
19	II-SUS-BW (T6)	55,174,172	53,899,574 (97.69 %)	42,222,134 (78.33 %)
20	III-SUS-BW (T7)	43,123,468	42,213,860 (97.89 %)	32,138,856 (76.13 %)
21	III-SUS-BW (T8)	53,099,410	52,385,034 (98.65 %)	40,650,674 (77.59 %)
Total reads for all RNA-seq		553,212,906	541,944,042 (97.96 %)	417,567,288 (77.05 %)

Note: *H* Healty sea cucumber, *BW* Body wall, *Int* Intestine, *RT* Respiratory tree, *C*Coelomocyte, *SUS* Skin ulceration syndrome, *I* SUS stage I, *II* SUS stage II, *III* SUS stage III,*N* Normal tissue

Table 2: Summary statistics of transcriptome assembly for *A. japonicus*

Category	Count
Total Clean Reads	
Illumina reads (First RNA-seq)	161,174,656
Illumina reads [30]	104,067,712
454 reads	1,076,411
Sanger reads	4,786
Assembly Results	
No.of unigenes	93,163
Mean Length	1,052
N50	1,575
Annotation	
Nr database	30,315
Nt database	11,771
Swiss-Prot	26,412
KEGG	21,316
GO	13,976
ALL	33,860

Analysis of the Gene Expression Profiles during SUS Progression

One of the primary goals of RNA-seq is to compare the gene expression levels from different samples. Analysis of the first 13 cDNA libraries revealed the sequential large-scale gene expression profiles of healthy and SUS-affected *A. japonicus* individuals, as well as the DEGs of ulcerative and normal BW in the same individuals. Transcripts with at least two-fold differences ($\log_2$Ratio ≥ 1, FDR ≤ 0.001) were classified as DEGs. As shown in Table 3, the largest number of DEGs was observed when the profiles of the Int, RT and C were compared to those of the BW in the healthy group (16,936, 17,164 and 16,692 DEGs, respectively), and the number of up-regulated DEGs was obviously higher than that of down-regulated DEGs. Interestingly, SUS caused a significant reduction in the number of DEGs between the BW and the other three tested tissues (Int, 8,647; RT, 6,888; and C, 6,072), with the number of down-regulated DEGs being greater than or similar to the number of up-regulated DEGs. It is noteworthy that the numbers of DEGs in the BW were higher in SUS-affected samples compared to those in healthy samples (I-SUS-BW vs. H-BW 9,854, II-SUS-BW vs. H-BW 10,182, III-SUS-BW vs. H-BW 9,726 and I-N-BW vs. H-BW 9,607, II-N-BW vs. H-BW 9,670, III-N-BW vs. H-BW 9,570 DEGs), but lower when SUS-affected samples obtained at the three infection stages were compared to each other (619, 514, and 946 in the SUS groups and 732, 499, and 961 in the normal groups). When the stage II SUS samples were compared to healthy samples, small numbers of DEGs were found in the Int (1,718), RT (3,279) and C (1,847), while more were found in the BW (II-SUS-BW vs. H-BW 10,182 and II-N-BW vs. H-BW 9,670). In addition, very few DEGs were found in the comparison of SUS and normal BW during the three SUS stages (646, 127 and 568 DEGs, respectively), and the lowest number of DEGs was screened at SUS stage II. These data demonstrated the following: (1) the differences in gene expression profiles between SUS-affected and healthy samples were larger in the BW than those in the Int, RT and C; (2) considerable differences in gene expression profiles were detected in the BW of SUS-affected individuals (including SUS stages I, II and III) compared to healthy individuals, while smaller differences in gene expression profiles were observed among the three stages of SUS progression; and (3) the number of DEGs between SUS and normal BW from the same individual exhibited minimum values at SUS stage II.

Table 3: The list of two-class DEGs among the first 13 libraries

Different tissues		Different SUS stages				SUS tissues vs normal tissues	
Libraries name	**No.of DEGs**	**Libraries name**	**No.of DEGs**	**Libraries name**	**No.of DEGs**	**Libraries name**	**No.of DEGs**
H-Int (D2) vs H-BW(D1)	↑12,639 ↓ 4,297 T 16,936	I-SUS-BW (D5) vs H-BW(D1)	↑7,401 ↓2,453 T 9,854	II-SUS-BW(D7) vs I-SUS-BW (D5)	↑340 ↓279 T 619	I-SUS-BW (D5) vs I-N-BW (D6)	↑442 ↓204 T 646
H-RT(D3) vs H-BW(D1)	↑ 13,923 ↓ 3,241 T 17,164	II-SUS-BW(D7) vs H-BW(D1)	↑7,705 ↓2,477 T 10,182	III-SUS-BW(D12) Vs II-SUS-BW(D7)	↑299 ↓215 T 514	II-SUS-BW (D7) vs II-N-BW (D8)	↑ 20 ↓107 T 127
H-C(D4) vs H-BW(D1)	↑ 12,294 ↓ 4,398 T 16,692	III-SUS-BW(D12) vs H-BW(D1)	↑7,229 ↓2,497 T 9,726	III-SUS-BW(D12) vs I-SUS-BW (D5)	↑493 ↓453 T 946	III-SUS-BW(D12) vs III-N-BW (D13)	↑307 ↓261 T 568
II-SUS-Int(D9) vs II-SUS-BW(D7)	↑3,758 ↓4,889 T 8,647	I-N-BW(D6) vs H-BW(D1)	↑7,805 ↓1,802 T 9,607	II-N-BW(D8) vs I-N-BW(D6)	↑437 ↓295 T 732	II-SUS-Int (D9) vs H-Int (D2)	↑ 581 ↓1,137 T 1,718
II-SUS-RT(D10) vs II-SUS-BW(D7)	↑3,523 ↓3,365 T 6,888	II-N-BW(D8) vs H-BW(D1)	↑7,662 ↓2,008 T 9,670	III-N-BW(D13) vs II-N-BW(D8)	↑205 ↓294 T 499	II-SUS-RT (D10) vs H-RT (D3)	↑ 871 ↓2,408 T 3,279

II-SUS-C(D11) vs II-SUS-BW(D7)	↑1,711 ↓4,361 T 6,072	III-N-BW(D13) vs H-BW(D1)	↑7,789 ↓1,781 T 9,570	III-N-BW(D13) vs I-N-BW(D6)	↑531 ↓430 T 961	II-SUS-C (D11) vs H-C (D4)	↑ 700 ↓1,147 T 1,847

Stage-Series Gene Expression Modeling during SUS Progression

The patterns of DEGs for SUS stages I, II and III were assigned as three serial points in the progression of SUS. To classify the dynamics of the SUS development transcriptome on a global scale, we performed gene expression profile clustering. The seven representative expression patterns of SUS progression are presented in Fig. 2. Visual inspection of these expression models suggested the existence of a diverse and complex pattern of regulation during SUS progression. The first model that represented the main expression pattern comprised Cluster 1, Cluster 4 and Cluster 7, containing 4,245 DEGs. These clusters showed higher levels of DEGs compared to healthy samples and were sustained stably with advancing SUS progression. In contrast, Cluster 2 showed sustained lower levels of DEG expression compared to healthy samples at the three SUS stages. Three expression models based on Cluster 3, Cluster 6, and Cluster 9, revealed a similar expression pattern at SUS stages I and III, while a distinct expression pattern was observed during the transition period at SUS stage II. Moreover, Cluster 5, Cluster 6 and Cluster 8 presented significantly higher levels of expression specific for stages I, II and III, respectively.

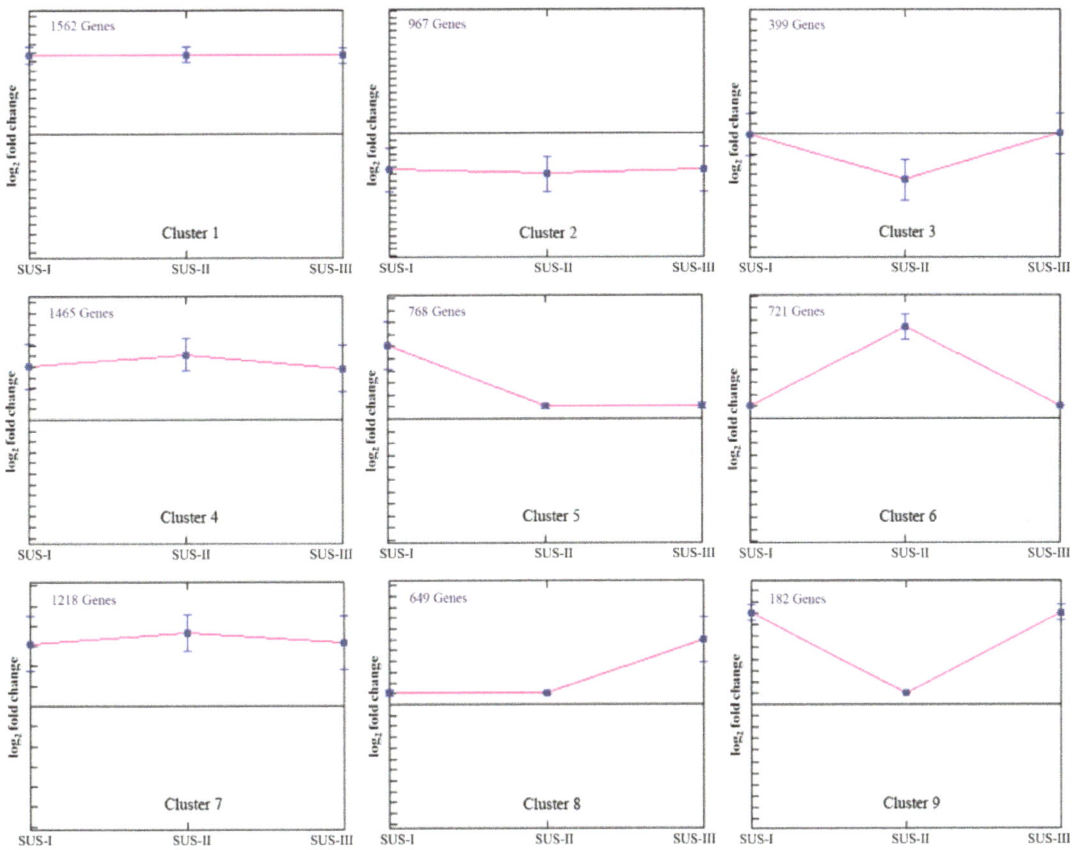

Figure 2: Dynamic expression patterns in *A. japonicus* during SUS progression. The expression profiles of the DEGs [the log_2Ratio ≥1 and the RPKM >2 at a minimum of one time-point] were determined over the three stages of SUS progression by clustering analysis based on the K-means method using the Euclidean distance algorithm. The three points along the x-axis represent I-SUS-BW/H-BW, II-SUS-BW/H-BW and III-SUS-BW/H-BW. Each tick on the y-axis represents a value of 1. Midline represents 0. "+1" represents up-regulated expression. "-1" represents down-regulated expression.

GO and KEGG enrichment analysis of the stably up-regulated expression pattern (4,245 DEGs) during SUS progression showed 175 GO terms (25 GO terms in the category "cellular component", 13 GO terms in the category "molecular function", and 137 GO terms in the category "biological process"). GO analysis demonstrated that these DEGs with positive regulation function were involved in a broad range of physiological processes. The most notably abundant GO terms in the stably up-regulated expression model were "protein binding", "signal transduction", "cell communication", "signaling" and "single organism signaling". This main stage-series gene expression model of SUS progression associated with biological pathways was evaluated by enrichment analysis of DEGs. Significantly enriched pathways were identified and are presented in Fig. 3. For the 35 significantly enriched pathways identified in this expression model, 14 were related to signal transduction. Eight significantly enriched pathways consisting of "natural killer cell mediated cytotoxicity", "leukocyte transendothelial migration", "Fc epsilon RI signaling pathway", "B cell receptor signaling pathway", "T cell receptor signaling pathway", "Toll-like receptor signaling pathway (TLR pathway)", "Fc gamma R-mediated phagocytosis" and "chemokine signaling pathway" were related to the immune system. Seven significantly enriched pathways were related to cellular processes, with apoptosis and regulation of autophagy pathways being noteworthy. Three significantly enriched pathways, "dorso-ventral axis formation", "osteoclast differentiation" and "axon guidance", were related to development.

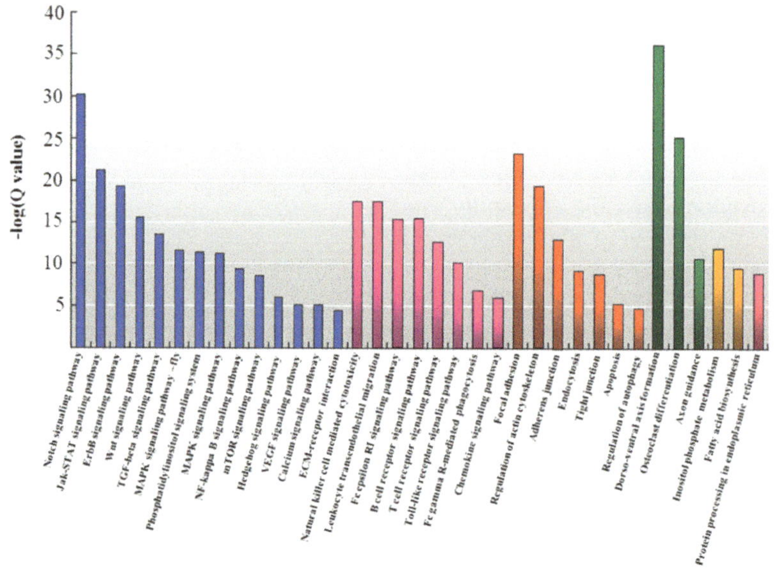

Figure 3: Significantly enriched pathway among the DEGs revealed by the up-regulated expression models in *A. japonicus* during SUS progression. Blue: Pathways related to signal transduction. Magenta: Pathways related to the immune system. Red: Pathways related to cellular processes. Green: Pathways related to development. Orange: Pathways related to metabolism. Pink: Pathways related to genetic information processing.

Investigation of DEGs during SUS Progression

Based on the gene expression profiles in the first 13 cDNA libraries, we focused on the changes in the transcriptome in the BW of sea cucumber during the SUS progression from stage I to stage III. Thereafter, we re-sequenced samples from ulcerative BW of the three SUS stages and healthy BW. Compared to healthy *A. japonicus*, the number of DEGs in the BW of SUS stages I, II and III were 5,478, 5,589 and 4,437, respectively. Combined with the results of the first 13 libraries, we further investigated the DEGs involved in the immune responses from extracellular interaction with bacteria to the activities within the nucleus. Based on GO and KEGG pathway annotations, manual blast and literature searches, 92 DEGs with Nr annotations were selected and divided into four main functional categories: (1) attachment/pathogen recognition (17 genes); (2) inflammatory reactions (38 genes); (3) oxidative stress response (7 genes); and (4) apoptosis (30 genes). A subset of these candidates is listed in Table 4. Among them, the vast majority of these DEGs were up-regulated in the process of SUS progression in *A. japonicus*. The comparison resulted in a trend concordance between the two sets of sequencing data.

Table 4: Special types of DEGs during SUS progression in *A. japonicus*

Gene name	First sequencing			Re-sequencing		
	SUS-I	SUS-II	SUS-III	SUS-I	SUS-II	SUS-III
Attachment/Pathogen recognition						
CLECT	−1.56	−1.79	−2.99	−3.27	−3.34	−3.77
CLECT isoform 3	−1.69	−2.01	−2.21	−3.27	−3.23	−3.64
CLEC19A	−1.71	−1.89	−2.12	−2.51	−2.68	−2.78
GalNAc-specific lectin	−1.21	−1.52	−1.91	−5.37	−6.05	−6.58
lactose-binding lectin l-2-like	a	a	−1.29	a	−2.96	−3.24
Fibrinogen-like protein A	−1.32	−1.25	−1.18	−2.74	−2.79	−2.18

SRCR protein	3.47	3.62	3.91	5.32	5.27	5.68
FNDC3A-like	7.49	7.51	6.39	5.09	5.08	4.91
Annexin A13	1.56	1.64	1.27	2.21	1.69	1.33
Mucin-2	6.21	5.49	7.66	8.78	8.38	8.64
Integrin alpha-1	2.14	1.85	1.8	3.80	3.20	3.23
Integrin alpha-9-like	4.6	4.02	5.08	5.45	4.77	4.65
Integrin Alpha-Lv1	5.88	6.23	7.08	6.23	5.31	5.5
Integrin beta-2	2.02	1.45	1.51	2.24	1.94	1.74
Integrin beta-C subunit	13.24	12.8	11.31	5.31	5.31	5.24
Integrin beta G subunit precursor	4.44	4.11	4.24	4.35	3.98	3.83
Tenascin-R-like	−1.67	−2.28	−2.04	−1.18	−1.77	−1.89
Inflammation Reactions						
Complement C3	2.82	2.93	3.15	3.68	3.55	3.39
Complement component 3-2	6.63	6.42	6.5	3.49	3.15	3.29
Complement Bf	1.29	1.63	1.64	3.3	3.45	3.41
Complement receptor type 2 isoform X1	2.71	3.43	2.87	3.42	4.25	3.51
IgGFc-binding protein	3.21	3.31	2.51	4.99	4.85	4.2
Collagen alpha-1 (XI) chain-like	5.12	5.27	5.11	5.99	5.91	5.81
Collagen alpha-2	7.11	6.34	6.03	5.93	5.06	5.13
Collagen alpha-5	4.51	3.62	3.31	5.02	4.52	4.06
MyD88	1.36	1.38	1.54	1.60	1.64	1.75
TRAF 1	2.29	2.63	2.69	4.02	4.1	4.42
TRAF5-like	5.46	7.65	5.94	2.44	2.93	2.57
IRAK4-like	3.65	3.57	3.91	3.7	3.09	2.86
NF-kB transcription factor Rel	3.25	3.78	3.48	3.64	4.11	3.91
NF-kB p105 subunit	3.45	3.54	3.82	4.17	4.22	4.4
MAPKKK	2.25	2.25	2.17	2.38	2.81	2.61
Serine/threonine-protein kinase TBK1-like	4.82	5.07	4.79	5.46	5.31	5.12
Stress-activated protein kinase JNK-like	4.24	4.21	4.33	3.89	3.63	4.02
TGF beta-activated kinase	2.89	2.98	2.96	4.35	4.21	4.53
TOLLIP	3.41	3.79	3.56	3.43	4.25	3.91
Zonadhesin-like	6.67	5.5	6.97	11.89	11.85	11.38

IFI27-like protein 2	−2.64	−2.47	−2.38	−3.02	−2.93	−3.17
Apolipoprotein B-100-like	3.47	3.83	4.22	2.93	3.18	3.56
Macrophage mannose receptor 1-like	3.01	3.25	3.18	4.3	4.63	4.72
Matrix metalloproteinase-9	5.72	4.84	5.42	8.41	7.71	8.15
Matrix metalloproteinase-24	3.91	3.71	4.01	2.41	1.41	1.57
Matrix metalloproteinase 24 preproprotein-like	4.52	3.43	3.59	3.91	3.71	4.01
Prostaglandin E synthase 2-like	1.26	1.42	1.23	1.51	1.77	1.85
Suppressor of cytokine signaling 2-like	5.33	4.89	5.18	4.95	4.45	4.81
Tumor protein p53-inducible protein 11-like	2.34	2.59	2.26	4.23	4.79	4.42
NFIL3	2.07	2.96	2.38	3.21	2.38	1.41
TMPRSS 5-like	10.34	10.37	10.79	10.39	10.41	10.83
LENG8 homolog	4.28	4.44	3.99	5.37	5.48	5.04
IGDCC4-like	3.76	3.08	3.15	1.87	1.95	1.88
LIG-3 isoform 2	2.22	2.51	2.41	3.45	3.94	3.63
Variable lymphocyte receptor	2.02	2.27	2.28	9.74	11.04	11.26
Thrombospondin-2	4.31	4.22	3.88	3.13	3.01	2.72
Thrombospondin-3	5.28	5.08	5.31	4.77	4.49	4.86
Major yolk protein	−2.63	−2.51	−2.13	−3.07	−3.03	−3.45
Oxidative Stress Response						
HSP 70	3.84	5.24	5.09	3.76	5.17	5.06
HSP 26	3.59	5.84	5.58	2.63	4.55	4.54
HSP 110	2.08	1.63	1.73	3.33	2.96	3.26
HSP 90	7.13	7.42	7.83	2.31	1.59	1.75
Dual oxidase 1	4.36	4.27	4.41	4.82	4.54	4.65
GPX1-like	2.09	1.92	1.62	2.72	2.50	2.65
Thioredoxin	2.33	2.19	2.32	2.91	2.7	3.38
Apoptosis/Autophagy/Lysosome/Phagosome						
Apoptosis regulator BAX-like	2.02	2.27	2.45	7.72	7.52	8.36
Apoptosis-inducing factor 2-like	2.08	2.14	1.99	2.21	2.48	2.01
Calpain-5	2.65	2.64	2.08	4.77	4.79	4.20
Calpain-7	1.96	1.94	1.99	3.06	3.12	3.08
Caspase-6-like	3.82	4.12	4.36	2.12	3.31	2.44

Cyclophilin B	1.97	1.96	1.91	1.90	2.03	2.30
Cytochrome c-like	−1.14	−1.32	−1.22	−1.65	−1.84	−1.75
Cytochrome c oxidase subunit 7C	−2.06	−1.95	−1.95	−2.57	−2.65	−2.76
BFAR -like	2.57	3.19	2.73	8.03	8.15	7.95
Bcl-2 protein	5.29	4.83	4.67	4.08	3.65	3.43
Bcl-2-like protein 1-like	2.25	2.58	2.32	4.04	4.34	4.14
BIRC2	3.28	2.9	3.01	3.8	3.52	3.42
BIRC 6-like	3.18	3.17	2.96	5.06	5.11	4.89
PDCD6IP -like	2.17	2.17	2.22	3.61	3.77	3.69
PDCD10 -like	3.79	3.47	3.54	4.41	4.29	4.32
DRAM 2-like	2.38	2.54	2.95	3.93	4.24	4.19
PDRG 1	−2.83	−2.83	−1.83	−3.81	−3.51	−3.13
Exportin-1-like	1.59	1.52	1.59	11.77	10.38	11.82
Exportin-T	1.37	1.49	1.61	1.64	1.42	1.59
Cathepsin D	7.13	7.47	7.44	6.13	7.16	8.09
Cathepsin L	1.72	1.83	1.79	3.18	3.84	4.71
Cathepsin B-like protease	3.08	3.19	3.25	3.23	3.37	3.57
Lysozyme-like	−1.51	a	−1.69	−3.38	a	−3.67
Autophagy-related protein 2 homolog B	6.46	6.31	5.79	3.82	4.29	3.75
Lysosome membrane protein 2-like	4.27	4.72	3.11	4.39	4.03	3.55
Lysosomal alpha-mannosidase-like	2.34	2.56	2.51	9.49	10.11	10.05
Lysosomal-trafficking regulator-like	1.92	1.51	1.43	4.85	4.68	4.7
ADP-ribosylation factor-like 1	1.72	1.42	1.58	6.12	6.09	6.43
ADP-ribosylation factor-like 3	2.32	1.51	2.68	4.25	4.61	4.52
Beta-galactosidase-like	1.66	1.42	1.78	4.11	4.01	3.61

[a]means no significantly different in gene expression. The numbers in the Table represent Log_2fold change of I-SUS-BW/H-BW, II-SUS-BW/H-BW and III-SUS-BW/H-BW

Expression Validation using qRT-PCR

To validate the reliability of the RNA-seq results, twenty DEGs from Table 4 were detected by qRT-PCR at three stages of SUS progression (SUS stage I, II and III) and H samples. The cytochrome b (Cyt b) gene was chosen as the reference gene [32]. The 20 DEGs were selected for their clear background information in the function of immune responses, and some of them were involved in the complement pathway, TLR signaling pathway and Apoptosis. The Pearson›s correlation coefficient (R) was used to assess the consistency of DEGs expression profiles between these two methods. All the DEGs at three stages of SUS progression showed a consistent expression pattern with R value ranging from 0.811 to 0.999 (Fig. 4).

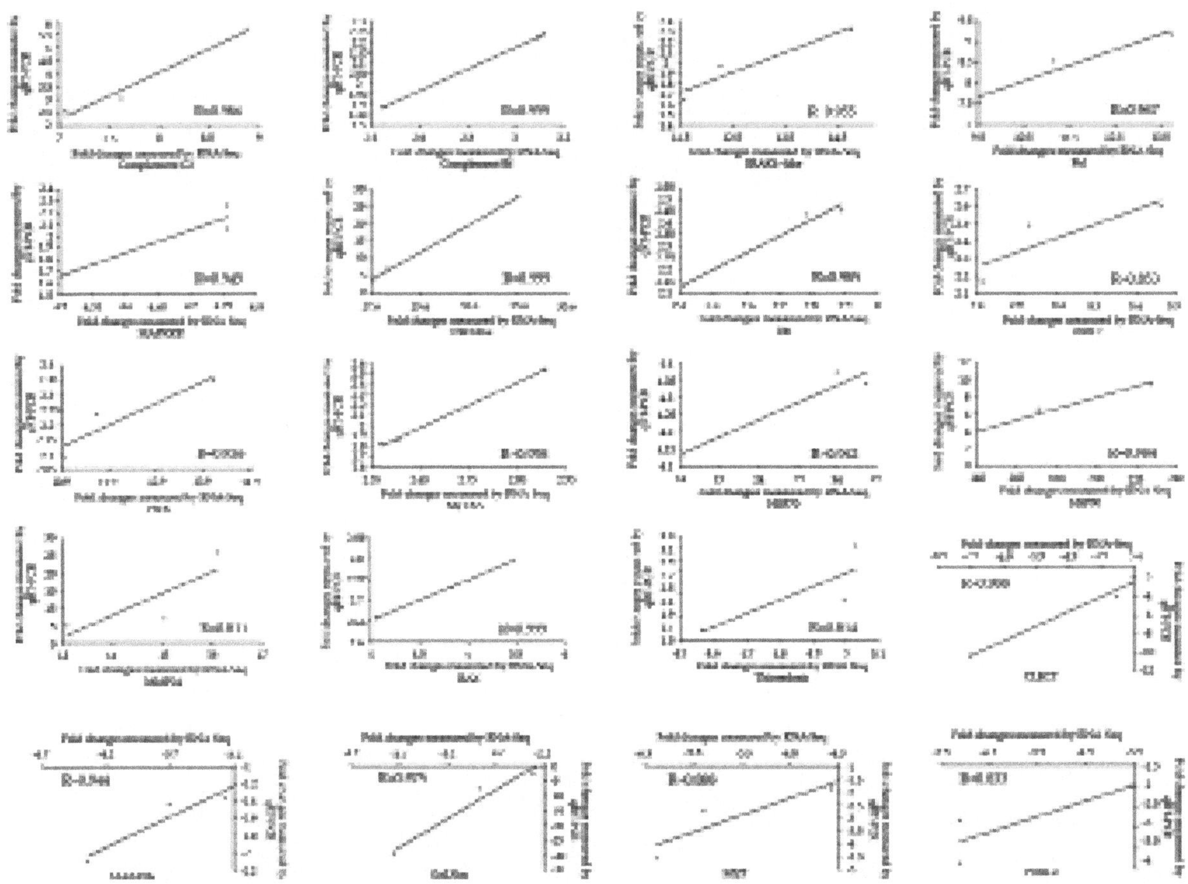

Figure 4: Validation of RNA-seq results using qRT-PCR. Twenty DEGs were selected and their relative fold changes were expressed as the ratio of gene expression in BW of *A. japonicus*at SUS stages I, II and III compared to H samples as normalized with the Cytb gene. The data obtained from the first RNA-seq and qRT-PCR were compared correspondingly and drawn as a Pearson correlation scatter plot.

DISCUSSION

Changes in Gene Expression Profiles during SUS Progression

Using RNA-seq we have generated an extensive transcriptome profile for SUS-affected*A. japonicus*. The transcriptome profile allowed us to look at the changes in gene expression associated with the progression of the disease. Through construction of the first 13 cDNA libraries and comparison of the DEGs in four different tissues taken from SUS-affected *A. japonicus* against those of healthy *A. japonicus* samples, as well as investigation of the transcriptome associated with the three disease stages, we acquired a broad understanding of the dynamic transcriptome during the progression of SUS. Recently, 4,858 DEGs were found in*A. japonicus* coelomocytes by comparisons of SUS diseased and healthy control sea cucumbers. These genes were significantly enriched in tyrosine transport (biological progression), melanosome membrane (cell component), and L-tyrosine transmembrane transporter activity and 1-3-beta-glycan binding (molecular function) [28]. Gao et al. identified 102 DEGs in *A. japonicus* coelomocytes of SUS-affected group by comparing with that in control group. Many DEGs were involved in immune related pathways, such as Endocytosis, Lysosome, MAPK, Chemokine and the ERBB signaling pathway [29]. However, our data showed that between the SUS and healthy individuals, the biggest difference in expression profiles was found in BW. White skin ulceration is the main symptom of SUS; therefore, we further analyzed the changes in the transcriptome in the BW of sea cucumber during SUS progression from stage I to stage III.

Pathways and DEGs during SUS Progression

During the progression of SUS caused by *V. splendidus* challenge, immune regulation was an important event in this host-pathogen interaction process. In fact, most of the significantly enriched pathways involved in the immune system were observed in the up-regulated expression model during SUS progression. Of the eight significantly enriched pathways that were related to immune system, the "TLR pathway" has been reported in *A. japonicus* [32–35] response to bacterial challenge and in sea star *Pycnopodia helianthoides* response to sea star wasting disease [36]. Several signal transduction pathways in the up-regulated expression model during SUS progression, including the insulin [37], TGF-beta [38], MAPK [29], NF-kappa B [39, 40], and Notch [41] signaling pathways, have been reported to involve in innate immunity in invertebrate. In the studies, we further investigated DEGs involved in attachment/pathogen recognition, inflammatory reactions and apoptosis.

Attachment/Pathogen Recognition

Immune responses against invasive pathogens depend primarily on the recognition of pathogen components by innate pattern recognition receptors (PRRs). C-type lectins, which are a type of PRRs involved in innate immunity, specifically recognize sugars on the surface of pathogens in the presence of Ca^{2+}, and cause a series of immune responses to resist pathogen invasion [42]. It is worth noting that four C-type lectins, including GalNAc-specific lectin [43], showed down-regulated expression and presented a downward trend during SUS progression (Table 4). Mannan-binding C-type lectin expression was also down-regulated at 48 h and 72 h after LPS challenge in *A. japonicus* [32]. Further studies are required to clarify the correlation between negative-regulation of this group of receptors and the occurrence of SUS in *A. japonicus*.

Cell adhesion molecules play important roles in many facets of the immune system. Integrins are the largest family of cell adhesion molecules that mediate cell-to-cell and cell-to-matrix interactions in a broad range of physiological and pathological processes [44]. The essential roles of integrins in developmental and innate immunity in invertebrates have been well studied, especially in phagocytosis [45, 46], encapsulation [47] and degranulation [48]. Integrin β1 is upregulated in hemocytes in response to various microbes in *Spodoptera exigua* [49] and integrin β is also involved as a cell adhesion receptor in the immune responses against microbe challenge in the shrimp *Litopenaeus vannamei* [50, 51]. Intriguingly, six integrin family genes were up-regulated during SUS progression in *A. japonicus*. The functions of integrins in *A. japonicus* response to SUS need to be further studied.

Inflammatory Reactions

The innate immune system is the host's first line of defense against infection. It triggers diverse humoral and cellular activities via signal transduction pathways which are conserved both in invertebrates and vertebrates [52]. In our studies, the DEGs included in inflammatory reactions were mainly involved in the complement pathway and TLR signaling pathway. The complement system is a major component of the innate immune system involved in defending against all the foreign pathogens [53]. The key genes of the complement system participating in this defense include complement components C3, C3-2, and factor B, all of which were up-regulated during SUS progression. In our earlier study, the expression of two complement component C3 genes in*A. japonicus* coelomocytes was also clearly up-regulated after the animal was challenged with LPS [54]. Zhong et al. reported that LPS challenge of *A. japonicus* induced significant up-regulation of the expression of a homologue of complement B factor (*AjBf-2*) in the coelomocyte and body wall [55]. C3 plays a pivotal role in the activation of both the classical and alternative complement pathways as well as the lectin pathways [56]. As the second complement component in the alternative pathway, Bf binds to the activated C3b and $C3(H_2O)$ [57]. In the sea urchin, these two complement components may be part of a simple complement system that is homologous to the alternative pathway in vertebrates [58].

The essential role of TLRs in the activation of innate immunity by recognizing specific patterns of microbial components is well established [59]. While individual TLR genes were not captured, downstream signaling components traditionally associated with TLR pathway were found. For up-regulated DEGs, the important molecules of this pathway, such as MyD88, IRAK4-like, TOLLIP, TAK, NF-kB p105 subunit, NF-kB Rel and stress-activated protein kinase JNK-like, were screened during SUS progression. It is noteworthy that TRAF1 and TRAF5-like, and serine/threonine-protein kinase TBK1-like were associated with the MyD88-independent TLR pathway. Two adaptor molecules in the *A. japonicus* Toll signaling cascade, MyD88 and TRAF6, have been isolated and characterized. The expression levels of these two genes have been shown to increase significantly during *V. splendidus* challenge [33]. The NF-kB homologues, *Aj-rel* and*Aj-p105,* have also been identified, and shown to be involved in LPS-induced immunity in *A. japonicus* [34]. The important molecules in this pathway such as MyD88, NF-kB Rel, TRAF6 and LPS-induced TNFα were all shown to be differentially expressed in *A. japonicus*coelomocytes after LPS challenge by RNA-seq analysis in our early studies [32]. As to the MyD88-independent pathway, two Toll-like receptor genes (*AjTLR3* and *AjToll*) were identified, which were functionally involved in the immune responses against Gram-positive and Gram-negative bacteria, fungi and double-stranded RNA (dsRNA) viruses [35].

Apoptosis

Apoptosis is an essential process in metazoans and is critical for the formation and function of tissues and organs [60]. Dysregulation of the apoptotic processes often leads to serious consequences in humans, such as neurodegenerative diseases [61], cancer [62], and autoimmunity [63]. Some apoptosis-related proteins, such as sarcoma oncogene (Src), vitronectin and vinculin, have been identified and displayed time-dependent depressed expression in the coelomocytes of *V. splendidus*-challenged *A. japonicus* [28]. In the present study, we identified the DEGs involved in the apoptosis pathway and determined their expression profiles during SUS progression. There are two major pathways leading to apoptosis: an extrinsic pathway initiated by death receptors and an intrinsic pathway that occurs through the mitochondria. In the extrinsic pathway, procaspase 8 is activated by receptors for FasL and TNF through the recruitment of intracellular death domain-containing proteins, such as FADD. In the intrinsic pathway, procaspase 9 is activated by cytochrome C released from the damaged mitochondria [64]. Our results showed down-regulated cytochrome C expression during SUS progression. Apoptosis regulator Bcl-2 is a family of evolutionarily related proteins that govern mitochondrial outer membrane permeabilization and perform either pro-apoptotic (such as BAX) or anti-apoptotic (such as Bcl-2) functions. Interestingly, our studies showed that these two opposing regulators were up-regulated expression during SUS progression. Another important apoptosis regulator, baculoviral inhibitor of apoptosis repeat-containing protein 2 (BIRC2), was identified with up-regulated expression during SUS progression. In addition, expressions of two transcription factors NF-kB p105 subunit and NF-kB Rel that suppress apoptosis were also up-regulated during SUS progression. In the study of human cancer, molecular targeting therapies have been focused on the regulation of apoptosis by Bcl-2 family proteins [65], IAPs [66] and NF-kB [67]. Since apoptotic signals are complicated and regulated at several levels, the mechanism underlying the regulation of apoptosis in SUS of sea cucumber is worthy of further exploration.

CONCLUSION

The development of SUS in sea cucumber is a complex process in which tens of thousands of genes showed significantly different expression during the progression of the disease. Systematic investigation of the polygenic regulation and multiple pathways involved in the multi-stage progression of SUS is required to elucidate the dynamic mechanism of SUS progression. The findings reported in this study will be useful for further studies into the origins and development of SUS in sea cucumber. In further investigations, we will focus on the network biology approach to comprehensively depict the miRNA-mRNA and mRNA-protein networks relevant to SUS in *A. japonicus*.

METHODS

Animals

Animals used in this research were obtained from commercial sea cucumber catches, therefore approval from any ethics committee or institutional review board was not necessary. Healthy *A. japonicus* (10–12 g) were collected from Zhuanghe, Liaoning Province (China). The animals were acclimated in the laboratory for 1 week, and then subjected to artificial infection. The sea water used in the experiment was filtered through sand, and then through 300-μm nylon sieves. Twenty-five percent of the seawater in the tank was exchanged daily. The animals were maintained at 12 °C, pH 8.1, with salinity of 32 and continuous aeration.

Artificial Infection and Sample Collection

V. splendidus used to infect *A. japonicus* was previously isolated and identified in our laboratory from SUS-affected *A. japonicus* according to the method reported by Deng et al. [22]. The bacteria were cultivated in 2216E medium at 28 °C for 24 h. The bacterial cells were then harvested by centrifugation at 1,000 × *g* for 5 min, and then re-suspended in 0.22-μm-filtered seawater. For the wounded immersion infection, 200 healthy *A. japonicus* individuals were cultured in five rectangular tanks (80 cm in length, 45 cm in width, and 45 cm in height). A cut (0.5 cm × 0.5 cm) was made in the body wall of the sea cucumbers using a scalpel and *V. splendidus* was added to the tank (final concentration, 5×10^9 CFU mL^{-1}), with 25 % seawater replenished daily; *V. splendidus* was supplemented and the bacterial concentrations were maintained. In the sample collection, white skin ulceration was considered to be the most important mark to distinguish diseased and healthy individuals [15, 28]. The progression of SUS was divided into three stages. In stage I, the animals showed one small white speck of skin ulceration (diameter <0.2 cm). The animals retained the ability to attach to the surface of the tank and did not eviscerate. In stage II, the animals exhibited 2 to 3 larger white specks (diameter >0.2 cm). The animals continued to exhibit the ability to attach to the surface of the tank and did not eviscerate. In stage III, the individuals showed several deep and extensive ulcerations, lost the ability to adhere to the tank and displayed evisceration. *A. japonicus* that cultured under normal conditions without the treatments of cut and bacterial challenge served as healthy controls. Thirty individuals were selected from each of the SUS stages (stage I, stage II, and stage III) and the ulcerative body wall and its distal normal tissue were sampled for RNA analysis (Fig. 1). In addition, samples of the Int, RT and C were also obtained at SUS stage II. The C were collected by centrifugation at 1,000 × *g* for 5 min. Samples of BW, Int, RT and C were separated from 30 healthy *A. japonicus* individuals. For each sampled individual, the size of BW tissues is about 0.2 cm × 0.2 cm × 0.2 cm. The quality of sampled Int and RT tissues is less than 100 ng. The volume of C is about 1 mL. All samples were frozen immediately in liquid nitrogen and then stored at −80 °C before RNA isolation.

RNA Extraction, cDNA Library Construction and Sequencing

Total RNA was extracted from the specimens with Trizol reagent (Invitrogen, Carlsbad,CA,USA) according to the manufacturer's instructions. The quality and concentration of total RNA were measured by Nanodrop 1000 (Thermo). The initial amount of high-quality total RNA was 1 μg per library (generated from 30 individuals). Subsequently, the mRNA in the total RNA was enriched using Oligo (dT) magnetic beads and sheared into short fragments by adding fragmentation buffer, followed by first- and second-strand cDNA synthesis using random hexamer primers. The cDNA fragments were subjected to an end-repair process, addition of "A" base, and ligation of sequencing adapters. After agarose gel electrophoresis, suitable fragments were selected and used as templates for the PCR amplification to create the final cDNA libraries. We first constructed 13 cDNA libraries (from D1 to D13), representing the ulcerative BW and their distal normal tissues at all three stages of SUS development (Fig. 1) and four tissues (BW, Int, RT and C) of SUS-affected and healthy samples. High-throughput sequencing was conducted using the Illumina HiSeq™ 2000 platform to generate 50-bp reads.

Data Processing, Assembly and Annotation

The original image data were converted to sequence data by base-calling and saved as fastq files. The raw reads were then cleaned by discarding adaptors, low-quality reads (quality scores <20), reads with unknown bases greater than 5 %, and reads of less than 20 nt. De novo transcriptome assembly combined with Illumina reads obtained from GenBank [28, 30] was carried out with the short-read assembly program, Trinity [68]. The assembled sequences from the Illumina reads, 454 sequences and ESTs were used in further processing of the assembly according to the method reported by Zhou et al. [30]. Blastx analysis of unigenes longer than 200 bp was conducted against the Non-Redundant (Nr) database, Swiss-Prot, COG and KEGG database with an *e*-value cutoff of 1E-5. GO classification was analyzed by Blast2GO software (v.2.5.0) based on best hits of Nr annotation. In addition, transcripts were also annotated with the NCBI non-redundant nucleotide (Nt) database using Blastn with an *e*-value cutoff of 1E-5.

Screening and Analysis of DEGs

The unigene expression was calculated using the RPKM (Reads Per kb per million reads) method [9]. To identify the

DEGs in the first 13 *A. japonicus* libraries, a rigorous algorithm was developed for statistical analysis according to "the significance of digital gene expression profiles" [69]. The FDR (false discovery rate) is used in multiple hypothesis testing to correct for*P*-values [70]. In our analysis, the FDR$\leq$0.001 and RPKM ratio larger than 2 ($|\log_2\text{Ratio}|\geq 1$) were used as the threshold to judge the significance of differences in gene expression. The dynamic expression profiles of DEGs during the three-stage SUS progression were visualized by clustering analysis, based on the K-means method using the Euclidean distance algorithm. The enrichment analysis was carried out based on an algorithm presented by KOBAS [71], with the entire transcriptome set as the background. The *P*-value was approximated by a hypergeometric distribution test and FDR multipletesting correctionwas used to corrected-*P*-value [70]. The GO enriched cutoff was a corrected-*P*-value$\leq$0.05, and the KEGG enriched cutoff was a Q value$\leq$0.05.

Transcriptome Sample Preparation, Re-Sequencing and DEG Identification

After analysis of the gene expression profiles in the first 13 cDNA libraries, we re-sequenced samples of the ulcerative BW of the three SUS stages and healthy BW. The three stages of SUS-affected samples were also infected with *V. splendidus*. Subsequently, ulcerative BW of individuals during the three SUS stages and healthy BW were separated for cDNA library construction (from T1 to T8) and RNA-seq. Two biological replicates of each sample were set up. High-throughput sequencing was conducted using the Illumina HiSeq™ 2000 platform to generate 100-bp paired-end reads. After quality control for raw reads, clean reads were mapped to the reference transcriptome and their abundance estimated using RSEM (RNA-seq by expectation maximization). DEGs were detected by using the DESEQ program from the R-Bioconductor package.

Expression Validation using qRT-PCR

To validate the RNA-seq results, 20 DEGs was employed for qRT-PCR (Mx3005p™ detection system, Agilent Stratagene, Santa Clara, CA, USA). Total RNA from samples used in the first RNA-seq was reverse-transcribed into cDNA templates with the PrimeScript™ RT reagent Kit (TaKaRa, Otsu, Japan) according to the manufacturer's instruction. Reactions were incubated at 37 °C for 15 min, and then 85 °C for 5 s. According to the sequence information in the transcriptome database, primers were designed using Primer 5.0 software according to rigorous criteria. The primer information was provided in Table 5. The cytochrome b (Cytb) gene was chosen as the reference gene [32]. Optimal primer pairs were examined by checking the melting curve at the end of each PCR reaction to confirm the specificity of PCR product. The qRT-PCR amplification was conducted in a volume of 20 μL containing 10 μL of 2× SYBR Premix Ex Taq™ II (*Tli* RNaseH Plus, TaKaRa, Otsu, Japan), 0.4 μL of ROX Reference Dye II, 2 μL of cDNA template, and 0.4 μM of each primer according to the introduction. Thermal cycling was as follows: 95 °C for 30 s, 40 cycles at 95 °C for 10 s, 56 °C for 25 s and 72 °C for 25 s. All reactions were run in triplicates. Relative Expression Software Tool 384 v.2 (REST) (Technical University of Munich, Munich, Germany) [72] was used to calculate the expression differences between control and SUS groups (SUS stage I, II and III). The correlation of DEGs expression profiles between RNA-seq and qRT-PCR was assessed by Pearson correlation coefficient (R). The nearer the scatter of points is to the straight line, the higher the strength of association between the variables. The value of R ranges from −1 to +1 [73, 74].

Table 5: Primers used for qRT-PCR validation

Gene	Primer Sequence (5′-3′)
Cytochrome b	Cytb-F: TGAGCCGCAACAGTAATC
	Cytb-R: AAGGGAAAAGGAAGTGAAAG
Complement component 3	AjC3-F: GCGTTGTTTCGTTCAACAAGGGGA
	AjC3-R: GCCATTCACTGGAGGTGTGCCA
Complement factor B	Bf-F: ATTATCTCGCAACAGCGATCC
	Bf-R: GGGCAACCACACCGGCTTCTCCA
IRAK4-like	IRAK4-F: TACACGTCAGATCGGGATGA
	IRAK4-R: TAAACGACGAGCGTACCACA
NF-kB transcription factor Rel	Rel-F: TGCGAAGCCACATCCATT
	Rel-R: AGGGCATCCTTTAAGTCAGC
MAPKKK	MAPKKK -F: GAATCAGAGGAGATAGATGTGGAGA
	MAPKKK -R: AGGAGGAGGAGGAAGACGAC
Serine/threonine-protein kinase TBK1-like	TBK1-F: AGATGATGTTGTCCATTCTCG
	TBK1-R: ACAGGAGGAAGTGATGTGCT
TGF beta-activated kinase	TAK -F: TCTCTGTAGCCTCCTTTGACG
	TAK -R: CTCGGTCTTCCAACCAACAC
BIRC2	BIRC2-F: TCAGGCACGAGTGACAAAGT
	BIRC2-R: GCATGAGCCATTCACATCTCA

NF-kB p105 subunit	p105-F: GCAACACACCCCTCCATCTT
	p105-R: TCTTCTTCGCTAACGTCACACC
MyD88	MyD88-F: CCGATGTAGGAGGATGGTAGTAG
	MyD88-R: CACAGTAAGGTGCTGAAGAATGC
HSP 70	HSP 70-F: AAGAGCACAGGCAAAGAG
	HSP 70-R: TGATGATGGGTTGGCACA
HSP 90	HSP 90-F: TATGAAAGCCTGACAGACGCAAGC
	HSP 90-R: TAACGCAGAGTAAAAGCCAACACC
Matrix metalloproteinase-24	MMP-F: CGATTCAGTCTTCCCTGGTG
	MMP-R: ACCGTCATCAACTTTCCTGGT
Apoptosis regulator BAX-like	BAX-F: GCCGTGGGACTGACTTTACA
	BAX-R: TCCATCTCGTAGTTCTCTCAACG
Thioredoxin	TRx-F: GCTGGTGACAAACTGGTGAT
	TRx -R: TGAGAAAGACAACGTCGGTA
CLECT	CLECT-F: GACGGCTTGTCCAGAGTT
	CLECT-R: AGGTCCATTGTTGGGTTC
CLEC19A	CLEC19A-F: ATGCAGCGAGAAGATGGAGT
	CLEC19A-R: TGGCAGGATATGCCCTAGAT
GalNAc-specific lectin	GalNAc -F: CCATCCTTCAGGGCAGATAA
	GalNAc -R: TTCATCGACCAAAATGCAGA
Major yolk protein	MYP-F: AGGAGGGAGACATTGCTT
	MYP-R: ATGATGCTTTCTGGGTTG
PDRG 1	PDRG1-F: AATTGGAGGAACTCGCTGAA
	PDRG1-R: TTGCTTATCGCCTTCTTGTG

ACKNOWLEDGEMENTS

This work was supported by grants from the National Natural Science Foundation of China (31272687), Science & Technology Project of Liaoning Province, China (2015103044) and the Natural Science Foundation of Liaoning Province, China (2015020786). The funding body had no role in study design, data collection, analysis and interpretation, decision to publish, or preparation of the manuscript.

AUTHORS' CONTRIBUTIONS

AY performed the experiments and wrote the manuscript. ZZ conceived the experiments and corrected the manuscript. YP conducted the data analysis. JJ, YD, XG, HS, SG and ZC were involved in one or more processes of material culture, samples collection, data analysis and manuscript preparation. All the authors have read and approved the final manuscript.

REFERENCES

1. Costa V, Angelini C, De Feis I, Ciccodicola A. Uncovering the complexity of transcriptomes with RNA-Seq. J Biomed Biotechnol. 2010;2010:853916.
2. Graveley BR, Brooks AN, Carlson JW, Duff MO, Landolin JM, Yang L, et al. The developmental transcriptome of *Drosophila melanogaster*. Nature. 2011;471:473–9.
3. Assou S, Boumela I, Haouzi D, Anahory T, Dechaud H, De Vos J, et al. Dynamic changes in gene expression during human early embryo development: from fundamental aspects to clinical applications. Hum Reprod Update. 2011;17:272–90.
4. Chen S, Yang P, Jiang F, Wei Y, Ma Z, Kang L. De novo analysis of transcriptome dynamics in the migratory locust during the development of phase traits. PLoS One. 2010;5, e15633.
5. Tan MH, Au KF, Yablonovitch AL, Wills AE, Chuang J, Baker JC, et al. RNA sequencing reveals a diverse and dynamic repertoire of the *Xenopus tropicalis* transcriptome over development. Genome Res. 2013;23:201–16.
6. Sun L, Yang H, Chen M, Ma D, Lin C. RNA-Seq Reveals Dynamic Changes of Gene Expression in Key Stages of Intestine Regeneration in the Sea Cucumber *Apostichopus japonicus*. PLoS One. 2013;8, e69441.
7. Yang H, Zhou Y, Gu J, Xie S, Xu Y, Zhu G, et al. Deep mRNA sequencing analysis to capture the transcriptome landscape of zebrafish embryos and larvae. PLoS One. 2013;8, e64058.
8. Wang Z, Gerstein M, Snyder M. RNA-Seq: a revolutionary tool for transcriptomics. Nat Rev Genet. 2009;10:57–63.
9. Mortazavi A, Williams BA, McCue K, Schaeffer L, Wold B. Mapping and quantifying mammalian transcriptomes by RNA-Seq. Nat Methods. 2008;5:621–8.

10. Gan Q, Chepelev I, Wei G, Tarayrah L, Cui KR, Zhao KJ, et al. Dynamic regulation of alternative splicing and chromatin structure in Drosophila gonads revealed by RNA-seq. Cell Res. 2010;20:763–83.

11. Zheng W, Wang Z, Collins JE, Andrews RM, Stemple D, Gong Z. Comparative transcriptome analyses indicate molecular homology of zebrafish swimbladder and mammalian lung. PLoS One. 2011;6, e24019.

12. Marioni JC, Mason CE, Mane SM, Stephens M, Gilad Y. RNA-seq: an assessment of technical reproducibility and comparison with gene expression arrays. Genome Res. 2008;18:1509–17.

13. Han X, Wu X, Chung WY, Li T, Nekrutenko A, Altman NS, et al. Transcriptome of embryonic and neonatal mouse cortex by high-throughput RNA sequencing. Proc Natl Acad Sci USA. 2009;106:12741–6.

14. Xiang LX, He D, Dong WR, Zhang YW, Shao JZ. Deep sequencing-based transcriptome profiling analysis of bacteria-challenged *Lateolabrax japonicus* reveals insight into the immune-relevant genes in marine fish. BMC Genomics. 2010;11:472.

15. Li C, Feng W, Qiu L, Xia C, Su X, Jin C, et al. Characterization of skin ulceration syndrome associated microRNAs in seacucumber *Apostichopus japonicus* by deep sequencing. Fish Shellfish Immunol. 2012;33:436–41.

16. Li P, Ponnala L, Gandotra N, Wang L, Si Y, Tausta SL, et al. The developmental dynamics of the maize leaf transcriptome. Nat Genet. 2010;42:1060–7.

17. Liu WY, Chang YM, Chen SC, Lu CH, Wu YH, Lu MY, et al. Anatomical and transcriptional dynamics of maize embryonic leaves during seed germination. Proc Natl Acad Sci USA. 2013;110:3979–84.

18. Chen J, Zeng B, Zhang M, Xie S, Wang G, Hauck A, et al. Dynamic transcriptome landscape of maize embryo and endosperm development. Plant Physiol. 2014;166:252–64.

19. Huang MY, Chen HC, Yang IP, Tsai HL, Wang TN, Juo SH, et al. Tumorigenesis and tumor progression related gene expression profiles in colorectal cancer. Cancer Biomark. 2013;13:269–79.

20. Zhao H, Li Y, Wang S, Yang Y, Wang J, Ruan X, et al. Whole transcriptome RNA-seq analysis: tumorigenesis and metastasis of melanoma. Gene. 2014;548:234–43.

21. Dong Y, Deng H, Sui X, Song L. Ulcer disease of farmed sea cucumber (*Apostichopus japonicus*). Fish Sci. 2005;24:4–6 .

22. Deng H, He C, Zhou Z, Liu C, Tan K, Wang N, et al. Isolation and pathogenicity of pathogens from skin ulceration disease and viscera ejection syndrome of the sea cucumber*Apostichopus japonicus*. Aquaculture. 2009;287:18–27.

23. Wang YG, Xu GR, Zhang CY, Sun SF. Main diseases of cultured *Apostichopus japonicus*: prevention and treatment. Mar Sci. 2005;29:1–7 .

24. Wang YG, Fang B, Zhang CY, Xu GR. Etiology of skin ulcer syndrome in cultured juveniles of *Apostichopus japonicus* and analysis of reservoir of the pathogens. J Fish Sci China. 2006;13:610–6 .

25. Ma YX, Xu GR, Chang YQ, Zhang EP, Zhou W, Song LS. Bacterial pathogens of skin ulceration disease in cultured sea cucumber *Apostichopus aponicus* (Selenka) juveniles. J Dalian Fish Univ. 2006;21:13–8 .

26. Deng H, Zhou ZC, Wang NB, Liu C. The syndrome of sea cucumber (*Apostichopus japonicus*) infected by virus and bacteria. Virologica Sinica. 2008;23:63–7.

27. Liu H, Zheng F, Sun X, Hong X, Dong S, Wang B, et al. Identification of the pathogens associated with skin ulceration and peristome tumescence in cultured sea cucumbers*Apostichopus japonicus* (Selenka). J Invertebr Pathol. 2010;105:236–42.

28. Zhang P, Li C, Zhu L, Su X, Li Y, Jin C, Li T. De novo assembly of the sea cucumber*Apostichopus japonicus* hemocytes transcriptome to identify miRNA targets associated with skin ulceration syndrome. PLoS One. 2013;8, e73506.

29. Gao Q, Liao M, Wang Y, Li B, Zhang Z, Rong X, et al. Transcriptome analysis and discovery of genes involved in immune pathways from coelomocytes of sea cucumber (*Apostichopus japonicus*) after *Vibrio splendidus* Challenge. Int J Mol Sci. 2015;16:16347–77.

30. Zhou ZC, Dong Y, Sun HJ, Yang AF, Chen Z, Gao S, et al. Transcriptome sequencing of sea cucumber (*Apostichopus japonicus*) and the identification of gene-associated markers. Mol Ecol Resour. 2014;14:127–38.

31. Du H, Bao Z, Hou R, Wang S, Su H, Yan J, et al. Transcriptome Sequencing and Characterization for the Sea Cucumber *Apostichopus japonicus* (Selenka, 1867). PLoS One. 2012;7, e33311.

32. Dong Y, Sun H, Zhou Z, Yang A, Chen Z, Guan X, et al. Expression analysis of immune related genes identified from the coelomocytes of *Apostichopus japonicus* in response to LPS challenge. Int J Mol Sci. 2014;15:19472–86.

33. Lu Y, Li C, Zhang P, Shao Y, Su X, Li Y, et al. Two adaptor molecules of MyD88 and TRAF6 in *Apostichopus japonicus* Toll signaling cascade: molecular cloning and expression analysis. Dev Comp Immunol. 2013;41:498–504.

34. Wang T, Sun Y, Jin L, Thacker P, Li S, Xu Y. *Aj-rel* and *Aj-p105*, two evolutionary conserved NF-kB homologues in sea cucumber (*Apostichopus japonicus*) and their involvement in LPS induced immunity. Fish Shellfish Immunol. 2013;34:17–22.

35. Sun H, Zhou Z, Dong Y, Yang A, Jiang B, Gao S, et al. Identification and expression analysis of two Toll-like receptor genes from sea cucumber (*Apostichopus japonicus*). Fish Shellfish Immunol. 2013;34:147–58.

36. Fuess LE, Eisenlord ME, Closek CJ, Tracy AM, Mauntz R, Gignoux-Wolfsohn S, et al. Up in Arms: Immune and Nervous System Response to Sea Star Wasting Disease. PLoS One. 2015;10, e0133053.

37. Kawli T, Tan MW. Neuroendocrine signals modulate the innate immunity of*Caenorhabditis elegans* through insulin signaling. Nat Immunol. 2008;9:1415–24.

38. Mallo GV, Kurz CL, Couillault C, Pujol N, Granjeaud S, Kohara Y, et al. Inducible antibacterial defense system in *C. elegans*. Curr Biol. 2002;12:1209–14.

39. Li H, Chen Y, Li M, Wang S, Zuo H, Xu X, et al. A C-type lectin (LvCTL4) from *Litopenaeus vannamei* is a downstream molecule of the NF-kB signaling pathway and participates in antibacterial immune response. Fish Shellfish Immunol. 2015;43:257–63.

40. Matova N, Anderson KV. Rel/NF-kappaB double mutants reveal that cellular immunity is central to Drosophila host defense. Proc Natl Acad Sci USA. 2006;103:16424–9.

41. Zacharioudaki E, Bray SJ. Tools and methods for studying Notch signaling in *Drosophila melanogaster*. Methods. 2014;68:173–82.

42. Zhang XH, Shi YH, Chen J. Molecular characterization of a transmemebrane C-type lectin receptor gene from ayu (*Plecoglossus altivelis*) and its effect on the recognition of different bacteria by monocytes/macrophages. Mol Immunol. 2015;66:439–50.

43. Kakiuchi M, Okino N, Sueyoshi N, Ichinose S, Omori A, Kawabata S, et al. Purification, characterization, and cDNA cloning of alpha-N-acetylgalactosamine -specific lectin from starfish, *Asterina pectinifera*. Glycobiology. 2002;12:85–94.

44. Zhang K, Xu M, Su J, Yu S, Sun Z, Li Y, et al. Characterization and identification of the integrin family in silkworm, *Bombyx mori*. Gene. 2014;549:149–55.

45. Mamali I, Lamprou I, Karagiannis F, Karakantza M, Lampropoulou M, Marmaras VJ. A beta integrin subunit regulates bacterial phagocytosis in medfly haemocytes. Dev Comp Immunol. 2009;33:858–66.

46. Moita LF, Vriend G, Mahairaki V, Louis C, Kafatos FC. Integrins of *Anopheles gambiae* and a putative role of a new beta integrin, BINT2, in phagocytosis of *E. coli*. Insect Biochem Mol Biol. 2006;36:282–90.

47. Hu J, Zhao H, Yu X, Liu J, Wang P, Chen J, et al. Integrin β1 subunit from *Ostrinia furnacalis* hemocytes: molecular characterization, expression, and effects on the spreading of plasmatocytes. J Insect Physiol. 2010;56:1846–56.

48. Johansson MW, Söderhäll K. A peptide containing the cell adhesion sequence RGD can mediate degranulation and cell adhesion of crayfish granular haemocytes in vitro. Insect Biochem. 1989;19:573–9.

49. Surakasi VP, Mohamed AA, Kim Y. RNA interference of β1 integrin subunit impairs development and immune responses of the beet armyworm, *Spodoptera exigua*. J Insect Physiol. 2011;57:1537–44.

50. Li DF, Zhang MC, Yang HJ, Zhu YB, Xu X. Beta-integrin mediates WSSV infection. Virology. 2007;368:122–32.

51. Zhang Y, Wang L, Wu N, Zhou Z, Song L. An integrin from shrimp *Litopenaeus vannamei*mediated microbial agglutination and cell proliferation. PLoS One. 2012;7, e40615.

52. Li F, Xiang J. Signaling pathways regulating innate immune responses in shrimp. Fish Shellfish Immunol. 2013;34:973–80.

53. Dunkelberger JR, Song WC. Complement and its role in innate and adaptive immune responses. Cell Res. 2010;20:34–50.

54. Zhou Z, Sun D, Yang A, Dong Y, Chen Z, Wang X, et al. Molecular characterization and expression analysis of a complement component 3 in the sea cucumber (*Apostichopus japonicus*). Fish Shellfish Immunol. 2011;31:540–7.

55. Zhong L, Zhang F, Chang Y. Gene cloning and function analysis of complement B factor-2 of *Apostichopus japonicus*. Fish shellfish immunol. 2012;33:504–13.

56. Fujita T. Evolution of the lectin-complement pathway and its role in innate immunity. Nat Rev Immunol. 2002;2:346–53.

57. Volanakis JE. Participation of C3 and its ligands in complement activation. Curr Top Microbiol Immunol. 1990;153:1–21.

58. Smith LC, Shih CS, Dachenhausen SG. Coelomocytes express SpBf, a homologue of factor B, the second component in the sea urchin complement system. J Immunol. 1998;161:6784–93.

59. Ulevitch RJ, Tobias PS. Receptor-dependent mechanisms of cell stimulation by bacterial endotoxin. Annu Rev Immunol. 1995;13:437–57.

60. Xu G, Shi Y. Apoptosis signaling pathways and lymphocyte homeostasis. Cell Res. 2007;17:759–71.

61. Ekshyyan O, Aw TY. Apoptosis: a key in neurodegenerative disorders. Curr Neurovasc Res. 2004;1:355–71.

62. Vermeulen K, Van Bockstaele DR, Berneman ZN. Apoptosis: mechanisms and relevance in cancer. Ann Hematol. 2005;84:627–39.

63. Mahoney JA, Rosen A. Apoptosis and autoimmunity. Curr Opin Immunol. 2005;17:583–8.

64. Danial NN, Korsmeyer SJ. Cell death: critical control points. Cell. 2004;116:205–19.

65. Leibowitz B, Yu J. Mitochondrial signaling in cell death via the Bcl-2 family. Cancer Biol Ther. 2010;9:417–22.

66. De Almagro MC, Vucic D. The inhibitor of apoptosis (IAP) proteins are critical regulators of signaling pathways and targets for anti-cancer therapy. Exp Oncol. 2012;34:200–11.

67. Baldwin AS. Control of oncogenesis and cancer therapy resistance by the transcription factor NF-kB. J Clin Invest. 2001;107:241–6.

68. Grabherr MG, Haas BJ, Yassour M, Levin JZ, Thompson DA, Amit I, et al. Full-length transcriptome assembly from RNA-Seq data without a reference genome. Nat Biotechnol. 2011;29:644–52.

69. Audic S, Claverie JM. The significance of digital gene expression profiles. Genome Res. 1997;7:986–95.

70. Benjamini Y, Drai D, Elmer G, Kafkafi N, Golani I. Controlling the false discovery rate in behavior genetics research. Behav Brain Res. 2001;125:279–84.

71. Xie C, Mao X, Huang J, Ding Y, Wu J, Dong S, et al. KOBAS 2.0: a web server for annotation and

identification of enriched pathways and diseases. Nucleic Acids Res. 2011;39:316–22.

72. Pfaffl MW, Horgan GW, Dempfle L. Relative expression software tool (REST©) for group-wise comparison and statistical analysis of relative expression results in real-time PCR. Nucleic Acids Res. 2002;30, e36.

73. Chen X, Li Q, Wang J, Guo X, Jiang X, Ren Z, et al. Identificaiton and characterization of novel amphioxus microRNAs by Solexa sequencing. Genome Biol. 2009;10:R78.

74. Sun H, Zhou Z, Dong Y, Yang A, Jiang J, Chen Z, et al. Expression analysis of microRNAs related to the *skin ulceration syndrome* of sea cucumber *Apostichopus japonicus*. Fish Shellfish Immunol. 2016;49:205–12.

CHAPTER 5

GENE EXPRESSION PROFILING OF INTESTINAL REGENERATION IN THE SEA CUCUMBER

Pablo A Ortiz-Pineda, Francisco Ramírez-Gómez, Judit Pérez-Ortiz, Sebastián González-Díaz, Francisco Santiago-De Jesús, Josue HernándezPasos, Cristina Del Valle-Avila, Carmencita Rojas-Cartagena, Edna C Suárez-Castillo, Karen Tossas, Ana T Méndez-Merced, José L RoigLópez, Humberto Ortiz-Zuazaga and José E García-Arrarás

BACKGROUND

All living organisms exhibit, to some extent, regenerative properties that allow them to deal with environmental events, physical trauma or diseases. Regenerative capacities have been studied in terms of stem cell recruitment, cell dedifferentiation, proliferation and migration, provision of specific regulatory/trophic factors, and expression or re-expression of the developmental program in adult animals [1]. However, the molecular/genetic basis of regeneration remains obscure.

Central to regeneration studies is the choice of the model organism, since metazoan species can show large variability in their regenerative capacities. In general, the ability to replace complex body parts decreases as one moves from the basal to the more highly derived taxa. However, even within the same phylum, not all animals are able to regenerate body parts, and not all tissues within a body can be equally repaired [2, 3]. One problem in elucidating the molecular basis of regeneration has been that organisms with high regenerative capacities do not lend themselves easily to traditional experimental genetics. Only recently has progress been made to make possible the use of some of these model systems to dissect the genetic basis of regeneration [4]. Among these, invertebrate systems such as Hydra and planaria, and vertebrates, such as ascidians and amphibian urodeles, have gained particular attention. Nonetheless, regeneration research remains underpopulated, and there are whole phyla of organisms, showing very interesting regenerative behaviors where little molecular research has been performed [5]

Among deuterostomes, the regenerative potential is maximally expressed in echinoderms, animals that can quickly replace most injured organs. In particular, sea cucumbers (holothurians) are excellent models for studying organ regeneration since they can regenerate many of their organs and appendages. Our laboratory has been using an echinoderm,*Holothuria glaberrima* as a model organism for the last decade to dissect the process of how the digestive tract regenerates once it is eliminated by auto-evisceration. We have shown that following evisceration the new intestine regenerates from the free end of the remaining mesentery [6]. Initially, the mesenterial tip thickens forming a continuous rod-like structure that extends from the esophagus to the cloaca. During the second regeneration week, luminal epithelial cells from the esophagus and from the cloaca migrate into this tube, forming the mucosal layer and giving rise to the intestinal lumen. We

Citation: Pablo A Ortiz-Pineda, Francisco Ramírez-Gómez, Judit Pérez-Ortiz, Sebastián González-Díaz, Francisco Santiago-De Jesús, Josue Hernández-Pasos, Cristina Del Valle-Avila, Carmencita Rojas-Cartagena, Edna C Suárez-Castillo, Karen Tossas, Ana T Méndez-Merced, José L Roig-López, Humberto Ortiz-Zuazaga, José E García-Arrarás, Gene expression profiling of intestinal regeneration in the sea cucumber, doi:10.1186/1471-2164-10-262.

have performed extensive studies at the cellular level showing the involvement of cell division, dedifferentiation, and migration in the regeneration of the intestine [6–11] as well as of events associated with the remodeling of the extracellular matrix [12].

We have also used the sea cucumber to explore the role of the genes that allow for such extraordinary regenerative processes [13]. Our approach has been to focus on target genes that have been associated with regenerative processes or identified in the regenerating tissues [14,15]. Similar gene by gene approaches have been used to study regeneration processes in other echinoderms, particularly in brittle stars and crinoids [16–20]. However, the regenerative capacities of the echinoderms have yet to be explored systematically using a large number of molecular tools [21]. In fact, one of the reasons that echinoderms have been sidelined in modern regeneration studies is the lack of genome-wide profiling approaches afforded by modern genomics [21]

We have now overcome this problem by determining the profile of gene activity during intestinal regeneration in *H. glaberrima* using microarray technology. For this, microarray slides were made using over 7000 ESTs that we previously identified from normal and regenerating intestine cDNA libraries [13]. Here we compare the gene expression profile of animals at 3, 7 and 14 days of regeneration, following the evisceration process typical of these organisms. Results show a large number of differentially expressed genes associated with intestinal regeneration. Some of these genes are homologues to metazoan genes associated with regenerative processes while many others might be novel sequences with little or no similarities to sequences in the databanks. To explore our results, we have focused on those sequences that are differentially expressed during early intestinal regeneration and that are associated with events previously shown to occur at the cellular level, namely developmental processes, cytoskeletal transformations and extracellular matrix remodeling. Our results make available the first overview of gene expression patterns in the regenerating intestine. More importantly, our results provide the basis for molecular studies aimed at exploring the molecular basis of regeneration in a group of animals that because of their remarkable properties and their key evolutionary position will provide important clues to the genetic basis of regenerative processes.

RESULTS

Technical Analyses

Technical analysis showed that the microarrays performed as expected. First, the intensities of the 536 internal controls of the Agilent array perfectly matched the expected ones as shown in the regression model. Second, no significant variation was observed between raw data and normalized MA plots, which means that there was no significant dye bias. Third, dye swap results showed extremely high similarity in their labeling pattern. Since the dye swaps were made with different animal samples (biological dye-swaps), this implies that not only did dyes produced similar results, but that results from animals at the same regeneration stage were equivalent. Fourth, analyses of genes spotted in duplicate within the same array (technical replicates) showed almost identical fold changes, evidencing the strong reproducibility of our data.

Gene Expression Profiles

A. Venn Diagrams

Microarray results showed dramatic differences in gene expression between regenerating and normal intestine and also among animals at different regeneration stages. Of the 14352 *H. glaberrima* probes in our microarray, 5915 showed similar expression in all groups, including normal (uneviscerated) and 3-, 7- and 14-days post-evisceration (dpe) regenerating intestines.

Most of the differential expression was observed when normal and regenerating intestines were compared. A large number of probes, 8437 (aprox 58%) showed differential expression at a significance level of $p < 0.01$ (Figure 1). As expected, the number of differentially expressed spots at $p < 0.05$ was larger (73%) but even at $p < 0.001$ a significant number of genes (39%) still were found to be differentially expressed (Figure 2). Differential gene expression was also observed when comparisons were made among regenerating stages. The largest differences were observed in probes that changed specifically in the 3-dpe stage, whether they were differentially expressed only at 3-dpe (2347), at both 3- and 7-dpe (3950) or at 3-, 7- and 14-dpe (1190) stages compared to normal. A smaller number of probes was differentially expressed only at 7-dpe (835).

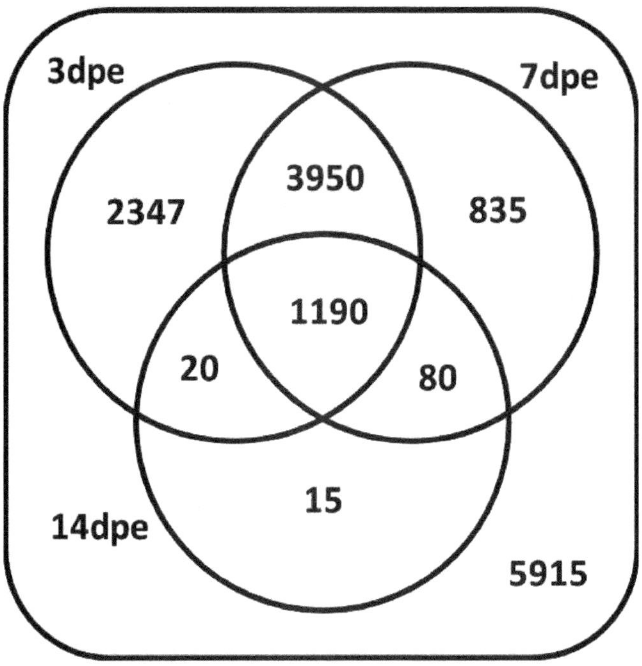

Figure 1: Venn Diagram of the EST distribution. 5915 probes do not show significant changes in expression when compared with normal tissues. The highest differential expression was found in both 3-dpe and 7-dpe stages (4439 differentially expressed spots); $p < 0.01$.

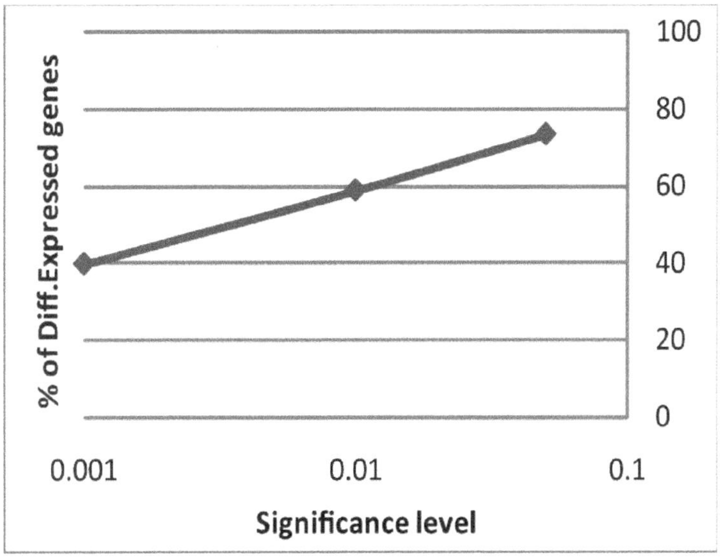

Figure 2: Percentage of differentially expressed sequences at $p < 0.001$, 0.01 and 0.05. At $p < 0.001$ only 39% of the sequences were significantly up or down regulated. *P*-values are shown in logarithmic scale.

B. Volcano Plots

In order to have a graphical representation of the gene expression levels and their significance, we analyzed our results using volcano plots (Figure 3).

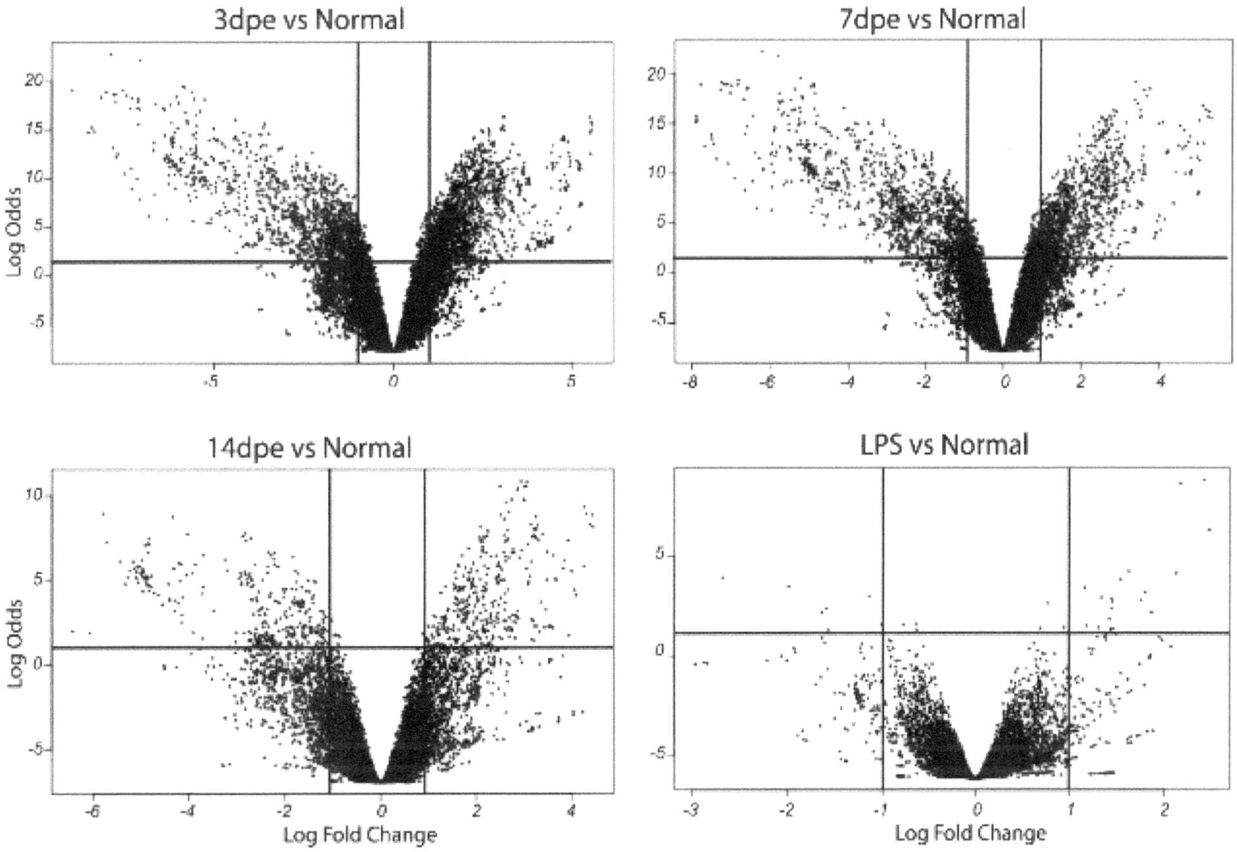

Figure 3: Microarrays volcano plots. Distribution of sequence expression between normal and normal injected with LPS, 3-, 7- and 14-dpe regenerating stages. Each plot shows the logarithm of the probability of the t-test as a function of the logarithm of fold change for each EST probe. Horizontal lines in each plot represent the nominal significant level of 0.001 for the *t*-student under the assumption that each gene has a unique variance. The vertical lines represent the limit of significance of the change in expression (fold change >2). Differentially expressed genes are located on the right (over-expressed) and left (under-expressed) top quadrants.

Results showed not only that a large number of genes are differentially expressed, but a high level of change and significance, mainly during the first week of regeneration. For example, at 3-dpe and 7-dpe the range of odds was significantly larger than at 14-dpe. Comparison with LPS showed a small number of genes being differentially expressed.

The top differentially expressed genes for 3-, 7- and 14-dpe stages of regeneration are shown on Tables 1, 2 and 3, respectively. Given that many probes have high significant p-values, we selected the top genes by two criteria: First, those probes with the highest p-value. Second, the fold change of the ratio of the intensities. The sequences were named either with their accession numbers (singlets) or with local identifications for multisequence contigs. All nucleotide sequences have been submitted to NCBI (taxonomy ID: 31192). Genes are shown in order of fold change and probability.

Table 1: Top 10 differentially expressed genes at 3dpe

Top	Gene ID	Lib.	BlastX*	ESTs	Exp.	logFC	P.Value
1	Contig 4863-1_			2	↑	5.541468	8.44E-11
2	ES726635	7d		s	↑	5.527128	6.06E-11
3	Contig 5300-1_			5	↑	4.879511	4.76E-10
4	Contig 3933-1_			2	↑	4.774727	3.03E-10
5	Contig 4702-1_			2	↑	3.091996	1.48E-10
6	Contig 798-1_			2	↑	3.080896	3.25E-11
7	ES725559	3d		s	↑	3.025192	1.2E-10
8	ES729465	N	TUBULIN A-1	s	↑	2.800781	5.86E-10
9	Contig 5344-1_		Cyclophilin	9	↑	2.797244	4.89E-10
10	ES725084	3d		s	↑	2.767621	5.84E-10
1	ES725883	7d		s	↓	-8.96457	1.67E-12
2	Contig 4571-1_			5	↓	-8.50941	1.83E-10
3	Contig 2794-1			2	↓	-7.88611	1.01E-14
4	ES728818	N		s	↓	-7.12955	1.32E-11
5	ES729355	N		s	↓	-6.66512	5.6E-10
6	ES729268	N		s	↓	-6.41638	1.32E-11
7	Contig 2489-1_			2	↓	-6.37315	1.7E-11
8	ES728532	N		s	↓	-6.18883	4E-10
9	Contig 5141-2_			6	↓	-6.17227	3.31E-10
10	Contig 5092-1_			2	↓	-6.00316	1.15E-10

Lib = cDNA library; FC = Fold Change; ↑: Up-regulated; ↓: Down-regulated; s:Singlet. *BlastX cut-off (1e-7)

Table 2: Top 10 differentially expressed genes at 7dpe

Top	Gene ID	Lib.	BlastX	ESTs	Exp.	logFC	P.Value
1	Contig 4863-1_			2	↑	5.355039	8.63E-11
2	Contig 5300-1_			5	↑	5.04046	1.11E-10
3	Contig 2092-1_			5	↑	4.098951	1.29E-10
4	ES725788	3d		s	↑	3.153745	1.31E-10
5	Contig 3636-1_			7	↑	2.913074	4.39E-11
6	ES725084	3d		s	↑	2.796475	1.71E-10
7	Contig 5564-1_		TENASCIN R	2	↑	2.64799	1.39E-10
8	Contig 798-1_			2	↑	2.542767	8.18E-11
9	ES729465	N	TUBULIN A-1	s	↑	2.450837	1.31E-10
10	Contig 4696-1_		MURINOGLOBULIN	2	↑	2.308496	2.07E-10
1	Contig 4571-1_			5	↓	-7.89392	1.27E-10
2	ES725883	7d		s	↓	-7.76893	2.30E-12
3	ES729355	N		s	↓	-6.61787	1.96E-10
4	Contig 2794-1_			2	↓	-6.23549	3.57E-14
5	ES728532	N		s	↓	-5.89441	2.13E-10

6	ES728818	N		s	↓	-5.80657	3.42E-11
7	Contig 2489-1_			2	↓	-5.79542	1.44E-11
8	Contig 5092-1_			2	↓	-5.72744	5.99E-11
9	ES729268	N		s	↓	-5.69887	1.42E-11
10	Contig 5131-1_			2	↓	-5.44947	3.85E-11

Lib = cDNA library; FC = Fold Change; ↑: Up-regulated; ↓: Down-regulated; s: Singlet.

Table 3: Top 10 differentially expressed genes at 14dpe

Top	Gene ID	Lib.	BlastX	ESTs	Exp.	logFC	P.Value
1	Contig 4863-1_			2	↑	4.432583	1.34E-07
2	Contig 5300-1_			5	↑	3.927607	3.05E-07
3	Contig 3636-1_			7	↑	3.386883	2.09E-07
4	Contig 5164-1_			5	↑	3.144439	1.11E-06
5	Contig 3791-1_		NM23 (NDK)	3	↑	2.953193	5.68E-07
6	Contig 4216-1_			2	↑	2.769543	5.21E-07
7	Contig 3895-1_			3	↑	2.735413	2.20E-08
8	ES727517	7d		s	↑	2.606467	8.29E-07
9	ES725788	3d		s	↑	2.576167	2.27E-07
10	ES726378	7d	LAMININ, A-4	s	↑	2.415549	7.96E-07
1	ES728532	N		s	↓	-5.80681	5.86E-08
2	Contig 5141-2_			6	↓	-5.16706	1.13E-06
3	ES725817	7d		s	↓	-4.33374	7.32E-08
4	ES725924	7d		s	↓	-4.21937	1.20E-06
5	ES725893	7d		s	↓	-4.15444	1.27E-06
6	ES726019	7d		s	↓	-4.02758	2.07E-07
7	ES729325	N		s	↓	-3.70124	7.48E-07
8	ES727944	7d	LEU-RICH REP.K-2	s	↓	-2.87982	2.40E-07
9	Contig 2398-1_			s	↓	-2.82164	3.09E-07
10	Contig 76-1_		SOMA FERRITIN	76	↓	-2.34468	6.22E-07

Lib = cDNA library; FC = Fold Change; ↑: Up-regulated; ↓: Down-regulated; s:Singlet.

Among the genes in this top 10 list that show known identities, were *cyclophilin, laminin, alpha tubulin, tenascin-R, ferritin. murinoglobulin* and *NM23*. Interestingly, most of the differentially expressed genes showed no significant similarity (e-value < 1.00E-07) to other sequences in publicly available nucleotide or protein databases. Unknown genes, such as Contig 4863-1 and 5300-1, were up-regulated at 3-, 7- and 14-dpe, while the singlet ES728532 showed down-regulation at all regeneration stages. On the other hand, some genes were specific for one stage such as Contig 3933-1 and Contig 4702-1 at 3-dpe. However, these genes were significantly differentially expressed but with higher p-values at other time points.

An alternate strategy to focus on gene groups that are differentially expressed at particular regeneration stages is to compare gene expression among regeneration stages (3-, 7- and 14dpe) and not directly to the normal intestine. To do this, we used the expression profile at 7-dpe as reference (up-regulated genes at 7dpe are considered as those whose expression is down-regulated at normal, 3 and 14dpe- stages; contrary, down-regulated genes are those whose expression is up-regulated at normal, 3 and 14 dpe). Seventeen genes (corresponding to 26 ESTs represented by 55 probes in the microarray) were found to be up-regulated at 3-dpe when compared with 7dpe (Table 4). The number of genes with up-regulated expression at 7- and 14-dpe stages was 24 and 13, respectively (Table 5 and 6).

Table 4: Upregulated genes specific to 3dpe

Gene ID	BlastX	p.value
Contig 739 2_		9.00E-05
Contig 66 1_		0.00016
Contig 5453-1_		0.00021

Contig 739 1_		0.00031
ES727789		0.00316
Contig 4797-1_		0.00334
Contig 4797-1_		0.00438
Contig 3640-1_		0.0058
Contig 798 1_		0.00669
Contig 3158-1_		0.00705
Contig 3640-1_		0.00784
Contig 3171-5_		0.00948
ES725288		0.01487
Contig 5652-1_	60S RIBOSOMAL L27	0.0157
ES727261		0.02239
Contig 5537-1_		0.02847
Contig 2300-1_		0.04107

Table 5: Upregulated genes specific to 7dpe

Gene ID	BlastX	p. alue
Contig 3895-1_		9.00E-05
ES729486		1.00E-04
ES727178		0.00016
Contig 771-1_	60S RIBOSOMAL PROTEIN L5	0.00032
ES726371		0.00044
Contig 5306-1_	APOLIPOPHORINS PREC.	6.00E-04
Contig 4810-1_		0.00062
ES727405		0.00065
Contig 3525-1_	EPENDYMIN REL. PROT.2	0.00073
Contig 89-1_6981		0.00169
ES727912	BONE MORPHOGENETIC PROT.1	0.00204
Contig 89-1_		0.00204
AY383544	EPENDYMIN REL. PROT. 1	0.00397
ES726186		0.00508
ES714731		0.00646
ES725228		0.00721
Contig 5342-1_		0.00906
ES728886		0.01851
ES729552	FAT (Cadherin)	0.01851
ES726575		0.02755
ES716626		0.02822
ES881026		0.03701
ES727189		0.04123
Contig 2092-1_		0.04342

Table 6: Upregulated genes specific to 14dpe

Gene ID	BlastX	p.value
Contig 2794-1_		6.00E-05
ES728818		0.00166
Contig 3191-1_		0.00176
ES725883		0.00551

ES729268		0.01215
ES728850		0.03556
Contig 4571-1_		0.04393
ES729033		0.0094
ES727539		0.02361
ES728538		0.00371
ES729914		0.0106
ES725898		0.03392
ES726811		0.02018

C. Validation of Microarray Results by RT-PCR

Gene-specific relative RT-PCRs were performed to validate the expression profiles of fifteen genes identified in the microarray as differentially expressed. The sequences of the primers used are shown on Table S1. Validation included genes with putative identity (*Wnt-9, Tensc-R, MMP-15, MMP-11, MMP-14, Actin 1, Actin 2, Actin 3* and *Hox12*). Most of these genes have been previously associated with wound-healing, development and regeneration in other animal species. The second group consisted of genes without identity (putative novel genes-Unknowns E, F and G, Contig 4874-1, Contig 5242-1 and Contig 5501-1). The validated genes include over-regulated and down-regulated sequences of the microarray.

The expression profiles of the 15 genes that were validated using RT-PCR confirmed the robustness of the microarray results (Figure 4 and 5). Eleven out of the 15 genes showed exact correlation in their expression profiles between RT-PCR and microarrays and very similar significance values in the statistical analyses. For example, unknown F showed up-regulation at 3-dpe and decreased gradually at 7- and 14-dpe. The unknowns 5501, 4874 and 5242 showed an exact match between the microarray and the PCR results. On the other hand, the correlation was not perfect for four genes (*actin 1, 2* and *3*, and *Hox 12*). These genes showed some correlations at particular stages, such as observed for *actin 1* and *2* and *Hox 12* at 3-dpe. At other stages some of the RT-PCR results although not significant do showed a trend that resembled the microarray results (see *actin 3* at 3-dpe). Overall the correlation of RT-PCR and microarray results was 84%, with only 7 of the 45 validated time points not showing the same profile between the microarrays and the RT-PCR. A direct comparison of the RT-PCR and the microarray results is shown on Table 7.

Table 7: Significance obtained from microarray and PCR at 3, 7 and 14dpe

Gene ID	3dpe Microarray	RT-PCR	7dpe Microarray	RT-PCR	14dpe Micro-array	RT-PCR
Wnt-9	↑ 7.00E-05	6.00E-05	↑ 4.00E-05	4.00E-03	↑ 0.013	0.009
Tensc-R	↑ 1.00E-05	5.00E-03	↑ <1.00E-05	4.00E-03	↑ 1.00E-4	8.00E-03
MMP-11	↑ 4.00E-05	5.00E-03	↑ <1.00E-05	6.00E-05	0.15	0.536
MMP-14	↑ 2.00E-03	1.00E-03	↑ 2.00E-04	2.00E-04	0.14	0.02
MMP-15	↑ 0.0485	0.0475	↑ 3.00E-03	4.00E-03	0.23	0.09
Unk- E	↓ 3.00E-05	4.00E-05	↓ 7.00E-04	1.00E-04	0.45	0.09
Unk- F	↑ 5.00E-04	3.00E-03	0.12	0.06	0.85	0.84
Unk- G	↓ <1.00E-05	5.00E-04	↓ <1.00E-05	2.00E-04	↑ 2.00E-4	4.00E-03
Unk- 4874-1	↑ 3.00E-05	1.00E-03	↑ 0.0011	0.02	0.1689	0.196
Unk- 5242-1	↑ 1.00E-04	4.00E-05	↑ 4.00E-04	1.00E-05	0.19	0.06
Unk- 5501-1	↑ 7.00E-05	6.00E-03	0.123	0.138	0.36	0.933
Actin-1	↑ <1.00E-05	0.04	↑ <1.00E-05	0.281	↑ 2.00E-4	0.276
Actin-2	↑ 4.00E-04	0.034	↑ 4.00E-04	0.116	0.05	0.09
Actin-3	↓ 1.00E-04	0.26	↓ 1.00E-04	0.005	↓ 0.013	0.005
Hox-12	↑ 2.00E-03	0.033	↑ 0.035	0.48	0.54	↓ 0.0016

↑: Up-regulated; ↓: Down-Regulated

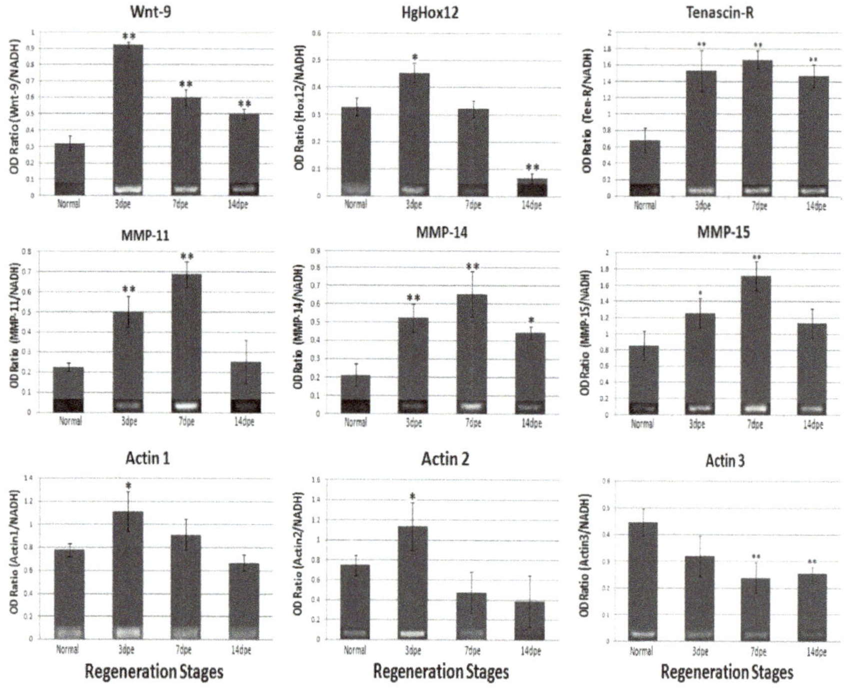

Figure 4: RT-PCR profiles of *Wnt-9, Hox12, Tenascin-R, MMP-11, MMP-14, MMP-15, Actin 1, Actin 2* and *Actin 3* in normal and regenerating intestines. The values were normalized against NADH. This set of genes showed up-regulation in the first week of regeneration except for *Actin 3* that showed down-regulation during regeneration. Interestingly, all extracellular matrix (ECM) related genes showed the peak of expression at 7-dpe. Moreover, *Wnt-9*, a developmental gene, was up-regulated during all stages. Values represent the mean ± S.E.M. of 7 biological samples (individual animals) per stage. Paired t-test for mean comparison was used. Significance levels were *:$p < 0.05$, and **:$p < 0.01$. Images at the base of the bar are RT-PCR pictures.

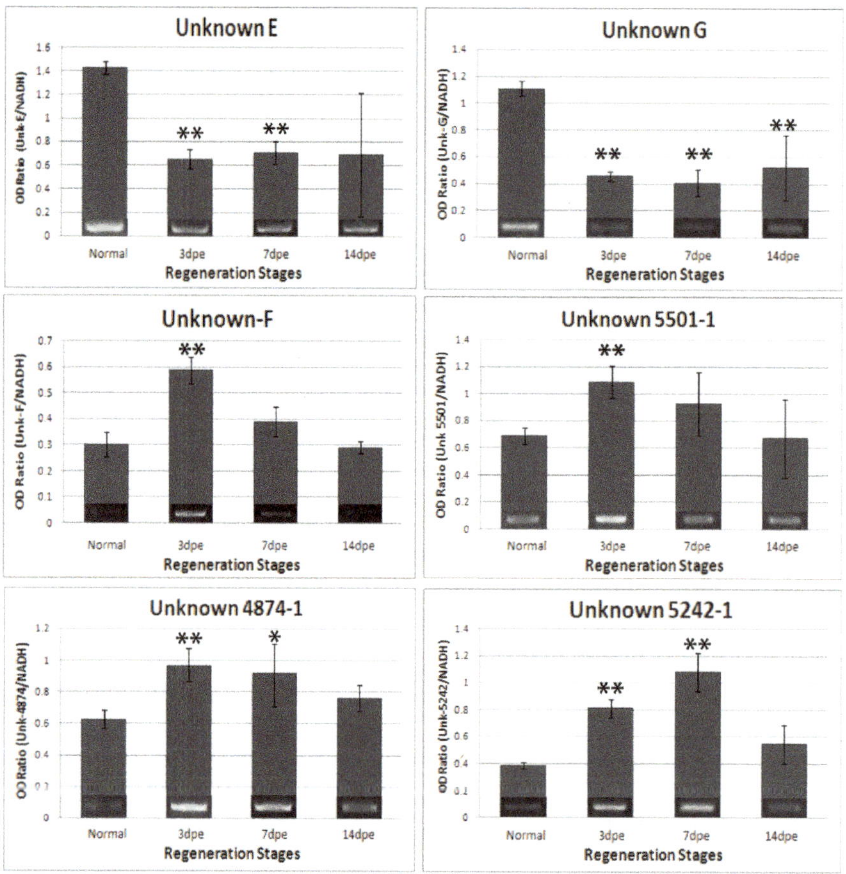

Figure 5: RT-PCR profiles of 6 novel gene sequences in normal and regenerating intestines. The values were normalized against NADH. This set of genes was mainly differentially expressed in the first week of regeneration, except by unknown-G that keeps down-regulated at 14dpe. Three of the four up-regulated genes showed the peak of expression at 3dpe and decrease the expression until 14dpe. Only unknown 5242-1 showed the peak at 7dpe but even at 3dpe the up-regulation was highly significant when compared with normal tissues. Values represent the mean ± S.E.M. of 7 samples per stage. Paired t-test for mean comparison was used. Significance levels were: *:$p < 0.05$, and **: $p < 0.01$. Images in the base of the bar are RT-PCR pictures.

D. Cluster Analysis

Cluster analysis was performed on the microarray results using hierarchical (not shown) and non-hierarchical methods in order to reveal several distinct profiles of expression (Figure 6).

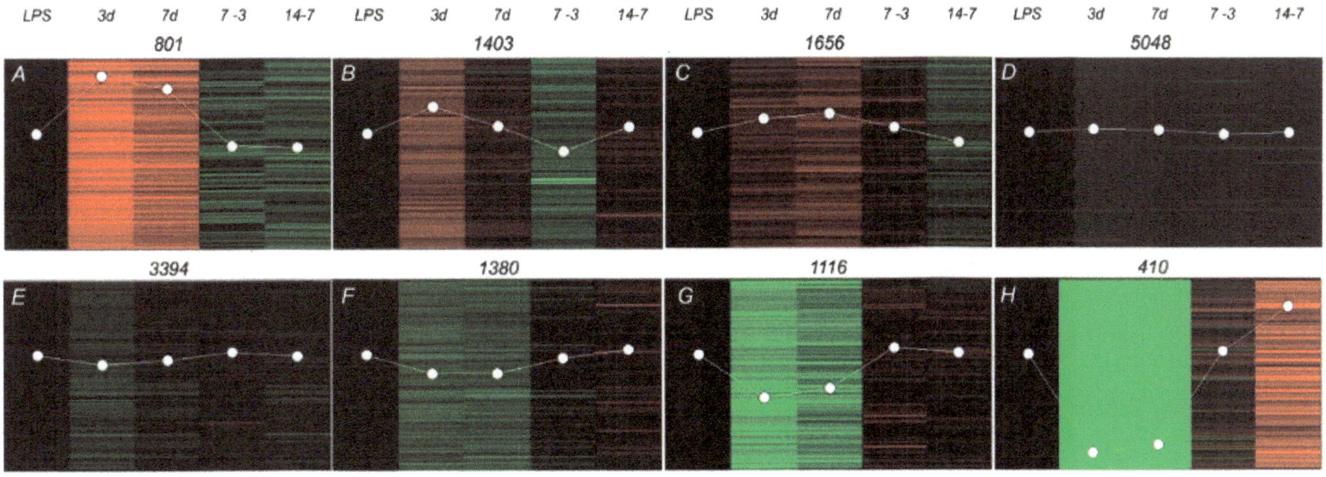

Figure 6: Cluster analysis of the microarray expression profiles for 15208 probes (All minus 536 controls). The 5 columns show the following comparisons: LPS (Red: "R") vs. Normal (Green: "G"): 3dpe(R) vs. Normal(G); 7dpe(R) vs. Normal(G); 7dpe(R) vs. 3dpe(G); 14dpe(R) vs. 7dpe(G). The number of genes in each cluster is shown above the bars. Since arrays were normalized against the LPS vs Normal results, the first column appears empty. Colors represent the level of expression of the gene, red for up-regulated and green for down-regulated; White spots represent the average of all the genes per cluster with "y" axis as a relative measure of expression levels.

Cluster # A contains probes that were highly over-expressed at 3- and 7-dpe when compared to normal intestine, and whose expression levels are slightly higher in 3- than in 7-dpe and are further decreased at 14-dpe.

Cluster # B contains probes that were moderately over-expressed at 3-dpe when compared to normal and whose expression levels decreased at 7-dpe. The expression levels of these probes at 14-dpe were similar to those at 7-dpe.

Cluster # C contains probes that were over-expressed at 3-dpe but whose expression peaked at 7-dpe, followed by a decrease in expression levels at 14-dpe.

Cluster # D contains probes that showed no significant differences in expression at any stage when compared to normal intestine.

Cluster # E contains probes that were slightly under-expressed at 3-dpe when compared to normal. The expression level of these probes was higher at 7- than at 3-dpe and somewhat similar between 7- and 14-dpe.

Cluster # F contains probes that were slightly under-expressed at 3- and 7-dpe when compared to normal and whose expression levels increase at 14-dpe.

Cluster # G contains probes that were moderately under-expressed at 3-dpe when compared to normal and whose expression levels slightly increase at 7- and 14-dpe.

Cluster # H contains probes that were highly under-expressed at 3- and 7-dpe and whose expression levels showed a moderate increase at 14-dpe.

E. Functional Category Analysis

We submitted all sea cucumber sequences to BlastX and parsed the results using a threshold of significance e-value < 1.00E-05. Approximately 58.1% of the sequences (3651) were "unknowns" that showed no similarity to subjects in the SwissProt database. On the other hand, 41.9% of the sequences (2636) had similarity with database sequences. Of these, 169 showed similarity with hypothetical proteins, most of them from the sea urchin *S. purpuratus*. GO searcheshttp://www.geneontology.org were performed with all the blast results and 1514 unique sequences were found with at least one reported ontology (Figure 7A).

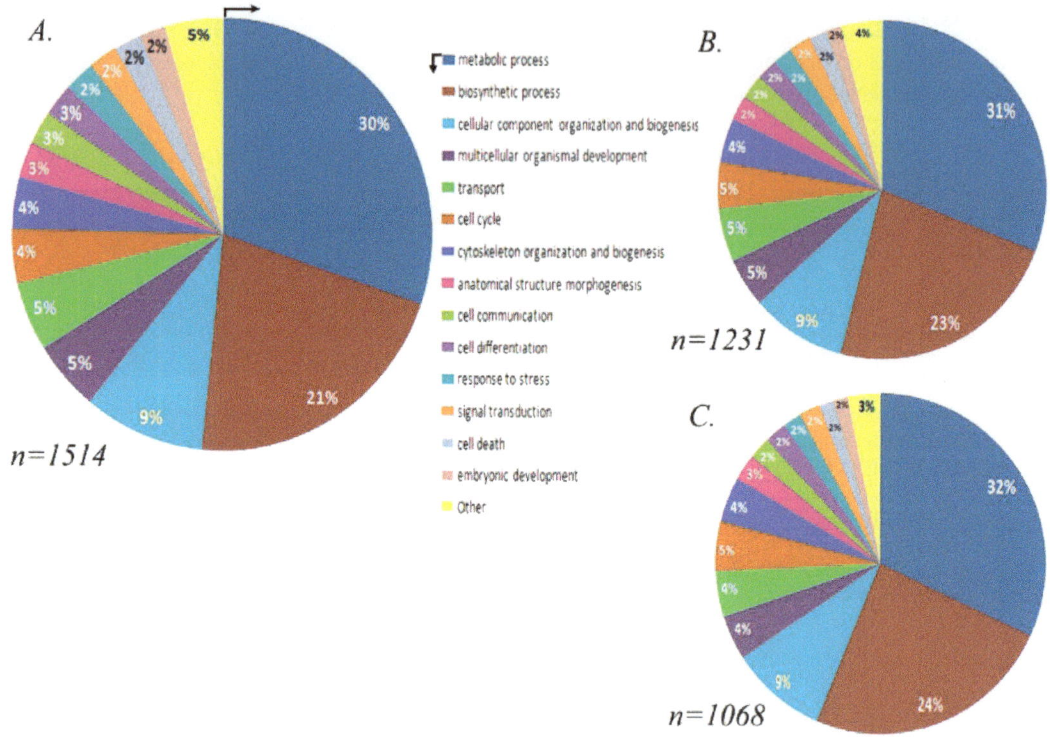

Figure 7: Distribution of gene function (GO) for ESTs analyzed in the microarrays. GO values of all ESTs printed on the microarray (A). Percentage of ESTs differentially expressed at 3-dpe (B) or 7-dpe (C). When compared with all the GOs, the 3- and 7-dpe distribution GOs showed no significant differences other than a small decrease in the number of GO results. The number of GOs results for each particular stage is shown in the bottom-left corner. The color code starts at 12 o'clock in the list order.

In order to determine if particular gene functions could be associated exclusively with specific regeneration stages, diagrams were done with only those genes that showed differential expression between 3-, 7- or 14-dpe and normal. At 3-dpe the GO search showed 1231 results while 1068 were found for 7-dpe (Figure 7B and 7C). The distribution of the GO was highly similar between all sequences and regeneration stages. For example, in all cases sequences associated with metabolic and biosynthetic processes corresponded to half of the GO results.

F. Individual gene expression

In addition to the global view of gene expression provided by microarray analyses, individual gene expression patterns can be determined. We have focused on various genes that have been associated with regenerative processes. For this we only chose those genes whose similarity to genes from other animals was sufficiently high (e-value < 1.00E-07) to be confident that they correspond to holothurian homologues. These genes are classified and presented in three main groups in relation with their function: a) Developmental genes, b) Cytoskeletal-related genes and c) Extracellular matrix-related genes.

a). Developmental genes

Many developmental gene pathways have been found to occur or be activated during regenerative processes [22, 23]. Thus, we have focused on the expression profile of development-associated genes to explore their roles during intestinal regeneration. Table 8shows the expression profile of well-known developmental genes in the microarray.

Table 8: Expression profile of developmental genes in the sea cucumber *H. glaberrima* during regeneration

Gene	3-dpe	7-dpe	14-dpe
Hox 1	-	-	-
Hox 2	-	-	-
Hox 3	-	-	-
Hox5	-	-	↑*
Hox 9	↑*	↑**	-
Hox10	↑**	↑**	-
Hox 11	-	-	-
Hox 12	↑**	↑*	-
Hox 13	-	-	-

TGFB-inducible LNR42	↑**	-	-
Krueppel-like	↓*	-	-
Wnt-9	↑**	↑**	↑*
BMP-1	-	↑**	-
Forkhead box prot. K1	↓*	-	-
Myotrophin	↓**	↓**	-

↑: up-regulated; ↓: down-regulated; *:$p < 0.05$; **:$p < 0.01$

In general, differential expression of these genes mainly occurred at early stages of regeneration. Most of the genes are activated or inhibited at 3-dpe (8 genes, being 5 of them highly significant) and 7-dpe (6 genes, being 5 highly significant). In contrast, at 14-dpe only two developmental genes were differentially expressed, both just at $p < 0.05$. Hox genes stand out among the group of development-associated genes, and our group has previously identified them in the holothurians [24]. Four of these genes appeared to be over-expressed, although their expression profile differed. *Hox 9, 10* and *12* were over-expressed at 3- and 7-dpe and returned to normal levels at 14-dpe. However, while *Hox 10* displayed the same levels of over-expression at 3- and 7-dpe, *Hox 12* and *9* showed peaks of expression at 3-dpe and 7-dpe, respectively. *Hox 5* showed a different pattern of expression altogether being over-expressed only at 14-dpe. Other development-associated genes were also found to be differentially expressed in our microarrays during intestinal regeneration. Once again the expression profile differed among them. For example, *Wnt 14* was found to be over-expressed at 3-, 7- and 14-dpe, while *BMP-1* was only up-regulated at 7-dpe. In contrast, *forkhead box K1* was down regulated at 3-dpe and, similarly, *myotrophin* was down-regulated at 3- and 7-dpe. Finally, not all development-associated genes showed changes in expression. For example, *Kruppel-like factor 2* remained at similar levels of expression during the first two weeks of regeneration.

To validate the developmental gene results we chose to study the expression of *Hox 12* and *Wnt 9* by RT-PCR. The sequences for these genes were obtained from the 7-dpe library and the complete sequence of *Hox 12* has been previously characterized by our laboratory [25] RT-PCR showed that genes were over expressed during early regeneration and their levels decreased during the second week of regeneration.

b). Cytoskeletal Proteins

Regeneration induces cellular changes that have been previously reported by our group. Among these are processes of dedifferentiation and migration that are associated with the formation of the new intestinal primordium during early regeneration [7–9]. Such processes are associated with changes in the expression of cytoskeletal proteins. Thus, it is expected that these changes would be reflected in the expression of mRNAs coding for cytoskeletal proteins. We have focused on the expression of seven cytoskeletal genes found within our EST database (Table 9). These include three actin family members, four tubulin family members, two myosin genes and one gelsolin gene. Although actin and tubulin isoforms are highly similar at the nucleotide level of the open reading frame (ORF), we identified ours based on differences in their 3'-unstranslated (UTR) region. The expression profile of these cytoskeletal mRNAs differed widely. Some were over-expressed during the first 2 weeks of regeneration (*Actin 1, Actin 2, Tubulin alfa and alfa-1*), others were under-expressed (*Actin 3, Myosin-11* and *gelsolin precursor*) while others showed no significant changes in their expression (*tubulin-beta*). Of particular interest were those sequences that showed a specific change at 3-dpe and appeared to only change slightly or not at all at other regeneration stages (*tubulin 3* and *myosin light 3F chain*), since these are more likely to play a role in specific processes occurring during intestinal regeneration.

Table 9: Expression profile of cytoskeletal genes in the sea cucumber *H. glaberrima* during regeneration

Gene	3-dpe	7-dpe	14-dpe
Beta-Tubilin	-	-	-
Alfa-Tubulin 1	↑**	↑**	↑**
Alfa-Tubulin 2	↑**	↑**	↑**
Alfa-Tubulin 3	↑**	↑**	-
Myosin-11	↓**	↓**	↓*
Myosin-Light 3F	-	↑*	-
Actin-1	↑**	↑**	↑**
Actin-2	↑**	↑**	↑*
Actin-3	↓**	↓**	↓*
Gelsolin-precurs	↓**	↓**	↓**

↑: up-regulated; ↓: down-regulated; *:$p < 0.05$; **:$p < 0.01$

Some of these sequences were validated by RT-PCR. In particular, the differential expression profiles of *actins 1* and *2* that are over-expressed during early regeneration stages, while *actin 3*is down-regulated, was clearly documented in our RT-PCRs.

c). ECM genes

We have previously described dramatic remodeling of the ECM and the involvement of MMPs during intestinal regeneration [12]. In this respect we searched for sequences in our database that might be associated with the ECM changes. Seven such genes were found, four of them corresponding to ECM molecules (*Echinonectin, Collagen alfa-1, Laminin alpha* and *Tenascin-R*) and three corresponding to matrix metalloproteases (MMPs) (15, 11, and 14) (Table 10). In this group, all analyzed genes showed up-regulation when compared with normal tissues. The mRNAs for the four ECM molecules showed over-expression during the first two regeneration weeks. Different to ECM molecules, MMPs were over-expressed at 3- and 7-dpe but not at 14-dpe, with *MMP 15* showing low significance at 3- but high at 7-dpe. The four genes from this group that were validated (Figure 5) showed strong correlation between RT-PCR and microarray results.

Table 10: Expression profile of ECM-related genes in the sea cucumber *H. glaberrima* during regeneration

Gene	3-dpe	7-dpe	14-dpe
Echinonectin	↑**	↑**	↑**
Collagen alfa-1(V)	↑**	↑**	↑**
tenascin-R	↑**	↑**	↑**
Laminin alfa 4 precurs	↑**	↑**	↑**
Laminin alpha 1	↑**	↑**	↑*
MMP 15	↑*	↑**	-
stromelysin-3 (MMP 11)	↑**	↑**	-
MMP-14	↑**	↑**	-

↑: up-regulated; ↓: down-regulated; *:$p < 0.05$; **: $p < 0.01$

d). Novel Sequences

As mentioned earlier many of the sequences in our microarray slides were unknowns. That is, sequences that showed no similarities to those in databanks. In many cases this was due to the sequence being either too short or being part of the mRNA's UTR. However, in some cases the sequence was long enough to contain a significant ORF that, in principle, should have shown similarities to other proteins when submitted to BLAST searches. These are the most interesting sequences to our research group, particularly those that were differentially expressed during regeneration, since they might correspond to novel holothurian-specific sequences associated with their striking regenerative capacities.

One strategy that we explored to study these unknowns was to focus on contigs. That is, those ESTs that shared overlapping sequence identity among them, and that together formed a composed sequence. The number of ESTs from a specific library (stage of regeneration) found to form a contig can be correlated with the expression level for that sequence at the stage from where the cDNA library was made. Therefore, an indirect measurement of the level of gene expression can be inferred from the number of identified ESTs from each library that formed contigs. We identified 596 multi-sequence contigs in our EST databank that had at least two assembled ESTs. Of these, 85.4% had ORFs larger than 100 nt (~33aa), and 46.5% had ORFs larger than 195 nt (~65aa). Only 27 contigs had no ORFs made up of at least 30 nt. We focused on contigs that showed over-expression at 3- and 7-dpe that were composed by at least three different ESTs and where all the EST probes showed similar results in the microarray. Sixteen of these unknown contigs are shown on Table 11. Their putative ORFs ranged from 98 to 794 nucleotides in length and they were represented by three to twenty-nine different ESTs. As expected, down-regulated contigs at 3-dpe appeared in the H and E clusters, while the up-regulated contigs at 3- and 7-dpe were located mostly in the A and B clusters. Moreover, some of these genes were validated with RT-PCR (Figure 6) showing high consistency in their expression profiles (Table 3).

Table 11: Novel genes differentially expressed during regeneration in H. glaberrima

ESTAP (cluster-assembly)	Val.	# Probes (ESTs)	3d D.E.	Orf Size (nt)	Orf-Start	Orf-end	G.O. Cluster
Contig 2748-1_Unk I		13 (5)	↓	371	719	1090	E
Contig 2794-1 _		5 (2)	↓	185	364	546	H
Contig 4277-8_Unk E	√	50 (26)	↓	110	1	111	E
Contig 4910-1_Unk H		9 (4)	↓	98	89	187	E
Contig 5141-2_Unk G	√	34 (13)	↓	209	144	353	H

Contig 293-1_		8 (3)	↑	737	46	783	D
Contig 4216-1_		7 (3)	↑	143	70	213	C
Contig 4766-1_Unk C	X	36 (18)	↑	365	91	456	A
Contig 4791-1_Unk F	√	35 (18)	↑	269	69	338	B
Contig 4860-1_Unk D	X	16 (8)	↑	242	48	290	B
Contig 4874-1_	√	16 (7)	↑	317	89	406	A
Contig 4911-1_Unk B		58 (29)	↑	218	1	219	A
Contig 5242-1_	√	9 (3)	↑	794	61	855	B
Contig 5501-1_	√	10 (4)	↑	125	1	126	B
Contig 5701-2_		6 (3)	↑	230	119	349	B

(√: validated genes in this paper; x: validated previously [13]

e). Other Sequences

Our microarray chip included the sequences of 75 ESTs from another sea cucumber *A. japonicus*, and 329 sequences from genes of the sea urchin *S. purpuratus*. Only five of the sea cucumber (*Collagen Alpha-1 (CA1); Cytoskeleton actin (cAct); Fibrinogen B precursor (Fib-Bp), Heat Shock protein 90b (HSP-90) and Senescence associated protein (SAP))* showed significant differences in expression in regenerating animals. Two of *A. japonicus* differentially expressed ESTs were underexpressed during regeneration (*Fib-Bp* and *SAP*), while the remaining three were overexpressed. Interestingly, four were similar to the *H. glaberrima* ESTs and followed the same expression profile (*Fib-Bp* was the different one). In all cases the probes of these other organisms showed similarity with the C-terminal region of the *H. glaberrima*sequences. The only sequence from the sea urchin that changed during regeneration was a hypothetical protein (Glean3-07946) showing a decrease in its expression at 3- and 7-dpe. The 24 sequences from zebra fish, human, rat, mice frog and axolotl behaved as expected; none showed any significant labeling or differential expression.

DISCUSSION

We have now shown by using microarray technology that a large number of genes, represented by EST sequences, are differentially expressed during sea cucumber intestinal regeneration. Nonetheless, microarray results are not fool-proof and must be validated by other methods. An indirect way of validating our results is to compare this microarray data to previous work from our laboratory showing individual gene expression patterns. Two examples of these are the expression of serum amyloid A (SAA) [14] and of ependymin [15], both of which have been well studied in the laboratory using a range of molecular techniques. SAA was shown by Northern blot analysis to be over-expressed during regeneration, with a peak during the second week[14]. Our microarray results also show SAA over-expression during intestinal regeneration, although the expression peak somewhat differs, occurring during the first week. This difference might be due to the fact that SAA is also associated with the immune status of the animals [26] therefore this status might differ between the animals used in the present experiments from those used for the Northern experiments performed years ago. On the other hand, the expression of ependymin, a gene associated with regeneration and plasticity, was previously shown to increase during intestinal regeneration [15]. These results were obtained with the highly accurate technique of quantitative Real Time PCR, showing a peak in expression at 7-dpe and no change at 3- or 14-dpe. These results are closely reproduced in our microarray analyses which show significant differences between 7-dpe animals and uneviscerated, but no differences at 3- or 14-dpe.

The comparison of RT-PCR results and microarray data obtained for fifteen selected ESTs provides a direct corroboration of their differential expression during intestinal regeneration. These results show that 11/15 of the sequences behave essentially the same whether measured by microarray or RT-PCR. Moreover, even the statistical significance of the result is within the same ballpark figure. The differences between RT-PCR and microarray results of the four remaining sequences (*Hox 12, actins 1, 2* and *3*), is mainly due to the RT-PCR results not being statistically significant, although in some cases they follow the same trend. It is interesting that these sequences are among the ones known to have closely related sequences in the genome (multiple actin genes and Hox genes), thus suggesting that the lack of correlation at some stages might be due to primers hybridizing with other sequences and increasing the background noise, making the RT-PCR results less sensitive than the microarray.

Why do so many Genes Show Differential Expression?

One of the most striking results from our microarray experiments, is that a high percentage of the spotted ESTs were found to be differentially expressed. Even when considering a very high level of significance ($p < 0.001$) more than a third of the ESTs showed differential expression. This contrasts with other regeneration studies where a smaller number of genes were found. For example, in studies of zebrafish heart regeneration, only about 5% of the assayed genes were found to be differentially regulated [27] and between 8 and 14% in studies of liver regeneration in rodents [28, 29].

The reasons for these differences might be due to two factors. First, most of our ESTs were obtained from cDNA libraries of regenerating intestine, thus the original pool of ESTs is probably biased toward those genes that are highly expressed in regenerating tissues. This was confirmed by the results themselves: the set of probes differentially expressed at 3dpe and 7dpe were 81.25% from the 3- and 7-dpe libraries and only 18.75% from the Normal library. In contrast, 42% of the down-regulated sequences come from the Normal library. Second, the process of intestinal regeneration that we are studying is complex and involves organogenesis, not merely wound healing. Furthermore, the intestine is an organ with multiple cell types and embryological origins. The intestine has components of the three embryonic germ layers, the enteric nervous system being an ectodermal derivative, the luminal epithelium an endodermal derivative and the muscular layer and submucosa being mesodermal derivatives. Therefore, when compared to a process such as heart or liver regeneration, where fewer cell types or more limited regeneration is involved, a larger number of genes might be necessary to be modulated to achieve the formation of the new intestine. Moreover, in the study of zebra fish heart, regeneration was limited to the ventricle and only to a portion of the organ since only 20% of the ventricle was removed. This contrasts largely from our studies in the sea cucumber where the new intestine must be completely formed from the remaining mesentery. In fact, a large number of differentially expressed genes was also found in another organism with high regenerative potential, the salamander *Ambyostoma mexicanum*. In this species when the tail, including the spinal cord segment is amputated, 76% of the genes in the microarray were differentially expressed [30]. It is interesting that they also found that even when increasing the stringency of the statistical analyses the number of differentially expressed genes (35%) was still significant. Thus, regeneration in deuterostomes imply the upregulation and downregulation of a large number of genes, making the search for "master regeneration genes", much more difficult than anticipated.

In view of this difficulty, we have developed certain strategies aimed at singling out those genes that might play important or crucial roles in regeneration. First, is to focus on genes that are differentially expressed at the 3-dpe stage. Previous experiments from our laboratory have shown that the cellular processes associated with intestinal regeneration occur early in intestinal regeneration. By 7-dpe, the blastema-like structure that will be the primordium for the new intestine is already formed [6]. Thus, genes associated with initiating regeneration must be regulated earlier than at 7-dpe. In this respect, some of the most likely candidates will be those genes that are differentially expressed between the 3- and 7-dpe stages. Second, focus on those genes that are most highly differentially expressed, such as those that were identified on our Top ten analyses. Third, look at sequences that form contigs. Although the presence of multiple contigs might be explained by several factors (some of them technique-related), they could be reflecting the abundance of gene expression in the original cDNA libraries, and therefore at the stage from where the library was made. Fourth, is to aim at those genes that have been previously associated with organogenesis in other species (see the following section). Finally, we expect that novel genes associated with regeneration will be characterized from our data. Thus, those sequences, and in particular those contigs that show ORFs with no similarity to sequences in databases are the strongest candidates to be pursued.

How do Microarray Results Correlate with Organogenesis Processes that occur during Embryonic Development?

Developmental Genes

The finding of genes associated with embryonic development being differentially expressed during regeneration is not surprising since developmental gene pathways have been found to occur or be activated during regenerative processes [22, 23]. Thus, these genes and signaling pathways provide excellent targets to study the molecular events of intestinal regeneration.

One of the most studied groups of genes is the Hox genes, which encode transcription factors that specify the body axis during embryonic development. Hox genes have been implicated in various regeneration processes, including the regeneration of amphibian tails and limbs [31–33], hydra [34], rat liver [35], and zebrafish tail [36] among others. This expression can be spatially or temporally specific in regenerating tissues [37, 38]. It is particularly interesting that *Hox12* had previously been found in our regenerating intestine cDNA library and that there is some evidence for homeobox gene involvement in echinoderm regeneration [20]. The finding that the most posterior genes (*Hox 9*, *10* and *12*) are the ones that showed an increase in expression during intestinal regeneration coincides with findings in the mammalian digestive tract where Hox genes are known to be expressed in a rostral-caudal gradient, and the Hox genes near the 5' end are expressed in the posterior digestive tract, i.e. intestine [39]. These are the same genes that when mutated cause defects in the development of the posterior digestive tract [40, 41].

Similarly, the finding that a Wnt homologue (*Wnt-9*) was overexpressed in the regenerating intestine during the first two weeks of regeneration coincides with published data on this molecular family. Wnt molecules are extracellular molecules that bind to their receptors and activate a signaling pathway that includes other known proteins, such as B-catenin. In recent years, Wnt pathways have been increasingly associated with regenerative phenomena. Wnt was found to be involved in blastema formation of the regenerating limbs of anuran tadpoles [42] and in lens regeneration in newts [43]. In mammals, Wnt has been studied in bone [44], hair follicle [45] and deer antler regeneration [46] among others. Liver regeneration was retarded in the absence of beta-catenin [47]. Wnt apparently plays a key role in the control of intestinal stem cell proliferation and differentiation [48]. Thus, it will be of great interest when the holothurian *Wnt-9* expression is localized to determine which intestinal cell types are expressing this protein. Wnt pathways have also been involved in other invertebrate regeneration models. In planaria, Wnt is necessary for proper brain pattern formation [49] and B-catenin for antero-posterior axis formation during regeneration [50].

Lastly, the family of bone morphogenetic proteins (BMPs) has been increasingly associated with regenerative phenomena. In newts, BMPs have been associated with the potential of the ventral iris to regenerate a new lens [51]. And in zebrafish, at least three members of the BMP family have been associated with fin regeneration [52]. Inhibition of BMP activity using the BMP antagonist, noggin, caused a reduction in blastema cell proliferation and a reduction in bone matrix deposition. Similarly, in experiments using transgenic Xenopus tadpole with increase noggin expression, both tail and limb regeneration were inhibited [53]. Finally, BMP has been shown to be essential for the correct establishment of the dorso-ventral axis during planarian regeneration [54, 55]. BMPs have also been associated with normal gastrointestinal tract development in vertebrates, and alterations in their expression caused morphological abnormalities in the intestinal tissues ([56, 57]). BMP have also been shown to be required for proper cloaca formation [58]. Therefore, the overexpression of BMP during intestinal regeneration observed in our system is consistent with its role in the proper formation of the digestive tract organs.

How do Microarray Results Correlate with Cellular Processes that occur during Intestinal Regeneration?

In order to understand how changes in gene expression are associated with changes at the tissue and organ levels it is important to have a clear understanding of the cellular events that underlie the regenerative process. For this reason, our laboratory has described the intestinal regeneration process in *H. glaberrima* from the cellular perspective [6–12]. Two events in particular are to be highlighted in view of their possible association with the differential gene expression profile, one is the change in cytoskeletal proteins and the second the remodeling of the extracellular matrix.

Cytoskeletal Proteins

Previous studies from our laboratory have shown dramatic changes in the cytoskeleton of cells in the regenerating intestine. Many of these changes were associated with de-differentiation of the mesenterial muscle during the initial regeneration stages and with the processes of myogenesis, innervation and luminal epithelial formation in later regenerative stages [6, 8, 10,11]. In fact, during the regenerative process muscle cells actively extrude their contractile apparatus as spindle-like structures (SLSs) [8]. This has also been shown in other echinoderm regenerative phenomena [1, 59]. We can speculate that the under-expression of some actin and myosin isoforms is associated with this dedifferentiation process. Similarly, the over expression of some actin isoforms could be associated with the migration of cells into the regenerating tissue [7, 9], or with the eventual differentiation of cells from committed precursors [11]. Nonetheless, these hypotheses will need to be verified in future experiments using *in situ*hybridization and immunochemical analyses.

One of the most interesting findings is the under expression of gelsolin in early regenerative stages. Results by Cowin and colleagues [60] have shown that gelsolin mRNA was upregulated and gelsolin protein was associated with actin filaments in the fetal skin of embryonic day 19 rats, but not at embryonic day 17. This is the age gap when fetal wounds properties change from being scar-free to scar-forming. In culture, skin from 17 day embryonic rats can epithelialize an excisional wound in culture, but this ability is lost by embryonic day 19. Thus, our finding of a correlation between gelsolin gene under-expression and early intestinal regeneration suggests that gelsolin might play an important inhibitory role in the process of wound healing and regeneration.

ECM

One of the events that is common to most regenerative processes is the remodeling of the extracellular matrix. These changes, which have been well studied particularly during epithelial wound healing, involve the replacement of the ECM by a transient ECM that facilitates the cellular events associated with the regenerative response [61]. In the holothurian intestinal system we have previously shown a dramatic extracellular remodeling during the early stages of regeneration [12]. During this time collagen and laminin immunoreactivities disappear from the mesentery that gives rise to the new organ. Similarly there is a rise in the activity of MMPs that coincides with the major changes in the ECM. Moreover, we showed that inhibition of MMPs at this time point causes an inhibition of the regeneration process [12]. At the microscopic level, the changes of ECM and MMP activity coincide with an increase presence of phagocytic amoebocytes in the regenerating areas [8]. Therefore, the differential expression of various molecules associated with ECM remodeling, particularly the MMPs provide specific targets to define their role in the regeneration process.

How do our Results Serve to Advance the Regeneration Field?

The cellular and molecular processes of organ regeneration have received increasing attention in recent years [5]. In fact, the question of what controls organ regeneration? has been highlighted as one of the top scientific issues that deserves to be studied with high priority [62]. However, organ regeneration remains one of the least understood biological processes, particularly at the molecular level. Progress toward characterizing the molecular basis of regenerative processes has been modest for at least three reasons [63, 64]. First, the regenerative capacity of most vertebrate animals is relatively limited [3], so popular biomedical models are not very useful. For example, some mammalian tissues (*e.g.*, intestinal epithelia) and organs (*e.g.*, skin and liver) can readily regenerate, but others, like heart and nervous tissues, have little or no regenerative capacity. Second, although development is a process closely akin to regeneration, most animals used extensively to study the molecular aspects of development (worms, fruitflies, zebrafish, chickens and mice) lack robust regenerative capacities. Third, those animals with remarkable regenerative capacities, such as coelenterates, flatworms, tunicates, newts and echinoderms have been little studied at the molecular/genetic level [1, 20,65].

This has started to change and ongoing projects promise a bright future for the identification of regeneration-associated genes. Among the studies/models that need to be highlighted in terms of EST/genomic are those in amphibians [30, 66, 67], zebrafish [27, 68, 69], ascidians [70,71] planaria [4, 55, 72–77] and Hydra [78–81]. However, the only echinoderm studies are those of the sea urchin genome [82], which ironically is the echinoderm group with the least regenerative capacities [83].

Analyses of regeneration-associated genes in multiple models will lead to important comparative studies among species with highly developed regenerative capacities or between closely related species where good and poor regenerators are found. The importance of comparative studies using a large range of animal models was highlighted by Sanchez Alvarado and Tsonis [4] who described the differences and similarities in the regeneration mechanisms among diverse animal species. Other investigators pointed out that the echinoderms were largely excluded from this analysis [84], a fact explained by one of the authors [21] as being due to the lack of modern molecular tools to study the regeneration phenomena. Nonetheless, all investigators agreed that the application of genome-wide profiling approaches (as well as other tools) to the echinoderms would provide significant contributions to the understanding of regeneration. Thus, the present results represent an important step toward the molecular study of the amazing regenerative capacities of echinoderms and more specifically toward their use to better understand how organs can be regenerated.

CONCLUSION

Despite the apparent simplicity of the echinoderms, here we show that the organ regeneration is a complex process that involves the up-regulation and down-regulation of thousands of genes. Consistent with previous findings, these genes have been reported as involved in regeneration, development, ECM rearrangement, cytoskeleton reorganization and wound healing such as Wnt-9, BMP-1, Hox12, Tubulin, Tens-R, MMPs and gelsolin respectively. Moreover, a large number of unknown genes were also found to be differentially expressed at different stages of regeneration and might represent novel genes. Finally, here we show that intestine regeneration in the sea cucumber is a novel and important model for studies to identify and characterize the molecular basis of regenerative processes.

METHODS

Microchip Preparation

We have used the Agilent platform to design, perform and analyze custom-made *H. glaberrima*arrays. For this we used the "eArray" server from Agilent to design 60-mer probes that were synthesized on the microarray. A design with 15744 spots on each array and 8 arrays per slide was selected. All sequences from the cDNA libraries of normal and regenerating intestines were represented at least twice. Thus, 14352 (91.2%) probes were from 6287 clean sequences from *H. glaberrima*. We also spotted other sequences. These sequences represented genes associated to development, regeneration, proliferation, wound healing and growth. This pool included 150 probes from the sea cucumber *Apostichopus japonicus* (reported by Zheng et al. 2006), 8 from other holothurians that could be found in the NCBI database and 658 from the sea urchin *S. purputatus* (whose genome was recently sequenced [82]). Finally, 40 probes were from vertebrate organisms including mice, zebrafish and axolotl. Technical controls were 536 (3.4%).

Animals and Treatments

Adult sea cucumbers (10~12 cm long) were collected from the north-eastern rocky shores of Puerto Rico. Animals were eviscerated at day 0, by intra-coelomic injection of 0.35 M KCl. They were left undisturbed in the aquaria to allow regeneration for 3, 7 and 14 days. At each time-point, animals were anesthetized by immersion in ice-cold sea water for 45 min. A dorsal incision was made to completely expose the internal cavity and allow the dissection of the intestinal primordium (3-, 7- or 14-dpe). After dissection, the regenerating intestines were placed in RNAlater® (Applied biosystems/Ambion, Austin TX) solution for RNA extraction. Uneviscerated (normal) animals were used as controls. One group was injected once with 1 mg lipopolysaccharides (LPS) diluted in 0.5 mL filtered sea water as reported by [85]. Intestines of the LPS treated animals were dissected 48 hrs after injection.

RNA Extraction

RNA was extracted using a combination of the [86] method using Tri-reagent® (N.93289, Sigma, St Louis, MO) and the RNAeasy mini kit from Quiagen (Valencia, CA) The concentration and integrity of RNA were determined using the NanoDrop-1000 Spectrophotometer (NanoDrop Technologies, Rockland, DE) and the 2100 Bioanalyzer (Agilent) with an RNA 6000 Nano LabChip® Kit at the Functional Genomics Research center (FGRc -UPR, Puerto Rico).

Synthesis and Labeling of cRNA

RNA samples were amplified and labeled using the Low Input Fluorescent Linear Amplification kit (Agilent). Briefly, cDNA was first synthesized combining 300 ng of RNA with T7 promoter primers, 5× First Strand Buffer, 0.1 M DTT, 10 mM dNTP mix, MMLV RT and RNaseOUT. This reaction mix was incubated at 40°C for 2 h. Then, cRNA synthesis and dye incorporation were performed as follows: cDNA of the first reaction was combined with cyanine-3-CTP (Cy3, 10 mM) or cyanine 5-CTP (Cy5,10 mM) and the Transcription Master Mix (4× transcription buffer, 0.1 M DTT, NTP mix, 50% PEG, RNAseOUT, inorganic pyrophosphase and T7 RNA polymerase) at 40°C for 2 h. The fluorescently labeled cRNA products were purified using RNeasy mini spin columns (QIAGEN), subsequently analyzed for yield and dye incorporation using the NanoDrop, and finally stored at -80°C until needed. Only samples that presented a concentration of more than 8 pmol/mg (measurement representing efficiency of dye incorporation) were used for hybridization as recommended by the manufacturer.

Hybridization & Scanning

Cy3 and Cy5-labeled cRNAs were combined, fragmented, and hybridized to an 8 × 15K 60-mer oligo microarray (Agilent) for 17.5 h at 60°C with continuing rotation at 4 rpm. After hybridization, slide arrays were subjected to two successive washes (wash solution 1: 6× SSPE, 0.005% N-Lauroylsarcosine, and wash solution 2: 0.06× SSPE, 0.005% N-Lauroylsarcosine) and dried with the Stabilization and Drying Solution (Agilent). Arrays were immediately scanned in a G250B Microarray Scanner (Agilent) to obtain fluorescence intensities and Cy5/Cy3 ratios for each gene on the array (see below). Scanning parameters were used as recommended by the manufacturer. The resolution settings for scanning were 5 μm using minimum (10%) and maximum (100%) Photo Multiplier Tube detection sensitivities.

Microarray Experimental Design

The experimental design consisted of comparisons between Normal (non-eviscerated) intestinal tissues and 3-, 7-, and 14-dpe regenerating intestines. To determine differences in gene expression between regenerating stages we also compared directly 3- vs. 7-dpe and 7- vs. 14-dpe (Figure 8). Comparisons were also made between the intestinal tissues of normal and immune activated animals (injected LPS). There were only 48 genes differentially expressed between normal and LPS-immune activated intestines and these genes will be discussed elsewhere. (F. Ramirez-Gomez et al. in prep.) These genes were subtracted from the regeneration profile to ensure that genes associated with the regeneration phenomenon were not differentially expressed due any possible immune activation (e.g., because of bacterial invasion during the evisceration process).

Technically, for each comparison, the microarray experiment was done twice, one using different biological samples, each consisting of a sample pool from 3 animals. Of the two biological replicates one was always performed as a dye-swap in order to correct any dye bias.

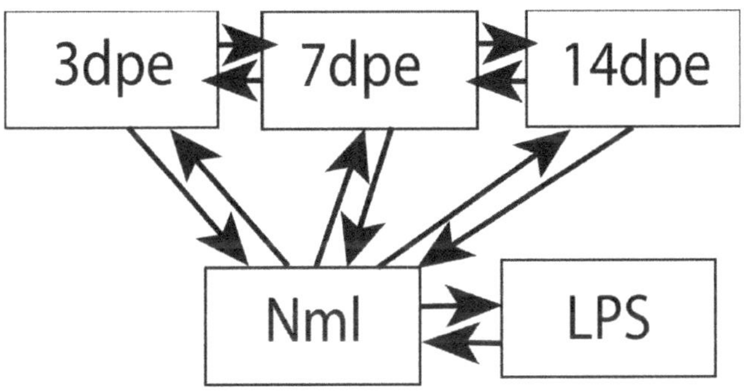

Figure 8: Microarray experimental design. Normal (Nml) and regenerating intestines at 3, 7 and 14 dpe (days post-evisceration) were compared among them. Each experiment was replicated as a dye swap with different biological samples. LPS treated tissues where used to normalize the expression for immune activated genes. Arrows show direct comparisons.

Microarray Data Analysis

Microarray data was extracted with Agilent's Feature Extraction software (version 5.1) and analyzed using the Limma package of Bioconductor. The data discussed in this publication have been deposited in NCBI's Gene Expression Omnibus [87] and are accessible through GEO Series accession number GSE16182 http://www.ncbi.nlm.nih.gov/geo/query/acc.cgi?acc=GSE16182. The analysis of the microarray data consisted of the following steps: 1) within-array and between-array normalizations; 2) fitting the data to a linear model; and 3) computing differential gene expression. For normalization purposes [88] MA-plots were generated representing the (R, G) data (R = red for Cy5 and G = green for Cy3), in which the log ratio of R versus G (M value = $\log_2$ R/G) was plotted against the overall intensity of each spot (A value = $\log_2$ (R + G)/2. Within-array normalization was first applied and M-values were normalized within each array using the Global Loess Normalization method. Aquantile normalization was then applied to the A-values as a method for between-array normalization, to assure that the intensities and log-ratios had similar distributions across arrays. To estimate the average M-value for each gene and assess differential gene expression, a simple linear model was fit to the data, and M-value averages and standard deviations for each gene were obtained. To find genes with significant expression changes between groups, empirical Bayes statistics were applied to the data by moderating the standard errors of the estimated M-values. P-values were obtained from the moderated t-statistic and corrected for multiple testing with the [89] method. The null hypothesis, that there is no differential expression of genes between regeneration stages compared with normal tissues, was rejected for p-values lower than 0.05. Thus, the change in expression is given by the fold change while the believability of the change is given by the odds. Alternative to this method, the GenePix Pro 6.0.1.26 and Acuity 4.0.0.60 software (from Molecular Devices, Sunnyvale CA) was also used to analyze gene expression values (log ratios) and perform clustering analyses. Methods used to visualize the expression profiles included hierarchical [90] and non-hierarchical clustering such as Self-Organizing Maps (SOM, [91]).

Gene Expression Analysis

The differential expression of the genes of interest was semi-quantitatively determined by gene-specific relative RT-PCR. Primers were designed for optimal performance using the Primer3 and Net primer web-servers. All primers were synthesized by AlphaDNA (Montreal, Canada). PCRs were performed in the Mastercycler ep (Eppendorf, Westbury, NY) using the Promega's Taq-Polimerase kit (Madison, USA). Cycling conditions for the amplified products were as follow: 94°C × 1 min, 55°C (± 3°C) × 1 min, 72°C × 1 min; all performed for 35~40 cycles. Optical densities of the PCR products were measured using Photoshop (Adobe). Values were extracted for background and then normalized against NADH-dehydrogenase unit 5 (NADH, 241 bp). Each stage of regeneration was analyzed with at least an N = 7, each sample represented the intestine of one individual. Statistical test of mean comparison was carried in R http://cran.r-project.org under the null hypothesis that there is no differential expression between different stages of regeneration and normal tissues. Significance of the paired t-test was assumed at $p < 0.05$.

ACKNOWLEDGEMENTS

This work was supported by NIH (1SC1GM084770-01), NSF (IBN-0110692) and NIH- (S06GM08102). We also acknowledge partial support from NIH INBRE (P20RR16470), DEGI-PBDT (doctoral dissertation fellowship to F. Ramírez-Gómez and E. Suarez-Castillo), the Sequencing and Genotyping facility of the UPR and the University of Puerto Rico.

AUTHORS' CONTRIBUTIONS

POP designed and carried out the study, analyzed and interpreted the data, and drafted the manuscript. FRG helped in the design and participated in the molecular experiments.

CRC, JRL, ESC, ATMM and KT participated in cDNA library construction, EST characterization and analysis and gene identification and sequencing. JPO, SGD, FSD, JH, and CDA performed most of the validation experiments and participated in the analysis of data. HOZ helped in bioinformatics analysis. JEGA conceived the study, participated in its design, data analyses and helped in the manuscript preparation. All authors have given final approval of the version to be published.

REFERENCES

1. Candia Carnevali MD, Bonasoro F: Introduction to the biology of regeneration in echinoderms. Microsc Res Tech. 2001, 55 (6): 365-368. 10.1002/jemt.1184.
2. Brockes JP: Amphibian limb regeneration: rebuilding a complex structure. Science. 1997, 276 (5309): 81-87. 10.1126/science.276.5309.81.
3. Tsonis PA: Regeneration in vertebrates. Dev Biol. 2000, 221 (2): 273-284. 10.1006/dbio.2000.9667.
4. Sanchez Alvarado A, Tsoni PA: Bridging the regeneration gap: genetic insights from diverse animal models. Nat Rev Genet. 2006, 7 (11): 873-884. 10.1038/nrg1923.
5. Slack JM: Regeneration research today. Dev Dyn. 2003, 226 (2): 162-166. 10.1002/dvdy.10232.
6. Garcia-Arraras JE, Estrada-Rodgers L, Santiago R, Torres , Diaz-Miranda L, Torres-Avillan I: Cellular mechanisms of intestine regeneration in the sea cucumber, Holothuria glaberrima Selenka (Holothuroidea:Echinodermata). J Exp Zool. 1998, 281 (4): 288-304. 10.1002/(SICI)1097-010X(19980701)281:4<288::AID-JEZ5>3.0.CO;2-K.
7. Cabrera-Serrano A, Garcia-Arraras JE: RGD-containing peptides inhibit intestinal regeneration in the sea cucumber Holothuria glaberrima. Dev Dyn. 2004, 231 (1): 171-178. 10.1002/dvdy.20112.
8. Candelaria AG, Murray G, File SK, Garcia-Arraras JE: Contribution of mesenterial muscle dedifferentiation to intestine regeneration in the sea cucumber Holothuria glaberrima. Cell Tissue Res. 2006, 325 (1): 55-65. 10.1007/s00441-006-0170-z.
9. Garcia-Arraras JE, Schenk C, Rodrigues-Ramirez R, Torres , Valentin G, Candelaria AG: Spherulocytes in the echinoderm Holothuria glaberrima and their involvement in intestinal regeneration. Dev Dyn. 2006, 235 (12): 3259-3267. 10.1002/dvdy.20983.
10. Garcia-Arraras JE, Torres-Avillan I: Developmental expression of galanin-like immunoreactivity by members of the avian sympathoadrenal cell lineage. Cell Tissue Res. 1999, 295 (1): 33-41. 10.1007/s004410051210.
11. Murray G, Garcia-Arraras JE: Myogenesis during holothurian intestinal regeneration. Cell Tissue Res. 2004, 318 (3): 515-524. 10.1007/s00441-004-0978-3
12. Quinones JL, Rosa R, Ruiz DL, Garcia-Arraras JE: Extracellular matrix remodeling and metalloproteinase involvement during intestine regeneration in the sea cucumber Holothuria glaberrima. Dev Biol. 2002, 250 (1): 181-197. 10.1006/dbio.2002.0778.
13. Rojas-Cartagena C, Ortiz-Pineda P, Ramirez-Gomez F, Suarez-Castillo EC, Matos-Cruz V, Rodriguez C, Ortiz-Zuazaga H, Garcia-Arraras JE: Distinct profiles of expressed sequence tags during intestinal regeneration in the sea cucumber Holothuria glaberrima. Physiol Genomics. 2007, 31 (2): 203-215. 10.1152/physiolgenomics.00228.2006.
14. Santiago P, Roig-Lopez JL, Santiago C, Garcia-Arraras JE: Serum amyloid A protein in an echinoderm: its primary structure and expression during intestinal regeneration in the sea cucumber Holothuria glaberrima. J Exp Zool. 2000, 288 (4): 335-344. 10.1002/1097-010X(20001215)288:4<335::AID-JEZ6>3.0.CO;2-1.
15. Suarez-Castillo EC, Medina-Ortiz WE, Roig-Lopez JL, Garcia-Arraras JE: Ependymin, a gene involved in regeneration and neuroplasticity in vertebrates, is overexpressed during regeneration in the echinoderm Holothuria glaberrima. Gene. 2004, 334: 133-143. 10.1016/j.gene.2004.03.023.
16. Bannister R, McGonnell IM, Graham A, Thorndyke MC, Beesley PW: Afuni, a novel transforming growth factor-beta gene is involved in arm regeneration by the brittle star Amphiura filiformis. Dev Genes Evol. 2005, 215 (8): 393-401. 10.1007/s00427-005-0487-8.
17. Bannister SR, Powell CA: Foreign body reaction to anorganic bovine bone and autogenous bone with platelet-rich plasma in guided bone regeneration. J Periodontol. 2008, 79 (6): 1116-1120. 10.1902/jop.2008.060475.
18. Patruno M, McGonnell I, Graham A, Beesley P, Candia Carnevali MD, Thorndyke M: Anbmp2/4 is a new member of the transforming growth factor-beta superfamily isolated from a crinoid and involved in regeneration. Proc Biol Sci. 2003, 270 (1522): 1341-1347. 10.1098/rspb.2003.2367.
19. Patruno M, Smertenko A, Candia Carnevali MD, Bonasoro F, Beesley PW, Thorndyke MC: Expression of transforming growth factor beta-like molecules in normal and regenerating arms of the crinoid Antedon mediterranea: immunocytochemical and biochemical evidence. Proc Biol Sci. 2002, 269 (1502): 1741-1747. 10.1098/rspb.2002.2010.
20. Thorndyke MC, Chen WC, Beesley PW, Patruno M: Molecular approach to echinoderm regeneration. Microsc Res Tech. 2001, 55 (6): 474-485. 10.1002/jemt.1192.
21. Sanchez Alvarado A: Stem cells and the Planarian Schmidtea mediterranea. C R Biol. 2007, 330 (67): 498-503. 10.1016/j.crvi.2007.05.005.
22. Carlson ME, O'Connor MS, Hsu M, Conboy IM: Notch signaling pathway and tissue engineering. Front Biosci. 2007, 12: 5143-5156. 10.2741/2554.
23. Stocum DL: Stem cells in CNS and cardiac regeneration. Adv Biochem Eng Biotechnol. 2005, 93: 135-159.
24. Mendez AT, Roig-Lopez JL, Santiago P, Santiago C, Garcia-Arraras JE: Identification of Hox Gene Sequences in the Sea Cucumber Holothuria glaberrima Selenka (Holothuroidea: Echinodermata). Mar Biotechnol (NY). 2000, 2 (3): 231-240.
25. Mendez AT: Characterization of homeobox genes in the sea cucumber, Holothuria glaberrima Selenka (holothuroidea:Echinodermata) and expression during

the regeneration of gastrointestinal tract. 2001, San Juan: University of Puerto Rico

26. Santiago-Cardona PG, Berrios CA, Ramirez F, Garcia-Arraras JE: Lipopolysaccharides induce intestinal serum amyloid A expression in the sea cucumber Holothuria glaberrima. Dev Comp Immunol. 2003, 27 (2): 105-110. 10.1016/S0145-305X(02)00068-X.
27. Lien CL, Schebesta M, Makino S, Weber GJ, Keating MT: Gene expression analysis of zebrafish heart regeneration. PLoS Biol. 2006, 4 (8): e260-10.1371/journal.pbio.0040260.
28. Arai M, Yokosuka O, Chiba T, Imazeki F, Kato M, Hashida J, Ueda Y, Sugano S, Hashimoto K, Saisho H, et al: Gene expression profiling reveals the mechanism and pathophysiology of mouse liver regeneration. J Biol Chem. 2003, 278 (32): 29813-29818. 10.1074/jbc.M212648200.
29. Fukuhara Y, Hirasawa A, Li XK, Kawasaki M, Fujino M, Funeshima N, Katsuma S, Shiojima S, Yamada M, Okuyama T, et al: Gene expression profile in the regenerating rat liver after partial hepatectomy. J Hepatol. 2003, 38 (6): 784-792. 10.1016/S0168-8278(03)00077-1.
30. Monaghan JR, Walker JA, Page RB, Putta S, Beachy CK, Voss SR: Early gene expression during natural spinal cord regeneration in the salamander Ambystoma mexicanum. J Neurochem. 2007, 101 (1): 27-40. 10.1111/j.1471-4159.2006.04344.x.
31. Beauchemin M, Noiseux N, Tremblay M, Savard P: Expression of Hox A11 in the limb and the regeneration blastema of adult newt. Int J Dev Biol. 1994, 38 (4): 641-649.
32. Carlson ER: Bone grafting the jaws in the 21st century: the use of platelet-rich plasma and bone morphogenetic protein. Alpha Omegan. 2000, 93 (3): 26-30.
33. Savard P, Gates PB, Brockes JP: Position dependent expression of a homeobox gene transcript in relation to amphibian limb regeneration. Embo J. 1988, 7 (13): 4275-4282.
34. Schummer M, Scheurlen I, Schaller C, Galliot B: HOM/HOX homeobox genes are present in hydra (Chlorohydra viridissima) and are differentially expressed during regeneration. Embo J. 1992, 11 (5): 1815-1823.
35. Mizuta I, Ogasawara N, Yoshikawa H, Sakoyama Y: Identification of homeobox genes expressed during the process of rat liver regeneration after partial hepatectomy. Biochem Genet. 1996, 34 (1–2): 1-15. 10.1007/BF02396236.
36. Thummel R, Ju M, Sarras MP, Godwin AR: Both Hoxc13 orthologs are functionally important for zebrafish tail fin regeneration. Dev Genes Evol. 2007, 217 (6): 413-420. 10.1007/s00427-007-0154-3.
37. Brown R, Brockes JP: Identification and expression of a regeneration-specific homeobox gene in the newt limb blastema. Development. 1991, 111 (2): 489-496.
38. Beck F, Tata F, Chawengsaksophak K: Homeobox genes and gut development. Bioessays. 2000, 22 (5): 431-441. 10.1002/(SICI)1521-1878(200005)22:5<431::AID-BIES5>3.0.CO;2-X.
39. Kawazoe Y, Sekimoto T, Araki M, Takagi K, Araki K, Yamamura K: Region-specific gastrointestinal Hox code during murine embryonal gut development. Dev Growth Differ. 2002, 44 (1): 77-84. 10.1046/j.1440-169x.2002.00623.x.
40. Wolgemuth DJ, Behringer RR, Mostoller MP, Brinster RL, Palmiter RD: Transgenic mice overexpressing the mouse homoeobox-containing gene Hox1.4 – exhibit abnormal gut development. Nature. 1989, 337 (6206): 464-467. 10.1038/337464a0.
41. Kondo T, Dolle P, Zakany J, Duboule D: Function of posterior HoxD genes in the morphogenesis of the anal sphincter. Development. 1996, 122 (9): 2651-2659.
42. Yokoyama H, Ogino H, Stoick-Cooper CL, Grainger RM, Moon RT: Wnt/beta-catenin signaling has an essential role in the initiation of limb regeneration. Dev Biol. 2007, 306 (1): 170-178. 10.1016/j.ydbio.2007.03.014.
43. Hayashi T, Mizuno N, Takada R, Takada S, Kondoh H: Determinative role of Wnt signals in dorsal iris-derived lens regeneration in newt eye. Mech Dev. 2006, 123 (11): 793-800. 10.1016/j.mod.2006.08.009.
44. Kim JB, Leucht P, Lam K, Luppen C, Ten Berge D, Nusse R, Helms JA: Bone regeneration is regulated by wnt signaling. J Bone Miner Res. 2007, 22 (12): 1913-1923. 10.1359/jbmr.070802.
45. Ito M, Yang Z, Andl T, Cui C, Kim N, Millar SE, Cotsarelis G: Wnt-dependent de novo hair follicle regeneration in adult mouse skin after wounding. Nature. 2007, 447 (7142): 316-320. 10.1038/nature05766.
46. Mount JG, Muzylak M, Allen S, Althnaian T, McGonnell IM, Price JS: Evidence that the canonical Wnt signalling pathway regulates deer antler regeneration. Dev Dyn. 2006, 235 (5): 1390-1399. 10.1002/dvdy.20742.
47. Tan X, Behari J, Cieply B, Michalopoulos GK, Monga SP: Conditional deletion of beta-catenin reveals its role in liver growth and regeneration. Gastroenterology. 2006, 131 (5): 1561-1572. 10.1053/j.gastro.2006.08.042.
48. Yen TH, Wright NA: The gastrointestinal tract stem cell niche. Stem Cell Rev. 2006, 2 (3): 203-212. 10.1007/s12015-006-0048-1.
49. Kobayashi C, Saito Y, Ogawa K, Agata K: Wnt signaling is required for antero-posterior patterning of the planarian brain. Dev Biol. 2007, 306 (2): 714-724. 10.1016/j.ydbio.2007.04.010.
50. Gurley KA, Rink JC, Sanchez Alvarado A: Beta-catenin defines head versus tail identity during planarian regeneration and homeostasis. Science. 2008, 319 (5861): 323-327. 10.1126/science.1150029.
51. Grogg MW, Call MK, Okamoto M, Vergara MN, Del Rio-Tsonis K, Tsonis PA: BMP inhibition-driven regulation of six-3 underlies induction of newt lens regeneration. Nature. 2005, 438 (7069): 858-862. 10.1038/nature04175.
52. Smith A, Avaron F, Guay D, Padhi BK, Akimenko MA: Inhibition of BMP signaling during zebrafish fin regeneration disrupts fin growth and scleroblasts

differentiation and function. Dev Biol. 2006, 299 (2): 438-454. 10.1016/j.ydbio.2006.08.016.

53. Beck CW, Christen B, Barker D, Slack JM: Temporal requirement for bone morphogenetic proteins in regeneration of the tail and limb of Xenopus tadpoles. Mech Dev. 2006, 123 (9): 674-688. 10.1016/j.mod.2006.07.001.

54. Molina MD, Salo E, Cebria F: The BMP pathway is essential for re-specification and maintenance of the dorsoventral axis in regenerating and intact planarians. Dev Biol. 2007, 311 (1): 79-94. 10.1016/j.ydbio.2007.08.019.

55. Reddien PW, Bermange AL, Kicza AM, Sanchez Alvarado A: BMP signaling regulates the dorsal planarian midline and is needed for asymmetric regeneration. Development. 2007, 134 (22): 4043-4051. 10.1242/dev.007138.

56. Batts LE, Polk DB, Dubois RN, Kulessa H: Bmp signaling is required for intestinal growth and morphogenesis. Dev Dyn. 2006, 235 (6): 1563-1570. 10.1002/dvdy.20741.

57. De Santa Barbara P, Williams J, Goldstein AM, Doyle AM, Nielsen C, Winfield S, Faure S, Roberts DJ: Bone morphogenetic protein signaling pathway plays multiple roles during gastrointestinal tract development. Dev Dyn. 2005, 234 (2): 312-322. 10.1002/dvdy.20554.

58. Pyati UJ, Cooper MS, Davidson AJ, Nechiporuk A, Kimelman D: Sustained Bmp signaling is essential for cloaca development in zebrafish. Development. 2006, 133 (11): 2275-2284. 10.1242/dev.02388.

59. Dolmatov I: Regeneration of the aquapharyngeal complex in the holothurian Eupentacta fraudatrix (Holothuroidea, Dendrochirota). Monogr Dev Biol. 1992, 23: 40-50.

60. Cowin AJ, Hatzirodos N, Teusner JT, Belford DA: Differential effect of wounding on actin and its associated proteins, paxillin and gelsolin, in fetal skin explants. J Invest Dermatol. 2003, 120 (6): 1118-1129. 10.1046/j.1523-1747.2003.12231.x.

61. Martin P: Wound healing – aiming for perfect skin regeneration. Science. 1997, 276 (5309): 75-81. 10.1126/science.276.5309.75.

62. Davenport RJ: What controls organ regeneration?. Science. 2005, 309 (5731): 84-10.1126/science.309.5731.84.

63. Sanchez Alvarado A: Regeneration in the metazoans: why does it happen?. Bioessays. 2000, 22 (6): 578-590. 10.1002/(SICI)1521-1878(200006)22:6<578::AID-BIES11>3.0.CO;2-#.

64. Sanchez Alvarado A: Regeneration and the need for simpler model organisms. Philos Trans R Soc Lond B Biol Sci. 2004, 359 (1445): 759-763. 10.1098/rstb.2004.1465.

65. Brockes JP, Kumar A: Plasticity and reprogramming of differentiated cells in amphibian regeneration. Nat Rev Mol Cell Biol. 2002, 3 (8): 566-574. 10.1038/nrm881.

66. Habermann B, Bebin AG, Herklotz S, Volkmer M, Eckelt K, Pehlke K, Epperlein HH, Schackert HK, Wiebe G, Tanaka EM: An Ambystoma mexicanum EST sequencing project: analysis of 17,352 expressed sequence tags from embryonic and regenerating blastema cDNA libraries. Genome Biol. 2004, 5 (9): R67-10.1186/gb-2004-5-9-r67.

67. Putta S, Smith JJ, Walker JA, Rondet M, Weisrock DW, Monaghan J, Samuels AK, Kump K, King DC, Maness NJ, et al: From biomedicine to natural history research: EST resources for ambystomatid salamanders. BMC Genomics. 2004, 5 (1): 54-10.1186/1471-2164-5-54.

68. Andreasen EA, Mathew LK, Lohr CV, Hasson R, Tanguay RL: Aryl hydrocarbon receptor activation impairs extracellular matrix remodeling during zebra fish fin regeneration. Toxicol Sci. 2007, 95 (1): 215-226. 10.1093/toxsci/kfl119.

69. Cameron DA, Gentile KL, Middleton FA, Yurco P: Gene expression profiles of intact and regenerating zebrafish retina. Mol Vis. 2005, 11: 775-791.

70. Azumi K, Takahashi H, Miki Y, Fujie M, Usami T, Ishikawa H, Kitayama A, Satou Y, Ueno N, Satoh N: Construction of a cDNA microarray derived from the ascidian Ciona intestinalis. Zoolog Sci. 2003, 20 (10): 1223-1229. 10.2108/zsj.20.1223.

71. Rinkevi Y, Douekch J, Haber O, Rinkevich B, Reshef R: Urochordate whole body regeneration inaugurates a diverse innate immune signaling profile. Dev Biol. 2007, 312 (1): 131-146. 10.1016/j.ydbio.2007.09.005.

72. Mineta K, Nakazawa M, Cebria F, Ikeo K, Agata K, Gojobori T: Origin and evolutionary process of the CNS elucidated by comparative genomics analysis of planarian ESTs. Proc Natl Acad Sci USA. 2003, 100 (13): 7666-7671. 10.1073/pnas.1332513100.

73. Nakazawa M, Cebria F, Mineta K, Ikeo K, Agata K, Gojobori T: Search for the evolutionary origin of a brain: planarian brain characterized by microarray. Mol Biol Evol. 2003, 20 (5): 784-791. 10.1093/molbev/msg086.

74. Newmark PA, Sanchez Alvarado A: Not your father's planarian: a classic model enters the era of functional genomics. Nat Rev Genet. 2002, 3 (3): 210-219. 10.1038/nrg759.

75. Rossi L, Salvetti A, Marincola FM, Lena A, Deri P, Mannini L, Batistoni R, Wang E, Gremigni V: Deciphering the molecular machinery of stem cells: a look at the neoblast gene expression profile. Genome Biol. 2007, 8 (4): R62-10.1186/gb-2007-8-4-r62.

76. Zayas RM, Hernandez A, Habermann B, Wang Y, Stary JM, Newmark PA: The planarian Schmidtea mediterranea as a model for epigenetic germ cell specification: analysis of ESTs from the hermaphroditic strain. Proc Natl Acad Sci USA. 2005, 102 (51): 18491-18496. 10.1073/pnas.0509507102.

77. Cebrla F, Kudome T, Nakazawa M, Mineta K, Ikeo K, Gojobori T, Agata K: The expression of neural-specific genes reveals the structural and molecular complexity of the planarian central nervous system. Mech Dev. 2002, 116 (1–2): 199-204. 10.1016/S0925-4773(02)00134-X.

78. Chera S, Kaloulis K, Galliot B: The cAMP response element binding protein (CREB) as an integrative HUB selector in metazoans: clues from the hydra model

system. Biosystems. 2007, 87 (2–3): 191-203. 10.1016/j.biosystems.2006.09.014.

79. Galliot B, Miljkovic-Licina M, de Rosa R, Chera S: Hydra, a niche for cell and developmental plasticity. Semin Cell Dev Biol. 2006, 17 (4): 492-502. 10.1016/j.semcdb.2006.05.005.

80. Galliot B, Miljkovic-Licina M, Ghila L, Chera S: RNAi gene silencing affects cell and developmental plasticity in hydra. C R Biol. 2007, 330 (6–7): 491-497. 10.1016/j.crvi.2007.01.008.

81. Stout T, McFarland T, Appukuttan B: Suppression subtractive hybridization identifies novel transcripts in regenerating Hydra littoralis. J Biochem Mol Biol. 2007, 40 (2): 286-289.

82. Sea-Urchin-Genome-Sequence-Consortium: The genome of the sea urchin Strongylocentrotus purpuratus. Science. 2006, 314 (5801): 941-952. 10.1126/science.1133609.

83. Dubois P, Ameye L: Regeneration of spines and pedicellariae in echinoderms: a review. Microsc Res Tech. 2001, 55 (6): 427-437. 10.1002/jemt.1188.

84. Dupont S, Thorndyke MC: Growth or differentiation? Adaptive regeneration in the brittlestar Amphiura filiformis. J Exp Biol. 2006, 209 (Pt 19): 3873-3881. 10.1242/jeb.02445.

85. Ramirez-Gomez F, Ortiz-Pineda PA, Rojas-Cartagena C, Suarez-Castillo EC, Garcia-Arraras JE: Immune-related genes associated with intestinal tissue in the sea cucumber Holothuria glaberrima. Immunogenetics. 2008, 60 (1): 57-71. 10.1007/s00251-007-0258-y.

86. Chomczynski P: A reagent for the single-step simultaneous isolation of RNA, DNA and proteins from cell and tissue samples. Biotechniques. 1993, 15 (3): 532-534. 536-537.

87. Edgar R, Domrachev M, Lash AE: Gene Expression Omnibus: NCBI gene expression and hybridization array data repository. Nucleic Acids Res. 2002, 30 (1): 207-210. 10.1093/nar/30.1.207.

88. Smyth GK, Speed T: Normalization of cDNA microarray data. Methods. 2003, 31 (4): 265-273. 10.1016/S1046-2023(03)00155-5.

89. Benjamini Y, Hochber Y: Controlling the false discovery rate: a practical and powerful approach to multiple testing. J R Stat Soc. 1995, 57: 289-300.

90. Eisen MB, Spellman PT, Brown PO, Botstein D: Cluster analysis and display of genome-wide expression patterns. Proc Natl Acad Sci USA. 1998, 95 (25): 14863-14868. 10.1073/pnas.95.25.14863.

91. Kohonen T: Cortical maps. Nature. 1990, 346 (6279): 24-10.1038/346024b0.

CHAPTER 6

DIETARY SEA CUCUMBER CEREBROSIDE ALLEVIATES OROTIC ACID-INDUCED EXCESS HEPATIC ADIPOPEXIS IN RATS

Bei Zhang, Changhu Xue, Xiaoqian Hu, Jie Xu, Zhaojie Li, Jingfeng Wang, Teruyoshi Yanagita, Yong Xue and Yuming Wang

Keywords: Sea cucumber cerebroside, Orotic acid, Fatty liver, Lipogenesis, Microsomal triglyceride transfer protein

BACKGROUND

The metabolic syndrome, a cluster of metabolic abnormalities such as hyperlipidemia, diabetes mellitus, and hypertension, is a widespread disease in industrialized countries and contributes to the increase in cardiovascular morbidity and mortality [1–3]. Nonalcoholic fatty liver disease (NAFLD), an increasingly prevalent chronic liver disease in many countries, has been associated with insulin resistance and metabolic syndrome. NAFLD is the preferred term to describe the spectrum of liver damage ranging from hepatic steatosis to steatohepatitis, liver fibrosis, and cirrhosis [4]. It has been reported that occurrence of NAFLD ranges from 10% to 24% in different populations, and even reaches 74% in obese individuals [5]. This prevalence prompted studies to focus on the prevention or therapy of NAFLD. Though the processes through which steatohepatitis evolves from hepatic steatosis are not fully understood, it is necessary to develop effective therapies for the treatment of NAFLD and to discover nutrients that will reduce the risk of NAFLD.

Orotic acid (OA), an intermediate in pyrimidine biosynthesis, is known to evoke fatty liver when fed over an appreciable amount in rats [6, 7]. The model established by OA-supplemented diet to rats reflects the natural etiologic setting in which NAFLD develops[8]. OA-induced NAFLD has been reported to be possibly generated, in part, by reduction of very low density lipoprotein (VLDL) secretion [9], enhancement of triglyceride (TG) synthesis [7, 10] and decrease in fatty acid β- oxidation [8, 11, 12]. The hepatic TG accumulation caused by OA which has been demonstrated before was likely to be alleviated by dietary nutrients. Buang Y has found that dietay phosphatidylcholine alleviated OA-induced hepatic steatosis and hepatomegaly through downregulation of fatty acid synthesis and up-regulation of β-oxidation in rats [13]. Dietary saponins of sea cucumber attenuating fatty liver partly associated with the promoted lipolysis via PPARα activation and inhibited SREBP-1c-mediated lipogenisis [12]. But no study was undertaken to ascertain the influence of dietary glycolipids against NAFLD.

The existence of glycolipids has been determined in wide range. Early in 2000, Suzuki H has detected glycolipids from several plants [14]. Recent growing interest in the physiological functions of glycolipid as a nutrient in food has generated. The antifungal and antibacterial activities in the three sphingolipids isolated from cucumber *Cucumis sativus L.* has been evaluated recently [15]. Glycolipids has been shown great anticancer potency through inducing Th1-biased cytokines and CD8/CD4

Citation: Zhang et al.: Dietary sea cucumber cerebroside alleviates orotic acid-induced excess hepatic adipopexis in rats. Lipids in Health and Disease 2012 11:48. doi:10.1186/1476-511X-11-48.

T cells in mice bearing breast and lung cancers [16]. Nakagawa R discovered that galactosylceramide activated NKT cells and then produced IFN-γ, which both increased the innate antitumor cytotoxicity of NK cells and expessed inhibition of tumor metastasis to liver through the adaptive antitumor response of CD8+ T cells in mice [17]. The effect of glycolipids on the immunomodulatory has also been investigated peviously[18]. The cerebroside, a type of endogenous glycolipid extensively existing in cell membranes of fungi, plants, animals and marine organism [19], has been proven to exert pharmacological effects, such as antibacterial, antihepatotoxic, anti-tumor and neuroprotective activities [20–23]. And the cerebroside was also reported to regulate lipid levels according to Cohn Jeffrey [24]. The marine cerebroside displayed strong bioactivity as a result of the special ocean circulation, although there are fewer classes and lower contents of cerebrosides in marine than living in the ground. Sea cucumber of Echinodermata is an important marine food and medical material in Asian countries. Researches on the influence of dietary whole sea cucumber on lipid metabolism have been much investigated. Tanaka K has found that dietary 5% black sea cucumber (*Stichophus japonicus*) revealed hypolipidemic effect in rats [25]. The results of Wang's study on sea cucumber (*Pearsonothuria graeffei* and *Apostichopus japonicus*) displayed its lipid-lowering effect in hypercholesteremic rats [26]. However, it is unknown whether the sea cucumber cerebroside (SCC) may prevent against lipid metabolism disorder in animal models.

Whereas the SCC may be expected to exhibit various physiological functions based on the results obtained from cerebrosides from other sources. Nonetheless, reports on biological evaluation of lipid metabolism on SCC are unavailable. Therefore, the present work was carried out to discuss the effect of SCC on the lipids of serum and liver in rats fed OA diet. To gain an insight into the mechanism of SCC on the NAFLD caused by OA, response of enzymatic activities and mRNA expression involved in hepatic lipid metabolism was also measured.

METHODS

Isolation and Purification of Cerebroside

The dried body wall of sea cucumber *Acaudina molpadioides* was grinded to powder and extracted three times with EtOH/H_2O (4:1). The combined extracts were concentrated in vacuo to give an aqueous solution. The aqueous solution was extracted with three portions of n-hexane. The organic layer was concentrated in vacuo and the residue was washed with cold acetone to give an acetone-insoluble fraction. The acetone-insoluble part was chromatographed on Silica gel column(solvent $CHCl_3$/MeOH 99:1 to $CHCl_3$/MeOH/H_2O 80:15:1) to afford cerebrosides (purity≥95%) [27]. The structure of the SCC is shown in Figure 1.

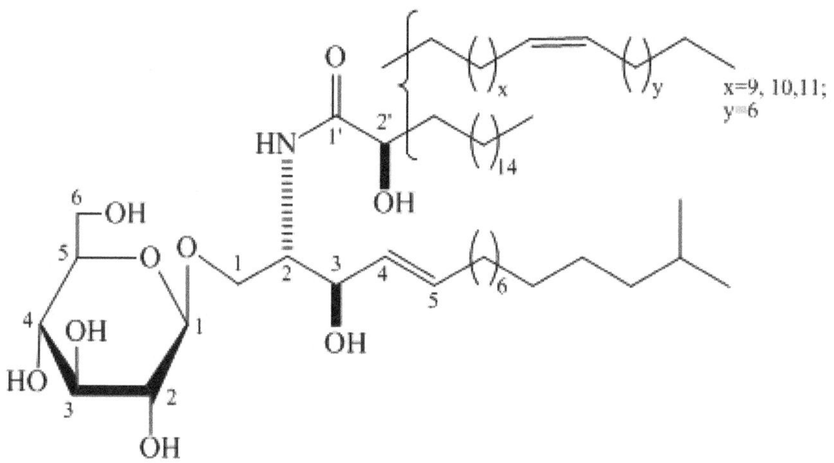

Figure 1: The structure of SCC from *Acaudina molpadioides.*

Animals and Diets

All aspects of the experiment were conducted according to guidelines provided by the ethical committee of experimental animal care at Ocean University of China (OUC, China).

Male Sprague–Dawley rats aged 5wk were purchased from Vital River (Beijing, China) and housed individually in an air-conditioned room with a 12-h light/dark cycle, a constant temperature of 24°C and relative humidity of 65±15%. After a 1-wk adaptation period, rats were randomly divided into four groups (n=6). The animals were then assigned to the following four different diets: 1) AIN-93G diet (Cont); 2) AIN-93G diet with 1% OA contained (OA); 3) 1% OA-contained AIN-93G diet plus 0.006%SCC (OA+0.006%); 4) 1% OA-contained AIN-93G diet plus 0.03% SCC (OA+0.03%). Basal experimental diets were prepared according to the recommendations of the American Institute of Nutrition (AIN). The composition of the diets and SCC concentration were shown in Table 1. Rats were allowed free access to water and food for 10 days. At the end of the feeding period, all rats were killed by bleeding from the abdominal aorta under diethyl ether anesthesia after overnight fasting. Serum was separated from the whole blood by centrifugation at 1500*g for 10 min at 4°C. Liver, perirenal adipose and epididymal adipose were immediately excised for analysis.

Table 1: Composition of experimental diets (g/kg diet)

Ingredients	Cont	OA	OA+0.006	OA+0.03%
Casein	200	200	200	200
Cornstarch	500	490	489.94	489.7
Sucrose	100	100	100	100
Corn oil	100	100	100	100
Mineral mix (AIN-93 G)	35	35	35	35
Vitamin mix(AIN-93 G)	10	10	10	10
Choline bitartrate	2	2	2	2
Cellulose	50	50	50	50
L-cystine	3	3	3	3
OA	—	10	10	10
SCC	—	—	0.06	0.3

Vitamin mixture (AIN-93VX; DyetsInc.).

Mineral mixture (AIN-93 G; DyetsInc.).

Serum Biochemistry Analysis

Serum total cholesterol (TC), TG and low density lipoprotein cholesterol (LDL-c) were measured using enzymatic reagent kits from Biosino (Beijing, China) according to the manufacturer's instructions. Serum ALT and AST activities were determined using enzymatic kits from Njjctech (Nanjing, China).

Measurement of Hepatic Lipid Levels

Hepatic lipids were extracted and purified according to the method of Folch [28] and then dissolved with 10% Triton X-100. The concentrations of TG and TC in liver were detected using enzymatic reagent kits from Biosino(Beijing, China), and the hepatic phospholipid (PL) levels were determined according to the methods of Bartlett [29].

Preparation of Liver Subcellular Fractions and Assays of Hepatic Enzyme Activities

Portions of the fresh livers from individual rats were homogenized in ice-cold 0.25 M sucrose solution containing 10 mM Tris–HCl buffer (pH7.4) and 1 mM EDTA. After precipitating the nuclei fraction, the supernatant was centrifuged at 10,000*g for 10 min at 4°C to obtain mitochondria, which was used for carnitinepalmitoyl transferase (CPT) activity as described previously [30]. The resulting supernatant was re-centrifuged at 125,000*g for 60 min,the supernatant (cytosol) was used for the assays of fatty acid synthase (FAS), malic enzyme (ME) and glucose-6-phosphatedehydrogenase (G6PDH) activities as described [30–32], and the precipitate(microsomes) was used for the assays of microsomal triglyceride transfer protein (MTP) activity with an MTP assay kit from Roar Biomedical (New York, US). The protein concentration was determined according to the method of Lowry [33], with bovine serum albumin used as the standard.

Determination of mRNA Expression of Hepatic Genes

For analysis of gene expression, total cellular RNA was extracted from 100 mg frozen liver samples using the Trizol reagent (Invitrogen, USA) according to the manufacturer's recommended procedures. RNA concentrations were determined spectrophotometrically and 1 μg total RNA was reverse transcribed using an MMLV reverse transcriptase kit (Promega, USA) and random primers (TOYOBO, Japan) in a reaction volume of 50uL for cDNA synthesis. The concentration of cDNA was analyzed by real-time detection PCR (ABI Prism 7500 Sequence Detection System, USA) using Sybr Green I Master Mix (TOYOBO, Japan). PCR was carried out with a final volume of 30 μL reaction mixture containing:15 μL master mix, 2 μL first-strand cDNA, 1 μL primer with each forward and reverse, and H_2O to make up 30 μL. A dilution curve from one cDNA source using dilutions of 1:2, 1:4, 1:8 and a no-template control was run for each gene including sterol- regulatory element binding protein (SREBP-1c). The gene expression was determined by relative quantification using the standard curve method. A final melting curve guaranteed the authenticity of the target product. The expression signal of the house keeping gene ß -actin served as an internal control for normalization. The primer sequences used for real-time PCR were shown in Table 2.

Table 2: Primers used in this study

Gene	Forward/reverse primer	Accession No.	Product length
β-actin	GCAGATGTGGATCAGCAAGC GTCAAAGAAAGGGTGTAAAACG	NM_031144	111 bp
FAS	GGAACTGAACGGCATTACTCG GCCCAAACCCCATTTTCTA	X62888	153 bp
ME	TCACCTGCCCTAATGTCCCT CATGCCGTTATCAACTTGTCC	NM_012600	185 bp
G6PDH	GTTTGGCAGCGGCAACTAA GGCATCACCCTGGTACAACTC	NM_031559	108 bp
SREBP-1c	CGCTACCGTTCCTCTATCAA TTCGCAGGGTCAGGTTCTC	AF286470	166 bp
CPT1a	GCTTCCCCTTACTGGTTCC AACTGGCAGGCAATGAGACT	NM_012930	115 bp

Gene-specific primers were designed by Primer Premier 5.0 software.

Statistical Analyses

All the values are expressed as mean±standard error of the mean of six rats. All statistical analyses were performed using PC SAS software. One-way ANOVA and Tukey's post hoc test with least significant difference were used to compare group means. Homogeneity of variances was tested by Levene's test and Welch's ANOVA was used to compare group means when the group variances were unequal. $P<0.05$ was considered statistically significant.

RESULTS

Body Weight Gain, Food Intake, Liver Weight and Adipose Weight

As described in Table 3, the initial body weight (BW) was similar among the four groups. The amount of food consumed and food intake were almost the same in the period of experiment. OA feeding markedly elevated the liver weight compared to the controls ($P<0.01$), while the rats fed with SCC resulted in a significant decrease in liver weight as compared with the OA-feeding rats ($P<0.05$). There was no statistical difference among the four groups in the adipose tissue weight although dietary SCC slightly reduced the weight of perirenal adipose.

Table 3: Effect of SCC on body weight gain, food intake, liver and adipose weights in rats

	Cont	OA	OA+0.006 %	OA+0.03 %
Initial BW, g	160±7	160±6	159±8	158±7
Final BW, g	248±9	242±8	238±5	236±9
Food intake, g/d	13.9±0.3	13.8±0.3	13.5±0.7	13.2±1.6
Liver index, g/100 g BW	3.77±0.3	4.65±0.4**	4.15±0.3#	4.05±0.2#
Perirenal adipose, g/100 g BW	0.85±0.11	0.89±0.15	0.84±0.17	0.76±0.21
Epididymal adipose,g/100 g BW	0.87±0.04	0.90±0.06	0.84±0.12	0.87±0.23

Rats were fed the experimental diets for 10d.

Values are expressed as mean±standard error of the mean of six rats. **$P<0.01$, different from the control group; #$P<0.05$, different from the OA group.

Serum and Hepatic Parameters

Table 4 summarized the lipid levels of the rats in serum and liver after 10 days of experiment. As compared with the controls, the serum TG and TC concentrations decreased ($P<0.05$, $P<0.05$) in the rats fed with OA, while the LDL-c level tended to decrease without significant difference. The rats in both SCC groups had higher serum TG content than the OA-fed rats ($P<0.05$, $P<0.05$), but the TC and the LDL-c concentrations, although tended to reduce, were of no statistic difference in the two groups contrasting the OA group.

Table 4: Effect of SCC on serum and hepatic parameters in rats

	Cont	OA	OA+0.006%	OA+0.03%
Serum lipids, mmol/L				
TG	1.44 ± 0.14	$0.69\pm0.12^{*}$	$1.25\pm0.09^{\#}$	$1.20\pm0.13^{\#}$
TC	1.75 ± 0.13	$1.47\pm0.16^{*}$	1.54 ± 0.12	1.52 ± 0.10
LDL-c	1.53 ± 0.11	1.07 ± 0.15	1.19 ± 0.12	1.21 ± 0.13
Hepatic lipids, umol/g				
TG	15.8 ± 1.3	$97.3\pm3.5^{**}$	$31.2\pm6.7^{\#}$	$15.7\pm2.9^{\#\#}$
TC	19.2 ± 0.5	$26.7\pm3.2^{*}$	$17.5\pm2.1^{\#}$	$14.3\pm0.7^{\#\#}$
PL	34.5 ± 1.2	32.7 ± 1.3	33.2 ± 1.3	32.7 ± 1.5
Enzyme activities, IU/L				
ALT	9.21 ± 2.01	$28.7\pm5.13^{**}$	$15.7\pm1.76^{\#}$	$15.9\pm1.93^{\#}$
AST	22.0 ± 1.76	$39.5\pm3.65^{**}$	$29.4\pm1.01^{\#}$	$28.1\pm1.65^{\#}$

Rats were fed the experimental diets for 10d.

Values are expressed as mean ± standard error of the mean of six rats.

$^{*}P<0.05$, $^{**}P<0.01$, different from the control group; $^{\#}P<0.05$, $^{\#\#}P<0.01$, different from the OA group.

OA-feeding rats had a hepatic TG level four times higher than that in the controls (P<0.01). But SCC supplementation at 0.006% and 0.03% in rats both attenuated the OA-induced TG accumulation in contrast to the OA feeding (P<0.05, P<0.01). The same tendency was observed in the hepatic TC level between the control and OA groups (P<0.05); dietary SCC at both doses reduced the TC concentration (P<0.05, P<0.01) in comparison with the OA diet. The PL content was of no marked difference among the four groups.

Besides, rats fed OA diet displayed severe NAFLD-accompanied liver injury. The activities of ALT increased about 2-fold in the OA group as compared with the controls (P<0.01) and the AST increased over 70% (P<0.01). Administration of SCC at both doses ameliorated the hepatic damage. Dietary SCC at 0.006% reduced the ALT and AST activities by 45.3% and 25.6% respectively (P<0.05, P<0.05) and SCC at 0.03% by 44.6% and 27.6% separately (P<0.05, P<0.05).

Hepatic Enzyme Activities Involved in Lipogenesis Metabolism

OA administration increased the activities of FAS (89.8%, P<0.01), ME (24.6%, P<0.05) and G6PDH (72.8%, P<0.05) in liver, while feeding SCC to NAFLD rats significantly abolished the OA-induced increase. Dietary SCC at 0.006% reduced the activity of FAS by 36.5% (P<0.01) and G6PDH by 26.0% (P<0.01), and tended to decreased the activity of ME. Addition of SCC at 0.03% significantly reduced the activities of FAS, ME and G6PDH by 59.1% (P<0.01), 42.8% (P<0.01) and 30.9% (P<0.01) (Figure 2).

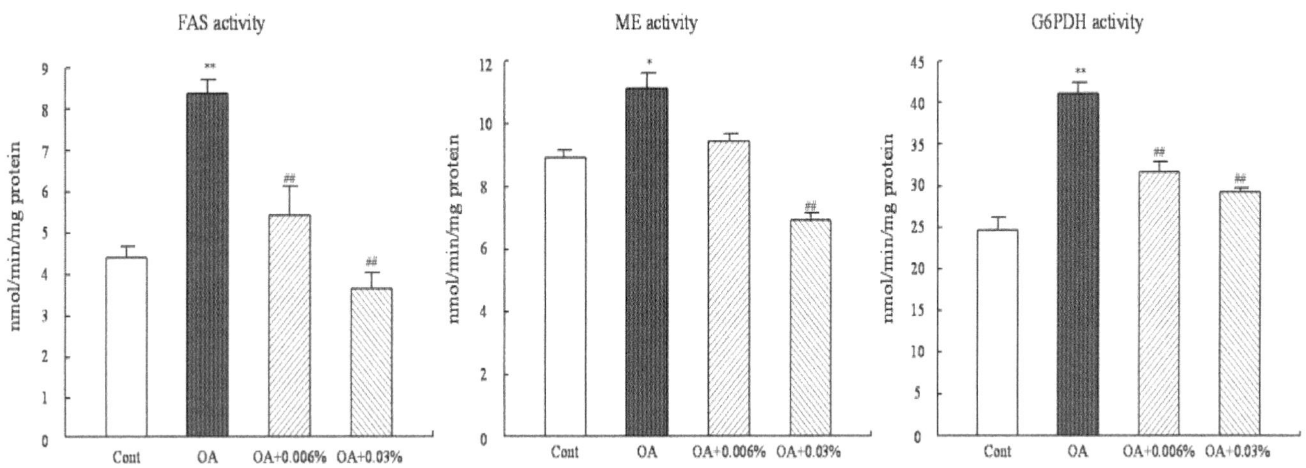

Figure 2: Effect of SCC on the activities of the enzymes related to fatty acid biosynthesis in rats. The activities of hepatic enzyme FAS, ME and G6PDH in rats fed Cont or OA-supplemented diet or the diet with OA+0.006% SCC or OA+0.03% SCC for 10d. Values are expressed as mean ± standard error of the mean of six rats. $^{*}P<0.05$, $^{**}P<0.01$, different from the control group; $^{\#\#}P<0.01$, different from the OA group.

Relative mRNA Expression of SREBP-1c and its Target Genes in Liver

In order to explore the molecular mechanism of the effect of SCC on the regulation of fatty acid lipogenesis, the expression of hepatic lipogenic genes was examined (Figure 3). Considering the integrated lipid-lowering effects of SCC, the group by dietary 0.03% SCC was chosen to access the changes of the mRNA expression for its most effective actions.

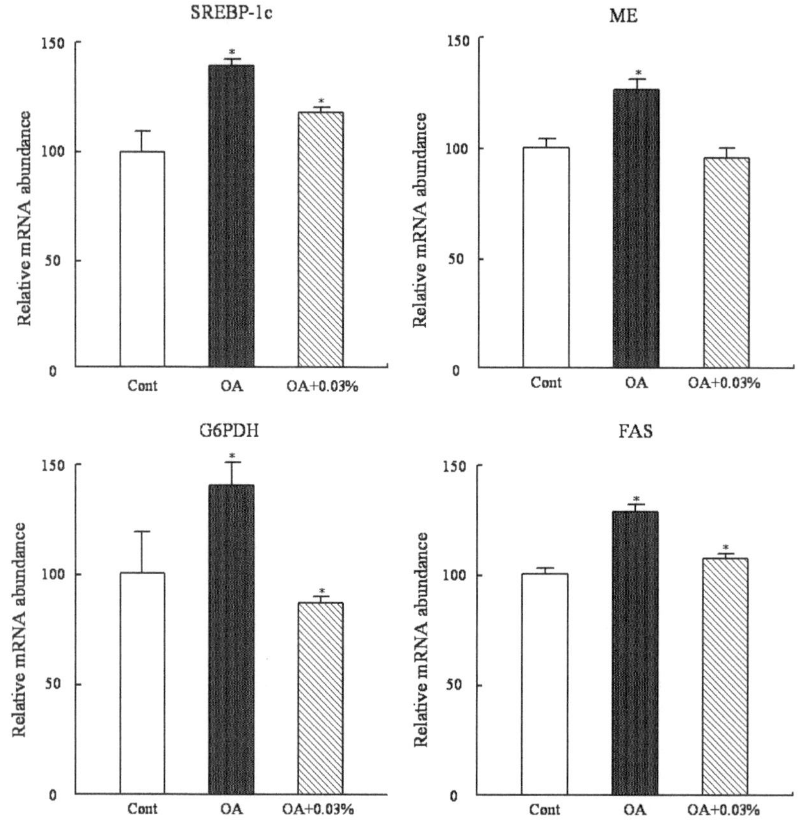

Figure 3: Effect of SCC on the mRNA expression of lipogenic genes in the liver of the rats. The mRNA expression of genes involved in lipogenesis SREBP-1c, FAS, ME and G6PDH in rats fed Cont or OA-supplemented diet or the diet with OA+0.03% SCC for 10d. Values are expressed as mean±standard error of the mean of six rats. *P<0.05 , different from the control group; #P<0.05, different from the OA group.

OA-supplemented diet to the rats up-regulated the mRNA expression of the lipogenic transcription factor SREBP-1c and its target genes (Figure 4). As compared with the controls, OA group exhibited a 1.4-fold SREBP-1c mRNA expreesion (P<0.05), 1.3-fold FAS (P<0.05) and 1.4-fold G6PDH (P<0.05); and the expression of ME mRNA tended to increase without significant difference. Dietary SCC at 0.03% prevented the OA-stimulated up-regulation of SREBP-1c and its response genes. SCC supplementation down-regulated the mRNA expression of FAS, G6PDH and ME by 17.2% (P<0.05), 43.0% (P<0.05) and 24%(P<0.05).

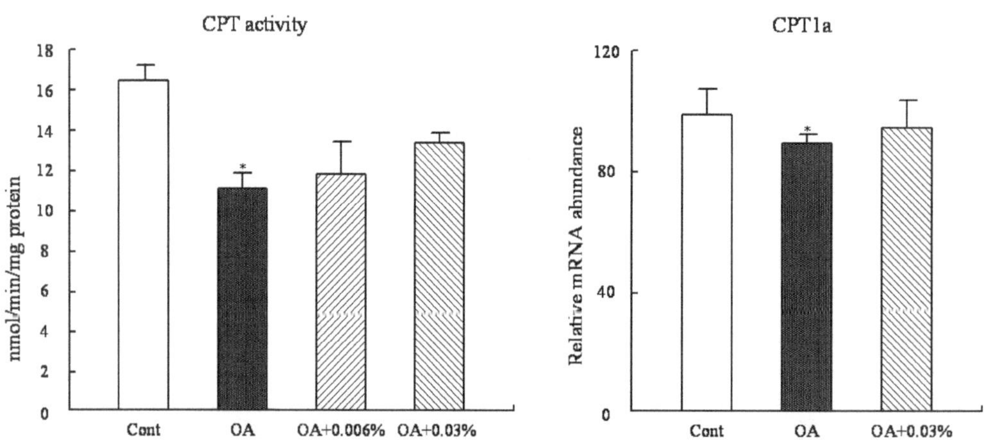

Figure 4: Effect of SCC on the CPT enzyme activity and mRNA expression in the liver of the rats. The activtiy of CPT in rats fed Cont or OA-supplemented diet or the diet with OA+0.006% SCC or OA+0.03% SCC and the mRNA expression of gene CPT1a in rats fed Cont or OA-supplemented diet or the diet withOA+0.03% SCC for 10d. Values are expressed as mean±standard error of the mean of six rats. *P<0.05 , different from the control group.

Hepatic CPT Activity and mRNA Expression

As depicted in Figure 4, 1% OA addition reduced the activity of CPT, a key enzyme in fatty acid β-oxidation, by 37.1%($P<0.05$) as compared with the basal diet; The SCC at both doses tended to attenuate the reduction with no statistical difference. In line with the previous results, SCC at 0.03% was selected for considering the relative mRNA concentration of CPT1a. And consistent with the change of CPT activity, the CPT1a mRNA expression reduced 11% ($P<0.05$) due to the supplemented-OA, but the SCC at 0.03% administration didn't affect its expression over the OA diet.

Hepatic MTP Activity

To gain further insight into the relationship of ipid metabolism between the liver and the serum, the activity of MTP was detected (Figure 5). OA feeding markedly inhibited the MTP activity (26.8%, $P<0.01$), and SCC at 0.03% significantly attenuated the inhibition by 30.2% ($P<0.05$) in contrast to the OA diet.

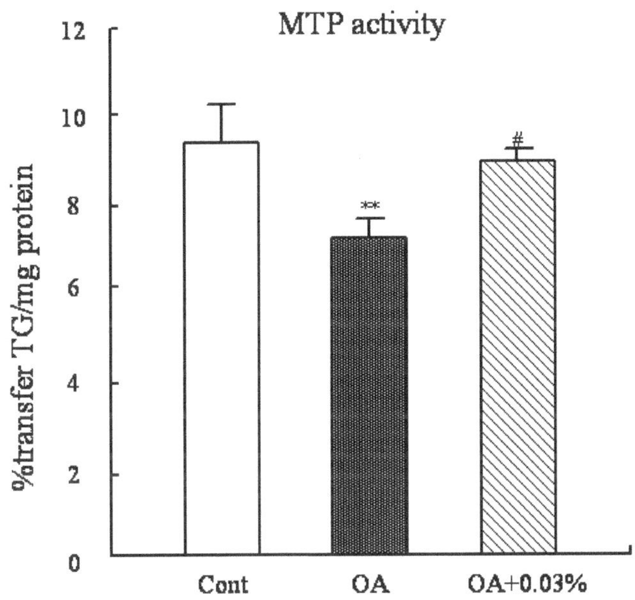

Figure 5: Effect of SCC on the hepatic MTP activity in rats. The activity of hepatic MTP in rats fed Cont or OA-supplemented diet or the diet with OA+0.03% SCC for 10d. Values are expressed as mean±standard error of the mean of six rats. $^{**}P<0.01$, different from the control group; $^{\#}P<0.05$, different from the OA group.

DISCUSSION

OA administration is widely used in studies of the mechanism of NAFLD. It is generally accepted that the enhancement of de novo lipid synthesis and access of hepatic fatty acids and inhibition of VLDL assembly and secretion result in OA-induced hepatic lipid accumulation [34]. The main objective of this study was to clarify if SCC contributed to NAFLD and to explore the potential mechanism in order to discover whether there is a new nutrient that prevents the development of NAFLD.

In the present study, dietary OA increased hepatic TG content by 4-fold, indicating the establishment of NAFLD model (Table4). However, the raise in the hepatic TG concentration in rat was markedly reduced by simultaneous feeding of SCC at the dose of 0.006% and 0.03%. The SCC at 0.03% decreased the lipid level to a much lower extent. The same tendency was observed in hepatic TC levels. The results above suggested that dietary SCC improved hepatic lipid accumulation in a dose–response manner.

In addition, OA resulted in hepatic lipid accumulation and the increase in the activity of serum ALT and AST (Table4). This indicated OA-induced NAFLD accompanied hepatic injury in rats. Nonetheless dietary SCC obviously attenuated the NAFLD-caused liver damage in rats via the declination of ALT and AST activities. Those results were in agreement with Zigmond E, who has found thatβ-glycosphingolipids alleviated liver damage through the immunomodulatory qualities inhibited local inflammation in the response organ in Cohen diabetic rats [35]. And to ourknowledg, this is the first report on the beneficial effect of SCC on NAFLD and concomitant liver injury in rats.

Several metabolic factors were involved in the hepatic lipid accumulation and the stimulation of hepatic fatty acid synthesis was the critical pathway of the NAFLD development. Therefore, the activities of the enzyme related to fatty acid biosynthesis were detected. And from Figure 1, a marked increase in the activity of FAS, a key enzyme involved in fatty acid synthesis, by OA diet was observed, and the activities of NADPH-generating enzymes required for FAS, ME and G6PDH, were also increasingly altered by OA-feeding, which was accordant with the study of Hu [12]. The present results demonstrated that SCC intake significantly suppressed the activities of those lipogenic enzymes, so one of the possible lipid-lowering mechanisms of SCC may be the inhibition of fatty acid biosynthesis.

As suggested in many researches, SREBP-1c acts a major part in the pathogenesis of NAFLD. The genes containing sterol-regulatory elements in their promoter regions in the lipogenic pathway are regulated by SREBP-1c including FAS, G6PDH and ME [36, 37]. Over expression of SREBP-1c produces a pronounced hepatic TG accumulation leading to the development of NAFLD [38, 39]. Thus, the depression of SREBP-1c expression in liver may potentially be responsible for the reduction of hepatic TG accumulation. As Figure 2 presented, SREBP-1c mRNA expression was strongly stimulated by OA diet, and supplementation of SCC inhibited its expression. And accordingly, the mRNA expressions of FAS, ME and G6PDH were concomitant markedly decreased in rats fed by simultaneously OA and SCC. Hence, the pathway of SCC-inhibition of lipogenis was supposed that SCC supplementation directly affected the SREBP-1c expression and then the alternation of these lipogenic gene transcriptions occur, thus reducing the activities of the responding enzymes. Another factor in charge of OA-induced hepatic TG accumulation is a declination of fatty acidβ-oxidation [12, 40]. Since the total β-oxidation critically depends on the mitochondrial oxidation, the activity of the rate-limiting enzyme of mitochondrial β-oxidation, CPT, was examined to reflect the β-oxidation ability (Figure 4). Consistent with Miyazawa S [41], dietary OA significantly decreased the hepaticβ-oxidation capacity. However, SCC intake showed no significantly effect on the key enzyme activities, although a slight increase was observed by dietary SCC at both doses, especially at 0.03%. The result of CPT1a mRNA expression was in accordance with the enzyme changes among the four groups. So the mechanism underlying the decrease in the activities of the lipolytic enzyme following SCC consumption was limited

Morifuji M has reported that the dietary OA decreased the level of serum TG [42]. In the present study, the dietary OA suppressed the serum TG and TC levels, but SCC intake significantly raised the depression (Table4). This result manifested that dietary SCC may contribute to regulating the assembly and secretion of lipoprotein in liver. MTP, a microsomal lumenal protein, is primarily responsible for transfer of TG and other lipids from their site of synthesis in the endoplasmic reticulum into the lumen during the assembly of VLDL [43–45]. VLDL produced by the liver are the major source of LDL in plasma. So the increase of MTP activity and expression accelerates the secretion of hepatic lipids in the form of lipoprotein. Evidence in this study displayed that OA supplementation significantly inhibited the MTP activity, which also explained the lipid accumulation in liver induced by dietary OA; While the SCC addition to the fatty liver rats elevated the MTP activity (Figure 5), which positively related to the increased serum lipid levels and the decreased hepatic TG concent in the SCC-feeding rats. Thus, this finding led to the suggestion that the enhancement of TG secretion from liver was another pathway of lipid-lowering mechanism of dietary SCC.

In Gao's study, the long chain base (LCB) of SCC exert as much effect as the cerebroside itself in lowering hepatic lipid levels in mice, which suggests that the LCB might take responsibility for the regulation of hepatic lipid levels [46]. This result provides elementary references for the relationship based on the structure and the activity of the cerebroside but it still deserves further investigation.

CONCLUSIONS

Supplementation with dietary SCC alleviated OA-induced fatty liver by attenuating TG synthesis through the down-regulation of fatty acid synthesis and improving TG secretion from liver. Findings of the present study also supported that SCC might be useful in the preservation of liver functions. But the relationship between the structure and the activity of the cerebroside still requires further investigation.

ACKNOWLEDGEMENTS

This work is supported by the National Natural Science Foundation of China (30972285) and International science & technology cooperation program of China (2010DFA31330).

AUTHORS' CONTRIBUTIONS

BZ , CX and YW contributed in design, experimental work, analysis and publication of results. BZ and YW were responsible for drafting the manuscript. CX , XH and ZL participated in the design of the study, animal studies and performed statistical analysis. JX and JFW were in charge of the sample preparation and discussing the results. YX participated in the design of the study and statistical analysis. CX and TY participated in drafting the manuscript, discussion of results and providing funding for the study. All authors have read and approved this manuscript.

REFERENCES

1. Farrell GC, Larter CZ: Nonalcoholic fatty liver disease: from steatosis to cirrhosis. Hepatology. 2006, 43: 99-112. 10.1002/hep.20973.CrossRef
2. Farres J, Pujol A, Coma M, Ruiz JL, Naval J, Mas JM, Molins A, Fondevila J, Aloy P: Revealing the molecular relationship between type 2 diabetes and the matabolic changes induced by a very-low-carbohydrated low-fat ketogenic diet. Nutr Metab. 2010, 7: 88-10.1186/1743-7075-7-88.CrossRef
3. Pagano G, Pacini G, Musso G, Gambino R, Mecca F, Depetris N, Cassader M, David E, Cavallo-Perin P, Rizzetto M: Non-alcoholic steatohepatitis, insulin resistance and metabolic syndrome: further evidence for an etiologic association. Hepatology. 2002, 35: 367-372. CrossRefPubMed
4. Nagao K, Inoue N, Wang YM, Shirouchi B, Yanagita T: Dietary conjugated linoleic acid alleviates nonalcoholic fatty liver disease in Zucker (fa/fa) rats. J Nutr. 2005, 135: 9-13.PubMed
5. Angulo P, Lindor K: Non-alcoholic fatty liver disease. J Gastroenterol Hepatol. 2002, 17: 186-190.CrossRef
6. Larson BL, Hegarty HM: Orotic acid in milks of various species and commercial dairy products. J Dairy Sci. 1979, 62: 1641-1644. 10.3168/jds.S0022-0302(79)83474-8. CrossRef
7. Cha JY, Mameda Y, Yamamoto K, Oogami K, Yanagita T: Association between hepatic triacylglycerol accumulation induced by administering orotic acid and enhanced phosphatidate phosphohydrolase activity in rats. Biosci Biotechnol Biochem. 1999, 62: 508-513. CrossRef
8. Wang YM, Hu XQ, Xue Y, Li ZJ, Yanagita T, Xue CH: Study on possible mechanism of orotic acid-induced fatty liver in rats. Nutrition. 2010, in press
9. Hamilton RL, Guo LSS, Felker T, Chao Y, Havel RJ: Nascent high density lipoprotein from liver perfusates of orotic acid fed rats. J Lipid Res. 1986, 27: 967-978. PubMed
10. Yanagita T, Oogami K, Yamamoto T, Cha JY, Nunez J: Triglyceride metabolism of fatty liver and the prevention by dietary n-3 fatty acid. Proc Jpn Conf Biochem Lipids. 1995, 38: 3-6.
11. Cha JY, Cho YS, Kim I, Anno T, Rahman SM, Yanagita T: Effect of hesperetin, a citrus flavonoid, on the liver triacylglycerol content and phosphatidate phosphohydrolase activity in orotic acid-fed rats. Plant Food Hum Nutr. 2001, 56: 349-358. 10.1023/A:1011884200848.CrossRef
12. Hu XQ, Wang YM, Wang JF, Xue Y, Li ZJ, Nagao K, Yanagita T, Xue CH: Dietary saponins of sea cucumber alleviate orotic acid-induced fatty liver in rats via PPARα and SREBP-1c signaling. Lipids Health Dis. 2010, 9: 25-PubMedCentralCrossRefPubMed

13. Buang Y, Wang YM, Cha JY, Nagao K, Yanagita T: Dietary phosphatidylcholine alleviates fatty liver induced by orotic acid. Nutr. 2005, 21: 867-873. 10.1016/j.nut.2004.11.019.CrossRef
14. Suzuki H, Park WK, Lim SY: Analyses of glycolipids in clove, red pepper, and nutmeg by high-performance liquid chromatography. J Food Sci. 2000, 65: 931-933. 10.1111/j.1365-2621.2000.tb09395.x.CrossRef
15. Tang J, Meng XJ, Liu H, Zhao JL, Zhou LG, Qiu MH, Zhang XM, Yu Z, Yang FY: Antimicrobial activity of sphingolipids isolated from the stems of cucumber (*cucumis sativus l.*). Molecules. 2010, 15: 9288-9297. CrossRefPubMed
16. Chang YJ, Huang JR, Tsai YC, Hung JT, Wu D, Fujio M, Wong CH, Yu AL: Potent immune-modulating and anticancer effects of NKT cell stimulatory glycolipids. Proceedings of the National Academy of Sciences of the United States of America. 2007, 104: 10299-10304. PubMedCentralCrossRefPubMed
17. Nakagawa R, Nagafune I, Tazunoki Y, Ehara H, Tomura H, Iijima R, Motoki K, Kamishohara M, Seki S: Mechanisms of the antimetastatic effect in the liver and of the hepatocyte injury induced by a-galactosylceramide in mice. J Immunol. 2001, 166: 6578-84.CrossRefPubMed
18. Wu D, Fujio M, Wong CH: Glycolipids as immunostimulating agents. Bioorgan Med Chem. 2008, 16: 1073-1083. 10.1016/j.bmc.2007.10.026.CrossRef
19. Rumiana K, Martin C: Phases and phase transitions of the sphingolipids. BBA. 1995, 1255: 213-236.
20. Mbosso EJ, Ngouela S, Nguedia JC, Penlap V, Rohmer M, Tsamo E: Spathoside, a cerebroside and other antibacterial constituents of the stem bark of Spathodea campanulata. Nat Prod Res. 2008, 22: 296-304. CrossRefPubMed
21. Kim SY, Choi YH, Huh H, Kim J, Kim YC, Lee HS: New bioactive cerebroside from Lycium chinense fruits. J Nat Prod. 1997, 60: 274-6.CrossRefPubMed
22. Hattori T, Kokura S, Okuda T, Okayama T, Takagi T, Handa O, Naito Y, Yoshida N, Yoshikawa T: Antitumor effect of whole body hyperthermia with alpha-galactosylceramide in a subcutaneous tumor model of colon cancer. Int J Hyperthermia. 2007, 23: 591-8. CrossRefPubMed
23. Boggs JM, Gao W, Hirahara Y: Myelin glycosphingolipids, galactosylceramide and sulfatide, participate in carbohydrated-carbohydrate interactions between apposed membranes and may form glycosynapses between oligodendrocyte and/or myelin membranes. Biochem Biophys Acta. 2008, 1780: 445-455.CrossRefPubMed
24. Jeffrey Cohn S, Wat Elaine, Kamili Alvin, Tandy Sally: Dietary phospholipids, hepatic lipid metabolism and cardiovascular disease. Current Opinion in Lipidology. 2008, 19: 257-262.CrossRef
25. Tanaka K, Nishizono S, Kase A, Ogura S, Kurita M, Murakami T, Kugino K, Matsumoto H, Ikeda I: Effects of dietary black sea cucumber on serum and liver lipid concentrations in rats. Journal of Japanese Society of Nutrition and Food Science. 2003, 56: 175-9. 10.4327/jsnfs.56.175.CrossRef
26. Wang JF, Pang L, Wang YM, Gao S, Xue CH: Studies on the Treatment effects of pearsonothuria graeffei and apostichopus japonicus on hyperlipidemia rats. Periodical of Ocean University of China. 2007, 4: 597-600.
27. Yamada K, Sasaki K, Harada Y, Isobe R, Higuchi R: Isolation and structure of glucocerebrosides from the sea cucumber *Holothuria pervicax*. Chem Pharm Bull. 2002, 50: 1467-1470.CrossRefPubMed
28. Folch J, Lees M, Slane-stanley GH: A simple method for the isolation and purification of total lipids from animal tissues. J Biol Chem. 1957, 226: 497-506.PubMed
29. Barlett GR: Colorimetric assay methods for free and phosphorylated glyceric acids. J Biol Chem. 1958, 234: 466-469.
30. Ikeda I, Cha JY, Yanagita T, Nakatani N, Oogami K, Imaizumi K, Yazawa K: Effects of dietary alpha-linolenic, eicosapentaenoic and docosahexaenoic acids on hepatic lipogenesis and beta-oxidation in rats. Biosci Biotechnol Biochem. 1998, 62: 675-80.CrossRefPubMed
31. Ochoa S, Mehler AH, Kornberg A: Biosynthesis of dicarboxylic acids by carbondioxide fixation. I. Isolation and properties of an enzyme from pigeon liver catalyzing the reversible oxidative decarboxylation of L-malicacid. J Biol Chem. 1948, 174: 979-1000.PubMed
32. Glock GE, Mclean P: Glucose 6-phosphate dehydrogenase activity of rat liver. Nature. 1952, 170: 119-20.CrossRefPubMed
33. Lowry OH, Rosebrough NJ, Farr AL, Randall RJ: Protein measurement with the Folin phenol reagent. J Biol Chem. 1951, 193: 265-75.PubMed
34. Valenti L, Fracanzani AL, Dongiovanni P, Santorelli G, Branchi A, Taioli E, Fiorelli G, Fargion S: Tumor necrosis factor alpha promoter polymorphisms and insulin resistance in nonalcoholic fatty liver disease. Gastroenterology. 2002, 2: 274-80.CrossRef
35. Zigmond E, Zangen SW, Pappo O, Sklair-Levy M, Lalazar G, Zolotaryova L, Raz I, Ilan Y: ß-Glycosphingolipids improve glucose intolerance and hepatic steatosis of the cohen diabetic rat. Am J Physiol Endocrinol Metab. 2009, 296: 72-78.CrossRef
36. Horton JD, Goldstein JL, Brown MS: SREBPs: activators of the complete program of cholesterol and fatty acid synthesis in the liver. J Clin Invest. 2002, 109: 1125-31. PubMedCentralCrossRefPubMed
37. Osborne TF: Sterol regulatory element-binding proteins (SREBPs): key regulators of nutritional homeostasis and insulin action. J Biol Chem. 2000, 275: 32379-82. CrossRefPubMed
38. Ahmed MH, Byrne CD: Modulation of sterol regulatory element binding proteins (SREBPs) as potential treatments for non-alcoholic fatty liver disease (NAFLD). Drug Discov Today. 2007, 12: 740-7.CrossRefPubMed
39. Rutledge AC, Adeli K: Fructose and the metabolic syndrome: pathophysiology and molecular mechanisms. Nutr Rev. 2007, 65: S13-23.CrossRefPubMed

40. Reddy JK, Hashimoto T: Peroxisomal beta-oxidation and peroxisome proliferator activated receptor alpha: an adaptive metabolic system. Annu Rev Nutr. 2001, 21: 193-230.CrossRefPubMed

41. Miyazawa S, Furuta S, Hashimoto T: Reduction of beta-oxidation capacity of rat liver mitochondria by feeding orotic acid. Biochim Biophys Acta. 1982, 711: 494-502. CrossRefPubMed

42. Morifuji M, Aoyama Y: Dietary orotic acid affects antioxidant enzyme mRNA levels and oxidative damage to lipids and proteins in rat liver. J Nutr Biochem. 2002, 13: 403-410.CrossRefPubMed

43. Hussain MM, Iqbal J, Anwar K, Rava P, Dai K: Microsomal triglyceride transfer protein: a multifunctional protein. Front Biosci. 2003, 8: 500-506. 10.2741/1071.CrossRef

44. Sundaram M, Yao Z: Recent progress in understanding protein and lipid factors affecting hepatic VLDL assembly and secretion. Nutr Metab. 2010, 7: 35-10.1186/1743-7075-7-35.CrossRef

45. Jun DW, Han JH, Jang EC, Kim SH, Kim SH, Jo YJ, Park YS, Chae JD: Polymorphisms of microsomal triglyceride transfer protein gene and phosphatidylethanolamine N-methyltransferase gene in alcoholic and nonalcoholic fatty liver disease in Koreans. Eur J Gastroenterol Hepatol. 2009, 6: 667-72.CrossRef

46. Gao Z, Zhou X, Hu XQ, Xue CH, Xu J, Wang YM: Effects of sea cucumber cerebroside and its long-chain base on lipid and glucose metabolism in obese mice. Zhejiang Da Xue Xue Bao Yi Xue Ban. 2012, 41: 60-64.

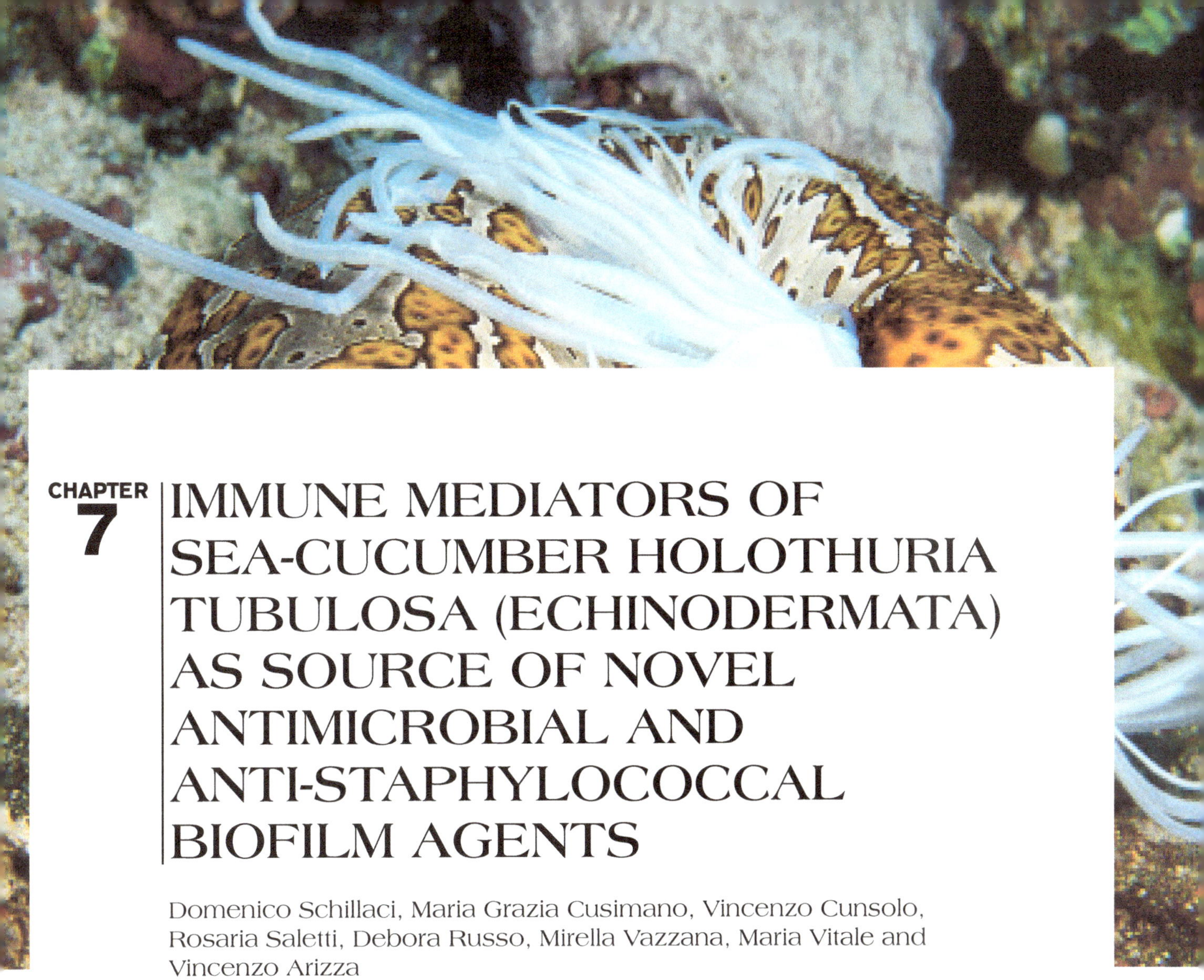

CHAPTER 7

IMMUNE MEDIATORS OF SEA-CUCUMBER HOLOTHURIA TUBULOSA (ECHINODERMATA) AS SOURCE OF NOVEL ANTIMICROBIAL AND ANTI-STAPHYLOCOCCAL BIOFILM AGENTS

Domenico Schillaci, Maria Grazia Cusimano, Vincenzo Cunsolo, Rosaria Saletti, Debora Russo, Mirella Vazzana, Maria Vitale and Vincenzo Arizza

Keywords: Biofilm, Staphylococci, Antimicrobial peptides (AMP), Innate immunity

INTRODUCTION

The survival and fitness of echinoderms in a marine environment, a habitat heavily populated by micro-organisms, suggest that they have developed a potent defensive mechanism that is part of an innate immune system (Tincu and Taylor 2004). The absence of fouling on many of these species indicates an effective defensive strategy to prevent the adhesion of bacteria, the first organisms that allow the formation of biofilm structures and fouling. With the aim of discovering new antimicrobial molecules, the echinoderm immune system was investigated as an unusual source of novel antimicrobial peptides (AMPs) with antibiofilm activity.

AMPs are small molecular weight proteins with broad spectrum antimicrobial activity against bacteria, viruses, and fungi. These evolutionarily conserved peptides are widely distributed in living organisms. Molecules with microbiocidal properties are found since in ameboid protozoa (Leippe 1999), prokaryotes (Pag and Sahl 2002), and plants (Garcia-Olmedo et al. 1998). Usually AMPs are positively charged and have both a hydrophobic and hydrophilic side that enables the molecule to be soluble in aqueous environments yet also enter lipid-rich membranes and kill target cells by using diverse mechanisms.

By focusing on the AMPs of echinoderms, many peptides have been discovered. In the coelomocytes (the immune mediators in echinoderms) of the sea star *Asterias rubens* a number of potential AMPs that are fragments of actin, histone HA2 and filamin were identified (Maltseva et al. 2004). In a recent study, the smallest peptide fragment of a beta-thymosin detected in the coelomocytes of Mediterranean sea-urchin *Paracentrotus lividus* showed the common chemical-physical characteristics of AMPs (Schillaci et al. 2012). In addition, two cystein-rich AMPs, named centrocyns, have been identified in the green sea urchin *Strongylocentrotus droebachiensis* (Li et al. 2010).

Recently Haug et al. (2002) described an antimicrobial peptide in the supernatant of the egg homogenate of *Cucumaria frondosa* with a high antibacterial activity, especially against Gram-positive bacteria. In *Cucumaria echinata* a peptide of 20 amino acids was described, corresponding to the sequence

---

Citation: Schillaci et al.: Immune mediators of sea-cucumber Holothuria tubulosa (Echinodermata) as source of novel antimicrobial and anti-staphylococcal biofilm agents. AMB Express 2013 3:35. doi:10.1186/2191-0855-3-35.

from residue 332 of hemolytic lectin CEL-III, which exhibited strong antibacterial activity toward Gram-positive bacteria. The antibacterial activity of this peptide was correlated with its membrane-permeabilizing activity that perturbs bacterial cell membranes, leading to enhancement of their permeability (Hatakeyama et al. 2004).

Over the last years, the study on natural AMPs has attracted considerable interest as a new class of antimicrobial drugs with antibiofilm activity (Batoni et al. 2011; Herrmann et al. 2010).

The growth of bacteria as biofilms is responsible for biomaterial and device-associated infections (Donlan 2001). Furthermore it is an important virulence factor of pathogens, like*Staphylococcus aureus* and *Pseudomonas aeruginosa*, in the development of the chronic form of their infections in human beings, such as native valve endocarditis, burn or wound infections, cystic fibrosis-associated infections, otolaryngological infections, etc. (Post et al. 2007). Biofilms show multi-factorial resistance to conventional antibiotics. Currently no effective therapies that target microbial biofilms exist; the prompt removal of the contaminated device or surgical intervention remain the most effective means to treat biofilm-associated infections (Brady et al.2008). Therefore novel anti-biofilm agents, treatments and strategies are needed (Projan and Youngman 2002).

With the aim of searching for new antimicrobial agents as alternatives to conventional antibiotics, we focused on the coelomocytes of the echinoderm *Holothuria tubulosa* (sea-cucumber). The antimicrobial activity of a 5kDa peptide fraction from coelomocyte cytosol of *H. tubulosa* (5-HCC) was evaluated against a group of important reference strains of well-known human pathogens. Nine peptides were detected in the content of 5-HCC, and two of them were chemically synthesized and tested for their antimicrobial and anti-biofilm activities.

MATERIALS AND METHODS

Animals and Bleeding Procedure

H. tubulosa (sea cucumber) were collected from the Gulf of Palermo (Mongerbino 38°06'58" nord, 13°30'26" est) at a depth of 10 m near a prairie of *Posidonia oceanica* and maintained until their use in a refrigerated aquarium at 15°C and fed with commercial invertebrate food (Algamac 2000 Bio-Marine CA USA). The coelomic fluid (CF) was collected by cutting the anterior-dorsal part of the animals with a scalpel and draining the fluid into a plastic cup containing an isosmotic anticoagulant solution (0.5 M NaCl, 20 mM Tris–HCl, 30 mM EDTA; pH 7.4) (ISO-EDTA). Coelomocytes were separated from plasma by centrifugation at 900 g for 10 min at 4°C, washed twice in ISO-EDTA and suspended in the same medium at 5×10^6cells/ml in ISO–EDTA. Total cell counts were determined using an improved Neubauer haemocytometer and dead cells were evaluated by using the eosin-y exclusion test (0.5% in ISO-EDTA).

Preparation of Coelomocyte Lysate Supernatant (CLS)

Acid-soluble protein extracts were prepared according to a previously described method (Mercado et al. 2005) with some slight modifications. Coelomocytes (1×10^7) were suspended in a solution of 10% acetic acid in ISO without EDTA, sonicated (Sonifier Branson, model B-15 Danbury, CT USA) for 1 min at 0°C (1 pulse per second, 70% duty cycle) and centrifuged at 27,000 × g for 30 min at 4°C to remove any precipitate. After centrifugation, the supernatants were filtered (0.45 μM, Millex™, Millipore Corp.) and freeze dried to be later re-dissolved in H_2O.

A 5 kDa peptide fraction of the cytosol from coelomocytes (5-HCC) was obtained by using a filter with a membrane with a nominal size of 5 kDa (Ultrafree-0.5 PBCC Centrifugal filter Unit, Amicon Millipore, MA, USA).

Identification of 5kDa Peptide Fraction (5-HCC) Components

Capillary RP-HPLC/nESI-MS/MS Analysis

HPLC-grade water and CH_3CN were provided by Carlo Erba (Milan, Italy). Capillary RP-HPLC/nESI-MS/MS analysis of the in-gel enzymatic digests was performed using an Ultimate 3000 LC system combined with an autosampler and a 1:100 flow splitter (Dionex Corporation, Sunnyvale, CA, USA), coupled on-line with a linear ion trap nano-electrospray mass spectrometer (LTQ, Thermo Fischer Scientific, San Jose, CA). Ionization was performed with liquid junction using an uncoated capillary probe (20 mm i.d.; New Objective, Woburn, MA, USA). 2 mg of 5-HCC was freeze dried and re-dissolved in 0.1% aqueous formic acid (FA) (1 mg/ml). 10 μl of this solution was directly loaded onto a C18 μ-precolumn cartridge (0.3 μm × 5 mm, 100 Å, 5 μm, PepMap, Dionex) equilibrated with 0.1% aqueous formic acid (FA) at a flow rate of 20 μl/min for 4 min. Then, the solution was switched onto a reversed-phase C18 column (0.18 × 150 mm, 300 Å, 5 μm, BioBasic, ThermoFisher Scientific) and peptides were separated by elution at room temperature with a linear gradient of solvent B (ACN+0.5% FA) in A (H_2O+0.5% FA) from 20% to 55% in 40 min at a flow rate of 1.5 μl/min. The nano-ESI source operated under the following conditions: capillary temperature 220°C, source voltage 2 kV, capillary voltage 45 V. Repetitive mass spectra were acquired in positive ion mode in the *m/z* range 450–2000. Characterization of peptide ions was performed by the data-dependent method as follows: 1) full MS scan (mass-to-charge ratio 450–2000); 2) zoom scan of the three most intense ions (isolation width: 2 Da) and 3) MS/MS of these three ions (activation Q 0.250; activation time 30 ms; normalized collision energy 27 a.u.). Mass calibration was made using a standard mixture of caffeine (Mr 194.1 Da), MRFA peptide (Mr 423.6 Da) and Ultramark (Mr 1621 Da). Data acquisition was performed using the Excalibur v. 1·4 software (ThermoFisher Scientific).

Database Search and Peptide Identification via MASCOT Engine

MS/MS data were used to perform peptide and protein identifications by searching in the NCBI non-redundant protein sequence database of *Echinodermata* (60697 entries, updated June 2011) using the MOWSE algorithm as implemented in the MS search engine MASCOT server (Matrix Science, London, UK, version 2.3). MASCOT was set up as follows:

i) assuming no enzyme as cleavage specificity; ii) with a fragment ion mass tolerance of ±0.6 Da and a parent ion mass (PIM) tolerance of± 1.6 Da; iii) specifying oxidation of methionine, transformation of N-terminal glutamine and N-terminal glutamic acid residue in the pyroglutamic acid form as variable modifications.

In MASCOT searches, proteins scores (as MudPIT score) were derived from ions scores as a non-probabilistic basis for ranking protein hits, disregarding all individual peptides with scores below 20. Only proteins that met the following criteria were accepted as unambiguously identified: 1) MASCOT score >76 (probability-based MOWSE score: -10*log(P), where P is the probability that the observed match is a random event; score>76 indicate identity or extensive homology (P<0·05)).

Because all protein hits derived from the MASCOT search did not satisfy the above reported criteria, all the MS/MS spectra were also subjected to manual interpretation assisted by the PepNovo software (http://proteomics.ucsd.edu/Software/PepNovo.html) in order to obtain peptide sequence information.

Synthetic Peptides

Two of the nine identified peptides, peptides H1 and H2, were purchased from CASLO, Lyngby, Denmark. The peptides were synthesized using Fmoc solid phase technology and the peptide content and purity were determined by high performance liquid chromatography (HPLC) and mass spectrometry (MS) analysis.

Antimicrobial Assays

Microbial Strains

The strains *Staphylococcus aureus* ATCC 25923, *Staphylococcus aureus* ATCC 29213, *Staphylococcus aureus* ATCC 6538, *Enterococcus faecalis* ATCC 29212, *Pseudomonas aeruginosa* ATCC 15442 were used in the assay. The good biofilm producer *Staphylococcus epidermidis* ATCC 35984 was also included in the analysis.

Minimum Inhibitory Concentrations (MIC)

MICs of 5-HCC and of chemically synthesized peptides were determined by a micro-method as previously described (Schillaci et al. 2008; Schillaci et al. 2005). Tryptic Soy Broth (TSB, Sigma) containing 2% glucose was used as medium.

Biofilm Capability Evaluation (Safranin Method)

The staphylococcal strains and *P. aeruginosa* ATCC15442 were tested for their ability to form biofilms. Briefly, bacteria were grown in TSB containing 2% glucose overnight at 37°C in a shaking bath and then diluted 1:200 to a suspension with optical density (OD) of about 0.040 at 570 nm (Pitts et al. 2003). Polystyrene 24-well tissue culture plates were filled with 1 ml of diluted suspension and incubated for 24 hours at 37°C. The wells were then washed three times with 1 ml of sterile phosphate-buffered saline (PBS) and stained with 1 ml of safranin 0.1% v/v for 1 min. The excess stain was removed by placing the plates under running tap water. Plates were dried overnight in an inverted position at 37°C. Safranin-stained adherent bacteria in each well were re-dissolved to homogeneity in 1 ml of 30% v/v glacial acetic acid, and the OD was read at 492 nm. Each assay was performed in triplicate and repeated at least twice.

Interference with Biofilm Formation Assay

The procedure described above was used to evaluate the activity of synthesized H1 and H2 in the prevention of biofilm formation. Polystyrene 24-well tissue culture plates were filled with 1 ml of diluted bacterial suspension, obtained and diluted as previously described, and sub-MIC concentrations (6.2 and 3.1 mg/ml) of peptides 7 and 8 were directly added to the bacterial suspension at time zero and incubated at 37°C for 24 hours. The wells were then washed and stained with safranin as per the biofilm forming assay. By comparing the average optical density (O.D.) of the growth control wells with the sample wells, the following formula was used to calculate the percentages of inhibition for each concentration of the sample:

inhibition(%)=ODgrowthcontrol−ODsampleODgrowthcontrol×100

$$inhibition(\%) = \frac{OD\ growth\ control - OD\ sample}{OD\ growth\ control} \times 100 \quad (1)$$

Hemolytic Assay

The assay was performed as described by (Arizza et al. 2013). In brief, freshly collected rabbit erythrocytes (RE) with heparin, kindly provided by the "Zooprophylaxis Institute of Sicily" (Palermo, Italy), were washed (400 g for 10 min at 4°C) to remove the buffy coat, and the erythrocytes obtained were washed three times with phosphate-buffered saline (PBS: 6 mM KH_2PO_4; 30 mM Na_2HPO_4; 0.11 M NaCl; pH 7.4) and suspended in 10 ml PBS to obtain an 80×10^6 cells/ml suspension. Aliquots of 200 µl of RE suspension were mixed with 200 µl of H1 and H2 concentrations (1.5, 3.2, 6.2 and 50 mg/ml) prepared in PBS, two-fold sequentially diluted v/v with a PBS. After a 1 h incubation at 37°C, the reaction mixture was centrifuged at 800 g for 15 min at 4°C to remove debris and residual erythrocytes. The O.D. of hemoglobin release was measured spectrophotometrically at a 541 nm wavelength. Spontaneous hemoglobin release (0%) was estimated incubating the RE with PBS while the complete hemolysis (100%) was assessed incubating the erythrocytes in a solution of 0.1% Triton-X 100 in distilled water and the hemolysis percentage was calculated according to the following equation:

Hemolysis(%)=O.D.Hbreleaseinthereactionmixture−O.D.spontaneousHbreleaseO.D.completeHbrelease×100

$$Hemolysis(\%) = \frac{O.D.Hb\ release\ in\ the\ reaction\ mixture - O.D.spontaneous\ Hb\ release}{O.D.complete\ Hb\ release} \times 100 \quad (2)$$

Statistical Analysis

Each experiment was performed in triplicate. The values were the mean of three assays ± SD. Significance was determined by using the Student's t-test and differences were considered significant at $P<0.05$.

RESULTS

Conformational Analysis of 5kDa Peptide Fraction (5-HCC) Components

With the aim of detecting the presence of novel antimicrobial peptides and determining their sequence, the 5-HCC fraction was subjected to capillary RP-HPLC-HPLC/nESI-MS/MS. The MS/MS data obtained were used to query the protein database of *Echinodermata* and to perform *de novo* peptide sequencing.

With this approach, nine principal peptides with molecular masses ranging from 805.5 to 2215.7 Da were identified. Their sequences are reported in Table 1.

Table 1: Chemical-physical characteristics and amino acid sequences of peptides detected in 5-HCC content

Peptides	ESI-MS calculated MH+Da	Total net charge (pH 7)	Total hydrophobic ratio (%)	pI	Sequences, and hydrophobic amino acids on the same face (underlined)
#1	805.5	-1.8	28.57	4.47	FTDESHA
#2	919.2	-0.6	50.00	6.48	LSELLHHA
#3	1003.4	-0.6	44.44	6.48	GVLSELLHH
#4	1017.4	-0.6	55.56	6.48	VLSELLHHA
#5	1074.4	-0.6	50.00	6.48	GVLSELLHHA
#6	1367.6	-0.4	41.67	6.77	HLGHHALDDLLK
#7	1389.5	+0.9	41.67	7.56	H L GHH AL DH L LK
#8	1547.6	+0.9	42.86	7.56	ASH L GHH AL DH L LK
#9	2111.3	-3.1	27.78	4.23	EVKPNLTEKIEDLSQQMD

By using an antimicrobial peptide database (Wang and Wang 2004), we observed that only two of the nine peptides, Holothuroidin 1 (H1) and Holothuroidin 2 (H2), identified as peptide #7 and #8, whose molecular weights are respectively 1389.5 and 1547.6 Da, were likely to be novel AMPs, due to their chemical-physical characteristics and similarity with already defined AMPs. They had a sequence of 12 and 14 amino acids (a.a.), respectively, and the H2 differed only for the first two a. a. (A^1 and S^2), which were not present in H1. They are cationic peptides mainly enriched by residues such as lysine hystidine and arginine, with a net charge of +0.9, calculated by using the protein calculator v3.3. (http://www.scripps.edu/~cdputnam/protcalc.html) and a deduced pI of 7.5 calculated by using a web tool (http://isoelectric.ovh.org/) (Table 1).

Both H1 and H2 peptides showed a similarity ≥ 35% respectively with protonectins, a peptide present in the venom of the neotropical social wasp *Agelaia pallipes pallipes*, with a potent antimicrobial action against both Gram-positive and Gram-negative bacteria (Mendes et al.2004) (Table 2) and signiferins, a naturally occurring cationic peptide produced by an Australian frog, *Crinia signifera*, that showed a wide spectrum of activity against Gram-positive and -negative bacteria including *B. cereus, E. faecalis, L. lactis, L. innocua, M. luteus, S. aureus, S. epidermidis* and *S. uberis* (Maselli et al. 2004) (Table 3). The other identified peptides (#1-#6 and #9) in *H. tubulosa* are negatively charged and have very little likelihood of being AMPs.

Table 2: Sequence comparation of Holothuroidin 1 with other AMPs

Peptides	Sequence	Similarity (%)	Species
Holothurioidin 1	H L **G** HHA **L** DH **L LK**		*Holothuria tubulosa*
Protonectin	I L **G**+TI **L**+G **L LK**GL	42.85	*Agelaia pallipes pallipes* (Mendes et al. 2004)
Polybia-CP	I L **G**+TI **L**+G **L LK**SL	42.85	*Polybia paulista* (Souza et al. 2005)
Macropin 1	G F **G**+MA **L**+K **L LK**KVL	40.00	*Macropis fulvipes* (Slaninova et al.2012)
Signiferin 2.2	IIGH LIKTA **L** GF **L** G **L**+	37.50	*Crinia signifera* (Maselli et al.2004)
Temporin-1DRa	HFL **G** ++T **L** VN **L** A **KK**IL	37.50	*Rana aurora draytonii* (Conlon et al. 2007)

Table 3: Sequence comparison of Holothuroidin 2 with other AMPs

Peptides	Sequence	Similarity (%)	Species
Holothurioidin 2	ASH **L** GHHA **L** DH**L** LK		*Holothuria tubulosa*
Frenatin 3	GLMSV **L** G+HAVGNV **L** GG**L** FK S	38.09	*Litoria infrafrenata* (Raftery et al. 1996)
Signiferin 2.2	IIGH **L** IKTA **L** GF **L**GL+	37.5	*Crinia signifera* (Maselli et al.2004)
Signiferin 2.1	IIGH **L** IKTA **L** GM **L**GL+	37.5	*Crinia signifera* (Maselli et al.2004)
Temporin 1Ja	ILP **L** VG+N L **L** ND+**L** L+	37.5	*Rana japonica* (Isaacson et al.2002)

A Schiffer-Edmundson helical wheel projection (Schiffer and Edmundson 1967) of H1 and H2, performed using Helical Wheel Projection (rzlab.ucr.edu/script/wheel/wheel.cgi), indicated that the α-helixes of both peptides had considerable amphipathic character, with the polar and mainly cationic residues D^8, H^1, H^4, H^5, H^9 and K^{12} segregated on one polar face and the hydrophobic or non-polar residues A^6, L^2, L^7, L^{10}, L^{11} on the opposite apolar face for H1 (Figure 1A), and D^{10}, H^3, H^6, H^7, H^{11}, and K^{14} segregated on the polar face and the hydrophobic or non-polar residues A^1, A^8, G^5, L^4, L^9, L^{12}, L^{13}, S^2 placed on the opposite apolar face for H2 Figure 1B). Their mean hydrophobic moments and hydrophobic arc size range were respectively 5.28 and 193.6° for H1 and 5.08 and 173.3° for H2 (Figure 1).

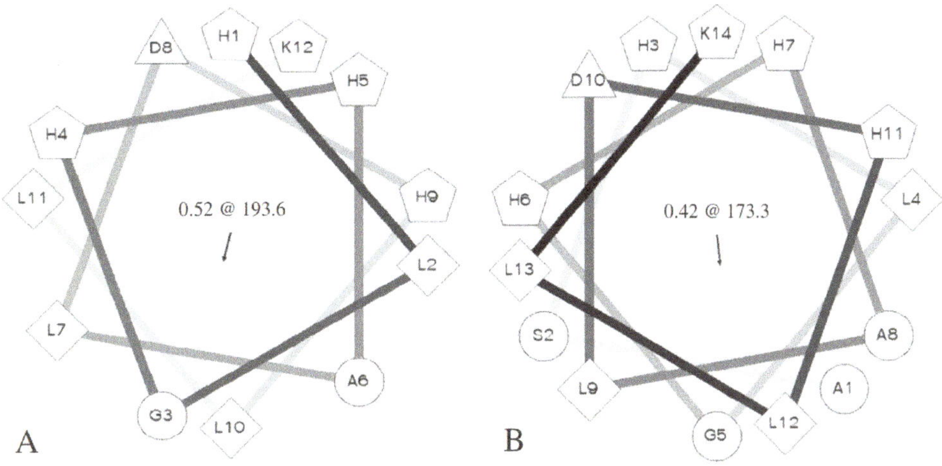

Figure 1: Edmundson wheel projection of (A) Holothuroidin 1 and (B) Holothuroidin 2 calculated by the Helical Wheel Projection web tool. Residues are schematically represented by geometric figures. Hydrophilic residues=circles, hydrophobic residues=diamonds, potentially negatively charged=triangles, and potentially positively charged=pentagons. In the centre mean relative hydrophobic moment and hydrophobic arc (@) are indicated.

The use of DeepView software (Swiss Institute of Bioinformatics - http://spdbv.vital-it.ch/) made it possible to resolve the secondary alpha-elical structure, which for both peptides was shown to be a periodic distribution of hydrophobic and hydrophilic residues, residues 4 – 10 for H1 and residues 2 – 12 for H2, in which the hydrophobic face is opposed to the hydrophilic one (Figure 2).

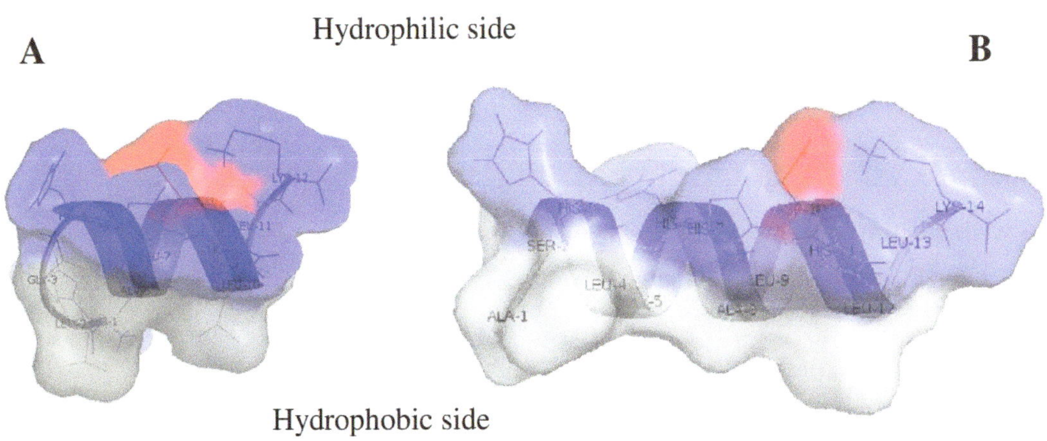

Figure 2: A ribbon representation of Holoturoidin 1 (A) and Holoturoidin 2 (B). The amphipathic nature of the peptide is shown in this representation with the hydrophilic side above and the hydrophobic side below the polypeptide backbone. The potential surface is superimposed. Color code: acidic residues in red, basic residues in blue, and hydrophobic residues in white.

Antimicrobial Activity of 5-HCC

The antimicrobial activity of the 5kDa peptide fraction from coelomocyte cytosol (5-HCC) of *H. tubulosa* expressed as minimum inhibitory concentrations (MICs) against microbial reference strains are listed in Table 4. The 5-HCC resulted active against all tested reference microbial strains with MIC values ranging from 125 to 500 mg/ml.

Table 4: Antibacterial activity of 5HCC assayed by MIC

Bacterial strains	**MIC**
	Values *in vitro*(mg/ml)
S. aureus ATCC 29213	**500**
S. aureus ATCC 25923	**250**
E. faecalis ATCC 29212	**500**
P. aeruginosa ATCC 15442	**500**

Antimicrobial Activity of Synthetic Peptides

The synthetic peptides H1 and H2 were screened for their antimicrobial activity against the same group used for the evaluation of biological activity of 5-HCC. The H2 resulted active at 12.5 mg/ml against planktonic Gram-positive (staphylococcal reference strains and *E. faecalis* ATCC 29212) and against *P. aeruginosa* ATCC 15442. A similar activity at the same concentration was observed for H1 but it resulted not active at the maximum tested concentration against *S. aureus* ATCC 29213, and *E. faecalis* ATCC 29212. See Table 5.

Table 5: Antibacterial activity of synthetic Holothuroidin 1 and Holothuroidin 2 assayed by MIC

Bacterial strains	**MIC (mg/ml)**	
	Holothuroidin 1	**Holothuroidin 2**
S. aureus ATCC 29213	>12.5	12.5
S. aureus ATCC 25923	12.5	12.5
S. aureus ATCC 6538	12.5	12.5
S. epidermidis ATCC 35984	12.5	12.5
E. faecalis ATCC 29212	>12.5	12.5
P. aeruginosa ATCC 15442	12.5	12.5

Interference with Biofilm Formation

The impact of H1 and H2 on the biofilm formation of two staphylococcal reference strains, *S. aureus* ATCC 25923, and *S. epidermidis* ATCC 35984, were evaluated at concentrations of 3.2 and 1.5 mg/ml (well below the MIC values obtained against the planktonic forms) by staining with safranin adherent bacteria. The H1 inhibited biofilm formation was 51.8% for *S. aureus*ATCC 25923 and 68.5% for *S. epidermidis* ATCC 35984 at 3.2 mg/ml; the inhibition was 37.9% for *S. aureus* and 58.2% for *S. epidermidis* at the lower concentration of 1.5 mg/ml. The same concentrations were used to analyze the effect of H2 on biofilm formation. Higher inhibitions of 57.7% and 73.8% at 3.2 mg/ml and 40.5% and 59.7% at 1.5 mg/ml were observed on *S. aureus*and *S. epidermidis* biofilms respectively.

H1 and H2 also interfere with the biofilm formation of *P. aeruginosa* ATCC 15442 at the sub-MIC concentrations of 6.2 and 3.1 mg/ml. Inhibition percentages of 69.9% and 62.7% for H1 and 64.3% and 43.8% for H2 were observed (Figure 3).

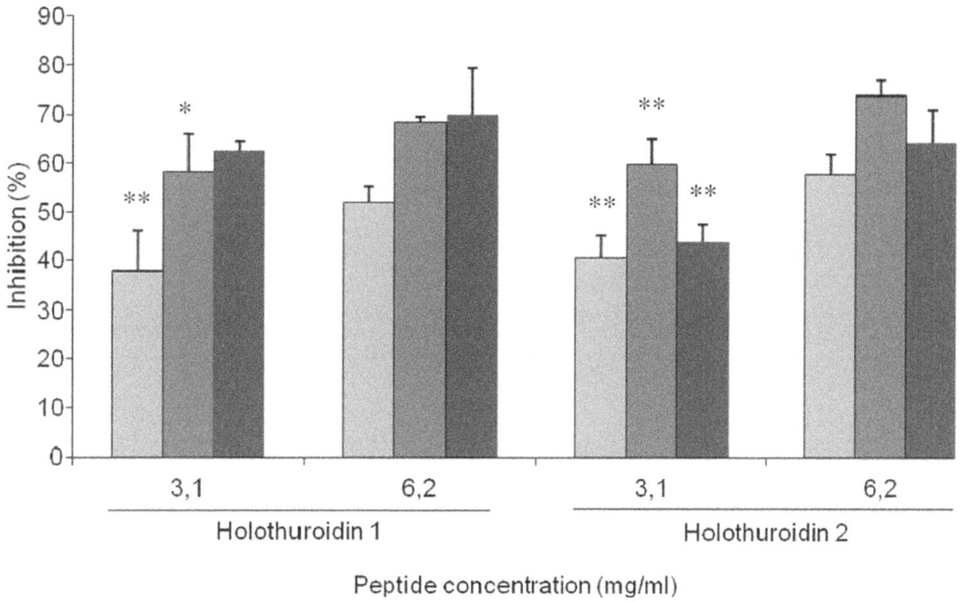

Figure 3: Inhibition activity of Holothuroidin 1 and 2 on *S. aureus* ATCC 25923, *S. epidermidis* ATCC 35984 (light gray square symbol) and *P. aeruginosa* ATCC 15442 (dark gray square symbol) biofilms. The data are expressed as the mean of three different experiments ± standard deviation. * = $p < 0.05$; ** = $p < 0.01$.

Hemolytic Assay

The hemolytic activity of antimicrobial peptides against mammalian erythrocytes is often used as to measure their cytotoxicity. The hemolytic experiment performed to evaluate the toxicity of the peptides H1 and H2 did not show a measurable toxic effect against RBC (~ 1%) at MIC concentrations (Figure 4). A slight hemolytic activity, about 12.5%, was evident only at the highest concentration (50 mg/ml) (Figure 4).

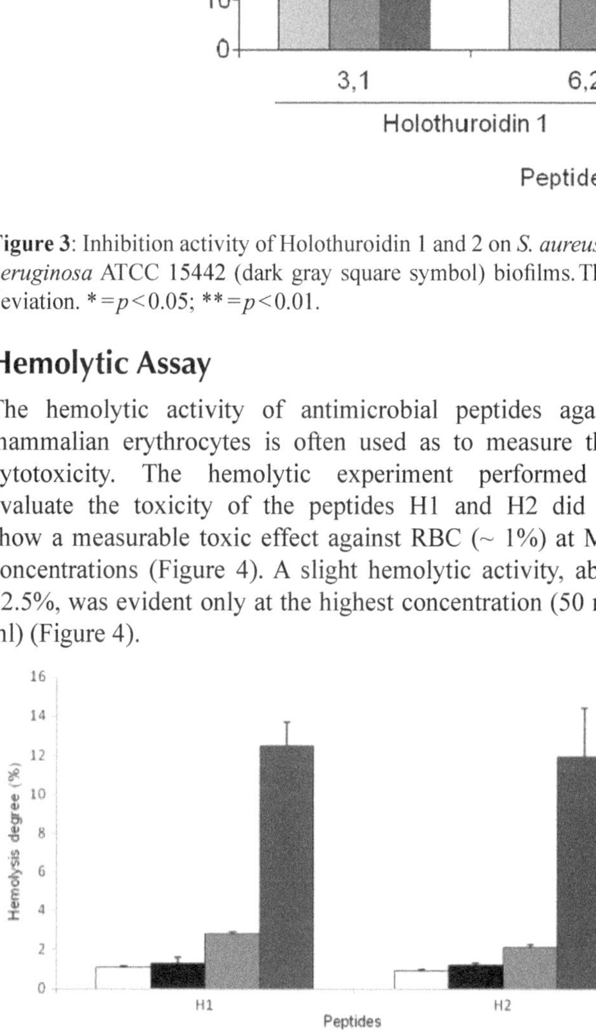

Figure 4: Hemolytic activity of H1 and H2 peptides from *Holothuria tubulosa*hemocytes against rabbit blood cells at different concentrations: white square symbol = 1.5 mg/ml; Black square symbol = 3.2; gray square symbol = 6.2; dark gray square symbol = 50 mg/ml. Data are the mean value of three separate experiments and expressed as percentage of hemolysis ± SD.

DISCUSSION

Antimicrobial peptides (AMPs) represent the first line of defense against pathogens in several organisms (Hancock and Lehrer 1998). The majority of AMPs, constituted by a short stretch of amino acids residues (10–100 maximum), present the following common features: broad antimicrobial spectrum, small sizes, a net positive charge due to an excess number of cationic residues, an over-representation of one or two amino acids (Hughes and Fenical 2010).

Recently, a variety of peptides with antimicrobial properties have been isolated from echinoderms (Andersson et al. 1983; Bryan et al. 1994; Ridzwan et al. 1995). Antimicrobial activity has also been reported in gonads of the asteroid *Marthasterias glacialis* (Stabili and Pagliara 1994). We have already demonstrated that echinoderms such as *Paracentrotus lividus*contain, in the low molecular weight fraction (<5kDa) of acid precipitate from their coelomocytes, peptides with antimicrobial activity against staphylococci antibiofilm (Schillaci et al. 2010; Schillaci et al. 2012).

In this study, the fraction of low molecular weight peptide (<5kDa) from coelomocytes of the *H. tubulosa* (5-HCC) showed broad antimicrobial activity against many Gram-positive and Gram-negative pathogens tested. The use of capillary RP-HPLC/nESI-MS/MS allowed the characterization in the amino acid sequence of nine principal peptides in the 5-HCC content. Only two peptides, #7 and #8, called respectively Holothuroidin 1 and Holothuroidin 2, of the total nine peptides found in 5-HCC, are likely to be AMPs (Wang and Wang 2004).

Structurally they share the characteristics of many antimicrobial peptides: they are cationic peptides and possess a significant proportion (~ 30%) of hydrophobic residues (Hancock and Lehrer 1998; Zasloff 2002). They may form, as secondary structures, an α-helical, which is the most common structure of AMPs in nature and which is essential, and often sufficient, for antimicrobial activity (Mor and Nicolas 1994; Skerlavaj et al. 1996; Storici et al. 1994; Tossi et al. 1994). The helixes of two peptides present an asymmetrical arrangement of the hydrophobic residues, specifically: four hydrophobic residues on the same face and a total hydrophobic ratio of 41.67% for H1, and for H2, a hydrophobic face opposite to the hydrophilic one and a total hydrophobic ratio of 42.86%. This characteristic conformation of the peptides, in the course of microbial membrane invasion, allows the non-polar face of their α-helical structure to interact with the membrane lipid core while at the same time permitting its hydrophilic face to engage in electrostatic interactions with the membrane lipid headgroup region (Phoenix et al. 2002).

Both peptides are not long enough (when in an alpha-helical conformation) to traverse the lipid bilayer of a bacterial cell. Indeed, the `barrel stave› mechanism requires a minimum of 20 residues (Bechinger 1997; Epand et al. 1995; Shai 1995), so probably some aggregation of Holothuroidin molecules on the membrane surface may be needed before the bacterial cell wall can be breached, e.g. via the so-called "carpet" mechanisms for the disruption of the bacterial membrane (Epand et al. 1995; Shai 1995).

In our study we observed that synthetic H1 and H2 did not show, in PBS, hemolytic activity, probably because the high ionic strength of PBS allows negatively charged sialic acid present on the erythrocyte membrane to neutralize the peptides. On the basis of this evidence, they could be classified as peptide antibiotics (Saberwal and Nagaraj 1995). The two peptides showed a modest broad-spectrum activity against planktonic strains (MIC equal to 12.5 mg/ml). However, they showed a good inhibition of biofilm formation *in vitro* at lower MIC concentrations when evaluated against the planktonic forms on two staphylococcal reference strains and *P. aeruginosa* ATCC 15442. Such additional antimicrobial effect could be due to different mechanisms of action during the biofilm formation: a) an interference with the initial adhesion of microbial cells to the surface could be due to modifications of the microbial membranes, b) the killing of the early bacterial colonizers, or c) the inhibition of Quorum Sensing (QS), i.e. the intercellular communication system involved in biofilm formation (Batoni et al. 2011).

Whatever the supposed mechanism of action, the prevention of biofilm formation – rather than its elimination – is an interesting strategy to contrast the growth, as a sessile community, of many pathogens.

The extensive clinical use of classical antibiotics has led to the growing emergence of many medically relevant resistant strains of bacteria biofilm (Patel 2005). Therefore, the development of a new class of antibiotics has become critical. The cationic antimicrobial peptides could represent such a new class of antibiotics (Andreu and Rivas 1998; Hancock 1997; Sitaram and Nagaraj 2002). The development of resistance to membrane active peptides whose sole target is the cytoplasmic membrane is not expected because this would require substantial changes in the lipid composition of cell membranes of microorganisms (Hancock 2001).

Novel AMPs derived from coelomocytes of *H. tubulosa* are attractive candidates for developing useful antimicrobial agents against surface adherent staphylococci and *P. aeruginosa* of clinical interest.

The anti-adhesion properties of the peptides H1 and H2 could be useful to contrast staphylococcal and *P. aeruginosa* biofilms involved in food spoilage and biofilm formation of food transmitted pathogens. Moreover, the tested peptides are a good starting point to design new synthetic derivatives with modified chemical-physical properties, with the aim to improve their antimicrobial activity against pathogens and their pharmaceutical potential (Brogden and Brogden 2011; Huang et al. 2010).

ACKNOWLEDGEMENTS

Financial support from Fondi di Ateneo 2007, Università degli Studi di Palermo (Italy), is gratefully acknowledged.

REFERENCES

1. Andersson L, Lidgren G, Bohlin L, Magni L, Ogren S, Afzelius L: Studies of Swedish marine organisms. I. Screening of biological activity. *Acta Pharm Suec* 1983,20(6):401–414.
2. Andreu D, Rivas L: Animal antimicrobial peptides: an overview. *Biopolymers*1998,47(6):415–433. 10.1002/(SICI)1097-0282(1998)47:6<415::AID-BIP2>3.0.CO;2-D
3. Arizza V, Vazzana M, Schillaci D, Russo D, Giaramita FT, Parrinello N: Gender differences in the immune system activities of sea urchin *Paracentrotus lividus*. Comparative biochemistry and physiology Part A. *Molecular & integrative physiology*2013,164(3):447–455. 10.1016/j.cbpa.2012.11.021
4. Batoni G, Maisetta G, Brancatisano FL, Esin S, Campa M: Use of antimicrobial peptides against microbial biofilms: advantages and limits. *Curr Med Chem* 2011,18(2):256–279. 10.2174/092986711794088399
5. Bechinger B: Structure and functions of channel-forming peptides: magainins, cecropins, melittin and alamethicin. *J Membrane Biol* 1997,156(3):197–211. 10.1007/s002329900201
6. Brady RA, Leid JG, Calhoun JH, Costerton JW, Shirtliff ME: Osteomyelitis and the role of biofilms in chronic infection. *FEMS Immunol Med Microbiol*2008,52(1):13–22. 10.1111/j.1574-695X.2007.00357.x
7. Brogden NK, Brogden KA: Will new generations of modified antimicrobial peptides improve their potential as pharmaceuticals? *Int J Antimicrob Ag* 2011,38(3):217–225. 10.1016/j.ijantimicag.2011.05.004
8. Bryan PJ, McClintock JB, Watts SA, Marion KR, Hopkins TS: Antimicrobial activity of ethanolic extracts of echinoderms from the northern Gulf of Mexico. In*Echinoderms Through Time*. Edited by: David B, Guille A, Feral J-P, Roux M. Balkema, Rotterdam; 1994:17–23.
9. Conlon JM, Coquet L, Leprince J, Jouenne T, Vaudry H, Kolodziejek J, Nowotny N, Bevier CR, Moler PE: Peptidomic analysis of skin secretions from *Rana heckscheri* and *Rana okaloosae* provides insight into phylogenetic relationships among frogs of the Aquarana species group. *Regul Pept* 2007,138(2–3):87–93. 10.1016/j.regpep.2006.08.007
10. Donlan RM: Biofilms and device-associated infections. *Emerg Infect Dis* 2001,7(2):277 281. 10.3201/eid0702.700277Central
11. Epand RM, Shai YC, Segrest JP, Anantharamaiah GM: Mechanisms for the modulation of membrane bilayer properties by amphipathic helical peptides. *Biopolymers*1995,37(5):319–338. 10.1002/bip.360370504

12. Garcia-Olmedo F, Molina A, Alamillo JM, Rodriguez-Palenzuela P: Plant defense peptides. *Biopolymers* 1998,47(6):479–491. 10.1002/(Sici)1097-0282(1998)47:6<479::Aid-Bip6>3.0.Co;2-K

13. Hancock REW: Peptide antibiotics. *Lancet* 1997,349(9049):418–422. 10.1016/S0140-6736(97)80051-7

14. Hancock RE: Cationic peptides: effectors in innate immunity and novel antimicrobials. *Lancet Infect Dis* 2001,1(3):156–164. 10.1016/S1473-3099(01)00092-5

15. Hancock RE, Lehrer R: Cationic peptides: a new source of antibiotics. *Trends Biotechnol* 1998,16(2):82–88. 10.1016/S0167-7799(97)01156-6

16. Hatakeyama T, Suenaga T, Eto S, Niidome T, Aoyagi H: Antibacterial activity of peptides derived from the C-terminal region of a hemolytic lectin, CEL-III, from the marine invertebrate *Cucumaria echinata* . *J Biochem* 2004,135(1):65–70. 10.1093/jb/mvh007

17. Haug T, Kjuul AK, Styrvold OB, Sandsdalen E, Olsen OM, Stensvag K: Antibacterial activity in *Strongylocentrotus droebachiensis* (*Echinoidea*), *Cucumaria frondosa* (*Holothuroidea*), and *Asterias rubens* (*Asteroidea*). *J Invertebr Pathol* 2002,81(2):94–102. 10.1016/S0022-2011(02)00153-2

18. Herrmann G, Yang L, Wu H, Song Z, Wang H, Hoiby N, Ulrich M, Molin S, Riethmuller J, Doring G: Colistin-tobramycin combinations are superior to monotherapy concerning the killing of biofilm *Pseudomonas aeruginosa* . *J Infect Dis* 2010,202(10):1585–1592. 10.1086/656788

19. Huang YB, Huang JF, Chen YX: Alpha-helical cationic antimicrobial peptides: relationships of structure and function. *Protein Cell* 2010,1(2):143–152. 10.1007/s13238-010-0004-3

20. Hughes CC, Fenical W: Antibacterials from the sea. *Chemistry (Weinheim an der Bergstrasse, Germany)* 2010,16(42):12512–12525. 10.1002/chem.201001279

21. Isaacson T, Soto A, Iwamuro S, Knoop FC, Conlon JM: Antimicrobial peptides with atypical structural features from the skin of the Japanese brown frog *Rana japonica* . *Peptides* 2002,23(3):419–25. 10.1016/S0196-9781(01)00634-9

22. Leippe M: Antimicrobial and cytolytic polypeptides of amoeboid protozoa--effector molecules of primitive phagocytes. *Dev Comp Immunol* 1999,23(4–5):267–279. 10.1016/S0145-305X(99)00010-5

23. Li C, Haug T, Moe MK, Styrvold OB, Stensvag K: Centrocins: isolation and characterization of novel dimeric antimicrobial peptides from the green sea urchin, *Strongylocentrotus droebachiensis* . *Dev Comp Immunol* 2010,34(9):959–968. 10.1016/j.dci.2010.04.004

24. Maltseva AL, Aleshina GM, Kokryakov VN, Krasnodembsky EG, Ovchinnikova TV: New antimicrobial peptides from coelomocytes of sea star *Asterias rubens* . *Biologiya* 2004, 4: 8.

25. Maselli VM, Brinkworth CS, Bowie JH, Tyler MJ: Host-defence skin peptides of the Australian Common Froglet *Crinia signifera* : sequence determination using positive and negative ion electrospray mass spectra. *Rapid Commun Mass Sp* 2004,18(18):2155–2161. 10.1002/Rcm.1602

26. Mendes MA, de Souza BM, Marques MR, Palma MS: Structural and biological characterization of two novel peptides from the venom of the neotropical social wasp *Agelaia pallipes pallipes* . *Toxicon* 2004,44(1):67–74. 10.1016/j.toxicon.2004.04.009

27. Mercado L, Schmitt P, Marshall SH, Arenas G: Gill tissues of the mussel *Mytilus edulis chilensis* : A new source for antimicrobial peptides. *Electron J Biotechn* 2005,8(3):284–290. 10.2225/vol8-issue3-fulltext-2

28. Mor A, Nicolas P: The Nh2-terminal alpha-helical domain 1–18 of Dermaseptin is responsible for antimicrobial activity. *J Biol Chem* 1994,269(3):1934–1939.

29. Pag U, Sahl HG: Lanthionine-containing bacterial peptides. In *Peptide Antibiotics: Discovery, Mode of Actions, and Applications*. Edited by: Dutton CJ, Haxell MA, McArthur HAI, Wax RG. Dekker, M, New York; 2002:47–80.

30. Patel R: Biofilms and antimicrobial resistance. *Clin Orthop Relat Res* 2005, (437):41–47.

31. Phoenix DA, Harris F, Daman OA, Wallace J: The prediction of amphiphilic alpha-helices. *Curr Protein Pept Sc* 2002,3(2):201–221. 10.2174/1389203024605368

32. Pitts B, Hamilton MA, Zelver N, Stewart PS: A microtiter-plate screening method for biofilm disinfection and removal. *J Microbiol Methods* 2003,54(2):269–276. 10.1016/S0167-7012(03)00034-4

33. Post JC, Hiller NL, Nistico L, Stoodley P, Ehrlich GD: The role of biofilms in otolaryngologic infections: update 2007. *Curr Opin Otolaryngol Head Neck Surg* 2007,15(5):347–351. 10.1097/MOO.0b013e3282b97327

34. Projan SJ, Youngman PJ: Antimicrobials: new solutions badly needed. *Curr Opin Microbiol* 2002,5(5):463–465. 10.1016/S1369-5274(02)00364-8

35. Raftery MJ, Waugh RJ, Bowie JH, Wallace JC, Tyler MJ: The structures of the frenatin **peptides from the skin secretion of the giant tree frog** *Litoria infrafrenata* . *J Pept Sci* 1996,2(2):117–24.

36. Ridzwan BH, Kaswandi MA, Azman Y, Fuad M: Screening for antibacterial agents in three species of sea cucumbers from coastal areas of Sabah. *Gen Pharmacol* 1995,26(7):1539–1543. 10.1016/0306-3623(95)00041-0

37. Saberwal G, Nagaraj R: Cell-Lytic and antibacterial peptides that act by perturbing the barrier function of membranes - facets of their conformational features, structure-function correlations and membrane-perturbing abilities. *Bba-Biomembranes* 1995,1235(1):159–159.

38. Schiffer M, Edmundson AB: Use of helical wheels to represent the structures of proteins and to identify segments with helical potential. *Biophys J* 1967,7(2):121–135. 10.1016/S0006-3495(67)86579-2

39. Schillaci D, Petruso S, Sciortino V: 3,4,5,3 ‹,5 ‹-Pentabromo-2-(2 ‹-hydroxybenzoyl)pyrrole: a potential lead compound as anti-Gram-positive and anti-biofilm agent. *Int J Antimicrob Ag* 2005,25(4):338–340. 10.1016/j.ijantimicag.2004.11.014

40. Schillaci D, Arizza V, Dayton T, Camarda L, Di Stefano V: In vitro anti-biofilm activity of*Boswellia* spp. oleogum resin essential oils. *Lett Appl Microbiol* 2008,47(5):433–438. 10.1111/j.1472-765X.2008.02469.x

41. Schillaci D, Arizza V, Parrinello N, Di Stefano V, Fanara S, Muccilli V, Cunsolo V, Haagensen JJ, Molin S: Antimicrobial and antistaphylococcal biofilm activity from the sea urchin*Paracentrotus lividus* . *J Appl Microbiol* 2010,108(1):17–24. 10.1111/j.1365-2672.2009.04394.x

42. Schillaci D, Vitale M, Cusimano MG, Arizza V: Fragments of beta-thymosin from the sea urchin *Paracentrotus lividus* as potential antimicrobial peptides against staphylococcal biofilms. *Ann N Y Acad Sci* 2012, 1270: 79–85. 10.1111/j.1749-6632.2012.06652.x

43. Shai Y: Molecular recognition between membrane-spanning polypeptides. *Trends Biochem Sci* 1995,20(11):460–464. 10.1016/S0968-0004(00)89101-X

44. Sitaram N, Nagaraj R: The therapeutic potential of host-defense antimicrobial peptides. *Curr Drug Targets*2002,3(3):259–267. 10.2174/1389450023347786

45. Skerlavaj B, Gennaro R, Bagella L, Merluzzi L, Risso A, Zanetti M: Biological characterization of two novel cathelicidin-derived peptides and identification of structural requirements for their antimicrobial and cell lytic activities. *J Biol Chem*1996,271(45):28375–28381. 10.1074/jbc.271.45.28375

46. Slaninová J, Mlsová V, Kroupová H, Alán L, Tůmová T, Monincová L, Borovičková L, Fučík V, Ceřovský V: Toxicity study of antimicrobial peptides from wild bee venom and their analogs toward mammalian normal and cancer cells. *Peptides* 2012,33(1):18–26. 10.1016/j.peptides.2011.11.002

47. Souza BM, Mendes MA, Santos LD, Marques MR, César LM, Almeida RN, Pagnocca FC, Konno K, Palma MS: Structural and functional characterization of two novel peptide toxins isolated from the venom of the social wasp Polybia paulista. *Peptides*2005,26(11):2157–64. 10.1016/j.peptides.2005.04.026

48. Stabili L, Pagliara P: Antibacterial protection in *Marthasterias glacialis* eggs: characterization of lysozyme-like activity. *Comp Biochem Physiol B Biochem Mol Biol*1994,109(4):709–713. 10.1016/0305-0491(94)90134-1

49. Storici P, Scocchi M, Tossi A, Gennaro R, Zanetti M: Chemical synthesis and biological-activity of a novel antibacterial peptide deduced from a pig myeloid cDNA. *FEBS Lett* 1994,337(3):303–307. 10.1016/0014-5793(94)80214-9

50. Tincu JA, Taylor SW: Antimicrobial peptides from marine invertebrates. *Antimicrob Agents Chemother* 2004,48(10):3645–3654. 10.1128/AAC.48.10.3645-3654.2004

51. Tossi A, Scocchi M, Skerlavaj B, Gennaro R: Identification and characterization of a primary antibacterial domain in Cap18, a lipopolysaccharide-binding protein from rabbit leukocytes. *FEBS Lett* 1994,339(1–2):108–112. 10.1016/0014-5793(94)80395-1

52. Wang Z, Wang G: APD: the antimicrobial peptide database. *Nucleic Acids Res*2004,32(Database issue):D590-D592.

53. Zasloff M: Antimicrobial peptides of multicellular organisms. *Nature*2002,415(6870):389–395. 10.1038/415389a

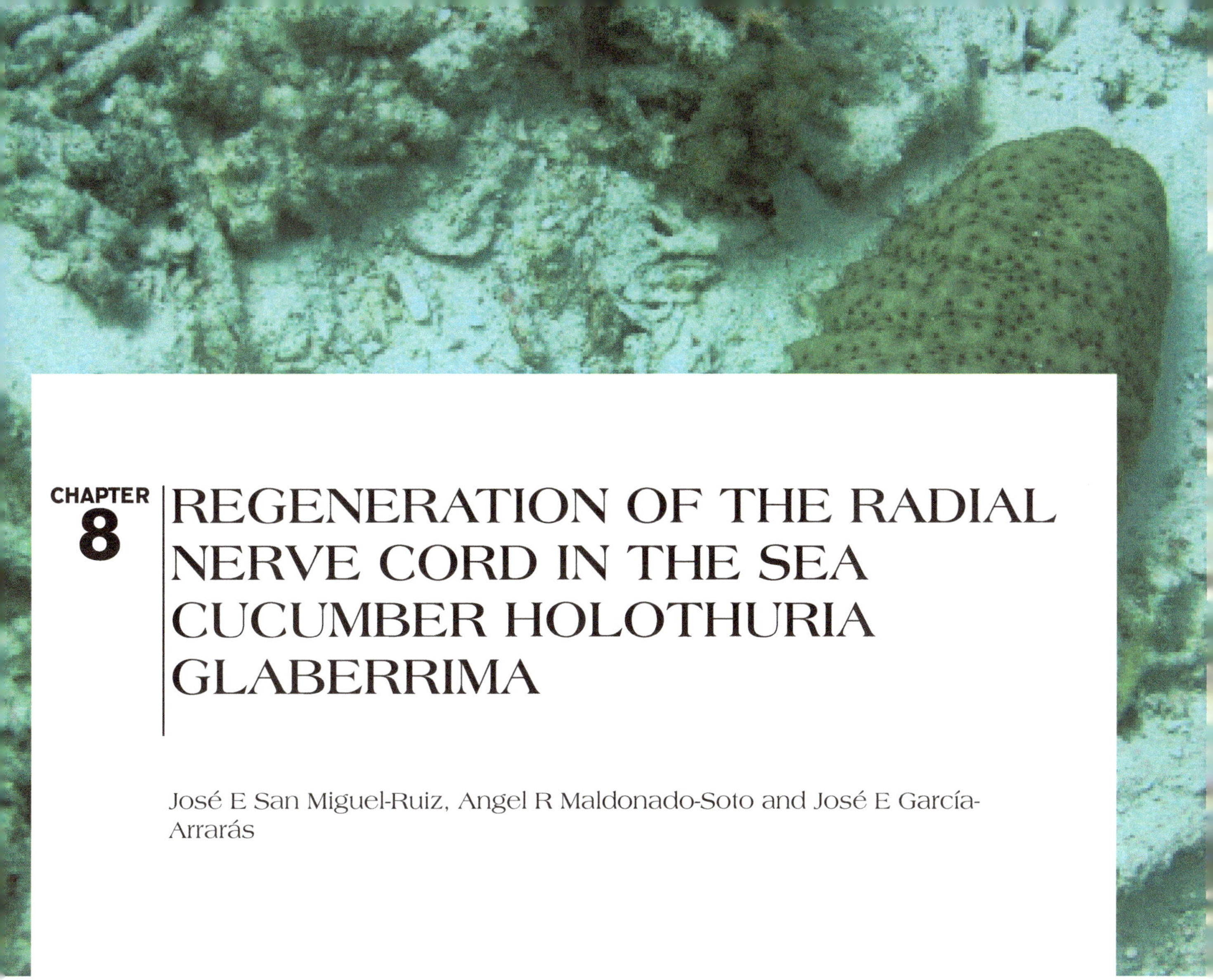

CHAPTER 8 REGENERATION OF THE RADIAL NERVE CORD IN THE SEA CUCUMBER HOLOTHURIA GLABERRIMA

José E San Miguel-Ruiz, Angel R Maldonado-Soto and José E García-Arrarás

BACKGROUND

Post-traumatic nervous regeneration is a rarely seen phenomenon in the central nervous system (CNS) of adult higher vertebrate animals. Organisms with CNS injuries can suffer the loss of motor activity and sensory perception and, as a result, develop limitations or disabilities. Some animal species, on the other hand, are capable of functional regeneration after an injury to their CNS. In these cases, severed neuronal fibers can find their targets through axonal path finding and neurogenesis can be activated. Efforts in different fields have generated a large amount of information as to what directs or limits the potential of neuronal cells to achieve proper functional regeneration. However, continued research and new approaches are still needed before we fully understand what governs the dynamics of injured nervous tissues and its regeneration.

One of the approaches that have been used to study nerve regeneration is the comparative analysis to determine the differences between those species that readily regenerate their CNS and those where regeneration is more limited [1–4]. The regeneration capabilities of the CNS have been said to decrease as we move higher in the phylogenetic tree [4, 5]. Thus, relatively simple organisms such as Hydra and Planaria, can regenerate their neuronal network after amputation of a body part [6]. Similarly, studies with mollusks and annelids (including leeches) have shown that they are capable of regeneration after a crush-type injury to one of the CNS tracts [5]. However, the assumption that organisms in the lower phylogenetic scale are more able to regenerate their nervous system is not always true. This is highlighted by the fact that crustaceans, as well as two of the most studied invertebrate model animals, *C. elegans* and*Drosophila*, show limited CNS regeneration [5, 7, 8].

Among deuterosotomes, most of the studies on CNS regeneration have been done on the spinal cord of vertebrates. Here again, organisms can be grouped into good regenerators and poor regenerators. Among the former are lampreys, fishes, and urodele amphibians, the latter include most amphibians, reptiles, birds and mammals [4, 9–11]. Some fish species have served as models to study axonal regeneration in the CNS of vertebrates, as these animals are capable of functional regeneration after a CNS injury [2, 11, 12]. Urodeles are among the most gifted to regenerate their CNS; following an injury they can regenerate a spinal cord with some functional recovery at any recovery time during their life cycle [3, 13, 14]. Other amphibians are also capable of regrowing spinal nerves after amputation of the tail, but only in an immature state [15, 16]. Even though these models have provided detailed descriptions of the problems and limitations that neuronal

Citation: José E San Miguel-Ruiz, Angel R Maldonado-Soto and José E García-Arrarás, Regeneration of the radial nerve cord in the sea cucumber Holothuria glaberrima, doi:10.1186/1471-213X-9-3.

tissues must overcome before functional regeneration, still a great deal of basic knowledge is missing to understand how this phenomenon could be regulated and neuronal regeneration enhanced.

CNS regeneration has been little studied in lower deuterostomes. In ascidians, the regeneration of the neural complex has been shown to occur after complete ablation [17, 18], but no study of nerve transection or partial injury is available. There is no information on CNS regeneration in hemichordates or cephalochordates. In echinoderms, organisms with an amazing regenerative capacity, a few publications provide some information on CNS regeneration, mainly describing what occurs following arm amputation in asteroids and crinoids [19–22]. However, these experiments provide a brief description of nerve regeneration that is part of an overall description of how the new arm forms or focus on some specific processes such as cell proliferation during regeneration. Only recently has a study on CNS regeneration in echinoderms been done, but it is limited to transmission electron microscopy [23]. Our work confirms some of their findings and greatly expands the analysis of CNS regeneration in echinoderms.

In this study we use the sea cucumber *Holothuria glaberrima* to study nervous system regeneration. *H. glaberrima*, is an echinoderm of the class Holothuroidea, and as such, is an invertebrate deuterostome, placing this organism in close evolutionary relationship with vertebrates. Moreover, these organisms have been known for a long time to possess tremendous capabilities to functionally regenerate most of their organs and tissues in a very short time [24,25]. We have previously used this model system to study intestinal regeneration [26], the regeneration of the enteric nervous system [27] and the wound healing of the body wall [28].

The holothurian CNS main component is an anterior nerve ring, at the base of the tentacles, from which 5 radial nerve cords (RNCs) exit and travel within the bodywall, spanning the length of the animal and ending blindly at the posterior end [24, 29, 30]. RNCs are subdivided into two components, the ectoneural (EN) or hyponeural (HN) band which are separated by a band of connective tissue (see Fig. 1A). In general terms the HN component is thought to be primarily motor, while the EN has both sensory and motor functions. RNCs are ganglionated nerves, meaning that they have neurons throughout their length. The neurons can be found in the periphery of both HN and EN components with the central portion, or neuropile, made up of nerve fibers. Innervation of organs occurs by peripheral nerves that emerge from the RNCs. We have developed a technique that allows for an accurate injury to be made to the RNC and are now able to study the regenerative potential of the holothurian CNS. We report here that following transection, re-construction of the nerve stumps occurs by 12 dpi, as a result of axonal growth; afterwards the RNC goes through a remodeling phase to achieve its original morphology. Other events associated with tissue development and regeneration were also observed, including cell division, migration and apoptosis. With this work we set the foundations for further studies on nervous regeneration by showing that this animal model carries out the two most important processes for CNS regeneration, explicitly neurogenesis and axonal pathfinding. More importantly, these occur as a natural response to the injury, allowing for future characterization of a permissive environment for possible therapy-related studies.

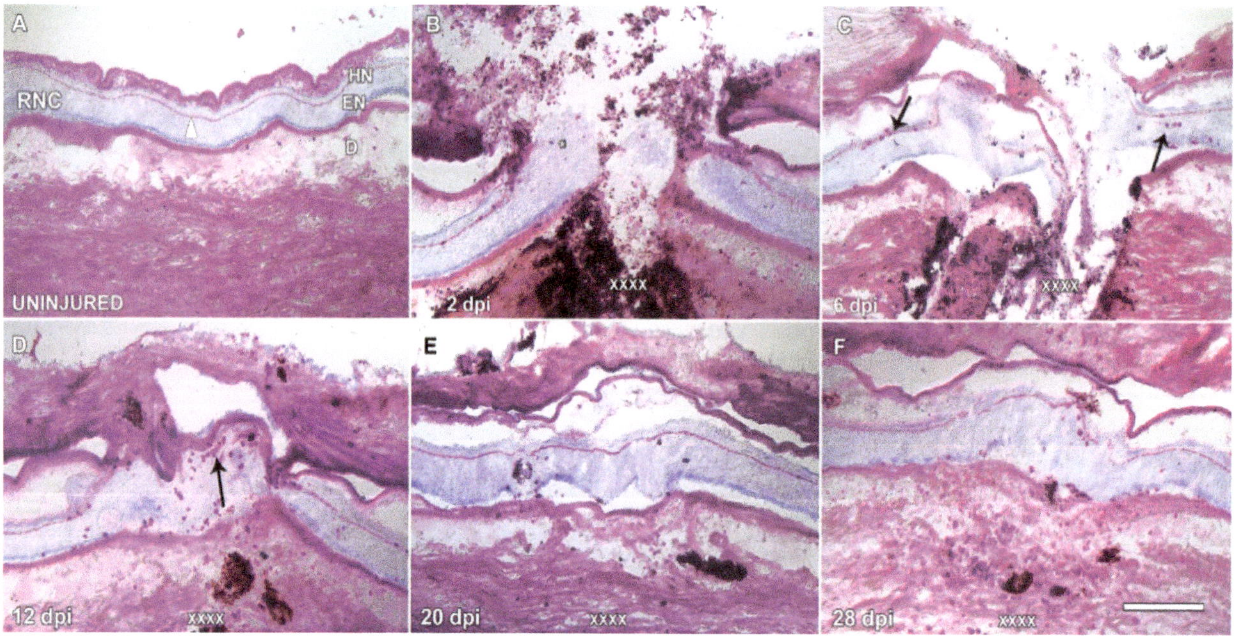

Figure 1: Uninjured and regenerating stages of the radial nerve cord (RNC) of *H. glaberrima*. Longitudinal tissue sections of (A) uninjured and regenerating RNC at (B) 2, (C) 6, (D) 12, (E) 20 and (F) 28 days post injury (dpi) were stained with Toluidene Blue. (A) The uninjured RNC clearly shows the large ectoneural (EN) component separated from the thinner hyponeural (HN) component by a connective tissue band. (B) At 2 dpi the area is filled with debris and the open end of the RNC is visible (arrowhead). (C) By 6 dpi, the RNC stumps have organized into club-shaped structures. (D) By 12 dpi the injury gap has been filled with nervous tissue that now forms a continuous extension that joins the stumps. (E) By 20 dpi the RNC still appears slightly disorganized and no connective tissue band is observed. (F) At 28 dpi, the RNC has recovered much of the structure and organization found within non-injured RNC including the separation between the EN and HN components by a connective tissue band. D-dermis, EN-epineural component, HN-hyponeural component, RNC-radial nerve cord. X's denote the injury site; Asterisks show the presence of tattoo ink used to label the injury site; arrows signal morula cells within the RNC. Bar = 300 μm.

METHODS

Animals

Holothuria glaberrima specimens, 10–16 cm long, were collected from the northeastern rocky shores of Puerto Rico and kept in an aquarium at 22–24°C with circulating seawater and constant oxygenation. All experiments conform to the regulations on animal research at the University of Puerto Rico.

Surgical Procedures

The technique used to injure the RNC has been described elsewhere [28]. Briefly, evisceration was induced by intracoelomic injection of 0.35 M KCl (3–5 ml). One hour following evisceration, animals were anesthetized by placing them for 10 min in 0.1% 1,1,1-Trichloro-2-methyl-2-propanol hydrate (chlorobutanol) diluted in seawater. Severing of the RNC was done by pushing the dorsal bodywall through the cloaca with a blunt rod, thus exposing the coelomic surface of the bodywall. A transection-type injury was made perpendicularly to the nerve using a scalpel, after which animals were reverted to their normal orientation and returned to their aquarium. By transecting the radial nerve from the inside we could obtain a reliable injury to the RNC with minimal damage to the surrounding tissues. Control animals were taken through the same protocol, except that no transection was made.

Tissue Harvesting and Histology

During the regeneration period, animals were sacrificed at different days post-injury (2, 6, 12, 20, and 28 dpi). Animals were anesthetized with chlorobutanol as before, slit open and pinned to a dissecting tray. Initially, they were directly fixed within the tray with 4% paraformaldehyde for 30 min and then the nerve-muscle complex, along with the bodywall, was excised. The tissue block was refixed overnight, washed in PBS and left in 30% sucrose until used.

Tissue blocks were embedded in OCT (Tissue Tek OCT; Miles Inc.), frozen at -35°C and longitudinally sectioned (20 μm) in a cryostat (Leica CM 1900). Sections were mounted on poly-L-lysine coated slides, left to dry for 1 hr at room temperature and stored in a dry chamber until needed. Since the RNCs are localized within the body wall, the tissue sections contained the regenerating RNC, the proximal regions that were 0.5 mm from the injury site, and the distal regions that were approximately 5 mm away.

Histological Analyses

Chemical Dyes – Toluidine Blue

To obtain a comprehensive view of the temporal modifications that the RNCs go through after an injury, longitudinal tissue sections were stained with toluidine blue, as described by Presnell and Schreibman [31]. In brief, sections were rinsed in PBS for 2 min, and then immersed in dye solution for 2 min. They were then rinsed with tap water, mounted in PBS-buffered glycerol and viewed under the microscope. Occasionally, staining was also done on tissue sections previously processed for immunohistochemistry (see below).

Morphometric Analysis

To evaluate changes in the RNCs following injury, tissue sections stained with toluidine blue were studied under light microscopy and the shortest distance between nerve stumps of injured animals was recorded. At least ten tissue sections were measured per animal and three to five animals were used for each stage. Sections were examined and photographed using a Nikon E600 microscope. Images were collected digitally with an RT Color Spot camera (Diagnostic Inst. Inc.) with 1520 × 1080 pixel resolution, and analyzed using the MetaVue software (version 6.0; Universal Imaging, Inc.) and Image J (version 1.37;NIH, USA).

Immunohistochemistry

Protocols for the immunohistochemical procedures performed for this study have been described elsewhere [28]. Tissue sections were permeabilized with 0.5% Triton X-100 in PBS, blocked with goat serum (1:50), and incubated at room temp overnight with the corresponding primary antibody: α-Bromodeoxyuridine (BrdU, 1:5, Amersham), Sph2 supernatant [32], α-GFSKLYFamide (α-GFS, 1:1000, [33]), α-Galanin (1:500, [34]), α-Nurr1 (1:1000, Sigma) or RN1 (1:50,000, [35]) in a humid chamber. The following day, sections were rinsed in PBS and treated with the FITC-labeled goat-anti-mouse secondary antibody (1:50, BioSource Int.). When double labeling was performed, the Cy3-labeled goat-anti-rabbit antibody (1:2,000) was used along with the FITC labeled antibody. Sections were rinsed again in PBS and mounted in buffered glycerol. Double or triple labeling with Hoechst nuclear dye (Sigma) was done by immersing slides in 1 μM Hoescht for 15 min after the secondary antibody [28]. Sections were examined and photographed using the Nikon E600 microscope equipped with FITC, R/DII and DAPI filters. Images were recorded using the MetaVue software as mentioned previously.

Cell Counts

Various cellular populations were labeled and quantified. These include (1) the overall number of cells within the RNC using Hoescht nuclear dye, (2) spherulocytes, cells that contained a large number of spherules within their cytoplasm and were labeled with monoclonal antibody Sph2, (see [32]), (3) morulas, cells that also contain vesicles or spherules and that are strongly stained with Toluidine blue (see [32]), (4) neuropeptide-containing neurons labeled with anti-GFS and (5) a neuronal population labeled with an antibody against the transcription factor NURR1. The number of cells was quantified by counting labeled cells in a 280 mm^2 area of the longitudinal RNC (this is the equivalent of the microscope field of view-FOW using the 40× objective). In order to be counted, cells had to be within the nerve or in close apposition with the somas in the nerve periphery, or with the fibers that had exited the nerve stumps. Cell counts were done in two fields of view chosen at random for each corresponding area in every animal and in at least five animals per regeneration stage.

TUNEL Assay

The TUNEL method was used to label apoptotic cells within the RNC. For this purpose we used the TdT-FragEl DNA

Fragmentation Detection kit (Calbiochem QIA33) and followed the manufacturer specifications with only minor modifications. Tissue sections were washed for 3 min in PBS to remove the embedding medium; afterwards they were refixed in 4% formaldehyde (diluted in 1× PBS) and washed in TBS for 15 min. Tissue sections were permeabilized with 50 μl proteinase K (5 μg/ml) for 15 min and then washed in TBS. Prior to labeling apoptotic cells, tissue sections were bathed with 1× TdT Equilibration Buffer (80 μl per section) for 30 min. For the labeling reaction, 60 μl of the mixture (57 μl Labeling Reaction Mix: 3 μl TdT enzyme) were applied to each section and incubated for 1 hr at room temperature. Slides were then washed 2× for 2 min in TBS, counterstained with the nuclear dye Hoescht for 10 min, washed for 5 min in TBS, and mounted with phosphate buffered glycerol.

Proliferation Assays

Pulse-Chase Experiments

Incorporation of BrdU into cell nuclei was used to study the spatio-temporal distribution of proliferating cells in regenerating animals. Animals were injected intracoelomically at 2, 6, 12, 20 and 28 dpi with 100 μl of a 1 mg/ml BrdU solution (approximately 0.01 mg/g wet weight) and sacrificed 24 hrs after. Tissues were fixed and treated as described above for immunohistochemistry. Cells undergoing replication were identified with an antibody against BrdU (Amersham). Prior to the incubation with the antibody, tissue sections were bathed with 0.05 N HCl for 1 h. This was done to improve the accessibility of the antibody to the BrdU epitope, followed by an additional 15 min PBS wash. Cell division index was obtained by dividing the number of BrdU-positive cells against the total number of Hoescht-labeled nuclei within a longitudinal section of the RNC in the microscope field of view. At least two measurements of corresponding areas chosen at random from each of at least five different animals were used for each stage.

Cellular Birth-Dating

To determine if dividing cells were giving rise to the neurons in the regenerating RNC, we injected animals with BrdU every other day for a period of 6 days (4 BrdU injections per animal in each group). There were a total of 6 groups: 0–6, 8–14, 16–22, 24–30, 32–38, and 40–46 dpi; all animals were sacrificed at 62 dpi. The rationale behind the multiple and consecutive doses of BrdU stems from the assumption that new neurons do not differentiate immediately after the division, but that a dividing neuronal precursor will not necessarily express the neuronal markers after the last division. Multiple injections were thought to be necessary to assure that at least some cells were caught at the time of division. Sections of RNCs from the various groups were double labeled against BrdU and markers for different subpopulations (GFS, Galanin, Nurr1) to establish the date of birth of the neurons that were repopulating the new RNC tissue.

Statistical Analyses

Student's t test and ANOVA's were used for statistical analyses. P values of < 0.05 were deemed to be significant.

RESULTS

Overview of Nerve Regeneration

To assess the morphological changes that the RNC underwent during the regeneration period, longitudinal sections of body wall were stained with Toluidine blue and studied under light microscopy (Fig. 1). The first stage (2 dpi) was characterized by an injured RNC with fan-shaped stumps (Fig. 1B). Remarkably, the pre-existing CT band in the stumps were practically unaffected; if anything, it lost a bit of cohesiveness at its most proximal end, adjacent to the injury. By 6 dpi, small bundles of fibers had extended from both nerve stumps, giving the structure an overall needle-like shape laying on the substratum provided by the body wall and pointing towards the opposing stump (Fig. 1C). At this stage, the origin of nerve bundles extending beyond the stump seemed to be, mostly, the result of an outgrowth in the nerve EN component, which lies closer to the body wall. Still, the lack of a substratum in the injury gap appeared to be the main limitation for scouting bundles. Support for this comes from observations of fiber bundles, which in the absence of a direct intervening matrix between the stumps, were found close to the trough of the injury gap. Nerve re-contact occurred by 12 dpi, concurrent with the appearance of a provisional matrix that filled the injury gap (Fig. 1D). The regenerated RNC spanned the injury gap but remained disorganized with empty spaces that could be seen. This was clearly observed by using the differential interference contrast (DIC) filter in the microscope. Although the nerve had reconnected by 12 dpi, the CT band that normally separates the HN and EN was not present. The RNC structure was much improved by 20 dpi (Fig. 1E). At this stage the number and localization of nuclei, both in the HN and EN, was similar to that of an uninjured nerve. The CT band separating the HN and EN was already perceptible at this stage in some, but not all animals, as traces of thin filaments along the length of the nerve. Approximately one month after the injury (28 dpi), the regenerated nerve had a normal architecture (Fig. 1F). The nuclear arrangement and density of fibers was now similar to that of a normal uninjured nerve and the CT band was already present and compact, although in some areas it still seemed diffuse.

Regeneration of Fiber Subpopulations

To obtain an in-depth view of the events that occur in the regenerating RNC, we studied the responses of different fiber subpopulations to the injury. Using immunohistochemistry we were able to correlate the gross appearance of the RNC, studied in light microscopy, with the growth of particular fiber populations across the injury site of the regenerated RNC.

RN1

This is a monoclonal antibody raised in our lab against *H. glaberrima* RNC tissue [35] that provides very specific labeling of what appears to be a large fiber population of the holothurian nervous system (Fig. 2A). RN1 does not label the CT band that separates the HN and EN component, thus, this band could be detected as a non-labeled distinct strip of tissue embedded within the high density of RN1 immunoreactive fibers. As a result, this antibody served as a two-edged marker, to measure progress during the process of nerve regeneration, namely the tracking of fibers as they elongate, and the appearance of the CT band as evidence of tissue organization into HN and EN components.

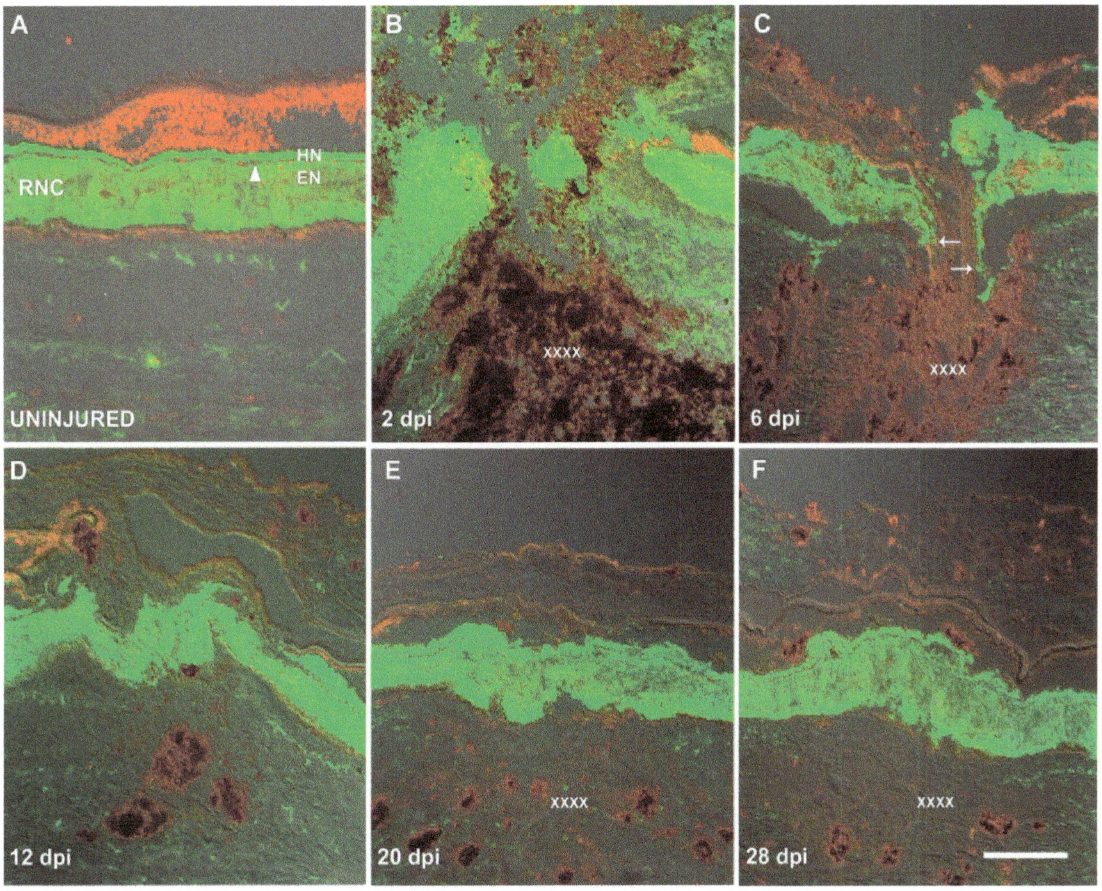

Figure 2: RN1 labeling of the regenerating radial nerve cord. Longitudinal tissue sections of (A) uninjured and regenerating radial nerve cords at (B) 2, (C) 6, (D) 12, (E) 20 and (F) 28 days post injury (dpi) were labeled with the monoclonal antibody RN1. (A) Labeling in the uninjured RNC is highly specific to the nervous components, both ectoneural (EN) and hyponeural (HN) component while the connective tissue band remains unlabeled (arrowhead). (B) At 2 dpi nerve fiber debris is visible around the injury site and the nerve stumps can be seen to be highly disorganized. (C) By 6 dpi, the RNC stumps have organized into club-shaped structures from which some nerve fibers can be observed to be extending (arrows). (D) By 12 dpi the injury gap has been filled with nervous tissue that now forms a continuous extension that joins the stumps. (E) By 20 dpi the RNC still appears slightly disorganized and the CT band is not present along the entire length of the regenerated RNC. (F) At 28 dpi, the RNC has recovered much of the structure and organization found within non-injured RNC including the connective tissue that separates the EN and HN components. EN-epineural component, HN-hyponeural component, RNC-radial nerve cord. X's denote the injury site; Bar = 300 μm.

Most of the data collected with this antibody confirm the observations made using the Toluidene blue dye; however, some interesting differences were found. The fan-shaped morphology of the nerve stumps at the injury site observed at 2 dpi was mostly the result of a swelling (enlargement) in the EN component, while the HN component was not severely affected (Fig.2B). At this stage, the HN component proximal to the injury appeared to retract in relation to the EN. At 6 dpi, the stumps acquired a club-shaped appearance and RN1-labeled fiber bundles could be observed protruding toward the opposite stump (Fig. 2C). Unfortunately, because of the high density of labeling, individual nerve fibers inside these pioneering bundles were hard to identify. However, the RN1 fiber bundles were always observed to be in direct contact with the substratum of the body wall and continuous with the injured stump. At this stage we observed the first difference in the rate of regeneration between the EN and HN components; the EN component provided the first scouting bundles that traverse the substratum towards the injury site, while RN1-labeled fibers of the HN appeared to be in a latent state. At 12 dpi, RN1 fibers from opposing nerve stumps came together, forming the newly regenerated RNC (Fig. 2D). At this stage, although RN1 fibers filled the area between the stumps, the separation of the RNC into its respective components had not occurred yet. During this stage RN1 labeled side branches could be observed originating from the same nerve bundles that form the connection between the stumps. These branches might be some that had gone astray into other injured tissues or they might be precursors to peripheral nerves. By 20 dpi the new RNC showed the first hint of connective tissue separation between the HN and EN components, albeit, rudimentary and irregular at times (Fig. 2E). The EN component was now structurally comparable to that of an uninjured RNC, as its wandering bundles of fibers were no longer present or at least apparent. Additionally, the regenerated RNC had a more cylindrical shape in accordance with the rest of the nerve; nonetheless, the staining intensity was now less than in uninjured areas. In spite of all these improvements in the EN component, the HN in most animals was still underdeveloped compared to uninjured counterparts in the same section. The new RNC was restored by 28 dpi with an HN and an EN component clearly separated from each other by the CT band between them (Fig. 2F).

α-GFS

α-GFS is a polyclonal antibody made in our lab against the neuropeptide GFSKLYF-amide that recognizes a subpopulation of neuronal fibers and somas in both HN and EN components of the RNC of *H. glaberrima* [33].

At 2 dpi, immunolabeling with GFS clearly showed the fan-shaped morphology of the cut RNC and the fraying of the nerve fibers close to the injury (Fig. 3A). The reduction in fiber density close to the injury was accompanied by an increase in labeling intensity close to the nerve stumps, which could be attributed to the appearance of α-GFS-labeled varicosities along the fibers. At 6 dpi, GFS-immunoreactive fibers had sprouted from the nerve stumps and the overall structure was clearly pointing towards the opposite stump (Fig. 1B). It is important to note that that the previously observed varicosities were also present in these migrating fibers. Similar to what was observed for RN1, fibers expressing the GFS neuropeptide also reconnected by 12 dpi (Fig. 3C). The labeling with GFS showed that immunoreactive fibers in the HN followed those of the EN; instead of using the body wall as a substrate. Even when the path taken by some fibers might be twisted and indirect, the overall axis of the fibers inside the regenerated EN was parallel to the long axis of the RNC. By 20 dpi both components were present in the regenerated RNC. No major differences were observed a month after transection (28 dpi) between regenerated RNC segment and its uninjured counterpart (Fig. 3D).

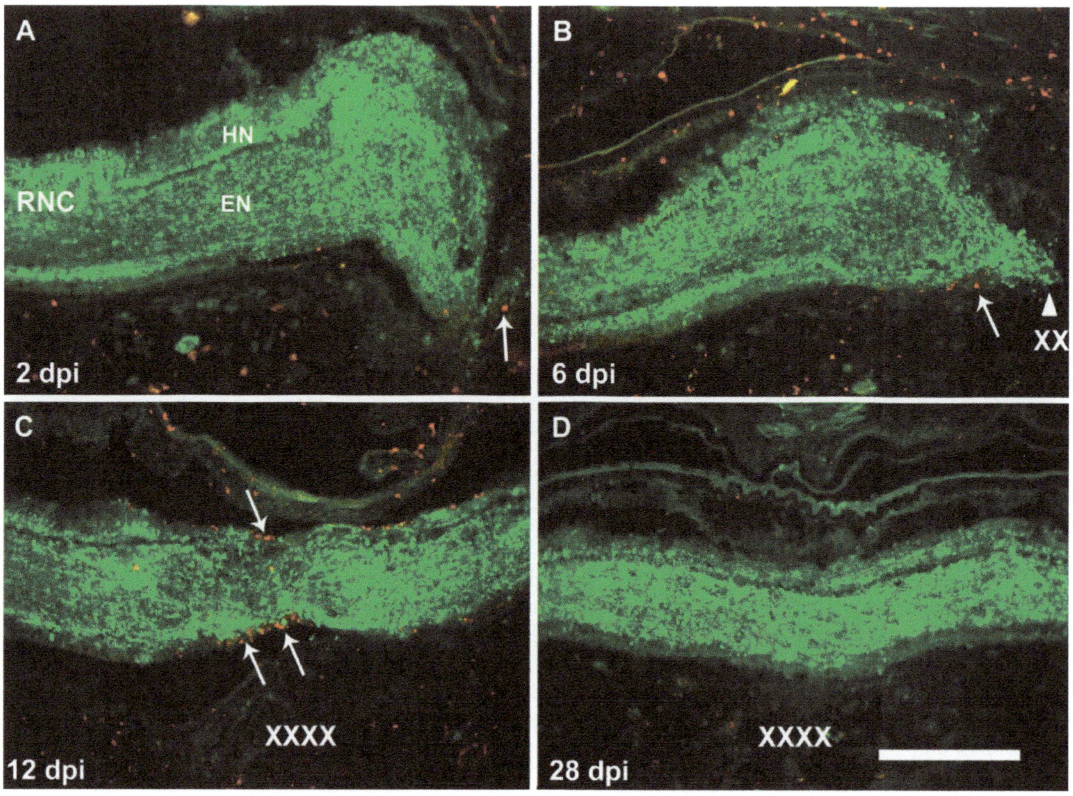

Figure 3: Double labeling of regenerating radial nerve cords for the fibers expressing the neuropeptide GFS and proliferating cells with BrdU. Longitudinal tissue sections of regenerating RNC at (A) 2, (B) 6, (C) 12, and (E) 28 days post injury (dpi) were labeled with anti-GFS (green) and anti-BrdU (red). Animals that had undergone a transection of their radial nerve were injected with BrdU 24 hrs prior to sacrifice. (B) GFS expressing fibers can be seen to elongate out of the nerve stump by 6 dpi (arrow). (D) Restoration of GFS-expressing fibers in the radial nerve cord is complete by 28 dpi. In terms of cell division, little cell division is observed at (A) 2 or (D) 28 dpi. Actively dividing cells were mainly observed in the periphery of the RNC where most of the nuclei are present (arrows). Some cell division is also observed among cells in the neuropile. Cell division begins around (B) 6 dpi and peaks around (C) 12 dpi in both hyponeural and ectoneural components (arrows). EN-epineural component, HN-hyponeural component, RNC-radial nerve cord. X's denote the injury site; arrows signal some of the BrdU-labeled cells within the RNC. Bar = 300 μm.

α-*Galanin*

In the RNC this antibody labels a relatively small population of fibers [34]. At 2 dpi, galanin immunoreactive fibers underwent a change in their morphology forming round varicosities with very intense labeling. This change was more conspicuous closer to the injury site, where the nerve fibers have became frayed; they were present in the HN and EN component of the RNC. At 6 dpi, varicosities were also found in the needle-shaped bundle of fibers exiting the nerve stump (Fig. 4A). Once the RNC reconnected at 12 dpi, the area of the injury could be easily identified by the presence of these large varicosities. At 20 dpi, galanin-positive fibers were present in both components of the RNC, although with a higher density in the EN (Fig. 4B). Galanin-positive fibers passing between these two components were also apparent at multiple points along the regenerated RNC, as is usual in an uninjured nerve. At 28 dpi, the large varicosities had disappeared and α-galanin produced a labeling that was virtually undistinguishable from that of a normal RNC.

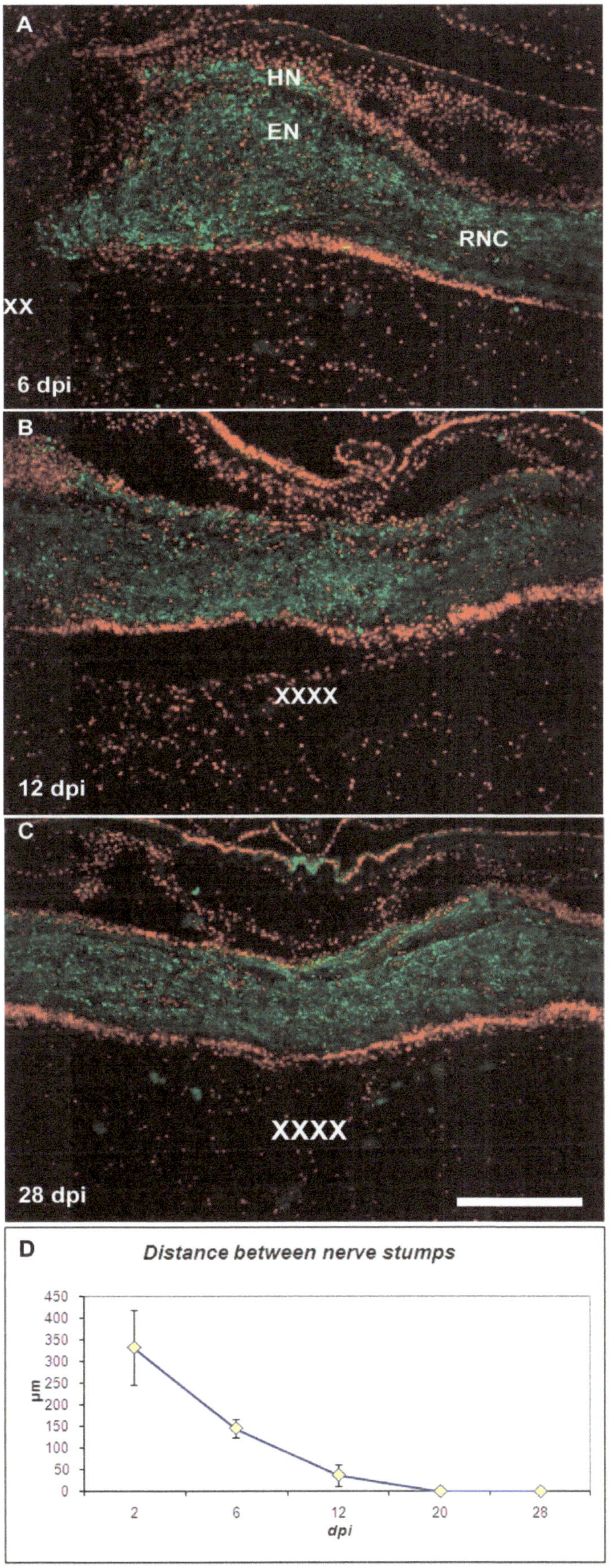

Figure 4: Galanin-like immunoreactivity in regenerating radial nerves. Longitudinal tissue sections of regenerating radial nerve cords at (A) 6, (B) 12 and (C) 28 days post injury (dpi) were labeled with anti-galanin (green) and counterstained with Hoescht (red) to label cell nuclei. (A) Galanin immunoreactivity disappears from the injury edge of the cord (asterisk) and the proximal stump acquires a punctuated labeling at 6 dpi. (B) At 12 dpi, the immunoreactive fibers have crossed the injury gap and labeling appears more homogeneous. (C) At 28 dpi labeling is similar to that in the uninjured cord (not shown). EN-ectoneural component, HN-hyponeural component, RNC-radial nerve cord. X's denote the injury site. Bar = 300 µm.

In summary, all neuronal markers used showed that the injury gap that separates the two RNC stumps is slowly filled by a growth of fibers that originate from the stump ends in the first two weeks of regeneration (Fig. 4C). This process is complete by 20 dpi and a RNC with high similarity to the uninjured RNC is in place by 28 dpi.

Cell Division

Pulse Chase

RNCs are ganglionated nerves where the neuronal soma and supporting cells can be mainly found in the periphery. To study the possibility that cell division is occurring in the RNC as a result of the injury, animals were injected with the thymidine analog, BrdU, and then checked for its incorporation into the nuclei. Initially, at 2 dpi few cells within the proximal stumps, both in the HN and EC bands, of the injured nerves were undergoing cell division; no cell division was seen distal to the injury site (Fig. 5). At 6 dpi, the highest numbers of dividing cells were seen in the proximal region, with about 6% of cells (6.81 ± 0.42 in HN and 5.36 ± 0.4 in EN) undergoing cell division. Cell division distal to the injury site never exceeded 2%. By 12 dpi, cell division in the proximal stumps remained comparatively similar (6.54 ± 0.23 in HN and 383 ± 0.21 in EN) to the previous stage. However, an impressive surge in cell division occurred in the regenerating RNC that is forming across the injury site, where close to 20% of the cells (22.2 ± 1.06 in HN and 14.9 ± 1.15 in EN) were undergoing cell division. Subsequently, at 20 dpi, a sharp drop in cell division was observed in the newly formed RNC. Proximal to the injured nerve, cell division did not exceed 2%. At the last stage studied (28 dpi), cell division was mainly restricted to the HN of the regenerated RNC, where cell division was 3.28 ± 0.7% in HN and 1.26 ± 0.3 in EN, everywhere else, cell division was less than 1%.

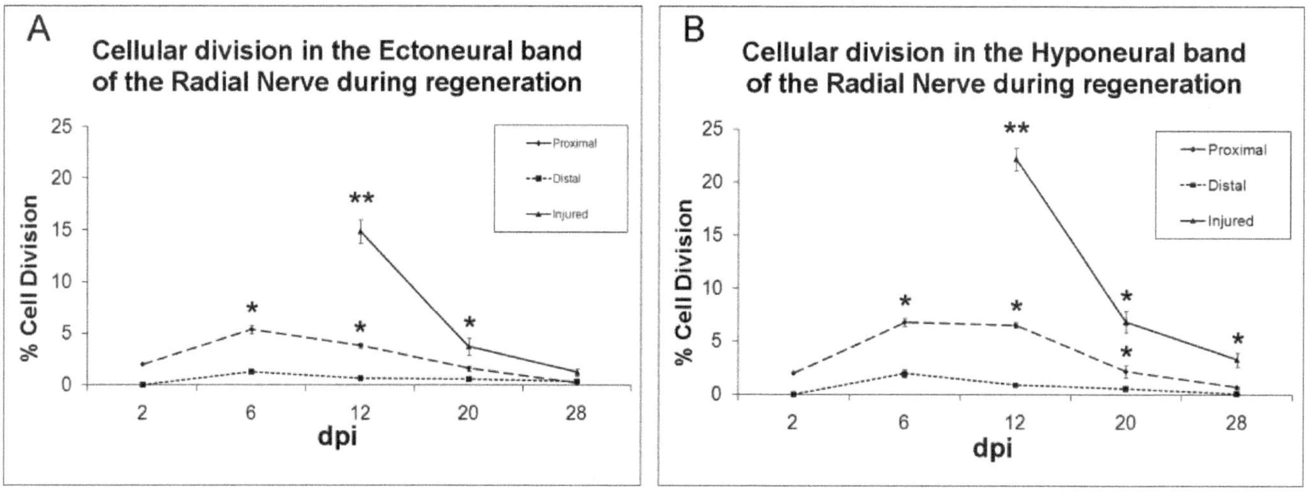

Figure 5: Patterns of cell proliferation in the radial nerve cord. The percentage of BrdU-labeled cells compared to Hoescht labeled nuclei in the (A) ectoneural and (B) hyponeural component of the radial nerve cord are shown. Measurements were done at the injury site and compared with sites proximal and distal to the injury. Animals were injected with BrdU 24 hrs before sacrificed. Quantification of cell division at the injury site shows a peak in cell proliferation at 12 dpi both in the hyponeural and ectoneural components. A smaller increase in cell division is also observed in the nerve stumps proximal to the injury site, but not so distal to the injury. Regenerated tissues in the injured area do not appear until 12-dpe. Each point represents the mean ± S.E. of at least five animals. When compared to distal, significant differences are observed at various stages in the proximal and injured areas *$p < .05$, **$p < .01$.

Cell Regeneration and Birth Dating

The next step was to determine whether the neurons in the regenerated RNC were originating from dividing precursors. For this we used two neuronal markers that label independent neuronal populations in the RNC, anti-GFS [33] and anti-Nurr1 (unp. obs.). These markers labeled neuronal cell bodies in both the EN and HN components of the RNC (Fig. 6A &6B). Initially, experiments were made by injecting animals with BrdU at different time periods and sacrificing animals at 30 dpi. However, these animals showed few if any GFS or Nurr1 neurons in their regenerated RNC. Thus, we decided to extend the experiment until day 62 dpi. For this, animals were separated into 6 groups and each group received BrdU injections during a particular 6-day period. All animals were sacrificed at 62 dpi. The rationale for this experiment was that only those cells that stop proliferation close to the last BrdU injections would retain the BrdU label until 62 dpi. Cells that kept proliferating would dilute out the BrdU and not be labeled. Thus, we would expect that dividing neuroblasts would stop dividing and then differentiate, so that most of the labeled cells that had stopped dividing within the RNC would be neurons.

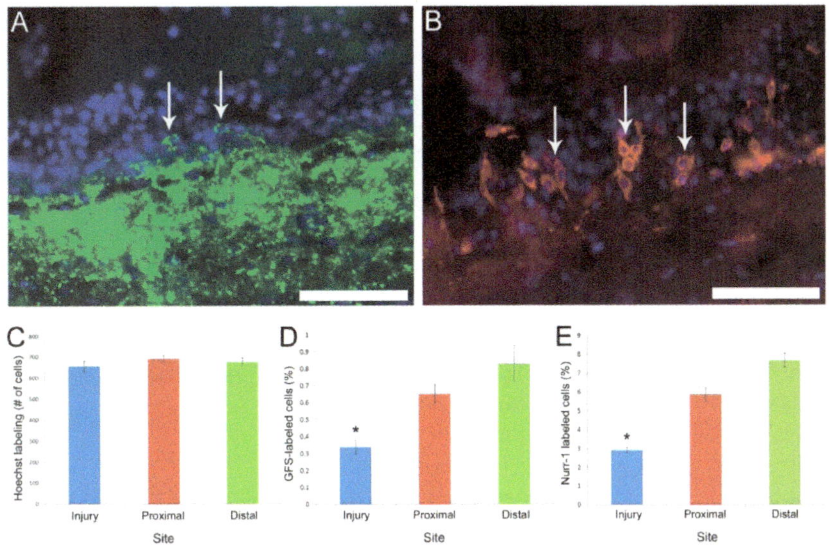

Figure 6: Neuronal labeling in the regenerating radial nerve cord. Neurons (arrows) within the regenerated radial nerves were observed 62 dpi using (A) anti-GFS and (B) anti-Nurr1 markers. (C) The total number of cells within the regenerated radial nerve is not significantly different from the number of cells at proximal or distal sites. In contrast, quantification of the percent of neurons at the three sites showed that the percentage of neurons at the injury site was observed to remain significantly lower than at the other two sites. Each point represents the mean ± S.E. of at least three animals. *p < .05, **p < .01. Scale bars = 50 μm.

Sections of the regenerated RNC were obtained and the cells in an area of equivalent length of RNC were determined on the microscope field of view (40×) were counted at 62 dpi. Results showed that the total number of cells (as measured by counting Hoescht-labeled nuclei) within the regenerated RNC did not differ from that in the proximal and distal RNC, suggesting that cell regeneration had occurred after 2 months (Fig. 6C). At this time, neurons expressing GFS or Nurr1 were also found within the regenerated RNC, albeit, in smaller numbers when compared to those within the proximal and distal RNC (Fig. 6D &6E)

Birthdating studies showed a peak during the second and third week of regeneration (groups 2 & 3, 8–22 dpi) when close to 3% of the cells (group 2 – 3.16 ± 0.54%, group 3 – 2.81 ± 0.59%) stopped dividing, and were still BrdU labeled at 62 dpi (Fig. 7A &7B). This would suggest that it is during this time period that most neurons have stopped dividing and started differentiation. To further verify that some of the dividing cells were neurons, we used double labeling with the neuronal markers GFS and NURR1. The results were somewhat unexpected. None of the cells that expressed GFS were double labeled with BrdU in any of the injected groups, even when over 300 GFS neurons were observed. In contrast, some cells labeled with Nurr1 were also labeled with BrdU (Fig. 7C). These cells were present only in the HN component of animals injected with BrdU during 8–22 dpi (which corresponds to the peak in cells that stopped proliferation). However, the number of neurons is rather small as only 2.5% of all NURR1 immunoreactive cells present in these groups were BrdU labeled (Fig. 7D).

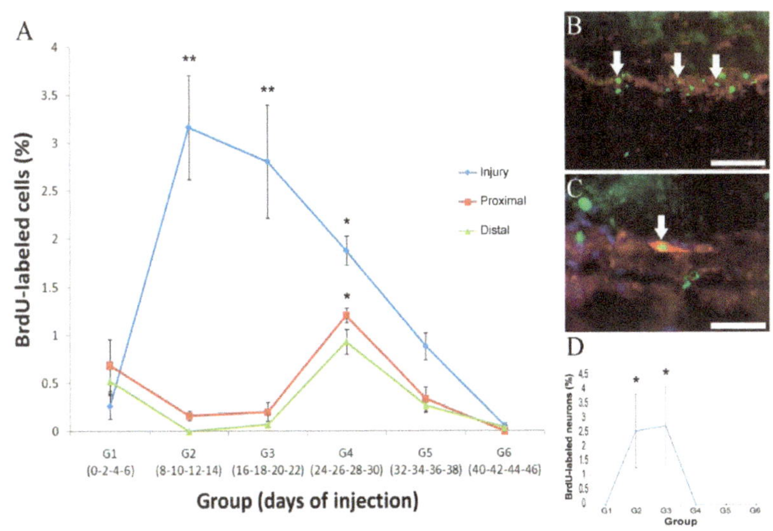

Figure 7: Cellular birth dating in the regenerating radial nerve cord. (A) Quantification of the percent of BrdU-labeled nuclei for each group of regenerating animals that were injected with BrdU at different time points and sacrificed at 62 dpi. (B) BrdU-labeled nuclei (green, arrows) among Hoechst-labeled nuclei (red) in the radial nerve cord at the injury site of an animal injected during the second week of regeneration. (C) Triple labeling showing Nurr1-expressing cell (red, arrow) with BrdU-labeled nuclei (green) in the hyponeural compartment at the injury site. Blue labeling is Hoechst dye. (D) Quantification of the percent of Nurr1 labeled neurons that were BrdU-labeled for each group at 62 dpi.

The only neurons labeled with BrdU at 62 dpi were from the animals injected during the second and third weeks of regeneration (groups 2 & 3). Each point represents the mean ± S.E. of at least three animals. When compared to distal, significant differences are observed at various stages in the injured area. Similarly, significant cell division occurs in the proximal and distal area at G4. *p < .05, **p < .01. Scale bars = (A) 50 μm, (B) 25 μm.

Apoptosis

To determine the cellular events that occurred during RNC regeneration following a transection-type injury, we decided to examine if programmed cell death was occurring (Fig. 8). As soon as 2 dpi, apoptotic cells were present in the HN component of the proximal stumps. The highest number of apoptotic cells in the proximal stumps occurred at 6 dpi, in both the HN and EC bands, although the magnitude of apoptotic cells was higher in the HN band. By 12 dpi and on to 28 dpi, few if any apoptotic cells were observed in the nerve proximal to the injury site. In contrast, at 12 dpi, a large wave of apoptosis occurred in the regenerated RNC that had just formed over the injury gap. The number of apoptotic cells in the regenerating RNC dropped in the two subsequent stages, until at 28 dpi, no apoptosis was observed in the EN and only in 1% of cells of the HN band.

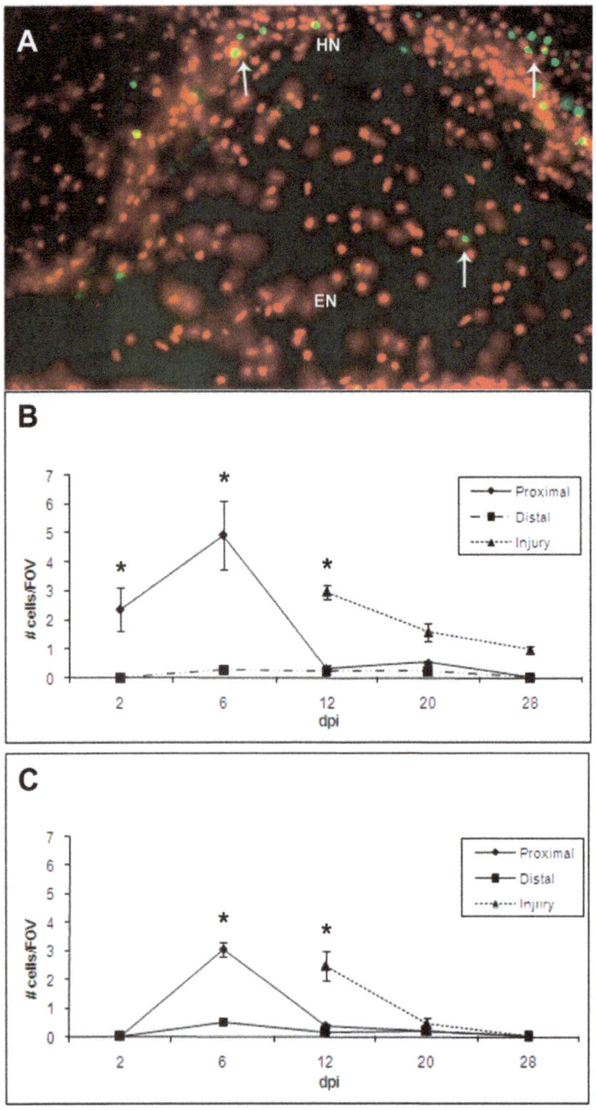

Figure 8: Apoptosis during radial nerve cord regeneration. (A.) Longitudinal section of regenerating RNC at 6 dpi showing TUNEL labeling (green) in cells undergoing apoptosis (arrows) (counter-stained with Hoescht-red). Apoptosis occurs in both (B) hyponeural and (C) ectoneural components of the regenerating radial nerve cord peaking at day 6 dpi in the proximal nerve stumps and at 12 dpi in the regenerating cord that is forming at the injury site. Little or no apoptosis occurs in areas distal to the injury. Regenerated tissues in the injured area do not appear until 12-dpe. Each point represents the mean ± S.E. of five animals. When compared to distal, significant differences are observed at various stages in the proximal and injured areas *p < .05, **p < .01. HN – hyponeural component, EN – ectoneural component. Bar = 25 μm.

Spherule-Containing Cells

Two types of spherule-containing cells have been described in our system: morulas, which are highly basophilic and stain dark purple with toluidine blue, and spherulocytes, which are recognized by a monoclonal antibody prepared against sea cucumber tissues [28, 32]. Only about 5% of spherule-containing cells are recognized by both markers.

Few (less than1 per visual field), if any, morula cells were found within the normal RNCs (see Fig. 1). (As mentioned earlier, fields of view, FOV, were obtained with the 40× objective and encompassed an equivalent area in all animals). Changes in their number within the RNC were already seen at 2 dpi, when they invaded the nerve stumps. They achieved their highest numbers at this stage, when the average number of cells per field of view within the proximal stumps was 6.9 + 0.2 cells per FOV (Fig. 9A, and see Fig. 1). Most cells were closely apposed to the CT band that separates the HN and EN components; this is the same location where they are normally found in an uninjured nerve. However, as a response to the injury, morulas were also found interspersed within the neuropile, more commonly where the nerve stumps were engorged, or fan-shaped. By 6 dpi, the number of cells in the proximal stump remained the same (6.4 + 0.24 cells per FOV) as in the previous stage. However, some of these cells were also associated with the RNC cell bodies, in both the HN and EN, adjacent to the coelomic side and also with the regrowing fibers that originated from the nerve stumps. At 12 dpi the number of morulas in the proximal RNC decreased to levels just slightly higher (1.6 + 0.1 cells per FOV) than the basal level and remained at this level until 28 dpi. The changes in the morula population only occurred proximally to the injury site since distal to the injury site the cell numbers were not significantly different from uninjured nerves at any stage studied. The number of cells was much higher in the new RNC bridging the injury site at 12 dpi, on average there were 7.12 ± 0.35 cells per FOV. At 20 dpi, the average number of cells in the regenerated RNC declined to 2.61 ± 0.25 cells per FOV; similar to the number of cells found at 28 dpi (2.89 ± 0.23 cells per FOV)(see Fig. 1).

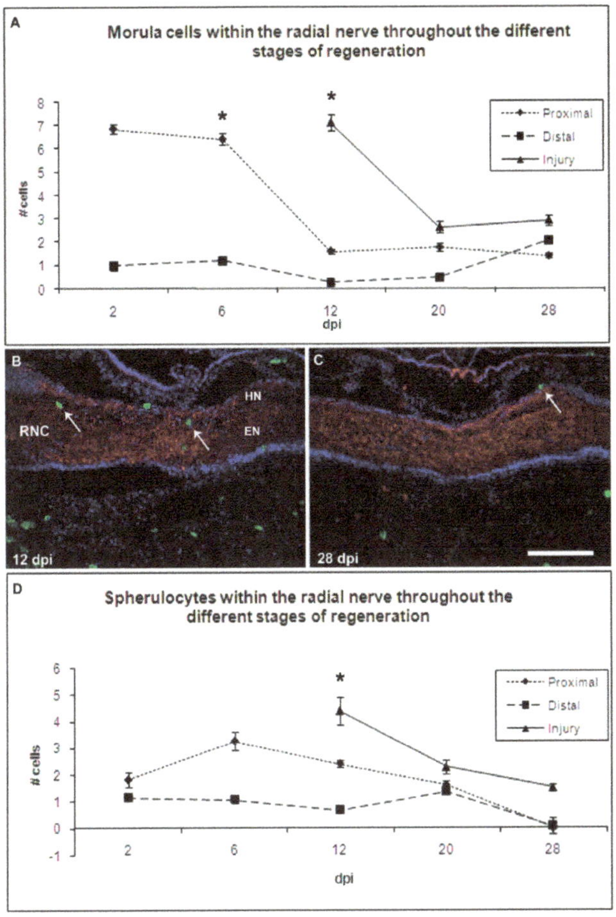

Figure 9: Changes in spherule-containing cells during radial nerve cord regeneration. The number of morula cells and spherulocytes increases in the regenerating radial nerve cord following transection. Morulas were detected using Toluidene blue (see Fig. 1). Quantitative analysis show that (A) the number of morulas peak between 2 and 6 dpi in the proximal nerve stumps and at 12 dpi in the regenerating cord that is forming at the injury site. Few changes are observed distal to the injury site. (B-C) Spherulocytes (green, arrows) were detected using the Sph2 antibody, a monoclonal antibody that recognizes the spherulocyte population. Longitudinal sections of animals at (B) 12 and (C) 28 dpi show the presence of spherulocytes within the nerve cord in early but not in late regenerating stages. Sections were also labeled with anti-galanin (red) and Hoeschst nuclear dye (blue). X's denote the injury site (D) Quantitative analysis show that the number of spherulocytes peaks at 6 dpi in the proximal stumps and at 12 dpi at the injury site, with little change at distal sites. Regenerated tissues in the injured area do not appear until 12-dpe. Each point represents the mean ± S.E. of five animals. When compared to distal, significant differences are observed at various stages in the proximal and injured areas *p < .05, **p < .01.

The number of spherulocytes within the RNCs also underwent changes during regeneration (Fig. 9D). These cells were not normally found in uninjured nerves, however, as soon as 2 dpi they were found within the proximal stumps (1.8 ± 0.27 cells per FOV); and to a lesser extent distal to the injury (1.1 ± 0. 07 cell per FOV). A consecutive increase was seen in the RNC stump by 6 dpi, when on average 3.23 ± 0.34 cells per FOV were observed in the RNC. By 12 dpi, the number of spherulocytes proximal to the regenerated RNC decreased, but still remained significantly higher than the basal level (see Fig. 9B). This was accompanied by a relatively high infiltration of these cells into the new RNC that had on average 4.34 ± 0.51 cells per FOV. The number of cells seen in the regenerated RNC by 20 dpi decreased to 2.24 ± 0.25 cells per FOV; the range of cells in proximal and distal parts relative to the injury site was 1.59 ± 0.17 to 1.34 ± 0.04 per FOV. Distally, the spherulocyte population remained unchanged from 3 dpi until 20 dpi at less than 1.4 cells per FOV. At 28 dpi, spherulocytes could only be detected in the regenerated portion of the RNC (see Fig. 9C), and although their levels were lower than those of previous stages, they were still significantly higher than in uninjured nerves.

DISCUSSION

The Regenerated RNC

Our results clearly show that the RNC of holothurians can regenerate within a month following a transection. The regenerated RNC looks amazingly similar to the non-injured RNC as judged by several histological and cellular parameters: It is well organized into ectoneural and hyponeural components, there is no scar or fibrotic tissue, the number of cells is similar to those found in uninjured tissue, neurons are localized to the nerve periphery, and spherule-containing cells that have infiltrated the injured cord have all but dissapeared. This is an amazing regeneration feat for an adult deuterostome, particularly if we consider that the RNC conforms the echinoderm CNS [24, 29, 30] and has been proposed to be evolutionary related to the vertebrate CNS [36–38].

Although we lack direct proof, a large amount of circumstantial evidence suggest that regeneration is not solely the joining back of the severed nerve stumps but that actually a new portion of the RNC is generated. First, from the moment it is cut, the wound forms a V-shaped structure where the nerve stumps are initially separated by a gap, and then by connective tissue. Second, most of the ongoing cellular division occurs within the regenerating RNC. Third, other tissues that surround the nerve and that also regenerate, do so by forming new tissues. This is particularly true of the surrounding muscle and connective tissues, which regenerate in parallel to the RNC and where very small differences in organization, cell density and texture distinguish the "new" tissues from the adjacent "old" [28]. Fourth, the initial and most visible regeneration event is the outgrowth of fibers from the RNC stumps. These outgrowths increase with time and eventually completely fill the space between the stumps. The latter retain their original morphology throughout regeneration, and it is the space in between that grows and becomes organized. Finally, recent work in another holothurian, *Eupentacta fraudatrix*, also shows regeneration of the RNC to be newly formed tissue [23].

In some ways, the formation of the new RNC can be compared to the "new" RNC segments that grow when the tip of the arm is severed in starfish or crinoids [19, 20, 22]. In all these cases the regenerated arm tip eventually grows to form new segments that include all the components found in the RNCs. Experimental evidence for the formation of a new RNC also comes from other closely related deuterostomes, the ascidians, that have an impressive ability to regenerate main components of their nervous system as has been shown by Bollner and colleagues [17, 18] in studies where the main neural ganglion regenerates afer ablation. Similarly, lower vertebrates are able to regenerate a lost segment of the spinal cord as has been shown by nerve cord injury in amphibians [39, 40] and fish [1].

Phylogenetic Analysis of Regeneration Following Transection

How does the regeneration of the holothurian RNC compares to that shown by other animals? To answer this question we will focus on those experiments where injury to the CNS is done by transection of the main nerve cord (or spinal cord). This includes studies in protostomes, such as mollusks and annelids and lower vertebrates, such as fish and urodele amphibians. It also includes developing vertebrates, since experiments have shown that spinal cord regeneration can occur at particular developmental stages of some higher vertebrates, such as chicks or marsupials, and then becomes restricted as development continues [41, 42].

RNC regeneration in *H. glaberrima* shows strong similarities to other animal species known to be capable of axonal regeneration and/or formation of a missing part of their CNS. In most species where CNS regeneration takes place, two events can be highlighed to occur: the outgrowth of fibers from the nerve stumps and the absence of fibrotic tissue or glial scar. The growth of fibers into the lessioned area provides the first indication of the regeneration process. As has been reported by TEM in the sea cucumber *E. fraudatrix* [23] and by immunohistochemistry in the starfish, *Asterias rubens* [22], these fibers originate from the lesioned stumps and move across the lesion eventually forming a bridge between the two nerve stumps. Similarly, in urodele larvae, fibers extending from each of the cut ends grow across the gap and eventually form a bridge connecting the two nerve ends [39]. New fibers are mainly attributed to the re-growth of fibers from lesioned cells as has been shown by retrograde labeling at the time of transection [43, 44]. Another common observation in regenerating animals, including holothurians, is the fact that there seems to be a directionality to the process, since fibers extending from one stump are directed toward the opposite stump and are rarely observed to be traveling toward another target, suggesting the participation of soluble factors originating from the severed stumps and/or guidance molecules in the surrounding environment.

The absence of a glial scar or fibrotic tissue in regenerating cords is another characteristic of those animals capable of regenerating their main nervous system. No scar tissue is detected or described in studies showing regeneration in urodeles, fish or echinoderms. Similarly, chick embryos that are capable of regeneration do not form scar tissue [41]. In fact, what investigators regularly show to occur during the regenerative response is an increase in cells or cell activity associated with the removal of the injury debris and of apoptotic cells.

There are also differences in the regeneration process among animals. One of these is the level or extent at which regeneration occurs. In holothurians, the regenerated RNC appears almost indistinct from the non-injured RNC. However, in other organisms where it has been studied, the regenerated cord is not fully equivalent to the normal one in spite of fibers crossing the lesion site. For example, in the regenerating chick embryo, fibers do cross the transection site and functional regeneration occurs, but the site of transection shows a slight disorganization of the gray matter [41]. Another example is the regenerating spinal cord of bullfrogs that have been transected as tadpoles, where the transection site is still visible 8 months after injury (and 6 after metamorphosis), showing a considerable constriction in size [15]. The differences can be striking, as occurs in the lizard where it has been shown that the regenerated spinal cord in the tail is considerably different from the uninjured cord and that it only contains descending fiber tracts from the old cord and scattered glial cells [45]. Other species show subtle differences, as in the case of regenerating 1 year-old goldfish, where the regenerated spinal cord appeared normal after 90 days post injury, but there were no motor horn cells in the regenerated area [46]. In *H. glaberrima*, the regenerated RNC has NURR 1 and GFS expressing neurons (albeit in lower numbers) and appears to be normal according to all anatomical parameters used. Nonetheless, more detailed experiments are needed to fully determine that all cell populations and fiber connections are present.

Finally, it is important to acknowledge that structural differences do exist in the regenerating nervous systems of animals. The echinoderm ganglionated nerve cord differs greatly from the regenerating nerves of some invertebrates where only fibers are found, and from the complex cellular and fiber organization of the vertebrate spinal cord. Thus, it remains unclear whether similar events occur during the regeneration of the different structures. Nonetheless, it will be by performing comparative studies, such as this one, that the commonalities and differences in nervous system regeneration among animals will be established.

Role of Cell division and Apoptosis

Two related and at the same time, opposite processes are occuring within the RNC during regeneration; cell proliferation and cell death. The two show a spatial and temporal profile that seems perplexing because they seem to be occurring simultaneously; both show a peak in the nerve stumps at 6 dpi and a peak within the regenerating RNC at 12 dpi.

Apoptosis

In vertebrates, apoptosis is associated with neurons that have been injured at the time of transection, but also with a period of secondary injury where the lesion size increases due to inflammation, immune-related events and hemorrhage [47–49]. Therefore, in our system, apoptosis within the nerve stumps seems to occur in cells whose fibers were damaged by the transection and that eventually failed to survive, undergoing apoptosis during the first week following injury. This was also observed at the TEM level in regenerating RNC of *E. fraudatrix*[23] and in regenerating fish CNS where apoptotic cells are seen for a long period of time following the lesion [50].

A different explanation might be involved in explaining the increase in apoptosis observed within the regenerating RNC in the second week following the transection. In this case the apoptosis event might be similar to what has been shown to occur during embryonic development in vertebrates, where about half of the neurons that are formed in the spinal cord undergo apoptosis [51]. Neuronal death in regenerating spinal cord has also been reported in the tail of the fish *Sternachus* [1, 52]. In these fishes, younger segments of the regenerated spinal cord have a larger number of neurons than older segments, once again suggesting that the regenerating spinal cord recapitulates the known embryonic developmental process of overproducing neurons that will eventually be discarded.

Cell Division

Cell division has been shown to occur in the regenerating arms of several echinoderms, mainly crinoids and starfish [20–22]. In crinoids, the nerve stump is also an active place of cell proliferation, although the dividing cells in the early stages appear to be non-neuronal [20]. In the starfish *Asterias rubens*, it was found that cell division in the regenerating nerve began 6–7 days following transection and increased during the next 2 weeks, gradually decreasing to control levels by day 60 post-injury [22]. Similarly, our results show cell division to occur within the ectoneural and hyponeural components of the nerve stump, but even more within the regenerating RNC. However, it is still unclear what type(s) of cells are proliferating since support cells, macrophages or some type of precursor cell could be accountable for the extent of cell division. This is a crucial issue that will be addressed below. Nonetheless, in recent studies Mashanov and colleagues [23] describe neurons and glial cells undergoing division in the regenerating RNC of *E. fraudatrix*.

The overall pattern of cell division within the regenerating RNC has some similarities to what has been shown to occur in the adjacent muscle and connective tissue, where cell division peaks between 6 and 12 days following injury [28]. These two weeks of extensive cell proliferation are undoubtly responsible for the finding that, at 30 dpi, animals have the same number of cells in the regenerated and in the uninjured RNCs.

Formation of new neurons has been shown to be an important process in animals where CNS regeneration is known to occur. In fish, BrdU labeling together with back-tracing techniques have shown that in the adult fish, neurons are regenerated from proliferating precursors following CNS injury [53]. In urodeles, new neurons are known to originate from proliferating ependymal cells [14]. Whether new neurons take part in the regeneration of the spinal tract observed in the chick embryo and the neonateal opposum is not known [54]. However, recent results suggest that stages where the ability to regenerate is still present, coincide with those stages where there is ongoing formation of new neurons [55]. On the other hand it is important to emphasize that some vertebrates can undergo functional regeneration with little or no contribution of neurogenesis, or that new neurons can originate from undifferentiated precursors.

In view of these findings, our results showing that neurons are present within the regenerated RNC and that some originate from dividing precursors provide for interesting comparisons to other systems. At the same time we must reconcile the finding that the densities of GFS and NURR1 neurons are lower in the regenerated RNC. One possible explanation is that the appearance of neurons in the regenerated RNC is a late event and that at 62 dpi some neuronal precursors have not fully differentiated. This would be consistent with experiments in starfish, where the first neurons observed in the regenerated segment following transection of the RNC were not observed until 28 days following injury [22]. These results might appear to contradict the observations by Mashanov and colleagues [23] in *E. fraudatrix*, who observed some neuronal division in the early stages of RNC regeneration, however, this might be explained when we take into account that early division was observed using TEM, while we used markers for differentiated neurons. Nonetheless, it is important to emphasize that in both *E. fraudatrix* and *H. glaberrima* neurons appear in the regenerated RNCs and that at least some of these neurons originate from dividing precursors. Similarly, in crinoid regenerating arms, some BrdU labeled neurons were recognized using electron microscopy, suggesting that in this echinoderm, neurons also originate from dividing precursors [21]. However, there are striking differences between neuronal populations since the GFS neurons were not observed to express BrdU, suggesting that they do not originate from dividing precursors. Once again this is consistent with studies in the transected starfish RNC where none of the neurons identified with the S1 neuropeptide (the starfish equivalent of GFS) showed BrdU labeling, even when enough time was allowed following the BrdU treatment for differentiaton to occur [22].

In order to fully identify the origin of neurons it will be important to determine the role of glia in holothurian RNC regeneration. Particularly in view that in *E. fraudatrix* Mashanov and colleagues propose that glial cells might be paving the path for neuron and fiber migration and even differentiating into neurons [23]. The role of the glial cells is not the only question that remains to be studied; several other questions remain to be answered. For example, does the observed anatomical regeneration provides for complete functional recovery? In many other model systems the extent of regeneration has been evidenced by backtracing neuronal connections and comparing the regenerated versus the uninjured connections. However, in view that little is known about the echinoderm radial nerve circuitry and nervous system physiology, it is not possible to determine at present to what extent there is morphological connectivity. Similarly, in other animals functional recovery can be documented by behavioral tests, that, at present, are not available for holothurians.

Comparison with other wound Healing/Regenerative Processes

Transection of the nervous system, whether experimentally induced or by accidental cause rarely, if ever, exclusively affect the nervous system. Thus, it is always important to view the regeneration of the nervous component as a process that occurs parallel to the regeneration of surrounding tissues. Granted, there are nerve-specific processes, such as axonal regeneration and synaptic specificity, that only occur in the nervous system. However, it is interesting that events that occur in the surrounding muscle and connective tissue previously published by our group [28] parallel those occuring within the nerve cord itself. One of them has already been addressed above: the pattern of cell division, showing that in muscle, coelomic epithelium and connective tissue the proliferation pattern coincides with that observed in the regenerating RNC.

Another event that provides for an interesting comparison is the presence of morula or spheruloytes cells within the RNC. These cells are rarely observed in uninjured nerves, however, their numbers increase after injury. This increase in cell numbers parallels what occurs within the regenerating intestine

[32] and following the transection injury in the adjacent tissues [28]. These cells have been proposed to play important roles in extracellular matrix remodeling, phagocytosis and antimicrobial activity, among others [32]. The presence of these cells within the regenerating nerve is quite intriguing and should be studied in further detail.

The extent of apoptosis observed after nerve transection can also be compared to what occurs during intestinal regeneration. We have recently completed a study of apoptosis during intestinal regeneration showing that high levels of apoptosis take place in the mesothelium 7 days following evisceration (unp. Observation). In both events, nerve regeneration and intestinal regeneration, apoptosis is rarely observed in the early days (2–5 days) following the injury or evisceration but occurs more likely during the second week (7–14 days). Moreover, apoptosis is more likely to take place within "new" tissues that have regenerated such as the newly formed intestinal blastema-like structure and the regenerated RNC, than in adjacent "old" tissues, such as the remaining nerve stumps or the intestinal mesentery.

For many investigators involved in studies of neurogenesis or neuronal regeneration, their focus has been strictly on the nervous system. However, our previous results show that the same events that occur during wound healing, also occur during intestinal regeneration [28]. And now we show that some of these events also occur during nervous system regeneration. Moreover, at the species level, the same species that can successfully regenerate their nervous systems are known to be exceptional regenerators of others tissues or organs, not just the nervous system. Thus, a major implication of our work is that the capacity to regenerate nervous system might be determined by the same processes and mechanisms in all tissues and that some animals have developed or retained these capabilities while in others they are absent. Although there are obvious exemptions to this statement, such as liver regeneration in vertebrates, the general rule remains that animals that have some capacity to regenerate their nervous system are, in general, good regenerators of many tissues or organs. If this is correct, then investigators interested in deciphering the mysteries of nervous system regeneration should also focus on regenerative or wound healing phenomena in other tissues and organs and on species where these regenerative processes are prominent.

ACKNOWLEDGEMENTS

The authors would like to thank Dr. V. Mashanov for helpful comments on the manuscript. Funded by NSF (IBN-0110692), NIH-MBRS (S06GM08102), NIH-1SC1GM084770-01), and the University of Puerto Rico. JSMR and ARMS are fellows of the NIH-MARC Program.

AUTHORS' CONTRIBUTIONS

JSMR carried out most of the experiments and helped in the analyses of the data and preparation of the manuscript. ARMS performed the birthdating experiments, analysed the data, and prepared the corresponding figures. JEGA conceived the study, participated in its design, data analyses and helped in the manuscript preparation. All authors read and approved the final manuscript.

REFERENCES

1. Anderson MJ, Waxman SG: Regeneration of spinal neurons in inframammalian vertebrates: morphological and developmental aspects. J Hirnforsch. 1983, 24: 371-398.
2. Blaugrund E, Lavie V, Cohen I, Solomon A, Schreyer DJ, Schwartz M: Axonal regeneration is associated with glial migration: comparison between the injured optic nerves of fish and rats. J Comp Neurol. 1993, 330: 105-112.
3. Chernoff EAG: Spinal cord regeneration: a phenomenon unique to urodeles?. Int J Dev Biol. 1996, 40: 823-831.
4. Ferretti P, Zhang F, O'Neill P: Changes in spinal cord regenerative ability through phylogenesis and development: lessons to be learnt. Dev Dyn. 2003, 226: 245-256.
5. Moffet SB: Nervous system regeneration in invertebrates. Zoophysiology. 1996, Springer NY, 34:
6. Sánchez Alvarado A, Tsonis PA: Bridging the regeneration gap: genetic insights from diverse animal models. Nat Rev Genet. 2006, 7: 873-884.
7. Wu Z, Ghosh-Roy A, Yanik MF, Zhang JZ, Jin Y, Chisholm A: Caenorhabditis elegans neuronal regeneration is influenced by life stage, eprhin signaling, and synaptic branching. Proc Natl Acad Sci. 2007, 104: 15132-15137.
8. Leyssen M, Hassan BA: A fruitfly's guide to keepin the brain wired. EMBO Rep. 2007, 8: 46-50.
9. Bernstein JJ, Gelderd JB: Synaptic reorganization following regeneration of goldfish spinal cord. Exp Neurol. 1973, 41: 402-410.
10. Lurie DI, Selzer ME: Axonal regeneration in the adult lamprey spinal cord. J Comp Neurol. 1991, 306: 409-416.
11. Sharma SC, Jadhao AG, Prasada Rao PD: Regeneration of supraspinal projection neurons in the adult goldfish. Brain Res. 1993, 620: 221-228.
12. Zupanc GKH: Adult neurogenesis and neuronal regeneration in the teleost fish brain: implications for the evolution of a primitive vertebrate trait. Evolution of Nervous Systems: A Comprehensive Reference. Edited by: Kaas JHJ. 2007, Non-mammalian Vertebrates, Acad. Press, 2: 485-519.
13. Nordlander RH, Singer M: The role of ependyma in regeneration of the spinal cord in the urodele amphibian tail. J Comp Neurol. 1978, 180: 349-374.
14. Chernoff EAG, Stocum DL, Nye HLD, Cameron JA: Urodele spinal cord regeneration and related processes. Dev Dyn. 2003, 226: 295-307.
15. Forehand CJ, Farel PB: Anatomical and behavioral recovery from the effects of spinal cord transection: dependence on metamorphosis in anuran larvae. J Neurosci. 1982, 2: 654-662.
16. Mochii M, Taniguchi Y, Shikata I: Tail regeneration in the Xenopus tadpole. Dev Growth Differ. 2007, 49: 155-161.

17. Bollner T, Beesley PW, Thorndyke MC: Pattern of substance P- and cholecystokinin-like immunoreactivity during regeneration of the neural complex in the ascidian Ciona intestinalis. J Comp Neurol. 1992, 325: 572-580.
18. Bollner T, Howalt S, Thorndyke MC, Beesley PW: Regeneration and post-metamorphic development of the nervous system in the protochordate Ciona intestinalis: a study with monoclonal antibodies. Cell Tissue Res. 1995, 279: 421-432.
19. Candia-Carnevali MD, Bonasoro F: Microscopic overview of crinoid regeneration. Microsc Res Tech. 2001, 55: 403-426.
20. Candia-Carnevali MD, Bonasoro F, Lucca E, Thorndyke MC: Pattern of cell proliferation in the early stages of arm regeneration in the feather star Antedon mediterranea. J Exp Zool. 1995, 272: 464-474.
21. Candia-Carnevali MD, Bonasoro F, Biale A: Pattern of bromodeoxyuridine incorporation in the advanced stages of arm regeneration in the feather star Antedon mediterranea. Cell Tissue Res. 1997, 289: 363-374.
22. Moss C, Hunter AJ, Thorndyke MC: Patterns of bromodeoxyuridine incorporation and neuropeptide immunoreactivity during arm regeneration in the starfish Asterias rubens. Philos Trans R Soc London B. 1998, 353: 421-436.
23. Mashanov VS, Zueva OR, Heinzeller T: Regeneration of the radial nerve cord in a holothurian: A promising new model system for studying post-traumatic recovery in the adult nervous system. Tissue Cell. 2008, 40: 351-372.
24. Hyman L: The Invertebrates: Echinodermata. 1995, New York: McGraw-Hill Book Co
25. García-Arrarás JE, Greenberg MC: Visceral regeneration in holothurians. Microsc Res Tech. 2001, 55 (6): 438-451.
26. García-Arrarás JE, Estrada-Rodgers L, Santiago R, Torres II, Díaz-Miranda L, Torres-Avillán I: Cellular mechanisms of intestine regeneration in the sea cucumber, *Holothuria glaberrima* Selenka (Holothuroidea:Echinodermata). J Exp Zool. 1998, 281: 288-304.
27. García-Arrarás JE, Díaz-Miranda L, Torres-Vázquez I, File S, Jiménez L, Rivera-Bermúdez K, Arroyo E, Cruz E: Regeneration of the enteric nervous system in the sea cucumber*Holothuria glaberrima*. J Comp Neurol. 1999, 406 (4): 461-475.
28. San Miguel-Ruiz JE, García-Arrarás JE: Common cellular events occur during wound healing and organ regeneration in the sea cucumber Holothuria glaberrima. BMC Dev Biol. 2007, 7: 115-
29. Cobb JLS: Neurobiology of the Echinodermata. Invertebrate Nervous Systems. Edited by: Ali MA. 1987, Plenum Press, New York, 483-525.
30. Pentreath VW, Cobb JLS: Neurobiology of the Echinodermata. Biol Rev. 1972, 47: 363-392.
31. Presnell JK, Schreibman MP: Humason's Animal Tissue Techniques. 1997, The Johns Hopkins Univ Press, Baltimore, 390-5
32. García-Arrarás JE, Schenk C, Rodrígues-Ramírez R, Torres II, Valentín G, Candelaria AG: Spherulocytes in the echinoderm Holothuria glaberrima and their involvement in intestinal regeneration. Dev Dyn. 2006, 235: 3259-3267.
33. Díaz-Miranda L, Blanco RE, García-Arrarás JE: Localization of the heptapeptide GFSKLYFamide in the sea cucumber *Holothuria glaberrima* (Echinodermata): a light and electron microscope study. J Comp Neurol. 1995, 352: 626-640.
34. Díaz-Miranda L, Pardo-Reoyo CF, Martínez R, García-Arrarás JE: Galanin-like immunoreactivity in the sea cucumber Holothuria glaberrima. Cell Tissue Res. 1996, 286: 385-391.
35. Diaz-Balzac CA, Santacana-Laffitte G, San Miguel-Ruiz JE, Tossas K, Valentin-Tirado G, Rives-Sanchez M, Mesleh A, Torres II, García-Arrarás JE: Identification of nerve plexi in connective tissues of the sea cucumber Holothuria glaberrima by using a novel nerve-specific antibody. Biol Bull. 2007, 213: 28-42.
36. Lacalli TC: Apical organs, epithelial domains, and the origin of the chordate nervous system. Am Zool. 1994, 34: 533-541.
37. Nielsen C: Origin of the chordate nervous system- and the origin of chordates. Dev Genes Evol. 1999, 209: 198-205.
38. Haag ES: Echinoderm rudiments, rudimentary bilaterians, and the origin of the chordates CNS. Evol Dev. 2005, 7: 280-281.
39. Butler EG, Ward MB: Reconstitution of the spinal cord following ablation in urodele larva. J Exp Zool. 1967, 160: 47-65.
40. Butler EG, Ward MB: Reconstitution of the spinal cord after ablation in adult Triturus. Dev Biol. 1967, 15: 464-486.
41. Shimizu I, Oppenheim RW, Obrien M, Shneiderman A: Anatomical and functional recovery following spinal cord transection in the chick embryo. J Neurobiol. 1990, 21: 918-937.
42. Saunders NR, Kitchener P, Knott GW, Nicholls JG, Potter A, Smith TJ: Development of walking, swimming and neuronal connections after complete spinal cord transection in the neonatal opossum Monodelphis domestica. J Neurosci. 1998, 18: 339-355.
43. Yamada H, Miyake T, Kitamura T: Regeneration of axons in transection of the carp spinal cord. Zoolog Sci. 1995, 12: 325-332.
44. Wang XM, Terman JR, Martin GF: Regeneration of supraspinal axons after transection of the thoracic spinal cord in the developing opossum, Didelphis virginiana. J Comp Neurol. 1998, 398: 83-97.
45. Simpson SB: Analysis of tail regeneration in the lizard *Lygosoma laterale*. J Morph. 1964, 114: 425-436.
46. Bernstein JJ: Relation of spinal cord regeneration to age in adult goldfish. Exp Neurol. 1964, 9: 161-174.
47. Beattie MS, Farooqui AA, Bresnahan JC: Review of current evidence for apoptosis after spinal cord injury. J Neurotrauma. 2000, 17: 915-925.

48. Beattie MS, Hermann GE, Rogers RC, Bresnahan JC: Cell death in models of spinal cord injury. Prog Brain Res. 2002, 137: 37-47.

49. Lu J, Ashwell KW, Waite P: Advances in secondary spinal cord injury: role of apoptosis. Spine. 2000, 25: 1859-1866.

50. Zupanc GK, Kompass KS, Horschke I, Ott R, Schwarz H: Apoptosis after injuries in the cerebellum of adult teleost fish. Exp Neurol. 1998, 152: 221-230.

51. Oppenheim RW: Cell death during development in the nervous system. Annu Rev Neurosci. 1991, 14: 453-501.

52. Anderson MJ, Waxman SG, Tadlock CH: Cell death of asynaptic neurons in regenerating spinal cord. Dev Biol. 1984, 103: 443-455.

53. Zupanc GK, Ott R: Cell proliferation after lesions in the cerebellum of adult teleost fish: time course, origin, and type of new cells produced. Exp Neurol. 1999, 160: 78-87.

54. Ferretti P: Neural stem cell plasticity: recruitment of endogenous populations for regeneration. Curr Neurovasc Res. 2004, 1: 215-229.

55. Ferretti P, Whalley K: Successful neural regeneration in amniotes: the developing chick spinal cord. Cell Mol Life Sci. 2008, 65: 45-53.

CHAPTER 9

VISCERAL REGENERATION IN A SEA CUCUMBER INVOLVES EXTENSIVE EXPRESSION OF SURVIVIN AND MORTALINHOMOLOGS IN THE MESOTHELIUM

Vladimir S Mashanov, Olga R Zueva, Carmencita Rojas-Catagena, Jose E Garcia-Arraras

BACKGROUND

The ability of echinoderms to repair their injured or autotomized body parts has been well known [1, 2]. One of the examples of such a remarkable capacity is rapid and complete regeneration of the digestive tube in holothurians (sea cucumbers) following induced or spontaneous evisceration (= autotomy of the viscera). Evisceration results in the loss of the entire digestive tube, except for small regions of the esophagus and cloaca (in some species, however, the esophagus and the pharynx are lost as well) [3, 4]. It has been shown that visceral regeneration in sea cucumbers is accomplished by massive remodeling of the remaining tissues of the mesentery and of the cloacal and esophageal stumps through a complex combination of morphogenetic events including de-differentiation of specialized cells, their migration, cell death, cell division, and re-differentiation [5–9]. Therefore, the regenerating digestive tube of sea cucumbers provides a unique experimental model for studying processes of extensive cell activation and proliferation without uncontrolled tumor formation. Another experimental advantage of this system is that the injury occurs by autotomy in pre-determined regions [3, 4], i.e., in a very consistent and repeatable manner, which excludes variation between animals in the extent and severity of the trauma. However, the molecular machinery underlying such an extraordinary plasticity in post-embryonic tissues remains largely unknown.

Post-traumatic regeneration, as any other developmental process, requires a tightly regulated interplay between cell proliferation and cell death. The present paper deals with two of the genes, *survivin and mortalin*, that are known to play a dual role in regulation of both programmed cell death and cell division in diverse groups of animals [10–14]. Although there are plenty of data on the involvement of *survivin* and *mortalin* in malignant diseases [12, 13,15–17], only rare studies have directly dealt with the functional significance of these two genes in post-traumatic regeneration [18–20].

Survivin (also known as BIRC5) is a small multifunctional protein, a member of the inhibitor of apoptosis (IAP) protein family [12]. IAPs are defined by the presence of the well-conserved baculovirus IAP repeat (BIR) domain [21, 22], which functions as a protein interaction module consisting of about 70 amino acids. This feature allows IAPs to control a wide variety of cellular pathways through cooperation with other polypeptides. Survivin interacts with many proteins that are important for regulation of both cell death and cell division.

Citation: Mashanov, V.S., Zueva, O.R., Rojas-Catagena, C. et al. Visceral regeneration in a sea cucumber involves extensive expression of survivin and mortalinhomologs in the mesothelium, BMC Dev Biol (2010) 10: 117. doi:10.1186/1471-213X-10-117.

For instance, like many other BIR-containing proteins, survivin suppresses apoptosis. After forming a complex with the co-factor protein HBXIP (hepatitis B X-interacting protein), survivin specifically binds pro-caspase 9, an initiator protease of the mitochondria-dependent apoptosis pathway [23]. Survivin also plays an important role in cell division and its expression was reported to overlap with several stem cell markers [24]. Through interaction with Ran, survivin protein regulates mitotic spindle formation [25]. In association with the proteins INCENP, aurora B, and borealin, survivin forms the multiprotein chromosomal passenger complex, which plays multiple roles in cell division, being involved, for instance, in correction of kinetochore attachment errors, assembly/stabilization of microtubules of the mitotic spindle, and completion of cytokinesis [26].

The *survivin* expression levels is usually high in most human cancers studied so far, but is largely absent from normal adult tissues, with a few notable exceptions, including the gastric mucosa, thymus, placenta, and testes [13, 27, 28]. Increased expression of *survivin* in cancer patients is considered an unfavorable prognostic marker correlating with decreased survival chances, risk of recurrence, metastasis, and resistance to anti-cancer drugs [13, 29]. During embryogenesis, *survivin* is prominently expressed in various (although not all) embryonic tissues, including the gastrointestinal tract, neural tube and blood vessels, in a developmentally regulated stage-dependent fashion, and knockdown experiments suggest that it is involved in regulation of neurogenesis, angiogenesis and hematopoiesis [30–32].

Mortalin (also known as Hspa9, Grp75, and PBP74) is a heat un-inducible member of Hsp70 family of proteins, which is cable of interacting with a variety of binding partners and performing various functions [14]. Like all Hsp70 family chaperones, mortalin is composed of two domains: an N-terminal nucleotide-binding (ATPase) domain and a C-terminal substrate binding domain [14]. As a chaperone, mortalin binds misfolded proteins and assists them to reach their functional configuration through ATP-dependent conformational change [33]. It is also involved in stress response and intracellular trafficking [14]. But most interestingly, mortalin is known to perform functions related to the control of cell proliferation and survival. It is known to bind the tumor suppressor protein p53 and therefore prevents the latter from inducing apoptosis and inhibiting cell division [34–36]. Elevated mortalin expression has been observed in many human tumors, with higher levels of mortalin expression corresponding to more aggressive tumor phenotypes [14, 16, 17]. Conversely, downregulation of mortalin expression was demonstrated to suppress the growth of human transformed cells [16].

Previous studies have demonstrated that successful visceral regeneration in sea cucumbers is accomplished through extensive proliferation of the tissues of the mesentery and the stumps of the gut [5, 7–9]. Cell death, however, has never been studied, although it is known to be equally important in regeneration [37, 38]. Here, we report that regeneration of the digestive tube in the sea cucumber *Holothuria glaberrima* involves a significant increase in cell death in comparison to non-injured animals and provide a detailed description of spatio-temporal expression pattern of *survivin* and *mortalin* transcripts.

METHODS

Animal Collection, Maintenance and Evisceration

Adult individuals of the sea cucumber *Holothuria glaberrima* Selenka, 1867 (Holothuroidea, Aspidochirota) were collected at low tide from the rocky coast adjacent to Old San Juan, Puerto Rico and immediately transported to the laboratory were they were allowed to adapt to laboratory conditions for 16 - 24 h in aerated seawater (brought from the sampling site) at room temperature. Evisceration was induced by injecting a few milliliters of 0.35 M KCl into the coelomic cavity. Eviscerated animals were kept in well-aerated in-door seawater tanks. The well-being of animals was ensured by keeping the density of holothurians in the tanks at approximately one individual per liter seawater. Seawater was changed on days 3, 7, 14, and 21 after evisceration.

Sequence Analysis

The homologs of *survivin* and *mortalin* were identified among 5173 EST sequences representing three cDNA libraries from the normal and regenerating digestive tube of *H. glaberrima* [39] by BLAST query against the non-redundant protein database of the NCBI. To obtain full-length cDNA sequences, we performed 5› and 3› RACE using SMARTer RACE cDNA Amplification kit (Clontech) following the manufacturer›s protocol. The resulting sequences of *survivin* and*mortalin* of *H. glaberrima* were deposited into the GenBank under accession numbers HQ174778 and HQ174779, respectively. Conserved domain search was performed by analyzing the predicted survivin and mortalin protein sequences with the online protein domain prediction program SMART [40] and InterProScan [41]. Coiled-coil regions were predicted using the COILS server at EMBnet (http://www.ch.embnet.org/software/COILS_form.html). Further sequence analysis was performed by aligning the predicted protein sequences of survivin and mortalin with the corresponding orthologs from other deuterostomes using ClustalX, version 2.0.10 [42]. Jalview version 2.4.0.b2 [43] was used for analysis of multiple sequence alignments.

Real-Time Quantitative RT-PCR

The non-eviscerated and regenerating individuals of *H. glaberrima* were anesthetized in 0.2% cholobutanol (1,1,1-trichloro-2-methyl-2-propanol hydrate) (Sigma) in seawater for 15 - 30 min at room temperature. In order to prevent possible RNA degradation during the subsequent dissection, the sedated holothurians were placed in ice-cold sea water and all manipulations were performed as quickly as possible. The animals were cut open along the dorsal interambulacrum, the normal gut or the anterior and posterior regenerates were excised, and immediately immersed in RNAlater (Ambion). Total RNA was isolated using TRI regent (Sigma) and treated with RNAse-free DNAse I (Qiagen) to minimize the noise due to possible genomic DNA contamination. First-strand cDNA was synthesized from 1 μg of the total RNA with random hexamer primers and ImPromt-II reverse transcriptase (Promega). PCR primers were designed using Primer Premier 5.0 software (PREMIER Biosoft International), and their sequences are shown in Additional File 1. qPCR reactions were set up in a reaction volume of 20 μl using PerfeCta SYBR Green Fast Mix (Quanta Biosciences) with the final concentration of the PCR primers of 200 nM and were then run on an iCycler iQ PCR Detection System (Bio-Rad) with the following parameters: 95°C for 10 min (denaturation step) followed by 45 cycles of 95°C for 30 sec, 55°C for 30 sec, and 72°C for 30 sec (amplification step). Fluorescence data were collected during the 72°C incubation phase. After amplification, melting curve analysis (55 - 95°C with a heating rate of 0.1°C/sec and a continuous fluorescence measurement) was performed for each of the PCR products to ensure the specificity of the reaction. Real-time PCR reactions were performed on three independent RNA samples purified from each of the regeneration stages as well as from the normal gut (biological replicates). All samples were analyzed in triplicate (technical replicates). The relative expression values of *survivin* and*mortalin* were normalized relative to the expression of the housekeeping gene *NADH dehydrogenase subunit 5* using equations from [44]. To determine the real-time PCR efficiencies, serial two-fold dilutions of cDNA templates were run in triplicates in a PCR reaction. The corresponding real-time PCR efficiencies (E) were calculated from the slope values produced by the iCycler software according to the equation: $E = 10^{(-1/slope)}$ [44]. The investigated transcripts showed amplification efficiencies of 2.00, 2.018, and 1.99, for *mortalin*,*survivin*, and *NADH dehydrogenase subunit 5*, respectively, with high linearity (correlation coefficient $R \geq 0.992$).

In Situ Hybridization

DIG-labeled riboprobes for in situ hybridization were synthesized from PCR-generated DNA templates. Briefly, RNA isolation and cDNA synthesis were performed as described above. The cDNA was amplified in a PCR reaction with gene-specific primers (without the RNA polymerase promoters at this stage) (Additional File 1) to generate what we call a pre-template for each of the genes of interest. The specificity of this PCR product was confirmed by direct sequencing. In the second set of PCR reactions, the templates were generated by amplifying the pre-templates with the appropriate primer (the reverse primer for the antisense probes, and the forward primer for the sense probes) containing the T7 RNA promoter sequence at the 5' end. The PCR products were then gel purified and used as templates to transcribe riboprobes with DIG RNA Labeling Kit (Roche) following the manufacturer's protocol. The *survivin* riboprobe targeted the last 20 nucleotides of the 5› UTR plus nucleotides 1 - 184 of the ORF. The *mortalin* riboprobe spanned nucleotides 1919 through 2353 of the ORF. Both antisense and sense probes were generated; the sense probes were used in the negative control reactions, and none of them showed detectable hybridization signal.

In situ hybridization staining was largely performed according to Holland et al. [45] with some modifications. Briefly, the animals were dissected as described above. Immediately upon excision, the tissue samples were briefly rinsed in ice-cold RNAse-free 0.01 M PBS (pH 7.4, 1030mOsm) and fixed overnight at 4°C in a freshly prepare mixture of 4% paraformaldehyde and 0.1% glutaraldehyde in PBS. After fixation, the samples were kept in 70% ethanol at -20°C. When needed, the samples were transferred to RNase-free 96-well cell culture plates, rinsed in PBS containing 0.1% Tween-20 (PBST), treated with 7.5 μg/ml proteinase K (Roche) for 10 min, acetylated sequentially in 0.25% and 0.5% acetic anhydride in 0.1 M triethanolamine, 5 min each. Prehybridization was performed at 60°C for 2 h or longer in hybridization buffer containing 5× SSC, 0.1% Tween-20, and 0.4 mg/ml salmon sperm DNA (Invitrogen). The riboprobes were diluted in warm (60°C) hybridization buffer to a final concentration of about 400 ng/ml and denatured at 80°C for 5 min. The hybridization was carried out at 58°C overnight in a hybridization oven equipped with a rocking platform. Stringency washes included four changes of 50% formamide in 5× SSC at 60°C, 50 min in 5× SSC at 37°C, 50 min in 2× SSC, and 15 min in 0.1× SSC at 50°C. The samples were then equilibrated in Washing Buffer (Roche) for 15 min at room temperature, followed by a 30 min incubation in Blocking Solution (Roche). Alkaline-phosphatase-conjugated anti-DIG antibodies (Roche) were diluted 1:2000 in Blocking Solution and applied overnight at 4°C. Excess antibody was removed by four washes in Washing Buffer, 20 min each. The samples were then equilibrated in four 10 min changes of a detection buffer containing 100 mM Tris-HCl pH 9.5, 100 mM NaCl and 0.1% Tween-20. Color reaction was performed at room temperature in the dark in a staining solution containing 4.5 μl of NBT stock solution (Roche) and 3.5 μl BCIP stock solution (Roche) per 1 ml of the detection buffer. The staining was stopped by two 10 min washes in a solution containing 10 mM Tris-HCl pH 8.1 and 1 mM EDTA. The samples were then equilibrated in PBST, postfixed in a mixture of 4% paraformaldehyde and 0.1% glutaraldehyde in PBS, cryoprotected in 10%, 20%, and 30% buffered sucrose, and incubated overnight in a 1:1 mixture of the 30% sucrose and the cryoembedding OCT medium (Takara) at 4°C. The samples were then frozen in the pure OCT medium. Serial cryosections were cut with a Leica CM1850 cryostat, collected onto gelatine-coated slides, dried overnight at 42°C and mounted in a mixture of 7.5% gelatine and 50% glycerol in 0.1 M PBS (pH 7.4). The preparations were analyzed and photographed with a Nikion Eclipse 600 microscope equipped with DIC optics and a SPOT RT3 digital camera (Diagnostic Instruments, Inc.).

All micrographs in the present paper represent transverse sections, which were cut orthogonal to the main axis of the organs. All figures are orientated with the ventral side of the animal to the bottom.

Quantification of Apoptosis

Tissue samples were obtained as described above and fixed overnight in 4% paraformaldehyde in 0.01 M PBS. Apoptosis was quantified by terminal deoxynucleotidyl transferase-mediated dUTP nick end labeling (TUNEL) with a Fluorescein FragEl DNA Fragmentation Detection kit (Calbiochem, Cat. QIA 39) on cryosections of the normal and regenerating gut. The percentage of TUNEL-positive cells was calculated by 'manually' counting FITC-positive cells and DAPI-positive nuclei on micrographs obtained with a 40× objective and imported into Fiji image processing software (http://pacific.mpi-cbg.de) with the Cell Counter plugin installed. Cell counting was performed on at least five 10 μm-thick cryosections per animal, and at least three animals were used per regeneration stage.

Double Labeling: In Situ Hybridization Combined with TUNEL Assay or BrdU Immunoistochemistry

In double labeling experiments, in situ hybridization was carried out first followed by either TUNEL assay of BrdU immunohistochemistry. The tissue processing was performed as described above, except that for the cell proliferation assay, the animals were injected with ~0.1 mg BrdU per animal 24 h before being sacrificed. To visualize BrdU incorporation, the sections of the in situ hybridization whole mounts were treated with 2 N HCl for 30 min at 37°C, the acid was neutralized with 0.1 M borate buffer (pH 8.5), followed by PBS washes, incubation in 0.1 M glycine (1 h) and 2% goat serum (1 h), and then the rat anti-BrdU antibody (GenWay, 20-783-71418) diluted at 1:400 were applied overnight at 4°C. Incubation in the goat anti-rat FITC conjugated secondary antibody (GenWay, 25-787-278232) diluted at 1:50 was performed for 1 h at RT.

Statistical Analysis

For evaluation of statistical differences between the non-eviscerated gut and the different stages visceral regeneration, we employed Welch's ANOVA and Welch's t-test, which do not assume samples having equal variances, and are, therefore, more suitable for biological samples of relatively small size than the ordinary Student's t-test [46]. All statistical analyses were performed in R package version 2.11.1 (http://cran.r-project.org/). All values are reported as mean ± standard error.

RESULTS

Orthologs of *Survivin* and *Mortalin* in *H. Glaberrima*

Sequences with significant (E value < 10^{-30}) similarity to database entries for deuterostome orthologs of *survivin* and *mortalin* were identified in the cDNA library derived from the regenerating gut of *H. glaberrima*. The deduced protein sequence of *H. glaberrima survivin*(143 amino acids long) was found to have a single highly conserved BIR domain (17-93 aa) (Figure 1, Additional File 2). Moreover, computer analysis strongly suggests that the survivin protein of *H. glaberrima* has a left-handed coiled-coil region at its C-terminus (121-143 aa) (Figure 1). The BIR domain is known to be essential both for apoptosis inhibition and mitosis-related functions, while the coiled-coil motif is thought to allow survivin protein to interact with microtubules of the mitotic spindle [13, 22]. The sea cucumber survivin protein exhibits 43-53% overall similarity with orthologs from other deuterostome species, with the greatest similarity (up to 69%) residing within the BIR domain (Additional files 2 and 3).

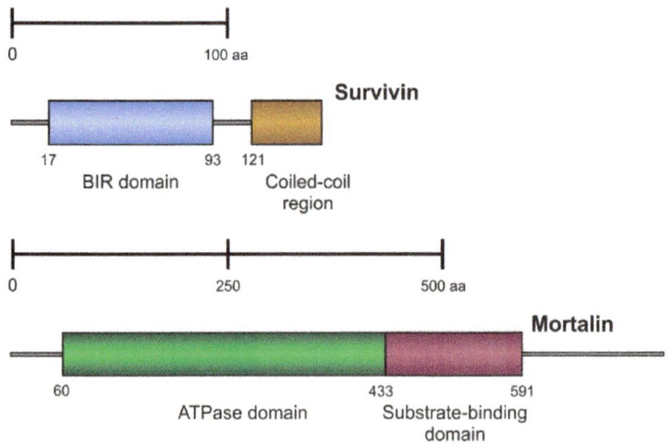

Figure 1: Domain organization of the predicted survivin and mortalin proteins of *H. glaberrima*.

The predicted sequence of mortalin is 752 amino acids long. As all Hsp70 family members [14], the deduced sequence of *H. glaberrima* contains an N-teminal ATPase domain followed by a substrate-binding domain (Figure 1). This coincides with the fact that the chaperoning functions of mortalin require multiple binding and release of the substrate peptide and are ATP-dependent [14, 33]. The sea cucumber mortalin protein shows a very high degree of identity (63.5 to 69.4%) with orthologs from other deuterostomes (Additional Files 4, 5, 6).

Overview of the Sea Cucumber Gut Organization and Evisceration Phenomenon

In order to make the reader familiar with the organization of the holothurian digestive system and the phenomenon of visceral regeneration, it may be helpful to provide a brief description here. For a more detailed reading, please refer to the previously published reviews and original papers (e.g., [1, 5, 8, 47–49]). As in other sea cucumber species [47, 49], the digestive tube of*H. glaberrima* consists of a pharynx, which lies within the so-called pharyngeal bulb, a short esophagus, an intestine, which is subdivided into the first descending, ascending and the second descending regions, and a cloaca (Figure 2A). The anterior regions of the digestive tube, including the esophagus and the first descending intestine, are suspended within the body cavity by the dorsal mesentery; the latter then continues into the lateral mesentery, which supports the ascending intestine, followed by the ventral mesentery attached to the second descending intestine in the posterior region of the body. The wall of the digestive tube consists of three histological layers: an inner digestive (luminal) epithelium, a connective tissue layer, and an outer mesothelium (also known as coelomic epithelium of the gut), which includes the gut musculature and a basiepithelial nervous plexus (Figure 2A). The gut mesothelium is continuous with the coelomic epithelia, which make up the mesentery.

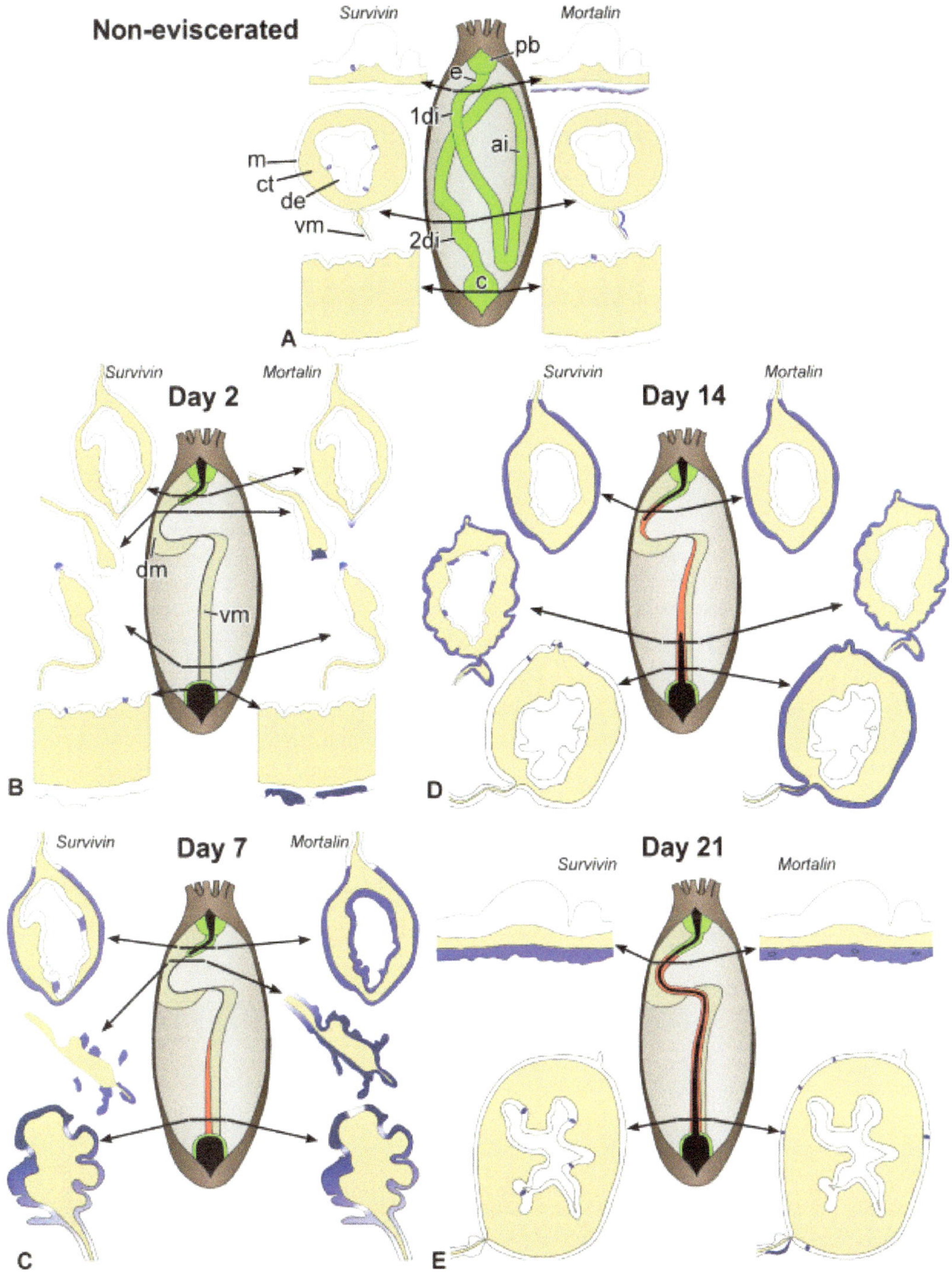

Figure 2: Diagram summarizing the anatomical features of the non-eviscerated (normal) and regenerating digestive tube and the expression patterns of*survivin* and *mortalin* in *H. glaberrima*. (A) Non-eviscerated animals (on the anatomical drawing, the gut mesenteria are not shown). (B) - (E) Regenerating animals at day 2, 7, 14, and 21, respectively. 1di - first descending intestine; 2di - second descending intestine; ai - ascending intestine; c - cloaca; ct - connective tissue layer; de - digestive (luminal) epithelium; dm - dorsal mesentery; e - esophagus; m - mesothelium; pb - pharyngeal bulb; vm - ventral mesentery. All anatomical drawings are positioned with the anterior to the top. The arrows indicate the position of the representative transverse sections. Colors indicate the following: blue - in situ hybridization signal; green - non-eviscerated (‹old›) tissues; red - regenerating (‹new›) tissues; black - lumen of the digestive tube. Not to scale.

Evisceration in *H. glaberrima* involves complete detachment of the intestine from the esophagus and the cloaca, and also from the mesentery [5]. The detached intestine, is then expelled through the rupture of the cloacal wall, and the wounds in the stumps of the esophagus and cloaca are initially sealed by muscular contraction before being healed.

Spatiotemporal pattern of *survivin*expression

In the normal gut, *survivin* transcripts are detected by in situ hybridization in scattered, but strongly labeled cells, which are often spherical in shape and are mostly localized in the basal region of the luminal epithelium. No labeling is seen in the mesothelium nor in the connective tissue layer of the gut wall (Figure 2A; 3A, B).

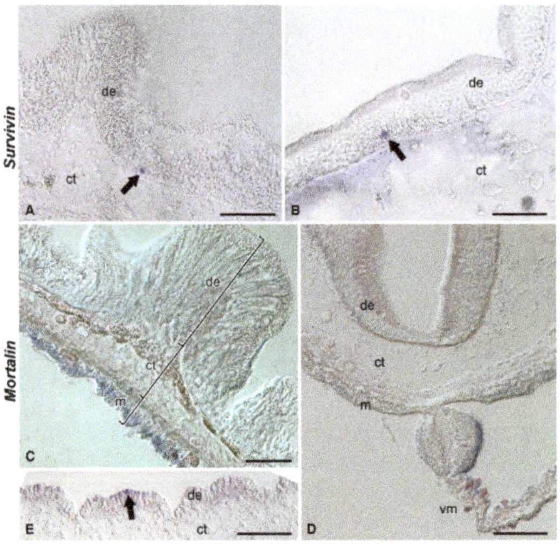

Figure 3: In situ hybridization. Expression of *survivin* (A and B) and *mortalin* (C - E) in the tissues of the digestive tube in non-eviscerated animals. (A) and (B) *survivin* expression in the scattered cells of the luminal epithelium in the esophagus (A) and the second descending intestine (B). (C) *mortalin* transcripts widely expressed in the apical region of the mesothelium in the esophagus. (D) Asymmetric distribution of *mortalin* transcripts in the distal region of the mesentery attached to the second descending intestine. (E)*mortalin*-expressing cell in the luminal epithelium of the cloaca. Arrows indicate rare cells showing in situ hybridization signal. de - digestive (luminal) epithelium; ct - connective tissue layer; m - mesothelium; vm - ventral mesentery. Scale bars = 50 µm in (A) - (C); 100 µm in (D); 200 µm in (E).

On days 2-3 following evisceration, the wound at the anterior end of the cloacal stump is healed. In the luminal epithelium of this region, strongly labeled scattered cells can be observed quite regularly (Figure 2B, 4A). Strong expression of *survivin* is also observed in the coelomic epithelium of the mesentery that runs forward from the cloaca. Interestingly, this expression is limited to a group of cells at the free margin of the mesentery and is absent from other areas (Figure 2B, 4B). The remnant (stump) of the esophagus at the anterior end of the animal also seals its wound at the autotomy plane, but no in situ hybridization signal is seen either in the stump, or in the anterior mesentery (Figure 2B, 4C).

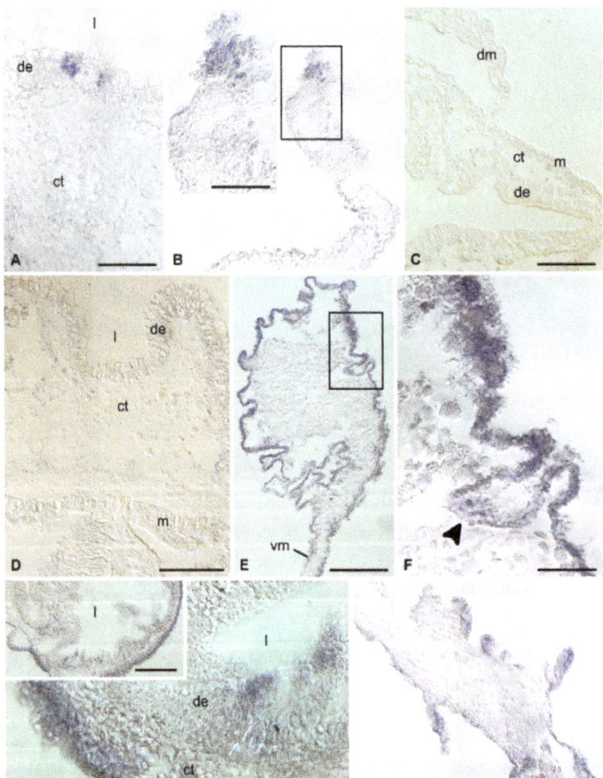

Figure 4: In situ hybridization. Expression of *survivin* at early stages of gut regeneration (days 2 - 6 after evisceration). (A) Digestive (luminal) epithelium of the cloaca on day 2. (B) Free distal edge of the posterior mesentery on day 2. The inset shows a higher magnification view of the boxed area. (C) The stump of the esophagus on day 2. (D) Wall of the cloaca on day 6. (E) A low-magnification view of the posterior regenerate on day 6. (F) Higher magnification of the boxed area on (E) showing a furrow (arrowhead) of the coelomic epithelium. (G) Esophageal stump on day 6. The inset shows a lower magnification view of the cross-section of the stump. (H) Free distal margin of the anterior mesentery on day 6. de - digestive (luminal) epithelium; dm - dorsal mesentery; ct - connective tissue layer; l - lumen of the gut; m - mesothelium; vm - ventral mesentery. Scale bars = 50 µm in (A), (B inset), (F) and (G); 100 µm in (B), (G inset), and (H); 200 µm in (C) - (E).

On days 6-7 after evisceration, no *survivin* transcripts are detected in the cloaca (Figure 4D). At this stage, a developing posterior rudiment can be clearly distinguished as a solid rod-like connective-tissue thickening running anteriorly from the cloaca along the free edge of the posterior mesentery. In cross sections, the thickening has a highly irregular shape and its covering mesothelium is thrown into numerous furrows of varying shape and size (Figure 2C, 4E, F). The mesothelium shows moderate to strong in situ signal, which is somewhat stronger at the antimesenterial side of the rudiment and gradually fades towards the mesenteric attachment (Figure 2C, 4E). It is worth noting that the expression is often weak or missing completely in the cells lining the bottom of the mesothelial folds (Figure 4F).

Moderate to strong *survivin* expression is also seen in the mesothelium and occasionally in the luminal epithelium of the healed posterior tip of the esophageal stump (Figure 2C, 4G) and also in the anterior mesentery that is attached to the tip of the esophageal stump. The hybridization signal is mostly confined to the tall folds of the mesothelium at the free distal edge of the mesentery (Figure 2C, 4H).

By day 12-14 after evisceration, the growing luminal epithelia of the esophagus and cloaca invade the connective tissue thickening of the anterior and posterior rudiments, respectively, thereby forming the inner tissue layer of the regenerating gut (Figure 2D). The posterior rudiment shows marked differences in *survivin* expression pattern along its length. In the posterior region, close to the cloaca, the in situ hybridization signal is almost completely absent from the tissues of the regenerate, with the exception of single cells or groups of a few cells in the mesothelium, predominantly on the anti-mesenterial side of the rudiment (Figure 2D, 5A, B). At more anterior levels of the posterior rudiment, *survivin* is expressed with varying intensity over most of the gut mesothelium with some cells showing particularly strong hybridization signal (Figure 2D, 5C). Some expression is also occasionally observed in the coelomic epithelium of the mesentery at its attachment to the posterior gut primordium (Figure 2D, 5C, D). In the luminal epithelium, weak to moderate in situ hybridization signal is often restricted to the apices of the irregularly shaped shallow folds, and this expression is evident mostly in the anti-mesenterial half of the rudiment (Figure 2D, 5C, E).

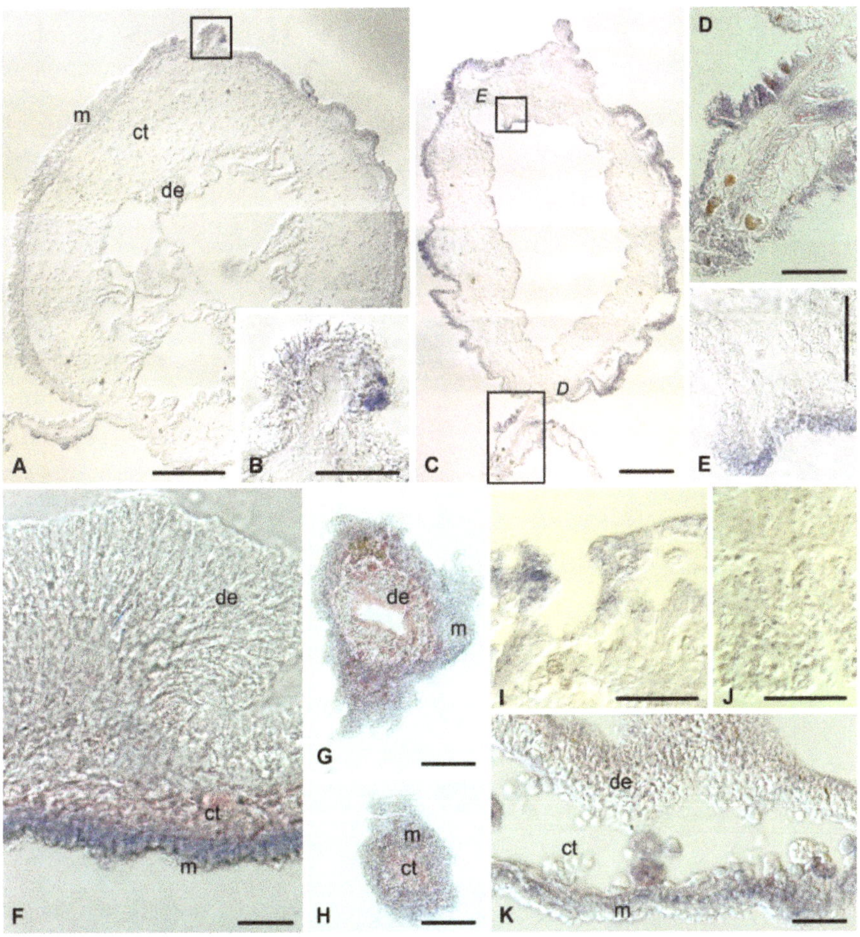

Figure 5: In situ hybridization. Expression of *survivin* at advanced stages of gut regeneration (days 12 - 21 after evisceration). (A) Section through the posterior rudiment at its attachment to the cloaca on day 12. (B) High magnification view of the boxed area on (A).(C) Section through the posterior regenerate at a more anterior level relative to (A). (D)and (E) show high magnification views of the boxed areas on (C), representing the mesenterial attachment and the digestive epithelium, respectively. (F) The wall of the esophageal stump on day 12. (G) and (H) Sequential sections trough the tip of the blindly ended anterior rudiment on day 12. (I) and (J) The wall of the posterior portion of the regenerated digestive tube on day 21, showing scatted labeled cells in the luminal (digestive) epithelium (I) and no labeling in the mesothelium (J). (K) The wall of the anterior region of the regenerated gut on day 21, showing expression of *survivin* in the mesothelium. de - digestive (luminal) epithelium; ct - connective tissue layer; m - mesothelium. Scale bars = 100 μm in (A), (D), (G), (H); 50 μm in (B), (E), (I), and (J); 200 μm in (C); 25 μm in (F) and (K).

In the anterior regenerate, *survivin* is broadly expressed at moderate to high levels in the mesothelium of both the esophageal stump and the newly created anterior rudiment, including its very tip (Figure 2D, 5F - H). However, no expression is detected in the luminal epithelium at any level along the anterior primordium.

By days 14 - 21 after evisceration, the continuous lumen is formed in the regenerating gut of eviscerated animals (Figure 2E). In cross sections, the organization of the gut is very similar to that of non-eviscerated animals. The expression pattern of *survivin* in the posterior portion of the newly regenerated gut (second descending region) strongly resembles that of the normal gut (Figure 2E, 5I, J), i.e., there are singly scattered strongly labeled cells in the digestive epithelium (Figure 5I), but no labeling in the mesothelium (Figure 5J). In the anterior portion of the gut, weak to moderate expression is still detected all over the mesothelium (Figure 2E, 5K).

Spatiotemporal Pattern of *Mortalin* Expression

As revealed by situ hybridization, some regions of the digestive tube, including the esophagus, second descending intestine, and cloaca express *mortalin* under normal conditions (Figure 2A,3C - E). In the esophagus, *mortalin* is widely expressed in the mesothelium of the gut wall, but within this epithelial layer the hybridization signal is restricted to the apical region, which is known to be occupied predominantly by cell bodies of peritoneocytes [47, 49] and where most of mesothelial cell division is observed (Additional File 7), and is absent from the basal region, where myoepithelial cells form the circular musculature of the gut (Figure 3C). In the second descending intestine, moderate to strong expression is observed in the apical region of the coelomic epithelium of the mesentery close to its attachment to the gut. This expression is highly asymmetrical with strong hybridization signal observed only on one side of the mesentery (Figure 2A, 3D). In the cloaca, very rare weakly labeled single cells or groups of a few cells are seen in the luminal epithelium (Figure 3E).

On days 2-3 after evisceration, moderate to strong in situ hybridization signal is seen in the mesothelium of the stump of the cloaca. This expression pattern is not continuous, but consists of patches of positive staining interspersed at irregular intervals (Figure 2B, 6A). Clear staining is also observed in the coelomic epithelial cells at the free margin of the posterior mesentery (Figure 2B, 6B, C). The level of *mortalin* expression in the mesothelium of the esophageal stump, when compared with that in the esophagus of non-eviscerated animals (Figure 3C), diminishes significantly, so that it mostly falls below the level of detection by in situ hybridization technique and only patches of very weak staining are occasionally observed in some areas of the mesothelium (Figure 2B, 6D). In contrast, the region of the mesentery that lies just posterior to the tip of the esophageal stump shows very intense labeling in some cells of the coelomic epithelium at its free edge (Figure 2B, 6E, F).

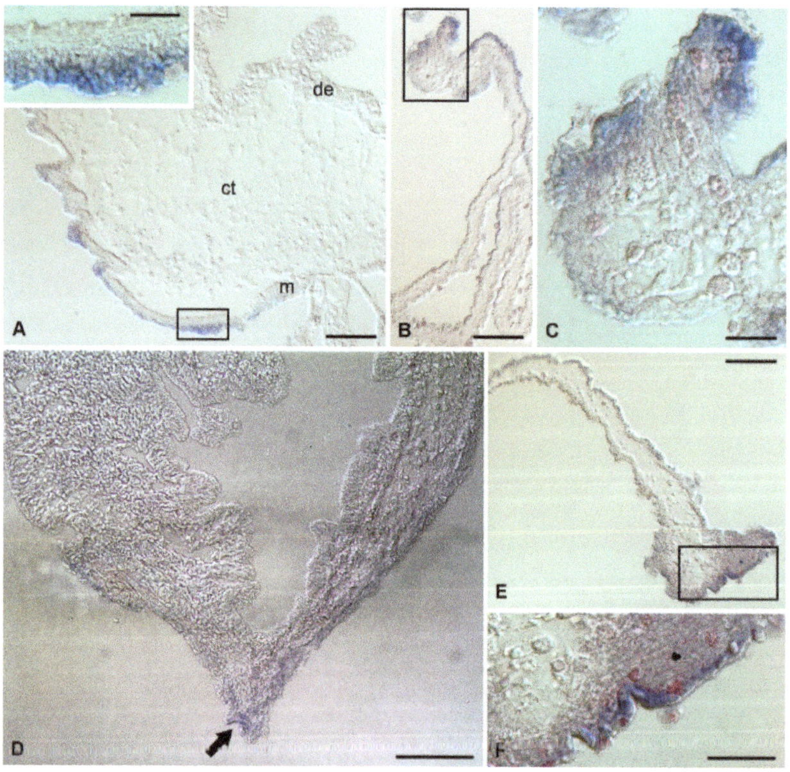

Figure 6: In situ hybridization. Expression of *mortalin* in gut tissues on day 2 after evisceration.(A) Patchy in situ hybridization signal in the mesothelium of the anterior region of the cloaca. The inset shows a detailed view of the mesothelium corresponding to the boxed area on the main image. (B) Low-magnification view of the posterior mesentery. (C) A higher magnification of the free distal edge of the posterior mesentery - boxed area in (B) - showing strong hybridization signal in some cells of the mesothelium. (D) Ventral (anti-mesenterial) region of the esophageal stump showing very weak and restricted in situ hybridization signal (arrow) in the mesothelium. (E) Low-magnification view of the anterior mesentery. (F) Magnified view of the boxed area in (E) showing strongly labeled cells in the mesothelium of the free distal edge of the mesentery. de - digestive (luminal) epithelium; ct - connective tissue layer; m - mesothelium. Scale bars = 100 µm in (A), (B),(D), and (E); 25 µm in (A inset) and (C); 50 µm in (F).

On days 6 - 7, *mortalin* is widely expressed in the mesothelium of the irregularly shaped posterior rudiment (Figure 2C, 7A). However, the expression is often weak or completely undetectable in the anti-mesenterial region of the rudiment and also at the bottom of the mesothelial furrows. In the stump of the esophagus and in the mesentery running posteriorly from the stump tip, *mortalin* is expressed all over the mesothelium at moderate to high levels (Figure 2C, 7B, C). At the sealed tip of the stump, where the lumen ends blindly, most of the cells of the digestive epithelium also show weak to moderate hybridization signal (Figure 7B).

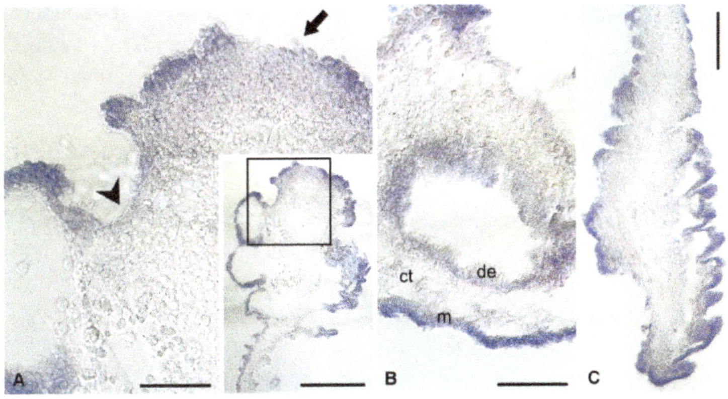

Figure 7: In situ hybridization. Expression of *mortalin* in gut tissues on day 6 after evisceration.(A) In situ hybridization signal in the mesothelium of the posterior rudiment. Note that the signal is weak or completely absent from the anti-mesenterial region of the rudiment (arrow) and from the bottom of mesothelial furrows (arrowhead). The inset shows a low-magnification view of the posterior rudiment with the boxed area corresponding to the main image. (B) The posterior tip of the esophageal stump. Note strong in situ hybridization signal in the mesothelium and moderate signal in the luminal (digestive) epithelium. (C) Strong expression of *mortalin* in the mesothelium of the anterior mesentery. de - digestive (luminal) epithelium; ct - connective tissue layer; m - mesothelium. Scale bars = 50 µm in (A); 200 µm in (A inset); 100 µm in (B) and (C).

On days 12-14 after evisceration, *mortalin* is strongly expressed all along the mesothelium of the posterior regenerate, including the anti-mesenterial region (Figure 2D, 8A). There are some local variations in the intensity of the signal between adjacent regions of the mesothelium in cross-sections, but those variations does not form any regular pattern. The localization of the hybridization signal shows no considerable differences along the rudiment either. No labeling is detected in the luminal epithelium of the posterior gut primordium.

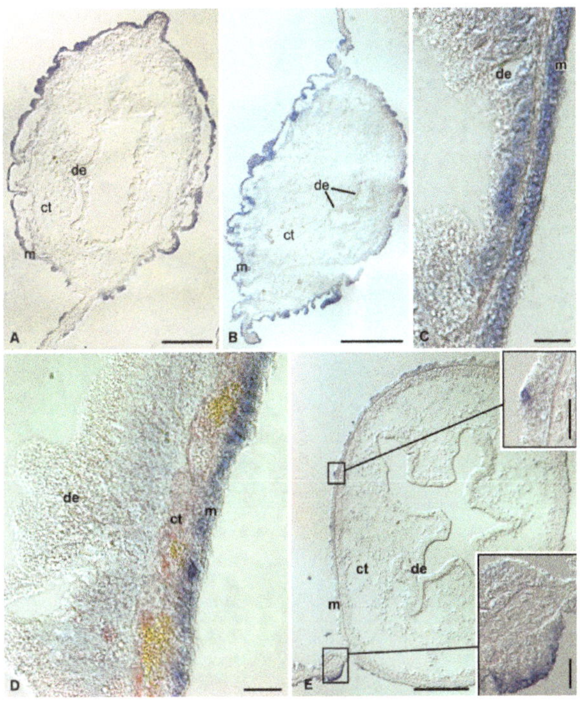

Figure 8: In situ hybridization. Expression of *mortalin* at advanced stages (days 12 - 21) of gut regeneration. (A) Posterior rudiment on day 12 after autotomy. (B) Growing posterior tip of the anterior rudiment on day 12 after evisceration. (C) The wall of the esophageal stump on day 12 after evisceration. (D) The wall of the newly regenerated posterior regions of the esophagus on day 21 after evisceration. (E) The second descending part of the newly regenerated intestine on day 21 after evisceration. The inserts show higher magnification view of the asymmetrical expression of *mortalin* in the mesothelium of the mesentery attachment and also strongly labeled singly scattered cells in other regions of the mesothelium. de - digestive (luminal) epithelium; ct - connective tissue layer; m - mesothelium. Scale bars = 200 µm in (A) and (E); 500 µm in (B); 25 µm in (C), (D) and(E insets).

The distribution of the *mortalin* in situ hybridization signal in the anterior rudiment is similar to that in the posterior rudiment, i.e., moderate to strong expression is seen mostly in the mesothelium (Figure 2D, 8B) with little or no variation between different regions in cross-sections and along the length of the primordium. Unlike at the previous stage, no expression is seen in the luminal epithelium at the growing tip of the anterior regenerate (Figure 8B), but*mortalin* transcripts are occasionally detected in the groups of cells in the luminal epithelium of the esophageal stump (Figure 8C).

On days 14 - 21, the spatial expression pattern of *mortalin* in the tissues of the newly regenerated digestive tube does not differ much from that of the non-eviscerated individuals (Figure 2E, 8D, E). For instance, moderate to strong hybridization signal is seen all over the mesothelium of the anterior region of the regenerate (esophagus) (Figure 8D) and on one of the sides of the mesentery attachment to the posterior regenerate (second descending part of the intestine) (Figure 8E). The only difference from the expression pattern in the intact gut is the presence of single scattered intensely labeled cells in the mesothelium at various levels along the regenerate. No significant hybridization signal is detected in the digestive (luminal) epithelium at this stage.

Qualitative Assessment of Transcript Abundance

The overall relative abundance of the *survivin* and *mortalin* transcripts in the tissue samples was assessed by real-time qualitative PCR. All data were compared to the normal gut and shown as fold change.

Both *survivin* (Figure 9A, B) and *mortalin* (Figure 9C, D) showed different temporal expression profiles in the anterior and posterior regenerates of the digestive tube. No significant difference in the overall *survivin* expression was found between the posterior regenerate at any stage of re-growth and the intact digestive tube ($p \geq 0.2269$) (Figure 9B), whereas the anterior regenerate shows a marginally insignificant increase in *survivin* mRNA level on days 7 ($p = 0.067$) and 14 ($p = 0.073$), which becomes a significant ($p = 0.037$) three-fold increase by day 21 after evisceration (Figure 9A).

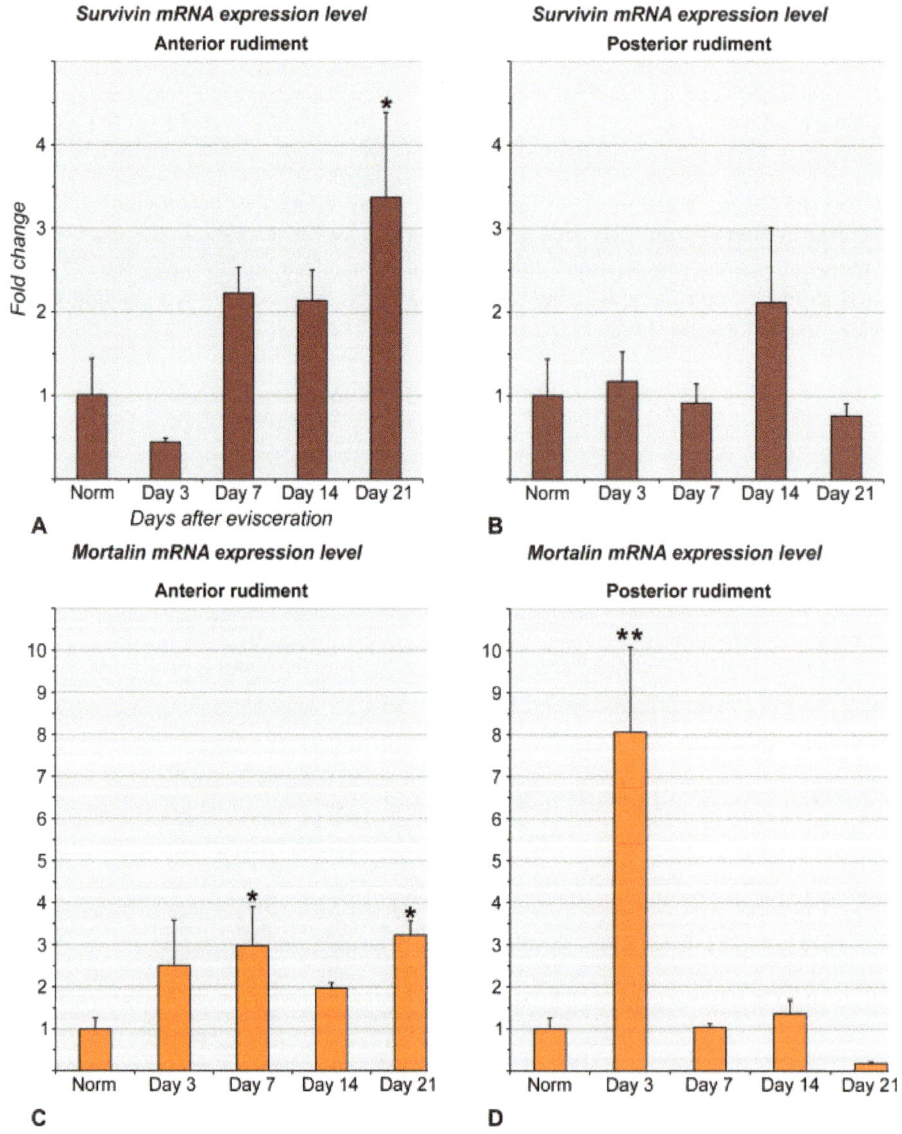

Figure 9: Real-time RT-qPCR. Overall abundance of *survivin* and *mortalin* transcripts in the regenerating digestive tube. (A) and (B) *Survivin* expression in the anterior and posterior regenerates, respectively. (C) and (D) *Mortalin* expression in the anterior and posterior regenerates, respectively. Transcript abundance is expressed as x-fold relative to the normal gut. Results are represented as mean ± standard error. $*P < 0.05$, $**P < 0.01$

Real-time RT-PCR analysis of *mortalin* transcript abundance in the anterior rudiment reveals two peaks of roughly 3-fold up-regulation, one on day 7 (p = 0.036) and another one on day 21 (p = 0.019) (Figure 9C), while, in the posterior rudiment, the overall *mortalin* expression level shows a highly significant increase as early as on day 3 after evisceration (p = 0.003) and then returns to the approximately normal level on days 7 to 21 (Figure 9D).

Apoptosis

Since both survivin and mortalin are known to act as anti-apoptotic proteins, we performed TUNEL assay to examine the extent of programmed cell death in the normal and regenerating digestive tube. Figure 10 shows the diagrams of the temporal changes in the percentage of TUNEL-positive cells in the regenerating digestive tube, and Figure 11 and 12 are representative micrographs used in cell counting assays.

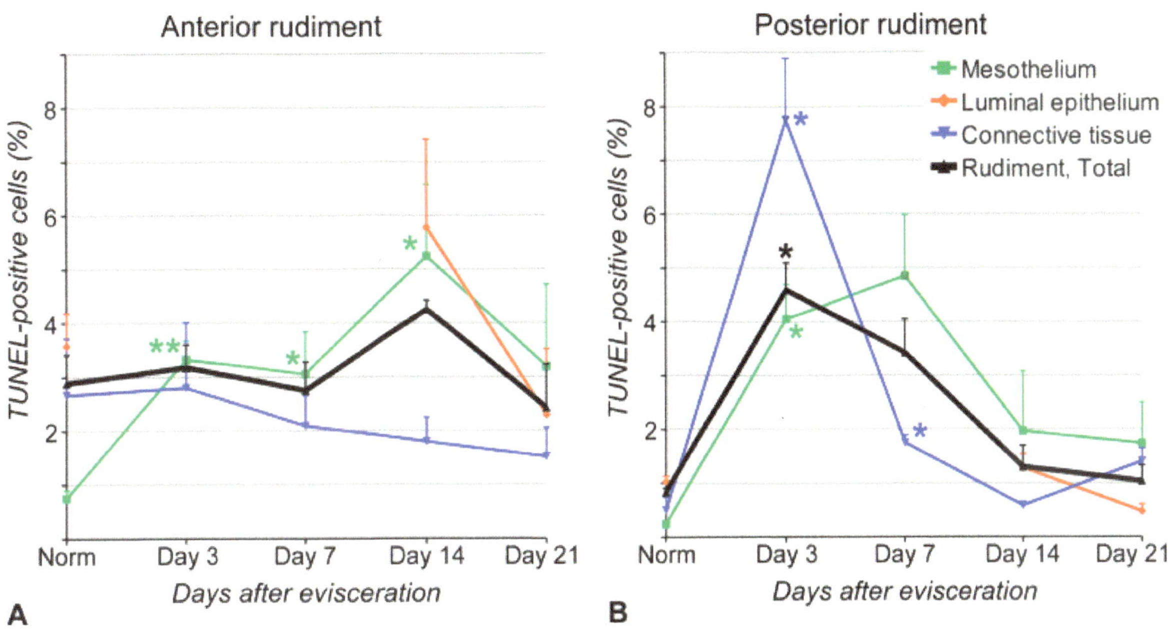

Figure 10: TUNEL assay. Percentage of apoptotic cells in tissue layers of the normal and regenerating digestive tube. (A) Cell death in the anterior regenerate. (B) Apoptosis in the posterior regenerate. Results are represented as mean ± standard error. *$P < 0.05$, **$P < 0.01$.

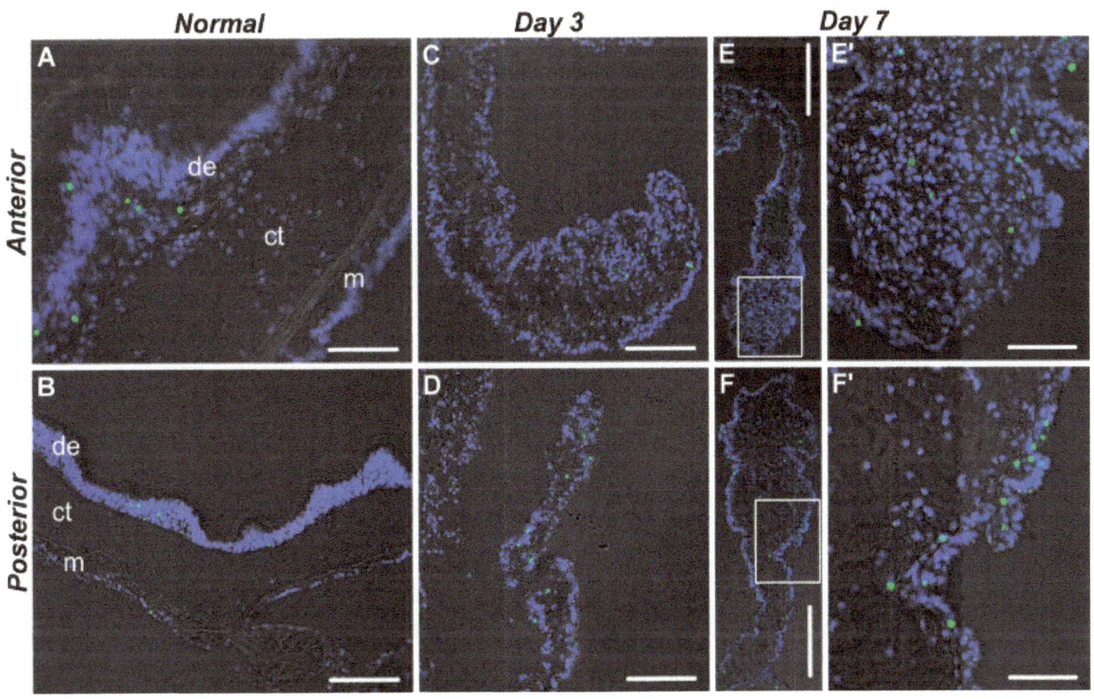

Figure 11: Representative micrographs of the distribution of TUNEL-positive cells (green) in the normal gut and in the early regenerates (days 3 - 7). (A) Wall of the esophagus in a non-eviscerated animal. (B) Wall of the second descending intestine in a non-eviscerated animal. (C) and (D) The anterior and posterior regenerates, respectively, on day 3. (E) and (F) General view of the anterior and posterior regenerates, respectively, on day 7. (E›) and (F›) Higher magnification of the boxed areas on (E) and(F), respectively. de - digestive (luminal) epithelium; ct - connective tissue layer; m - mesothelium. TUNEL-positive cells are green; nuclei were stained with DAPI and are shown in blue. Scale bars = 50 µm in (A), (E›), and (F›); 100 µm in (B) - (D); 200 µm in(E) and (F).

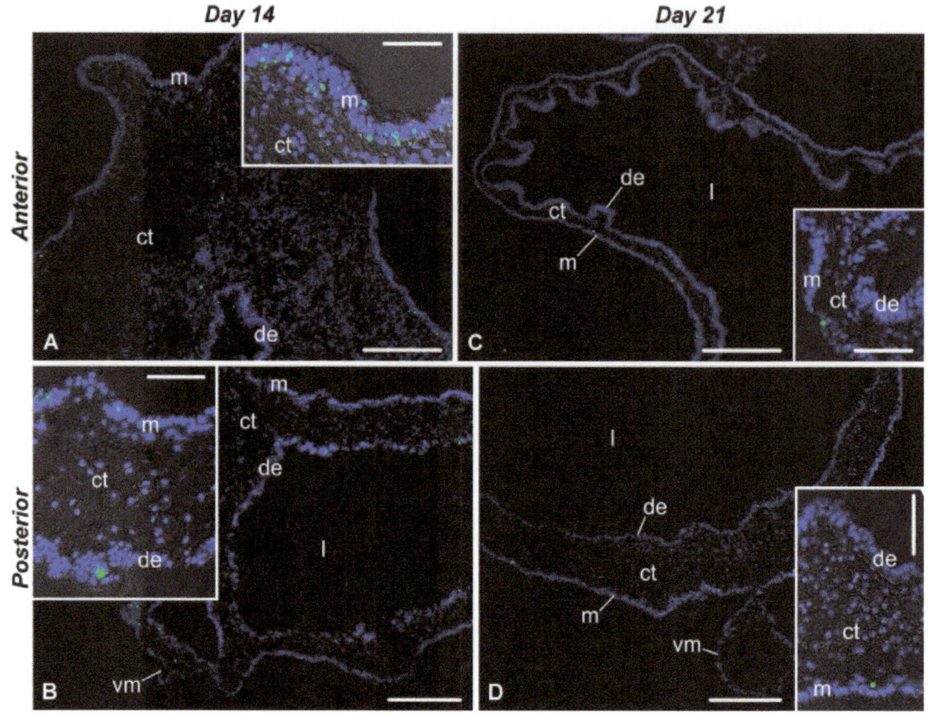

Figure 12: Representative micrographs of the distribution of TUNEL-positive cells during the late phase (days 14 and 21) of visceral regeneration. (A) and (B) The anterior and posterior regenerates, respectively on day 14. (C) and (D) The newly regenerated posterior region of the esophagus and the second descending intestine on day 21 after evisceration. Insets show higher magnification views of the gut wall. de - digestive (luminal) epithelium; ct - connective tissue layer; l - gut lumen; m - mesothelium; vm - visceral mesentery. TUNEL-positive cells are green; nuclei were stained with DAPI and are shown in blue. Scale bars = 200 μm in (A) - (D); 50 μm in all insets.

As could be expected, some cell death occurs even in the tissues of the non-eviscerated digestive tube, the dying cells being mostly restricted to the luminal epithelium (Figure 10, 11A, B). The apoptotic cells are significantly (Welch›s t-test, $p = 0.03$) more abundant in the anterior regions (esophagus) of the digestive tube ($2.88 \pm 0.53\%$ of the total cell number), than in the posterior part ($0.83 \pm 0.08\%$). As early as on day 3 after evisceration, the mesothelium of both the anterior and posterior regenerates shows a significant increase (4-fold and almost 17-fold, respectively) in percentage of TUNEL-positive cells (with the corresponding p-values of 0.009 and 0.03, respectively) (Figure 10, 11C, D). In the mesothelium of the anterior regenerate, this elevated level of cells death persists until day 14 and then declines at later stages (Figure 10A). In the mesothelium of the posterior rudiment, the increase in percentage of TUNEL-positive cells remains almost significant until day 7 ($p = 0.056$), before declining later on (Figure 10).

Cells in the connective tissue of the anterior and posterior rudiments respond differently to injury. In the anterior rudiment, no significant changes in percentage of TUNEL-positive cells were observed (Welch›s ANOVA, $F(4, 5.9) = 0.32$, $p = 0.86$) at any of the regeneration stages studied, whereas the connective tissue layer of the posterior rudiment shows a sharp increase in cell death on day 3 after autotomy ($p = 0.02$), which is followed by a rapid decline, and, by day 14 of regeneration, the percentage of TUNEL-positive cells resumes its normal values (Figure10). As the new luminal epithelium develops on days 14 - 21 after evisceration, it does not show any stage-dependent variation in the overall rate of cell death (anterior rudiment: $F(2, 3.5) = 1.22$, $p = 0.4$; posterior rudiment: $F(2, 3.6) = 6.24$, $p = 0.07$, Welch›s ANOVA) (Figure 10, 12).

Multiple Labeling

To obtain some insight into possible function(s) of *survivin* and *mortalin* in sea cucumber gut regeneration, we performed double labeling experiments by subjecting the samples of the posterior gut rudiment at the stage of 7 days after evisceration (extensive expression of both genes, elevated levels of both cell death and proliferation [5]) to in situ hybridization followed by either TUNEL assay or BrdU immunohistochemistry. In the lateral and anti-mesenterial regions of the regenerate, where *survivin* transcripts are most abundant, TUNEL-positive cells are rare. However, the region where the newly developing gut attaches to the mesentery is characterized both by a weaker *survivin* mRNA hybridization signal and an increased abundance of apoptotic cells in the mesothelium (Figure 13A-C). Therefore, there is a negative correlation between the survivin level and the extent of cell death in the coelomic epithelial cells. The spatial relationship between *mortalin* expression and the cell death is less straightforward, because, although *mortalin* transcripts are seen in the lateral regions of the regenerate, where cell death is less extensive than in the mesenterial attachment, they are often absent from the mesothelium covering the anti-mesenterial region of the rudiment, where the apoptotic cells are scarcely seen (Figure 13D - F). Combined in situ hybridization and BrdU immunohistochemistry shows that the localization of the cell division in the regenerating mesothelium largely coincides with the expression domains of *survivin* and *mortalin* (Figure 14).

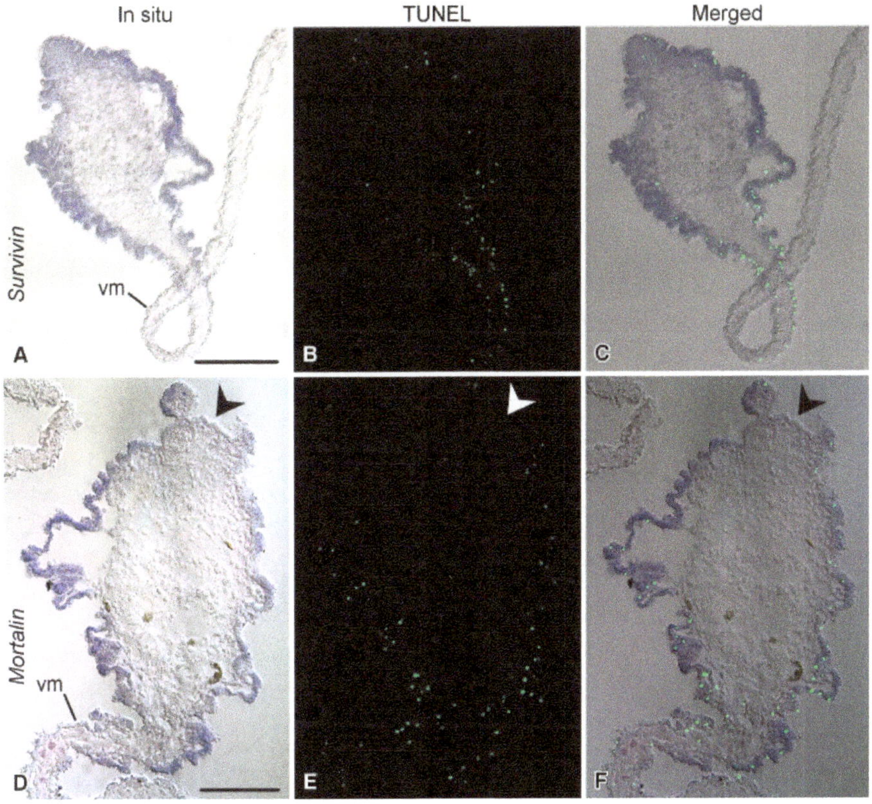

Figure 13: Double labeling with riboprobes for *survivin* and *mortalin* (blue) and TUNEL assay (green) on the posterior regenerate on day 7. (A) - (C) *Survivin* riboprobe and TUNEL assay. (D) - (F) *Mortalin* riboprobe and TUNEL assay. vm - ventral mesentery. Arrowhead on (D) - (F) marks the anti-mesenterial region of the rudiment, where *mortalin* transcript are absent. Note a negative correlation between the localization of *survivin* in situ hybridization signal and the density of the TUNEL-positive cells (A) -(C). Scale bars = 100 μm.

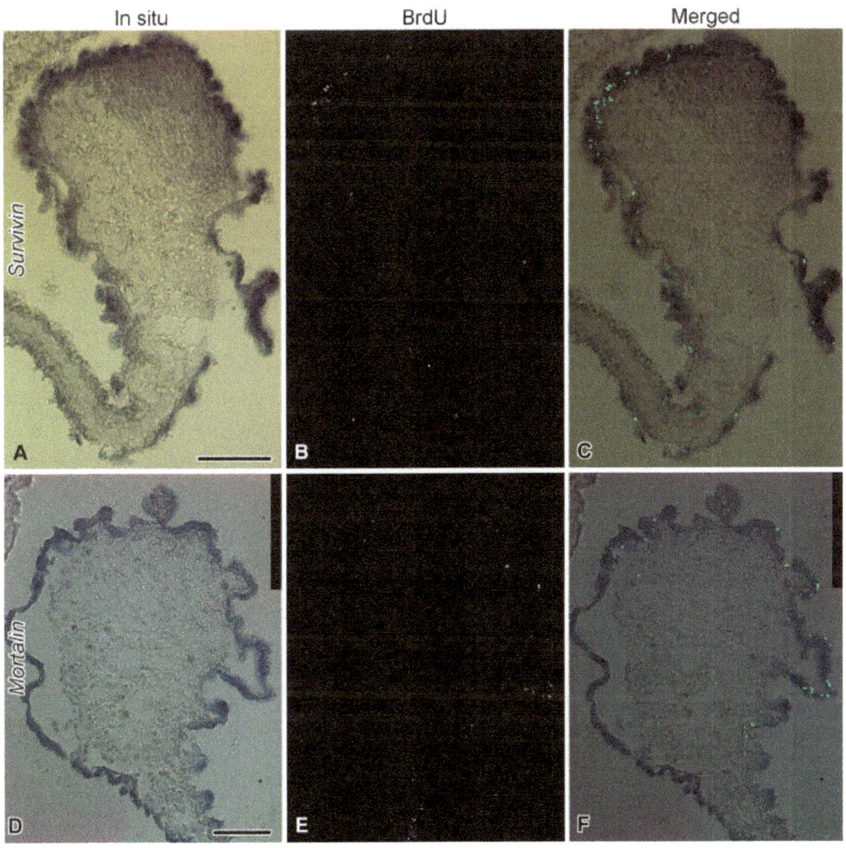

Figure 14: Double labeling with riboprobes for *survivin* and *mortalin* (blue) and BrdU immunocytochemistry (green) on the posterior regenerate on day 7. (A) - (C)*Survivin* riboprobe and BrdU immunohistochemstry. (D) - (F) *Mortalin* riboprobe and BrdU immunohistochemistry. Note that BrdU-incorporating cells are mostly distributed within the expression domains of *survivin* (A) - (C) and *mortalin* (D) - (F). Scale bars = 100 μm.

DISCUSSION

Cell death and cell division are the two fundamental processes that create tissue homeostasis. The ability to tightly control them is of vital importance for any multicellular organism. Although one can intuitively perceive that regeneration will certainly shift the balance in favor of cell proliferation, induction of apoptosis is nevertheless known to be equally important for successful regeneration, and, moreover, can be absolutely required to trigger tissue repair [37,38]. Visceral regeneration in the sea cucumber *H. glaberrima* involves both extensive cell proliferation, as was documented earlier [5], and a transient increase in programmed cell death in the regrowing tissues, as we have shown here. The present study examines the expression pattern of *survivin* and *mortalin*, known to be involved in both mitosis and apoptosis in various animal taxa, in the normal and regenerating gut of the sea cucumber.

Both genes show certain basal levels of expression in the digestive tube of non-eviscerated sea cucumber individuals (Figure 2A, 9). Survivin protein is believed to be absent from most of the adult tissues of vertebrates. Notable known exceptions include organs with high rate of physiological cell turnover, such as thymus and gastric mucosa [13, 27, 28]. In our study, we detected *survivin* transcripts in single cells widely scattered throughout the luminal (digestive) epithelium of the sea cucumber. Unlike *survivin*, *mortalin* is almost entirely absent from the luminal epithelium of the normal gut with the exception of vary rare cells in the digestive epithelium of the cloaca. However, *mortalin* is widely expressed in the mesothelium (coelomic epithelium) of the esophagus and also shows an interesting asymmetric expression pattern at the attachment of the mesentery to the posterior region of the intestine. Functional significance of the expression of the two genes in the normal digestive tube is not yet clear, but since both epithelia of the holothurian gut are known to slowly self-renew [49] (Additional File 7), involvement in cell turnover could be a possible explanation.

Visceral autotomy (evisceration) in sea cucumbers triggers a series of developmental events, such as apoptosis, cell division, migration of individual cells and epithelial sheets, connective tissue remodeling, de-differentiation, trans-differentiation, and re-differentiation [5, 8, 48, 50]. These events, although partially overlapping in time and space, unfold in a certain temporal order to eventually result in the successful regeneration of the digestive tube. Likewise, both*survivin* and *mortalin* show time-dependent changes in their expression in the regenerating digestive tube of the sea cucumber. These changes do not necessarily manifest themselves in increased or decreased numeric values of the overall relative abundance of mRNA transcripts, but can involve mostly changes in the spatial distribution of the transcripts instead. For instance, although the overall quantity of *survivin* transcripts in the posterior gut regenerate of*H. glaberrima* does not change significantly (relative to the non-eviscerated gut), in situ hybridization shows marked spatial alterations of the expression pattern as regeneration progresses.

In situ hybridization revealed that both *survivin* and *mortalin* are much more abundantly expressed in the mesothelium of the regenerating gut, than in the luminal epithelium, and that the two genes are completely absent from the cells of the connective tissue layer. The visceral mesothelium of echinoderms shows a complex histological organization [8, 47, 49]. It is mostly made up of specialized peritoneocytes and myoepithelial cells, which are assembled in a highly organized architecture, and also contains a basiepithelial nerve plexus. In spite of this high level of histological complexity, the mesothelium of sea cucumbers is known to possess an extraordinary histogenetic potential even in adult animals. Shortly after evisceration, the mesothelia of the mesentery and the remaining portions of the digestive tube (stumps) undergo drastic de-differentiation, which transforms the highly specialized tissue into a layer of greatly simplified peritoneal and myoepithelial cells, which lose their characteristic properties, such as long basal processes and myofilaments, respectively [6–8, 51]. In this dedifferentiated condition, the mesothelium undergoes extensive cell division and expands to accommodate the connective tissue swelling, which is being developed along the free edge of the mesentery. The combination of the connective tissue thickening with the surrounding dedifferentiated mesothelium constitutes the early regenerate of the digestive tube. As regeneration progresses, the mesothelium gradually resumes its normal organization, i.e., undergoes re-differentiation.

In the regenerating digestive tube of *H. glaberrima*, the most extensive expression of *survivin*and *mortalin* in the mesothelium occurs on days 3 through 14 after evisceration. This is exactly the time when the mesothelium shows a significant increase in cell death (this study) and cell division [5], and undergoes the dedifferentiation - expansion - redifferentiation cycle [5, 7, 8]. Therefore, since the activation of *survivin* and *mortalin* expression coincides in time with the major morphogenic processes, the two genes are likely to be involved somehow in the remodeling of the gut mesothelium during regeneration. For instance, the peak in cell proliferation in the regenerating gut mesothelium of *H. glaberrima* occurs on days 7 though 14 after evisceration [5], and since both genes are widely expressed in the same regions where proliferation occurs (this study) at these stages, it is possible that they are involved in regulation of cell division.

There is a discrepancy in the literature on the anti-apoptotic function of *survivin* in different regenerating tissues. In human and rodent liver regeneration, *survivin* up-regulation was related to cell proliferation, but not to apoptosis inhibition [18, 19]. On the other hand, a role for*survivin* in suppression of the programmed cell death was demonstrated in traumatic brain injury in rats [52]. Combined in situ hybridization and TUNEL labeling of the regenerating gut of *H. glaberrima* showed that, although *survivin* expression and the increase in cell death rate occurred concomitantly in the mesothelium, the strongest *survivin* hybridization signal and the highest abundance of apoptotic cells were mostly localized to different territories (the basal plus lateral surfaces of the rudiment and the mesenterial attachment, respectively), suggesting an anti-apoptotic role for *survivin* in the regenerating coelomic epithelium of the holothurian gut. No such clear relationship exists between the cell death and *mortalin* expression, which can be explained by a variety of other functions that the mortalin protein is known to perform [14].

In terms of cell sources of regeneration, there are two major groups of events: those that involve some kind of undifferentiated reserve/stem cells (such as neoblasts in planarians) [53] and those that rely on the plasticity of the existing differentiated cells (as in case of mammalian pancreatic beta-cells and liver regeneration after acute injury) [54, 55]. Previous electron microscopy studies [6–8] clearly demonstrated that both the luminal epithelium and the mesothelium of the sea cucumber digestive tube regenerate through induction of extensive proliferation of the differentiated cells resulting in expansion of the tissue layers of the gut stumps into the regenerate. The peritoneal and myoepithelial cells of the mesothelium undergo drastic dedifferentiation by losing their characteristic features and enter the cell cycle, but remain nevertheless connected to each other by intercellular junctions within the epithelial sheet. The present study shows that the mesothelium in this dedifferentiated condition expresses*mortalin* and *survivin*. In this regard, it is important to note that the expression of these two genes is known to be associated with stem cells. *Mortalin*, for instance, is constitutively expressed by planarian neoblasts and its knockdown results in inability to regenerate and maintain normal cell turnover [20]. *Survivin* is known to be expressed in stem cells of a variety of tissues undergoing cell turnover [24, 56], where it is though to contribute to stem cell maintenance and protection from cell death. Therefore, the results of the present study combined with the data obtained earlier, suggest that, although the mesothelium of the sea cucumber gut is devoid of resident stem cells, most of the mesothelial cells themselves temporarily acquire some stem cell properties through reversible dedifferentiation. Those properties include the absence of specialized cytoplasmic features, ability to go through cell divisions, and expression of *survivin* and *mortalin*. It is worth mentioning here that in vitro studies of the cells derived from sea cucumber visceral regenerates showed that only the cells obtained during the phase of extensive dedifferentiation and proliferation, were capable of sustained growth in culture [9].

It is not clear why the transcripts of both *survivin* and *mortalin* are much less abundant in the regenerating luminal epithelium, than in the mesothelium. The luminal epithelium *H. glaberrima*, as in other members of the order Aspidochirota, also regenerates via proliferation of the enterocytes that remain in the esophageal and cloacal stumps after evisceration [5, 7, 9], i.e. employs the same basic mechanisms, as the mesothelium. The obvious explanation is that regeneration of the luminal epithelium may employ additional pathways, besides inducing extensive expression of *survivin* and *mortalin*, to coordinate cell death and/or proliferation. This conclusion is in line with findings that, contrary to a previous belief, the survivin protein is not absolutely required to prevent cell death during mitosis [57].

The differences in regeneration mechanisms of the same tissue between different species or between different developmental stages of the same animal are not uncommon and are not surprising. However, visceral regeneration in holothurians provides an intriguing example of how different cellular and/or molecular mechanisms can be employed simultaneously in the same organ of the same individual. One of the most extreme studied cases is gut regeneration in a dendrochirotid holothurian *Eupentacta fraudatrix* [8]. In this species, the luminal epithelium in the posterior gut rudiment, as could be expected, develops from the luminal epithelium of the cloacal stump, whereas the luminal epithelium of the anterior regenerate develops from the mesodermally derived cells of the mesothelium. In spite of different origin, the anterior and posterior luminal epithelia are indistinguishable from each other in histological and ultrastructural organization once regeneration is completed. In *H. glaberrima*, the species used in the present study, the differences in mechanisms between the anterior and posterior rudiments are not as prominent, but still are present (Figure 2, 9, 10). They include different timing of *survivin* and *mortalin* expression peaks, somewhat different pattern of spatial distribution of the transcripts of the two genes, as well as some differences in the programmed cell death dynamics, particularly in the connective tissue layer. It is not clear yet whether the differences in regeneration mechanisms between the anterior and posterior regenerates are related to the oral-aboral polarity of the animal, reflect the evolutionary history of the regenerative mechanisms, or have any adaptive significance.

CONCLUSIONS

All developmental events (broadly defined) including embryogenesis, postnatal cell turnover, tumor formation, and regeneration rely on the balance between cell division and cell death. Understanding the basic mechanisms that regulate these two processes is not only of great academic interest but also holds promise for medical advances. The present study examines the expression pattern of *survivin* and *mortalin*, two genes known to be involved in regulation of both cell division and apoptosis, in the regenerating viscera of the sea cucumber *Holothuria glaberrima*. In response to injury, both genes show changes in the spatial distribution of the transcripts and/or in the overall abundance of the transcripts in the gut regenerates. Although the two genes show some expression in the regenerating luminal epithelium (at certain stages, in certain regions), the most extensive expression is seen in the mesothelium (the outer layer of the gut) at days 6 through 14, the stage, at which the mesothelial cells are known to be dedifferentiated and engaged in extensive proliferation [5, 7–9]. Our data also show elevated levels of cell death in the regenerating mesothelium. Double labeling experiments suggest that both genes are likely to support cell proliferation in the regenerating gut, while *survivin* might also be involved in apoptosis suppression. It also cannot be ruled out that the two genes play some other additional functions in the regenerating tissues.

The very fact that *survivin* and *mortalin* are expressed in the sea cucumber digestive tube raises an interesting question. Since both genes are known to be involved in carcinogenesis [14,58], why is it that tumor formation has never been reported in studies of visceral regeneration in holothurians or documented in animals captured in the wild? In metazoans (multicellular organisms) with a relatively long life span, the ability to replace worn-out cells under normal conditions and/or replenish the cell mass lost to injury strongly correlates with the presence of potent tumor suppression mechanisms that keep the rate of cell division within secure limits to match the interests of the organism as a whole [59]. Sea cucumbers are characterized by a relatively long life span, estimated at about four to ten years [60], they constantly renew cells in their adult tissues, including the digestive tube [49] and, most interestingly, they can quickly regrow most of their tissues after traumatic injury, autotomy, or seasonal atrophy [1, 2] and regenerate the same structure multiple times over their lifetime. Therefore, sea cucumbers, and echinoderms in general, must have evolved a particularly strong set of antitumor mechanisms, further studies of which could improve our understanding of relationships between embryogenesis, cancer and regeneration, might help us to devise more effective cancer treatment strategies.

ACKNOWLEDGEMENTS

The authors thank Dr. A. Vazquez for her advice on the real-time qPCR technique and the Sequencing and Genotyping Facility of the University of Puerto Rico (and personally Mrs. D. Rodriguez) for providing sequencing services in an efficient and timely manner. Special thanks are to Dr. A.M. Najakshin for his invaluable advice on molecular biology techniques. This research was funded by NIH (grant number: 1SC1GM084770-01), NSF (grant number: IOS-0842870), and the University of Puerto Rico.

AUTHORS' CONTRIBUTIONS

VSM, ORZ, CRC, and JGA conceived the study and interpreted the results. CRC and JGA performed the EST data analysis and identified the genes. VSM and ORZ carried out the experimental procedures and analyzed the data. VSM drafted the manuscript. VSM, ORZ, and JGA finalized the manuscript. All authors read and approved the final version of the manuscript.

REFERENCES

1. Hyman L: The Invertebrates. IV. Echinodermata. The Celomate Bilateria. 1955, McGraw-Hill Book Co. Inc., New York
2. Candia Carnevali M: Regeneration in echinoderms: repair, regrowth, cloning. ISJ. 2006, 3: 64-76.
3. Byrne M: The morphology of autotomy structures in the sea cucumber *Eupentacta quinquesemita* before and during evisceration. J Exp Biol. 2001, 204: 849-863. PubMed
4. Wilkie IC: Autotomy as a prelude to regeneration in echinoderms. Microsc Res Tech. 2001, 55: 369-396. 10.1002/jemt.1185.CrossRefPubMed
5. García-Arrarás JE, Estrada-Rodgers L, Santiago R, Torres II, Díaz-Miranda L, Torres-Avillán I: Cellular mechanisms of intestine regeneration in the sea cucumber, *Holothuria glaberrima* Selenka (Holothuroidea:Echinodermata). J Exp Zool. 1998, 281: 288-304. 10.1002/(SICI)1097-010X(19980701)281:4<288::AID-JEZ5>3.0.CO;2-K. CrossRefPubMed
6. Mashanov V, Dolmatov I: Regeneration of the digestive tract in the pentactulae of the far-eastern holothurian *Eupentacta fraudatrix* (Holothuroidea, Dendrochirota). Invertebrate Reproduction and Development. 2001, 39: 143-151.CrossRef
7. Shukalyuk A, Dolmatov I: Regeneration of the digestive tube in the holothurian*Apostichopus japonicus* after evisceration. Russian Journal of Marine Biology. 2001, 27: 168-173. 10.1023/A:1016717502616.CrossRef
8. Mashanov VS, Dolmatov IY, Heinzeller T: Transdifferentiation in holothurian gut regeneration. Biol Bull. 2005, 209: 184-193. 10.2307/3593108. CrossRefPubMed
9. Odintsova N, Dolmatov I, Mashanov V: Regenerating holothurian tissues as a source of cells for long-time cell culture. Marine Biology. 2005, 146: 915-921. 10.1007/s00227-004-1495-3.CrossRef
10. Fortugno P, Wall NR, Giodini A, O›Connor DS, Plescia J, Padgett KM, Tognin S, Marchisio PC, Altieri DC: Survivin exists in immunochemically distinct subcellular pools and is involved in spindle microtubule function. J Cell Sci. 2002, 115: 575-585.PubMed
11. Johnson ME, Howerth EW: Survivin: a bifunctional inhibitor of apoptosis protein. Vet Pathol. 2004, 41: 599-607. 10.1354/vp.41-6-599.CrossRefPubMed
12. Altieri DC: New wirings in the survivin networks. Oncogene. 2008, 27: 6276-6284. 10.1038/onc.2008.303. PubMedCentralCrossRefPubMed
13. Mita AC, Mita MM, Nawrocki ST, Giles FJ: Survivin: key regulator of mitosis and apoptosis and novel target for cancer therapeutics. Clin Cancer Res. 2008, 14: 5000-5005. 10.1158/1078-0432.CCR-08-0746. CrossRefPubMed
14. Kaul SC, Deocaris CC, Wadhwa R: Three faces of mortalin: a housekeeper, guardian and killer. Exp Gerontol. 2007, 42: 263-274. 10.1016/j.exger.2006.10.020.CrossRefPubMed
15. Gianani R, Jarboe E, Orlicky D, Frost M, Bobak J, Lehner R, Shroyer KR: Expression of survivin in normal, hyperplastic, and neoplastic colonic mucosa. Hum Pathol. 2001, 32: 119-125. 10.1053/hupa.2001.21897. CrossRefPubMed
16. Wadhwa R, Takano S, Kaur K, Deocaris CC, Pereira-Smith OM, Reddel RR, Kaul SC: Upregulation of mortalin/mthsp70/Grp75 contributes to human carcinogenesis. Int J Cancer. 2006, 118: 2973-2980. 10.1002/ijc.21773.CrossRefPubMed
17. Yi X, Luk JM, Lee NP, Peng J, Leng X, Guan X, Lau GK, Beretta L, Fan S: Association of mortalin (HSPA9) with liver cancer metastasis and prediction for early tumor recurrence. Mol Cell Proteomics. 2008, 7: 315-325.CrossRefPubMed
18. Deguchi M, Shiraki K, Inoue H, Okano H, Ito T, Yamanaka T, Sugimoto K, Sakai T, Ohmori S, Murata K, Furusaka A, Hisatomi H, Nakano T: Expression of survivin during liver regeneration. Biochem Biophys Res Commun. 2002, 297: 59-64. 10.1016/S0006-291X(02)02128-9.CrossRefPubMed
19. Baba HA, Wohlschlaeger J, Schmitz KJ, Nadalin S, Lang H, Benesch A, Gu Y, Biglarnia A, Sotiropoulos GC, Takeda A, Takeda N, von Wnuck Lipinski K, Levkau B: Survivin is upregulated during liver regeneration in rats and humans and is associated with hepatocyte proliferation. Liver Int. 2009, 29: 585-592. 10.1111/j.1478-3231.2008.01911.x.CrossRefPubMed
20. Conte M, Deri P, Isolani ME, Mannini L, Batistoni R: A mortalin-like gene is crucial for planarian stem cell viability. Dev Biol. 2009, 334: 109-118. 10.1016/j.ydbio.2009.07.010.CrossRefPubMed

21. Dubrez-Daloz L, Dupoux A, Cartier J: IAPs: more than just inhibitors of apoptosis proteins. Cell Cycle. 2008, 7: 1036-1046.CrossRefPubMed

22. Yue Z, Carvalho A, Xu Z, Yuan X, Cardinale S, Ribeiro S, Lai F, Ogawa H, Gudmundsdottir E, Gassmann R, Morrison CG, Ruchaud S, Earnshaw WC: Deconstructing Survivin: comprehensive genetic analysis of Survivin function by conditional knockout in a vertebrate cell line. J Cell Biol. 2008, 183: 279-296. 10.1083/jcb.200806118. PubMedCentralCrossRefPubMed

23. Marusawa H, Matsuzawa S, Welsh K, Zou H, Armstrong R, Tamm I, Reed JC: HBXIP functions as a cofactor of survivin in apoptosis suppression. EMBO J. 2003, 22: 2729-2740. 10.1093/emboj/cdg263. PubMedCentralCrossRefPubMed

24. Li F, Cheng Q, Ling X, Stablewski A, Tang L, Foster BA, Johnson CS, Rustum YM, Porter CW: Generation of a Novel Transgenic Mouse Model for Bioluminescent Monitoring of Survivin Gene Activity in Vivo at Various Pathophysiological Processes. Survivin Expression Overlaps with Stem Cell Markers. Am J Pathol. 2010

25. Xia F, Canovas PM, Guadagno TM, Altieri DC: A survivin-ran complex regulates spindle formation in tumor cells. Mol Cell Biol. 2008, 28: 5299-5311. 10.1128/MCB.02039-07.PubMedCentralCrossRefPubMed

26. Ruchaud S, Carmena M, Earnshaw WC: The chromosomal passenger complex: one for all and all for one. Cell. 2007, 131: 230-231. 10.1016/j.cell.2007.10.002.CrossRefPubMed

27. Ambrosini G, Adida C, Altieri DC: A novel anti-apoptosis gene, survivin, expressed in cancer and lymphoma. Nat Med. 1997, 3: 917-921. 10.1038/nm0897-917. CrossRefPubMed

28. Li F, Brattain MG: Role of the Survivin gene in pathophysiology. Am J Pathol. 2006, 169: 1-11. 10.2353/ajpath.2006.060121.PubMedCentralCrossRefPubMed

29. Li F: Survivin study: what is the next wave?. J Cell Physiol. 2003, 197: 8-29. 10.1002/jcp.10327. CrossRefPubMed

30. Adida C, Crotty PL, McGrath J, Berrebi D, Diebold J, Altieri DC: Developmentally regulated expression of the novel cancer anti-apoptosis gene survivin in human and mouse differentiation. Am J Pathol. 1998, 152: 43-49. PubMedCentralPubMed

31. Ma AC, Lin R, Chan P, Leung JCK, Chan LYY, Meng A, Verfaillie CM, Liang R, Leung AYH: The role of survivin in angiogenesis during zebrafish embryonic development. BMC Dev Biol. 2007, 7: 50-10.1186/1471-213X-7-50.PubMedCentralCrossRefPubMed

32. Delvaeye M, Devriese A, Zwerts F, Betz I, Moons M, Autiero M, Conway EM: Role of the 2 zebrafish survivin genes in vasculo-angiogenesis, neurogenesis, cardiogenesis and hematopoiesis. BMC Dev Biol. 2009, 9: 25-10.1186/1471-213X-9-25. PubMedCentralCrossRefPubMed

33. Bukau B, Horwich AL: The Hsp70 and Hsp60 chaperone machines. Cell. 1998, 92: 351-366. 10.1016/S0092-8674(00)80928-9.CrossRefPubMed

34. Wadhwa R, Yaguchi T, Hasan MK, Mitsui Y, Reddel RR, Kaul SC: Hsp70 family member, mot-2/mthsp70/GRP75, binds to the cytoplasmic sequestration domain of the p53 protein. Exp Cell Res. 2002, 274: 246-253. 10.1006/excr.2002.5468.CrossRefPubMed

35. Ma Z, Izumi H, Kanai M, Kabuyama Y, Ahn NG, Fukasawa K: Mortalin controls centrosome duplication via modulating centrosomal localization of p53. Oncogene. 2006, 25: 5377-5390. 10.1038/sj.onc.1209543.CrossRefPubMed

36. Iosefson O, Azem A: Reconstitution of the mitochondrial Hsp70 (mortalin)-p53 interaction using purified proteins--identification of additional interacting regions. FEBS Lett. 2010, 584: 1080-1084. 10.1016/j.febslet.2010.02.019.CrossRefPubMed

37. Tseng A, Adams DS, Qiu D, Koustubhan P, Levin M: Apoptosis is required during early stages of tail regeneration in *Xenopus laevis*. Dev Biol. 2007, 301: 62-69. 10.1016/j.ydbio.2006.10.048. PubMedCentralCrossRefPubMed

38. Li F, Huang Q, Chen J, Peng Y, Roop DR, Bedford JS, Li C: Apoptotic cells activate the "phoenix rising" pathway to promote wound healing and tissue regeneration. Sci Signal. 2010, 3: ra13-10.1126/scisignal.2000634. PubMedCentralPubMed

39. Rojas-Cartagena C, Ortíz-Pineda P, Ramírez-Gómez F, Suárez-Castillo EC, Matos-Cruz V, Rodríguez C, Ortíz-Zuazaga H, García-Arrarás JE: Distinct profiles of expressed sequence tags during intestinal regeneration in the sea cucumber *Holothuria glaberrima*. Physiol Genomics. 2007, 31: 203-215. 10.1152/physiolgenomics.00228.2006. PubMedCentralCrossRefPubMed

40. Letunic I, Doerks T, Bork P: SMART 6: recent updates and new developments. Nucleic Acids Res. 2009, 37: D229-32. 10.1093/nar/gkn808. PubMedCentralCrossRefPubMed

41. Zdobnov EM, Apweiler R: InterProScan--an integration platform for the signature-recognition methods in InterPro. Bioinformatics. 2001, 17: 847-848. 10.1093/bioinformatics/17.9.847.CrossRefPubMed

42. Larkin MA, Blackshields G, Brown NP, Chenna R, McGettigan PA, McWilliam H, Valentin F, Wallace IM, Wilm A, Lopez R, Thompson JD, Gibson TJ, Higgins DG: Clustal W and Clustal × version 2.0. Bioinformatics. 2007, 23: 2947-2948. 10.1093/bioinformatics/btm404. CrossRefPubMed

43. Waterhouse AM, Procter JB, Martin DMA, Clamp M, Barton GJ: Jalview Version 2--a multiple sequence alignment editor and analysis workbench. Bioinformatics. 2009, 25: 1189-1191. 10.1093/bioinformatics/btp033. PubMedCentralCrossRefPubMed

44. Pfaffl MW: A new mathematical model for relative quantification in real-time RT-PCR. Nucleic Acids Res. 2001, 29: e45-10.1093/nar/29.9.e45. PubMedCentralCrossRefPubMed

45. Holland L, Holland P, Holland N: Whole mount in situ hybridization applicable to Amphioxus and other small larvae. Molecular zoology: advances, strategies and

protocols. Edited by: Ferraris J, Palumbi S. 1996, 476-483.

46. Ruxton G: The unequal variance t-test as an underused alternative to Student's t-test and the Mann-Whitney U test. Behavioral Ecology. 2006, 17: 688-690. 10.1093/beheco/ark016.CrossRef
47. Feral J, Massin C: Digestive system: Holothuroidea. Echinoderm nutrition. Edited by: Jangoux M, Lawrence J, Balkema. 1982, 192-212.
48. García-Arrarás JE, Greenberg MJ: Visceral regeneration in holothurians. Microsc Res Tech. 2001, 55: 438-451. 10.1002/jemt.1189.CrossRefPubMed
49. Mashanov V, Frolova L, Dolmatov I: Structure of the digestive tube in the holothurian*Eupentacta fraudatrix* (Holothuroidea:Dendrochirota). Russian Journal of Marine Biology. 2004, 30: 314-322. 10.1023/B:RUMB.0000046550.23343.b4.CrossRef
50. Quiñones JL, Rosa R, Ruiz DL, García-Arrarás JE: Extracellular matrix remodeling and metalloproteinase involvement during intestine regeneration in the sea cucumber*Holothuria glaberrima*. Dev Biol. 2002, 250: 181-197. 10.1006/dbio.2002.0778.CrossRefPubMed
51. García-Arrarás JE, Dolmatov IY: Echinoderms: potential model systems for studies on muscle regeneration. Curr Pharm Des. 2010, 16: 942-955. 10.2174/138161210790883426. PubMedCentralCrossRefPubMed
52. Johnson EA, Svetlov SI, Wang KKW, Hayes RL, Pineda JA: Cell-specific DNA fragmentation may be attenuated by a survivin-dependent mechanism after traumatic brain injury in rats. Exp Brain Res. 2005, 167: 17-26. 10.1007/s00221-005-2362-2.CrossRefPubMed
53. Gurley K, Sanchez Alvarado A: Stem cells in animal models of regeneration. Stembook. 2008, The Stem Cell Research Community
54. Dor Y, Melton DA: How important are adult stem cells for tissue maintenance?. Cell Cycle. 2004, 3: 1104-1106. CrossRefPubMed
55. Kung JWC, Currie IS, Forbes SJ, Ross JA: Liver development, regeneration, and carcinogenesis. J Biomed Biotechnol. 2010, 984248-
56. Marconi A, Dallaglio K, Lotti R, Vaschieri C, Truzzi F, Fantini F, Pincelli C: Survivin identifies keratinocyte stem cells and is downregulated by anti-beta1 integrin during anoikis. Stem Cells. 2007, 25: 149-155. 10.1634/stemcells.2006-0165.CrossRefPubMed
57. Yang D, Welm A, Bishop JM: Cell division and cell survival in the absence of survivin. Proc Natl Acad Sci USA. 2004, 101: 15100-15105. 10.1073/pnas.0406665101.PubMedCentralCrossRefPubMed
58. Ryan BM, O'Donovan N, Duffy MJ: Survivin: a new target for anti-cancer therapy. Cancer Treat Rev. 2009, 35: 553-562. 10.1016/j.ctrv.2009.05.003.CrossRefPubMed
59. Pearson B, Sanchez Alvarado A: Regeneration, stem cells, and the Evolution of Tumor Supression. Cold Spring Harbor Symposia on Quantitative Biology. 2008, 1-8. LXXIII
60. Storelli M, Storelli A, Marcotrigiano G: Heavy metals in the aquatic environment of the Southern Adriatic Sea, Italy. Macroalgae, sediments, and benthic species. Environment International. 2001, 26: 505-509. 10.1016/S0160-4120(01)00034-4.

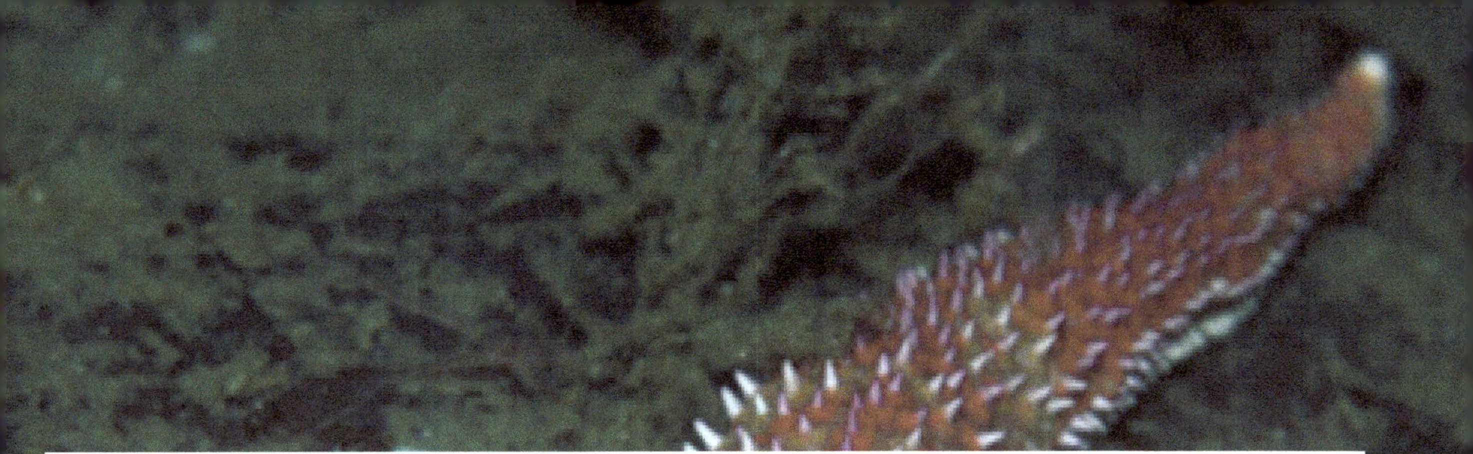

CHAPTER 10 SYNERGY OF MICROCAPSULE POLYSACCHARIDES ANDBACILLUS SUBTILIS ON THE GROWTH, IMMUNITY AND RESISTANCE OF SEA CUCUMBER APOSTICHOPUS JAPONICUSAGAINST VIBRIO SPLENDIDUS INFECTION

Ying Fan, Xiaoqing Yu, La Xu, Shuxian Wang, Haibin Ye, Jing Diao, Xiusheng Yang, Tianbao Li

Keywords: Apostichopus japonicas, Polysaccharide, Bacillus subtilis, Growth, Immunity, Vibrio splendidus

INTRODUCTION

As a traditional food and invigorant, sea cucumber is nowadays extensively cultivated in China, which also brings severe diseases. That's why antibiotics and chemotherapeutics are used today. However, abuse of these results in the spread of drug-resistant pathogens, environmental pollution and unexpected residues in aquaculture [1–4]. As an echinoderm species, sea cucumbers lack an adaptive immune system, and its key defenses against different substances are cellular and humoral immune responses. So the most promising method for controlling sea cucumber disease in aquaculture is to strengthen its defense mechanisms by prophylactic administration of immunostimulants [5].

Many Chinese herbs, such as *Astragalus membranaceus* and tuckahoe, have been used as immune boosters for nearly 2000 years [6–9]. Many feeding trials and other tests have shown that polysaccharides have significant immunostimulatory effects through different mechanisms, for example by activating mouse B cells and macrophages, regulating the intestinal microbiota to improve productive performance, as well as enhancing the nonspecific immunity of some animals, such as lysozyme, superoxide, and alkaline phosphatase activities in fish or *A. japonicus* [10–13]. But application of microcapsule technology in polysaccharides has not been investigated, and has not been applied in *A. japonicus*, either.

Most researchers applied only Chinese herb or *Bacillus* as immunostimulants in their studies [13–15] and a few of them have combined different kinds of Chinese herbs or active polysaccharides or *Bacillus* alone in order to amplify the immune response of aquatic animals [16–18], but their synergistic effects have not been confirmed. In previous work we found that there was better function by injecting *Astragalus* polysaccharide (APS) and tuckahoe polysaccharide (TPS) into sea cucumber, and APS enhanced the immunity of *A. japonicus* [19]. In the present study, microcapsule technology can decrease the problem of dissolving of polysaccharides, and polysaccharides can be directly fed in aquaculture. To the best of our knowledge, effects of (APS + TPS) microcapsules + *B. subtilis* supplementation on growth, immune responses and disease resistance of sea cucumber have not been defined.

Citation: Fan, Y., Yu, X., Xu, L. et al. Fish Sci (2013) 79: 807, Synergy of microcapsule polysaccharides and Bacillus subtilis on the growth, immunity and resistance of sea cucumber Apostichopus japonicus against Vibrio splendidus infection. doi:10.1007/s12562-013-0644-3.

MATERIALS AND METHODS

Experimental Animals and Culture Condition

Healthy sea cucumbers (initial weight 40.2 ± 2.0 g, mean ± SE) were obtained from a farm in Qingdao (China) and kept in cylindrical 6 l tanks with recirculating seawater for a 2-week conditioning period. During the experiment, the seawater temperature was 15–18 °C, pH was 7.8-8.2, salinity was 31–32 PSU, dissolved oxygen was >5 mg/l. One-half of the seawater in the recirculating system was replaced by fresh seawater once per day and all was replaced once per week to maintain the water quality.

Experimental Design and Diets

Sea cucumbers were randomly divided into six groups, three replicates per group and 12 sea cucumbers per replicate. The basal diet with dried seaweed *Sargassum thunbergii* meal (group 1) was fed superfluously at a rate of 2 % body weight; the basal diet was supplemented with blank microcapsules (group 2), APS microcapsules (group 3), TPS microcapsules (group 4), (APS + TPS) microcapsules (group 5), (APS + TPS) microcapsules + *B. subtilis* (group 6). These supplements were fed at a rate of 3 % basal diet. All experimental animals were fed with their different diets at 16:00 hours.

APS and TPS were extracted from Chinese herbs *Astragalus membranaceus* and tuckahoe by the method of water decoction. The polysaccharide content of APS was up to 37 % and TPS was 50 %. Microcapsules were prepared through the spray method of "tiny hole and solidifying bath", the encapsulation rate was up to 85 % and the loading dose was 17 %. For *B. subtilis,* the number of living bacteria was 10^{11} CFU/g, bought from Qingdao Biocom Biology Technology Company, and added at a rate of 2×10^7 CFU/g body weight.

Experimental Procedure and Sampling Procedure

Three individuals in each replicate were randomly sampled for coelomic fluid on 7th, 14th, 21st, and 28th days. Coelomic fluid was collected with a 1-ml sterile syringe through the body wall and frozen until required. For serum separation, the collected coelomic fluid was spun down at 4,000 rpm for 10 min at 4 °C. The supernatant was stored in sterile microcentrifuge tubes at −20 °C for use. At the end of the 4-week feeding trial, the remaining sea cucumbers were weighed to monitor growth, and challenged against *V. splendidus*.

Specific Growth Rate (SGR)

The SGR is an important factor for measuring growth. We monitored it by collectively weighing the sea cucumbers in each group. The growth was calculated by the formula: specific growth rate (SGR) = $(\ln W_t - \ln W_0) \times 100/t$; where W_t and W_0 were final and initial sea cucumber weights respectively; t was duration of experiment in days.

Lysozyme Activity

Lysozyme activity was measured by a blank control method using a LSZ detection kit (Nanjing Jiancheng Bioengineering Institute, China). The increase in the transmittance of the sample at 530 nm was determined after 15 min of incubation at 37 °C. One unit of LSZ activity was defined as the amount of enzyme causing a reduction in absorbance of 0.001 per min.

Superoxide Dismutase Activity

Superoxide dismutase was measured by its ability to inhibit superoxide anions generated by xanthine and the xanthine oxidase reaction system with the SOD assay kit (Nanjing Jiancheng Bioengineering Institute, China). The optical density was measured at 550 nm. One unit of SOD was defined as the amount required for inhibiting the rate of xanthine reduction by 50 % in a 1-ml reaction system.

Alkaline Phosphatase Activity

Alkaline phosphatase activity was determined by using disodium phenyl phosphate as a substrate with a chemical detection kit (Nanjing Jiancheng, Bioengineering Institute, China). The unit definitions of AKP enzymatic activity correspond to the degradation of 1 mg phenol per 100 ml serum at 37 °C within 15 min.

Content of Complement Component 3 (C3)

The content of C3 was measured using a detection kit made by Zhejiang Elikan Biological Technology. The increase in the absorbance of the sample at 340 nm was determined after 10 min of incubation at 37 °C. The content of C3 was proportional to the amount of added antibody.

Vibrio Splendidus Challenge

The virulent strain was originally isolated from sea cucumbers diagnosed with skin ulceration disease, which were provided by Yellow-sea Fishery Research Institute, Chinese Academy of Fishery Sciences (Qingdao, China) [20]. The LD50 for 7 days determined before challenge was 5×10^8 CFU/ml. *V. splendidus* was grown in tryptic soy broth (TSB) medium with 1.5 % NaCl at 28 °C for 24 h, and then was adjusted to10^9 CFU/ml. At the end of the feeding trial, 12 sea cucumbers from each group were injected twice with 0.1 ml live *V. splendidus.* Mortality was monitored for 14 days.

Statistical Analysis

All statistical analyses were performed with SPSS version 17.0. The results were presented as mean ± SE (standard error of the means). Data were analyzed by one-way analysis of variance (ANOVA). When overall differences were significant at less than the 5 % level, Tukey's multiple range tests were used to compare the means among individual treatments.

RESULTS

Growth Performance

After 4 weeks of feeding, the SGRs of the sea cucumbers were significantly affected by the different dietary supplements (Fig. 1). There were no significant differences between group 2 (0.34 %) and the control (0.31 %) ($P > 0.05$), other groups had significantly higher SGRs compared with the control ($P < 0.05$), especially group 6 (0.97 %).

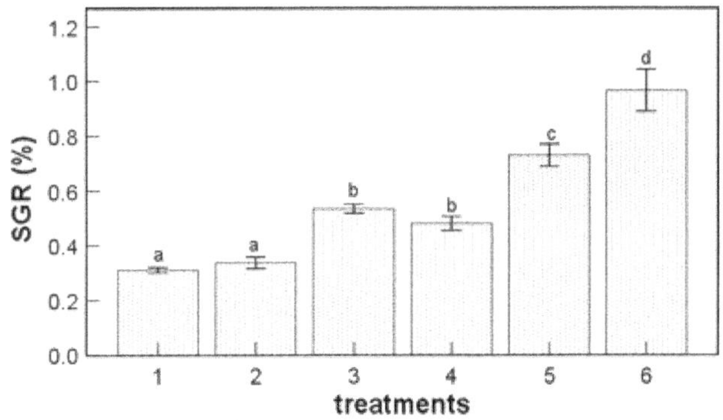

Figure 1: Effect of different treatments on the growth of *Apostichopus japonicus*. *1* control, *2* blank microcapsule, *3 Astragalus* polysaccharide (APS) microcapsule, *4* tuckahoe (TPS) microcapsule, *5* APS + TPS microcapsule, *6* (APS + TPS) microcapsule + *Bacillus subtilis.* Values are means and standard errors of three replicates (mean ± SE; *n* = 3). Treatments with different letters are significantly different ($P < 0.05$).

LSZ Activity

After feeding for 28 days, LSZ activity in the sea cucumbers had increased noticeably with different dietary supplements (Fig. 2a). The sea cucumbers fed a diet supplemented with (APS + TPS) microcapsules + *B. subtilis* exhibited the highest LSZ activity on the 7th day ($P < 0.05$), at 246 U/ml, but there was no significant difference compared with group 5 ($P > 0.05$). The activity reduced with the time, being lower on the 14th and 28th days.

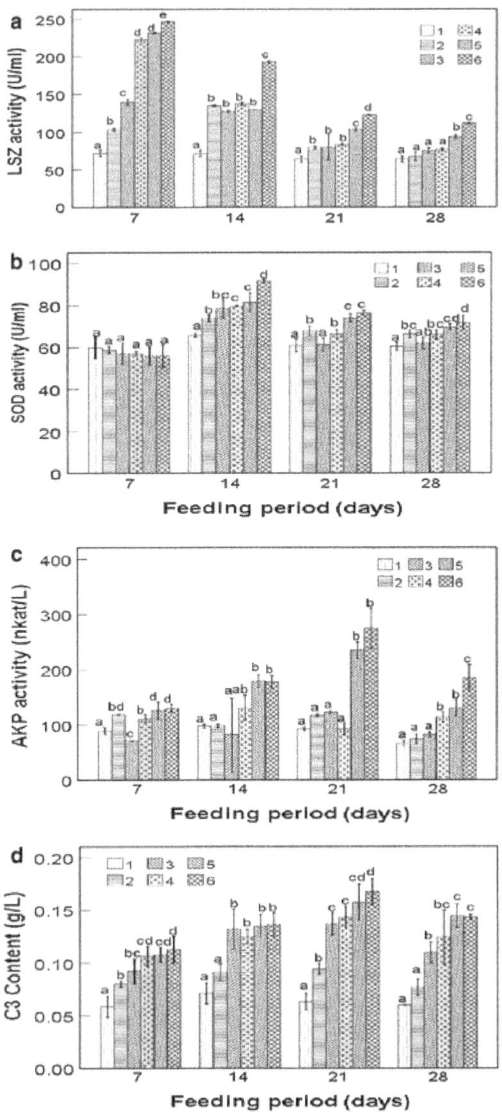

Figure 2: Lysozyme activity (a), superoxide dismutase activity(b), alkaline phosphatase activity(c), and complement 3 content (d) of *Apostichopus japonicus* fed with different diets for 4 weeks. *1* control, *2* blank microcapsule, *3 Astragalus* polysaccharide (APS) microcapsule, *4* tuckahoe (TPS) microcapsule, *5* APS + TPS microcapsule, *6* (APS + TPS) microcapsule + *Bacillus subtilis.* Values are means and standard errors of three replicates (mean ± SE; *n* = 3). Treatments with different letters are significantly different ($P < 0.05$).

SOD Activity

From Fig. 2b we can see that the SOD activity of sea cucumbers fed a diet supplemented with (APS + TPS) microcapsules + *B. subtilis* was at its highest ($P < 0.05$) on the 14th day, up to 91.7 U/ml, and there were irregular changes among the other groups. However, as time progressed, the activities decreased, and there were no significant differences between groups ($P > 0.05$).

AKP Activity

The AKP activity value in the coelomic fluid of sea cucumbers is shown in Fig. 2c. Dietary supplementation with (APS + TPS) microcapsules + *B. subtilis* significantly influenced activity, and achieved the highest value (273.8 nkat/l) on the 21st day. This was significant compared with the control ($P < 0.05$). Group 5, fed a diet supplemented with (APS + TPS) microcapsules, showed an obvious advantage, increasing to 234.9 nkat/l on the 21st day ($P < 0.05$). Apart from groups 5 and 6, there were no significant increases between other groups and the control ($P > 0.05$).

C3 Content

C3 content was significantly affected by different dietary supplements, and the result is shown in Fig. 2d. The combination of (APS + TPS) microcapsules and *B. subtilis* gave the best effect, which was 0.17 g/l on the 21st day ($P < 0.05$), but the control group produced the lowest value.

Vibrio Splendidus Challenge

The cumulative mortality rate within 14 days of sea cucumbers fed with (APS + TPS) microcapsules + *B. subtilis* was only 8.3 %, which was significantly lower than that of sea cucumbers fed with the control diet (49.6 %) ($P < 0.05$). Death of sea cucumbers fed with the control diet was relatively earlier than in other groups (Fig. 3). The challenge test showed that oral administration of (APS + TPS) microcapsules + *B. subtilis* for 4 weeks significantly enhanced protection against *V. splendidus* infection.

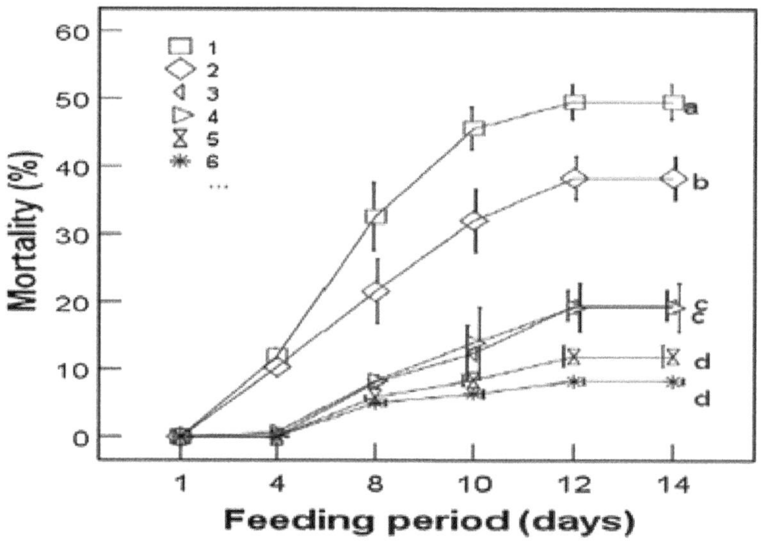

Figure 3: Cumulative morbidity during a 14-day *Vibrio splendidus* challenge of *Apostichopus japonicus* fed with different diets. *1* control, *2* blank microcapsule, *3* *Astragalus* polysaccharide (APS) microcapsule, *4* tuckahoe (TPS) microcapsule, *5* APS + TPS microcapsule, *6* (APS + TPS) microcapsule + *Bacillus subtilis*. Values are means and standard errors of three replicates (mean ± SE; $n = 3$). Treatments with different letters are significantly different ($P < 0.05$).

DISCUSSION

Growth Performance

The effects of Chinese herbs and their polysaccharides have been studied before, successfully used in aquaculture and have been verified to affect growth and non-specific immunity [9, 14,21, 22]. Wang et al. [14] showed that supplements of 3 % conventional fine powder (CP) or superfine powder (SP) of *Astragalus membranaceus* root or 0.3 % APS, given over a period of 60 days could enhance the immune responses of *A. japonicus*. Some studies showed that feeding Chinese herb polysaccharides or their combinations could improve the growth of animals [18, 23]. Gu et al. [15] demonstrated that dietary β-glucan, manna oligosaccharide (MOS) and their combinations for 4 weeks had obvious effects on the growth performance of the sea cucumber *A. japonicus*. However, little work has been undertaken to study the application of microcapsule polysaccharides in aquaculture; neither the application of microencapsulation technology products in aquaculture, nor the effect on *A. japonicus*. In the present work, it is suggested that feeding combinations of (APS + TPS) microcapsules and *B. subtilis* significantly increased the SGR of sea cucumbers ($P < 0.05$). It reached 0.97 %, higher than (APS + TPS) microcapsules or APS microcapsules or TPS microcapsules. It could be suggested that among the different active substances in Chinese herbs there are significant synergies, for example between APS and TPS, and the *B. subtilis* could further enhance their effect, perhaps due to their regulatory role in the environment or ecology.

Probiotics (such as *Bacillus*), defined as micro-organisms whose products are healthy to the host, are known for their antagonism towards pathogens, enhancement of growth, immune response and feeding efficiency, and improvement of micro-flora balance in humans and animals [24–27]. The genus *Bacillus* has been used extensively as a feed additive that can be resistant to high temperature and high pressure [28–30]. Some studies showed that dietary *B. subtilis* can improve the growth of aquatic animals, but suitable doses in the diet are needed [31–33], based on differences in strains, animals (species and sizes) and experimental conditions [13, 17, 34, 35]. Zhang et al. [17] demonstrated that a commercial *B. subtilis* could significantly increase growth of *A. japonicus* on a 1.82×10^7 CFU/g diet for 56 days. Compared to those reports, in the present study, the dietary level of *B. subtilis* was near 2×10^7 CFU/g, which could enhance the synergistic effect on the SGR of *A. japonicus*. But Zhao et al. [13] showed that a dietary level of *B. subtilis* T13 at a much higher level (10^9 CFU/g) significantly improved the SGR of sea cucumbers.

Moreover, in the current study, APS and TPS microcapsules could be mixed directly with the basal diet and then the mixture could be sprayed evenly into aquaria, to decrease the loss of polysaccharides. Microcapsules are a new dosage form of immune-enhancement agents, applied in aquaculture, especially suitable for animals with certain feeding characteristics, such as sea cucumbers.

Immune Response

Modulation of the immune system is one of the ordinary benefits of the Chinese herb active substances and probiotics. The present study also demonstrated that APS, TPS and *B. subtilis*could significantly stimulate the immune response of sea cucumbers. As we all know, sea cucumbers lack an adaptive immune system; their humoral immune responses are the second line of defense against infections and injuries [5, 36]. And in the humoral immune response, lysosomal enzymes, superoxide dismutase, alkaline phosphatase, and complement together participate in the destruction of external substances, so that they can play a protective role. The single Chinese herb of *A. membranaceus* or APS has been shown to increase lysozyme activity in fish blood and in sea cucumbers [8, 9, 14, 19]. However, the combination of polysaccharides and *B. subtilis* has not been previously reported, or the microcapsule technology. In the present study, APS microcapsules or TPS microcapsules could increase lysozyme values ($P < 0.05$) during the experiment to different extents; on the 7th day the lysozyme values reached the highest point, showing a similar course over time as groups 5 and 6. Moreover, combination of APS and TPS microcapsules, and combination of (APS + TPS) microcapsules and *B. subtilis*indicated a stronger effect; the combined effect among different Chinese herbs could have a complementary synergy, which is usually called compatibility. Furthermore, the role of *B. subtilis* in regulating micro-ecological balance and water quality indirectly enhanced the growth and immune effect of sea cucumber, as seen in the performance of LSZ activity.

SOD catalyses the dismutation of the extra bactericidal highly reactive O^{2-} to O_2 and less reactive H_2O_2, and it is an important component of the antioxidant defense system of the organism [37, 38]. In our laboratory, the tests by injecting APS and TPS into *A. japonicus*suggested that APS and TPS could significantly increase SOD activity in the coelomic fluid of the sea cucumber, and enhance antioxidant capacity, which could be related to the role that flavonoids and saponins play in *A. membranaceus* to eliminate oxygen free radicals. The effect of dietary administration of APS, TPS, and *B. subtilis* on SOD activity of the sea cucumber had different results in the present study. And a significant difference was observed in group 6 on the 14th day; 91.7 U/ml ($P < 0.05$). Other treatments also gave higher values on the 14th day; there was a rise-drop trend between the time and the effect. This could be associated with the structure of the polysaccharide and immunomodulatory mechanisms, as well as the immune system of sea cucumbers. The results of the present study confirmed that the APS and TPS microcapsules could be fed directly to the sea cucumbers, the size of the microcapsules were right for them, while at the same time polysaccharide loss was reduced in the seawater; the microcapsule in actual application is suitable for breeding and easily accepted by farmers.

Alkaline phosphatase is a marker enzyme of the phagocytes lysosomal, especially in the mollusks. Sun et al. [39] pointed out that AKP in the sea cucumber immune system plays an important role in the process of swallow the foreign substances. The experimental results indicated that AKP activity showed a sensitive response to the combination of (APS + TPS) microcapsules and *B. subtilis*, reaching the highest value (273.8 nkat/l) on the 21st day ($P < 0.05$). However, the activity had decreased by the 28th day. So we deduced that there was a certain degree of correlation between the dietary time and activity.

At present the complement system is a highly sophisticated defense system against common pathogens acting in the innate immunity of invertebrates and vertebrates, which is induced by antigen–antibody interactions in the traditional pathway. As a central component in the complement system, component 3 is an intermediary between the innate and the adaptive immune systems [40, 41]. During the past few years, the homologs of C3 have been identified, from higher vertebrates to lower protostomes including human, fish, amphioxus, sea squirt, sea urchin, horseshoe crab, coral, and sea anemone [42–44]. Past studies have shown that there are analogs of complement in *Asterias forbesi*, and in its coelomic cells there were C3b and C3bi complement receptors [45, 46]. Using enzyme-linked chemiluminescence immune detection (chemiluminescent immunoassay, CLIA), they found complement analogs in the coelomic fluid of the sea cucumber: the content of C3 was 6.58 (± 1.4) μg/ml, and C4 was 0.67 (± 0.3) μg/ml. This is an important observation about *A. japonicus* complement, and may provide a theoretical basis for the development of new immunostimulants [47, 48]. Zhou et al. [49] studied the molecular characterization and expression analysis of C3 in the sea cucumber (*A. japonicus*), it was suggested that AjC3-2 and AjC3 genes play a pivotal role in immune responses to bacterial infection in sea cucumbers. The results showed that complement of *A. japonicus* played an important role in the immune system. In the present study, C3 content was highest on the 21st day in group 6, reaching 0.17 g/l; it was significantly different to the control group and groups 2, 3, and 4 ($P < 0.05$). Sea cucumber C3 needs further research. C3 content and activities of other enzymes reached their highest values at different times. This may be related to the regulation mechanism of immunostimulants, since different immune indices have different responses.

Challenge Assay

The challenge assay is the direct way to reflect the effect of dietary immunostimulants. *V.splendidus* is the pathogen causing skin ulcer syndrome, which can cause serious harm to sea cucumber aquaculture. Dong et al. [3] have shown that exposure of *A. japonicus* to *V. splendidus* at a concentration of 10^6 cells/ml for 6 days could result in the occurrence of disease. To determine the efficacy of different dietary supplements, it has been shown that the survival rate of *A. japonicus* and its resistance to *V. splendidus* could be enhanced by administration of *Astragalus* and APS [14]. The present study showed that the oral administration of (APS + TPS) microcapsules and *B. subtilis* reduced the mortality of sea cucumber after being challenged by *V. splendidus*. The improved resistance of sea cucumbers may be partly attributable to the increased activities of different enzymes. There was significant difference between blank microcapsules and the control, which could be due to the effects of sodium alginate. Sodium alginate is alginate extracted from seaweed, and other factors could cause the errors. The resistance of sea cucumbers was further improved after adding APS and other microcapsules, however, there was no significant difference between groups 5 and 6, but group 6 also reflected an increase under the action of *B. subtilis,* that probably was due to the improvement of water quality. Zhao et al. [13] reported that *A. japonicus* fed with probiotic *B. subtilis* T13 show significant improved resistance against *V. splendidus*.

As we all know, the role of probiotics could selectively stimulate the growth and activity of one or more bacteria, and so result in beneficial or harmful effects on the host [24, 50]. This may be the reason that APS, TPS, and *B. subtilis* could improve the resistance to disease of sea cucumbers in the present study. Polysaccharides could play the role of probiotics and stimulate the secretion of cytokines to improve immune function and increase glucose tolerance.

ACKNOWLEDGMENTS

We would like to thank the Fund Project of Shandong Province modern agricultural technology system for sea cucumber innovation team; shandong province policy project (No. 2012YD10016); marine public research fund project (No. 201305005). We also thank Drs Diao and Chen for their contribution on checking the manuscript.

REFERENCES

1. Reilly A, Kaferstein F (1997) Food safety hazards and the application of the principles of the hazard analysis and critical control point (HACCP) system for their control in aquaculture production. Aquac Res 28:735–752
2. Bachère E (2003) Anti-infectious immune effectors in marine invertebrates: potential tools for disease control in larviculture. Aquaculture 227:427–438
3. Dong Y, Deng H, Sui XL, Song L (2005) Ulcer disease of farmed sea cucumber (*Apostichopus japonicus*). Fish Sci (China) 24:4–6 (in Chinese, with English abstract)
4. Deng H, He CB, Zhou ZC, Liu C, Tan KF, Wang NB, Jiang B (2009) Isolation and pathogenicity of pathogens from skin ulceration disease and viscera ejection syndrome of the sea cucumber *Apostichopus japonicus*. Aquaculture 287:18–27
5. Sun YX, Qang JQ, Wang TT, Xue JP, Liu G, You JH, Xu YP (2007) A review: defense mechanism in sea cucumber. Fish Sci (China) 26:358–361
6. Bedir E, Pugh N, Calis I, Pasco DS, Khan IA (2000) Immunostimulatory effects of cycloartane-type triterpene glycosides from *Astragalus* species. Biol Pharm Bull 23:834–837
7. Tan BK, Vanitha J (2004) Immunomodulatory and antimicrobial effects of some traditional Chinese medicinal herbs: a review. Curr Med Chem 11:1423–1430
8. Chansue N, Ponpornpisit A, Endo M, Sakai M, Satoshi Y (2000) Improved immunity of tilapia *Oreochromis niloticus* by CUP III, a herb medicine. Fish Pathol 35:89–90
9. Jian JC, Wu ZH (2004) Influences of traditional Chinese medicine on non-specific immunity of Jian carp (*Cyprinus carpio* var. *Jian*). Fish Shellfish Immunol 16:185–191
10. Shao BM, Xu W, Dai H, Tu P, Li Z, Gao XM (2004) A study on the immune receptors for polysaccharides from the roots of *Astragalus membranaceus*, a Chinese medicinal herb. Biochem Biophys Res Commun 320:1103–1111
11. Seguin-Devaux C, Hanriot D, Dailloux M, Latger-Cannard V, Zannad F, Mertes PM, Longrosis D, Devaux Y (2005) Retinoic acid amplifies the host immune response to LPS through increased T lymphocytes number and LPS binding protein expression. Mol Cell Endocrinol 245:67–76
12. Dalmo RA, Bøgwald J (2008) β-glucans as conductors of immune symphonies. Fish Shellfish Immunol 5:384–396
13. Zhao YC, Zhang WB, Xu W, Mai KS, Zhang YJ, Liufu ZG (2012) Effects of potential probiotic *Bacillus subtilis* T13 on growth, immunity and disease resistance against *Vibrio splendidus* infection in juvenile sea cucumber *Apostichopus japonicus*. Fish Shellfish Immunol 32:750–755
14. Wang TT, Sun YX, Jin LJ, Xu YP, Wang L, Ren TJ, Wang KL (2009) Enhancement of non-specific immune response in sea cucumber (*Apostichopus japonicus*) by *Astragalus membranaceus* and its polysaccharides. Fish Shellfish Immunol 27:757–762
15. Gu M, Ma HM, Mai KS, Zhang WB, Bai N, Wang XJ (2011) Effects of dietary β-glucan, mannan oligosaccharide and their combinations on growth performance, immunity and resistance against *Vibrio splendidus* of sea cucumber, *Apostichopus japonicus*. Fish Shellfish Immunol 31:303–309
16. Ortuño J, Cuesta A, Esteban A, Meseguer J (2001) Effect of oral administration of high vitamin C and E dosages on the gilthead sea bream (*Sparus aurata* L.) innate immune system. Vet Immunol Immunopathol 79:167–180
17. Zhang Q, Ma HM, Mai KS, Zhang WB, Liufu ZG, Xu W (2010) Interaction of dietary*Bacillus subtilis* and fructooligo saccharide on the growth performance,

nonspecific immunity of sea cucumber, *Apostichopus japonicus*. Fish Shellfish Immunol 29:204–211

18. Selvaraj V, Sampath K, Sekar V (2006) Adjuvant and immunostimulatory effects of β-glucan administration in combination with lipopolysaccharide enhances survival and some immune parameters in carp challenged with *Aeromonas hydrophila*. Vet Immun Immunopathol 114:15–24
19. Fan Y, Wang SX, Ye HB, Xu L, Zhu AC, Yang XS, Li TB (2010) Effects of polysaccharides from *Astragalus membranaceus* on non-specific immune in sea cucumber *Apostichopus japonicus*. Fish Sci (China) 29:321–324
20. Zhang CY, Wang YG, Rong XJ (2006) Isolation and identification of causative pathogen for skin ulcerative syndrome in *Apostichopus japonicus*. J Fish China 30:106–118
21. Yin GJ, Jeney G, Racz T, Xu P, Jun X, Jeney Z (2006) Effect of two Chinese herbs (*Astragalus radix* and *Scutellaria radix*) on nonspecific immune response of tilapia,*Oreochromis niloticus*. Aquaculture 253:39–47
22. Huang XX, Zhou HQ, Zhang H (2006) The effect of *Sargassum fusiforme* polysaccharide extracts on vibriosis resistance and immune activity of the shrimp, *Fenneropenaeus chinensis*. Fish Shellfish Immunol 20:750–757
23. Gu M, Mai KS Ma HM, Zhang WB, Ai QH, Wang XJ, Bai N (2010) Immune response of sea cucumber *Apostichopus japonicus* coelomocytes to several immunostimulants in vitro. Aquaculture 306:49–56
24. Suzer C, Çoban D, Kamaci HO, Saka Ş, Firat K, Özge Otgucuoğlu, Küçüksari H (2008)*Lactobacillus* spp. bacteria as probiotics in gilthead sea bream (*Sparus aurata*, L.) larvae: effects on growth performance and digestive enzyme activities. Aquaculture 280:140–145
25. Wang YB, Li JR, Lin JD (2008) Probiotics in aquaculture: challenges and outlook. Aquaculture 281:1–4
26. Gatesoupe FJ (1999) The use of probiotics in aquaculture. Aquaculture 180:147–165
27. Li JQ, Tan BP, Mai KS (2009) Dietary probiotic *Bacillus* OJ and isomalto oligosaccharides influence the intestine microbial populations, immune responses and resistance to white spot syndrome virus in shrimp (*Litopenaeus vannamei*). Aquaculture 291:35–40
28. Rengpipat S, Rukpratanporn S, Piyatiratitivorakul S, Menasaveta P (2000) Immunity enhancement in black tiger shrimp (*Penaeus monodon*) by a probiont bacterium (*Bacillus*S11). Aquaculture 191:271–288
29. Rengpipat S, Phianphak W, Piyatiratitivorakul S, Menasveta P (1998) Effects of a probiotic bacterium on black tigers shrimp *Penaeus monodon* survival and growth. Aquaculture 167:301–313
30. Ochoa-Solano JL, Olmos-Soto J (2006) The functional property of *Bacillus* for shrimp feeds. Food Microbiol 23:519–525
31. Liu CH, Chiu CS, Ho PL, Wang SW (2009) Improvement in the growth performance of white shrimp, *Litopenaeus vannamei*, by a protease-producing probiotic, *Bacillus subtilis*E20, from natto. J Appl Microbiol 107:1031–1041
32. Tseng DY, Ho PL, Huang SY, Cheng SC, Shiu YL, Chiu CS, Liu CH (2009) Enhancement of immunity and disease resistance in the white shrimp, *Litopenaeus vannamei*, by the probiotic, *Bacillus subtilis* E20. Fish Shellfish Immunol 26:339–344
33. Aly SM, Mohamed MF, John G (2008) Effect of probiotics on the survival, growth and challenge infection in *Tilapia nilotica* (*Oreochromis niloticus*). Aquac Res 39:647–656
34. Nayak SK (2010) Probiotics and immunity: a fish perspective. Fish Shellfish Immunol 29:2–14
35. Sun YZ, Yang HL, Ma RL, Lin WY (2010) Probiotic applications of two dominant gut*Bacillus* strains with antagonistic activity improved the growth performance and immune responses of grouper *Epinephelus coioides*. Fish Shellfish Immunol 29:803–809
36. Canicatti C, Parrinello N (1985) Hemaglutinin and hemolysin level in coelomic fluid from*Holothuria polii* (Echinodermata) following sheep erythrocyte injection. Biol Bull 168:175–182
37. Fu LL, Shuai JB, Xu ZR, Li JR, Li WF (2010) Immune responses of *Fenneropenaeus chinensis* against white spot syndrome virus after oral delivery of VP28 using *Bacillus subtilis* as vehicles. Fish Shellfish Immunol 28:49–55
38. Castex M, Lemaire P, Wabete N, Chim L (2010) Effect of probiotic *Pediococcus acidilactici*on antioxidant defences and oxidative stress of *Litopenaeus stylirostris* under *Vibrio nigripulchritudo* challenge. Fish Shellfish Immunol 28:622–631
39. Sun YX, Jin LJ, Wang TT, Xue JP, Liu G, Li XY, You JS, Li SY, Xu YP (2008) Polysaccharides from *Astragalus membranaceus* promote phagocytosis and superoxide anion (O_2^-) production by coelomocytes from sea cucumber *Apostichopus japonicus* in vitro. Comp Biochem Phys Part C 147:293–298
40. Dunkelberger JR, Song WC (2010) Complement and its role in innate and adaptive immune responses. Cell Res 20:34–50
41. Fujita T (2002) Evolution of the lectin-complement pathway and its role in innate immunity. Nat Rev Immunol 2:346–353
42. Boshra H, Li J, Sunyer JO (2006) Recent advances on the complement system of teleost fish. Fish Shellfish Immunol 20:239–262
43. Dishaw LJ, Smith SL, Bigger CH (2005) Characterization of a C3-like cDNA in a coral: phylogenetic implications. Immunogenetics 57:535–548
44. Fujito NT, Sugimoto S, Nonaka M (2010) Evolution of thioester-containing proteins revealed by cloning and characterization of their genes from a cnidarian sea anemone,*Haliplanella lineate*. Dev Comp Immunol 34:775–784
45. Berthuessen K, Seljelid R (1982) Receptors for complement on echinoid phagocytes. The opsonic effect

of vertebrae sera on echinoid phagocytosis. Dev Comp Immunol 6:423–431

46. Leonard LA, Strandberg JD, Winkelstein JA (1990) Complement-like activity in the sea star, *Asterias forbesi*. Dev Comp Immunol 14:19–30

47. Zhang F, Gong J, Wang HF, Zhao X, Li XY (2007) The detection of complement analogues AjC3, and AjC4 in the sea cucumber *Apostichopus japonicus*. J Dalian Fish Univ 22:246–248

48. Zhang F, Wang HF, Gong J, Chang SJ (2007) Detection method of complement analogues of coelomic fluid in *Apostichopus japonicus* by chemiluminescent immunoassay. J Nuclear Agric Sci (JNAS) 21:413–416

49. Zhou ZC, Sun DP, Yang AF, Dong Y, Chen Zh, Wang XY, Guan XY, Jiang B, Wang B (2011) Molecular characterization and expression analysis of a complement component 3 in the sea cucumber (*Apostichopus japonicus*). Fish Shellfish Immunol 31:540–547

50. Vine N, Leukes W, Kaiser H (2006) Probiotics in marine larviculture. FEMS Microbiol Rev 30:404–427

CHAPTER 11

A DETAILED STAGING SCHEME FOR LATE LARVAL DEVELOPMENT IN STRONGYLOCENTROTUS PURPURATUS

Andreas Heyland and Jason Hodin

Keywords: Sea urchin, Echinoidea, Echinodermata, Skeletogenesis, Rudiment, Metamorphosis, Morphogenesis, Growth, Development

BACKGROUND

The purple sea urchin, *Strongylocentrotus purpuratus*, has been the focus of developmental and ecological studies for over a hundred years. Its widespread distribution, large population sizes, active fishery, dramatic high energy habitat, ease of obtaining adults and gametes and hardiness as a laboratory study organism has contributed to this long term usage [1]. More recently, *S. purpuratus* was the first free-living, non-chordate marine invertebrate with a fully sequenced genome [2], and the application of modern gene manipulation techniques to the early sea urchin embryo has reinvigorated its use in understanding the basic mechanisms of cell biology and development.

Purple urchins develop through a feeding larval stage that, depending on latitude and food supply [3], spends 1–3 months feeding on phytoplankton before settling to the sea floor and completing metamorphosis into a feeding juvenile. This complex life history determines the nature of dispersal from parental sites and recruitment patterns of juveniles into adult populations [4]. Moreover, the transformation from a larva to a juvenile, in which the bilaterally symmetric larva transforms into the pentameral juvenile urchin, is a fascinating ontogenetic event unto itself.

In contrast to insects and amphibians, relatively little is known about the mechanisms of echinoderm metamorphosis in general, and purple urchin metamorphosis specifically (for what is known see [5–16]). Comparative morphology and functional studies in a phylogenetic context suggest that echinoderm metamorphosis evolved independently from that in insects and amphibians, and that many other marine invertebrate groups themselves likely evolved metamorphosis independently [12, 17–24]. Still, a more complete understanding of the mechanisms underlying echinoderm metamorphosis can shed new light on the origins of metamorphosis in general [20]. Furthermore, a better mechanistic appreciation of echinoderm metamorphosis will provide insights into body plan evolution within the echinoderms and ultimately the deuterostomes [25–28]. Finally, the metamorphic transition is notable as one in which multiple external environmental signals are integrated by the developing larvae to control the proper timing and location of settlement [12], which subsequently impacts the successful recruitment of larvae into benthic populations. Therefore, there is a fascinating interplay between ecology, evolution and development that occurs at the metamorphic transition in marine invertebrates that makes it an under-appreciated archetype for eco-evo-devo studies (sensu [29, 30]).

Citation: Heyland and Hodin: A detailed staging scheme for late larval development in Strongylocentrotus purpuratus. BMC Developmental Biology 2014 14:22. doi:10.1186/1471-213X-14-22.

One impediment to using echinoderms as subjects for detailed metamorphic studies is the lack of an agreed upon, simple, yet appropriately detailed scheme for describing the stages of juvenile morphogenesis leading up to settlement (except for brittle stars; [31, 32]). This information gap exists despite many careful studies over more than a hundred years into the ontogeny of the structures in echinoderm larvae that are fated to form the juvenile (i.e., juvenile structures in contrast to larval ones – see e.g., [33–36]).

Two explicit staging schemes have been proposed for echinoids that include the development of the juvenile rudiment through settlement: one for purple urchins [37], the other encompassing three other regular echinoids [38]. Unfortunately, these schemes provide little detail on the events of juvenile morphogenesis that occur after the onset of juvenile calcification, a period of complex ontogeny which represents approximately the last 1/3 of the total larval phase, and during which most of the definitive structures of the juvenile first appear. Indeed, these are the very stages where we expect to find species-specific differences in ontogeny underlying the substantial variation in juvenile morphology across echinoids [39]. Furthermore, the proposed purple urchin staging scheme of Smith et al. [37] mixes larval and juvenile characters. This is a considerable shortcoming, as sea urchin larvae demonstrate marked phenotypic plasticity with respect to the timing of appearance of larval versus juvenile structures [38, 40, 41].

In an effort to build on these existing schemes for sea urchin larval development, we here propose the first detailed staging scheme in any echinoid for the morphogenesis of juvenile structures alone, from rudiment invagination until settlement. We have divided our scheme into soft tissue and skeletogenic stages, in part for clarity, and in part due to our observations of variation within and among clutches in the timing of appearance of skeleton relative to our defined soft tissue stages. In fact, such heterochronies (differences among individuals in the relative timing of developmental events), occur commonly in the formation of different juvenile characters, and we have endeavored to take account of this variation in our proposed staging scheme. Above all, we have only included characters that are readily visible in live larvae using even the most basic compound microscope equipped with cross-polarized light at about 100× total magnification. Our intention is that this staging scheme will not only facilitate a broader range of studies into late larval development in purple sea urchins, but will also be the starting point for similar staging schemes across echinoids and other echinoderms.

RESULTS

Staging Scheme

In Table 1 & Figure 1 (for soft-tissue stages) and in Table 2 & Figure 2 (for skeletogenic stages) we present a new staging scheme for the development of juvenile structures in the purple sea urchin, *Strongylocentrotus purpuratus*, beginning with the invagination of the larval epithelium that forms the juvenile ectoderm (="echinus rudiment"), and ending with well-developed tube foot end plates and the elongation of adult spines (reviewed in [42]). This latter stage (our Stage 10 in Table 2, Figure 2) is approximately the point at which larvae become competent to settle when reared in the laboratory under controlled conditions (see Figure 2M & O), and occurs at approximately 4 to 4.5 weeks after fertilization at 14°C under abundant food conditions.

Table 1: Soft tissue stages in rudiment

Stage	Shorthand	Days	Drawings	Smith et al. 2008 [[37]]	Chino et al. 1994 [[38]]	Description
-	pre invagination	< 13		I-II	a-c	6-8 arm larva, no invagination forming on the left side yet.
i	invagination	14	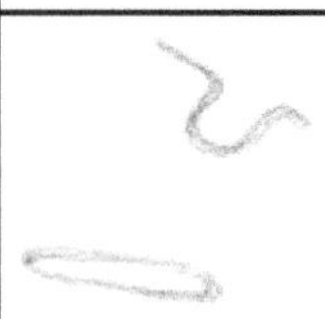	III	c-d	Rudiment invaginating on left side, not yet contacting hydrocoel; this can be further subdivided as "% invagination" (i.e., the depth of the invagination measured from the ectoderm through the axis of the invagination to the hydrocoel).
ii	contact	15	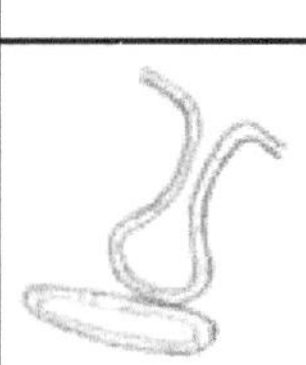	IV	e-f	Rudiment has contacted the hydrocoel, but has not flattened alongside it yet.
iii	contact flattened	16		IV	f-g	Invaginated rudiment has now flattened alongside the hydrocoel, but there is as yet no 5-fold symmetry apparent in the mesoderm or the invaginated ectoderm.

iv	5-fold meso-derm	17		V	h	Hydrocoel showing first visible signs of 5-fold symmetry (5 bumps), but the ecto-derm (invaginated rudiment) not yet show-ing any sign of 5-fold symmetry.
v	5-fold ecto-derm	18-19	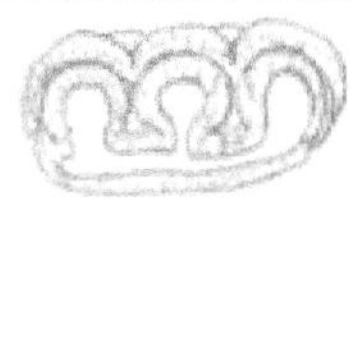	V	i	Ectoderm now also showing 5-fold sym-metry as the primordia of the 5 podia begin to push through the floor of the vestibular ectoderm. At this point, the interior of the 5 incipient podia are spherical in shape or shorter than wide.
vi	primary podia (pp)	19-21		V	i-j	Interior of 5 primary podia are now taller than wide, but the podia are not yet folding in towards one another.
vii	pp-folded	21-23		VI?	k	The 5 podia are now folding towards one another (i.e. towards the center of the oral field of the forming juvenile), but the tips of adjacent podia are not yet touching one another. This soft tissue stage usually co-incides with skeletogenic Stage 1 or 2 (see Table 2).
viii	pp-touching	23	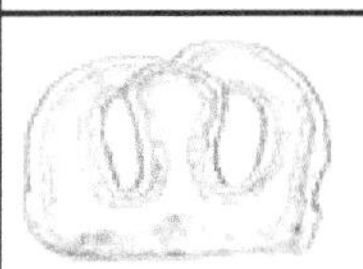	VI	k-l	The tips of at least 2 of the 5 tube feet are now touching. This soft tissue stage usually coincides with skeletogenic Stage 2+ (see Table 2).

Description and illustration of stages presented in the new staging scheme as well as approximate timing in days after fertilization at 14°C. Note that roman numerals in the column “Smith et al. [37]” and the letters in the column “Chino et al. [38]” are the stages proposed in those studies (see Figure eleven in [37], and Figure two in [38]) as they relate approximately to the stages used in our new soft tissue staging scheme here. Question marks indicate ambiguities in the comparisons between staging schemes.

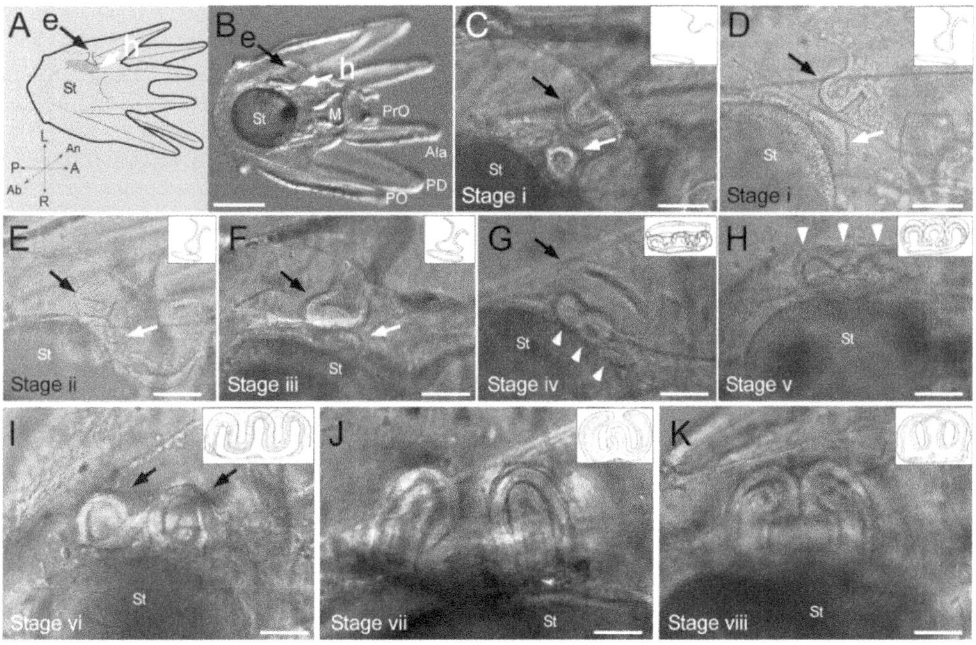

Figure 1: Examples of soft tissue developmental stages of *S. purpuratus* larvae as defined in Table1(roman numerals). All images (except as indicated) are close-up views of the rudiment in the same orientation as indicated in A, i.e., an abanal view of the larvae (sensu [37]). Insets in C-K show schematic views of the stages in question; see also Table 1. (A) Overview diagram of larva with corresponding DIC image of actual larva (B – anal view). (C-F) Black arrow- invaginating ectoderm (e); white arrow- hydrocoel (h). (C)Stage i, approximately 60% ectodermal invagination. (D) Stage i, approximately 90% ectodermal invagination (anal view). (E) Stage ii, contact of invaginating ectoderm with hydrocoel (anal view). (F) Stage iii, ectoderm flattening alongside hydrocoel (anal view).(G) Stage iv, 5-fold mesoderm (first visible sign of 5-fold symmetry); white arrowheads- 3 of the 5 primary podia anlage are visible in this view; black arrow- invaginating ectoderm.(H) Stage v, 5-fold ectoderm; arrowheads as in E. (I) Stage vi, primary podia stage, arrowheads indicate two of the forming podia. (J) Soft tissue Stage vii with folded primary podia. (K) Soft tissue Stage viii with primary podia bending at the tip and touching each other (anal view). Note that schematic in A does not show the pair of preoral arms for simplicity. Abbreviations: Ala – anterolateral arms; M – Mouth; St – stomach; PD – postdorsal arms; PO – postoral arms, PrO – preoral arms; h – hydrocoel; e – invaginating ectoderm (=vestibule); L – Left larval side; R – Right larval side; A – anterior; P – posterior; An – anal; Ab – abanal. Scale bars: B- 200 μm C, D, E – 35 μm; F - 25 μm; G,H, I - 40 μm, J - 30 μm; K - 40 μm.

Table 2: Skeletogenic stages in rudiment

Stage	Shorthand	Days	Drawings	Smith et al. 2008 [[37]]	Chino et al. 1994 [[38]]	Description
0	no skeleton	< 21		I-V	a-j	No spicule dot (or any other rudiment skeleton) is visible; "soft tissue" Stage vi or earlier (see Table 1).
1	spicule dot	22		VI	k	At least one spicule dot is visible in the rudiment (first spicule dot is usually at the anterior edge of the rudiment, near the stomach/esophagus border); this occurs at about "soft tissue" Stage vi-vii (see Table 1).
2	spicule	22-23		VI	k	At least one triradiate spicule present in the rudiment.
3	MB/TF spic dots	23	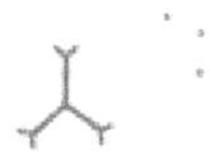	VI	k	Multi-branched spicule (MB) and/or tube foot (TF) spicule dots present in the rudiment. The first MB spicules seen are the de novo ocular plate primordia of the urchin test, as well as the interambulacral plate primordia that will articulate with adult spines after Stage 8.
4	TF spicules	24		VI	k	TF triradiate spicule present in at least one primary podium.
5	sp primord/incomplete ring	25	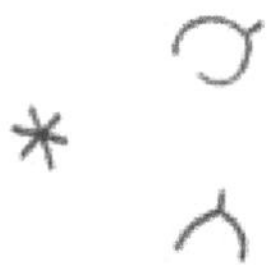	VI	k-l	Adult spine primordium (6-prong spicule) present and/or TF spicules elongating but not forming complete ring in any primary podia.
6	spine primord+base/1st TF ring complete	26	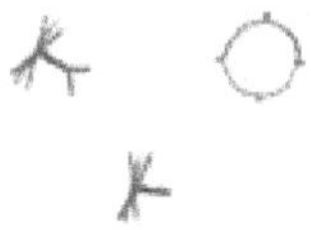	VI	l	Adult spine primordium element elongating in direction orthogonal to 6-sided primordium (i.e., along the axis of the incipient spine) and/or 1st TF ring has fused to form a complete ring in at least one primary podium. 2nd TF ring may be beginning to form as bifurcating extensions from the 1st TF ring, but none of these bifurcations have yet started to fuse back with the 1st TF ring.
7	pre-spine/2nd TF ring < ½ complete	27	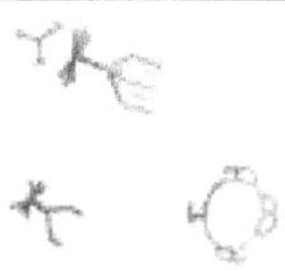	VI	l	At least one adult spine primordium has a complete base+6 fronds, but no cross hatches have formed between the fronds and/or the 2nd TF ring has bifurcating spicules that may have started to fuse in at least one primary podium, forming part of the incipient second ring, but this ring is at most half complete.
8	spines	28	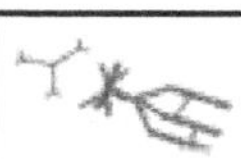	VI	l	At least one cross hatch is now present in at least one forming adult (6-sided) spine. Spine growth can be further subdivided by counting the average and maximum number of cross hatches (see Figure 5).

9	2nd TF ring>½ complete	29		VI?	l	Bifurcating bits of the 2nd TF ring have continued to fuse in at least one primary podium, so that more than half or the 2nd TF ring is complete, but the 2nd TF ring is not yet complete in any podium.
10	2nd TF ring complete	30		VII?	m	2nd TF ring is now complete in at least one primary podium. Larvae start becoming competent to settle at around this stage, though timing of competence seems to vary with skeletogenic stage.

Description and illustration of stages presented in the new staging scheme as well as approximate timing in days after fertilization at 14°C. Note that roman numerals in the column "Smith et al. [37]" and the letters in the column "Chino et al. [38]" are the stage designations proposed in those studies (see Figure eleven in Smith et al. [37], and Figure two in Chino et al. [38]) as they relate approximately to the stages used in our new skeletogenic staging scheme here. Question marks indicate ambiguities in the comparison of staging schemes.

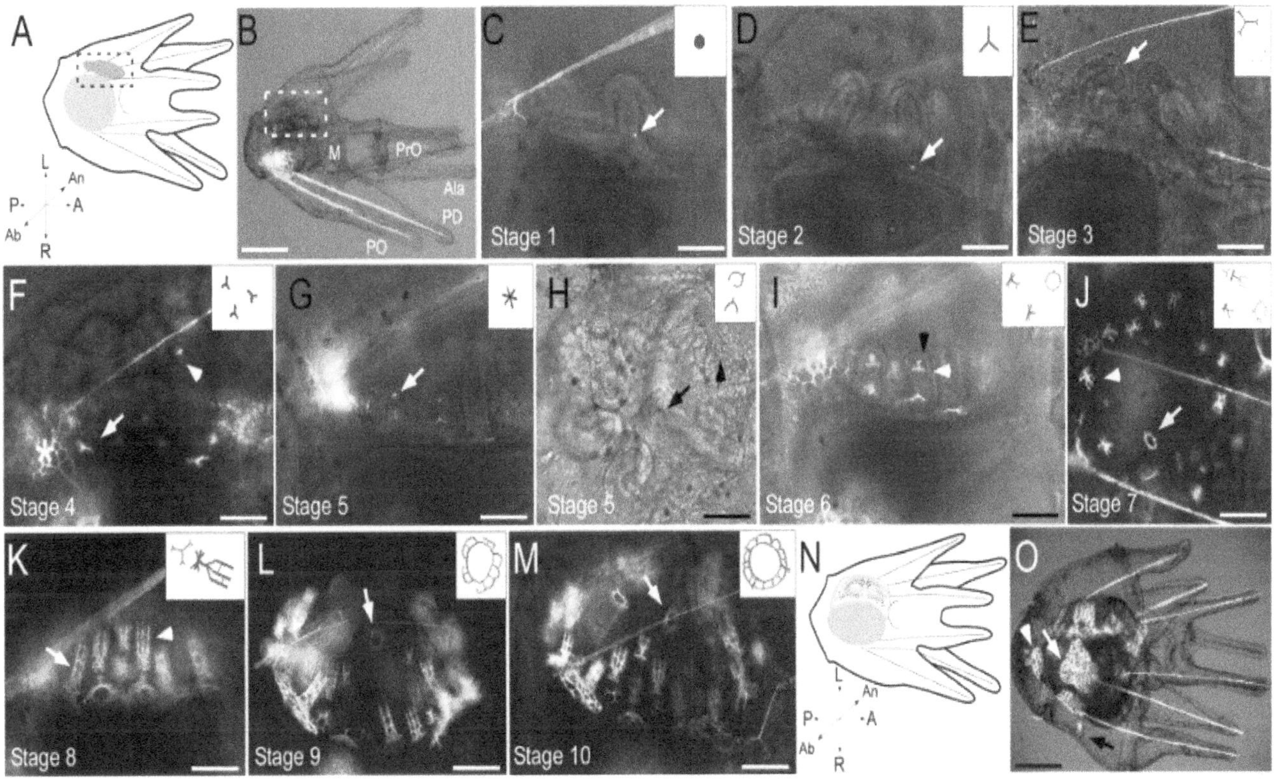

Figure 2: Examples of skeletogenic developmental stages of *S. purpuratus* larvae as defined in Table2(arabic numerals). All images (except as indicated) are rudiment close-up views in same orientation (i.e., abanal view, sensu [37]) as indicated in A (earlier stages) and N (later stages). E, F, G, K, M are anal views. Insets in C-M: schematic views of defining skeletogenic features for stage in question, not drawn to scale; see also Table 2. Drawing (A) and corresponding live image (B) of representative early skeletogenic stage larva. (C) Stage 1: spicule dot (white arrow). (D) Stage 2: spicule (white arrow). (E) Stage 3: tube foot spicule dot (white arrow). (F) Stage 4: tri-radiate tube foot spicule (white arrowhead); multi-branched spicule (white arrow), fated to form ocular plate 1 (see [34,35]). White asterisk indicates ocular plate 5 –a non-rudiment skeletal character not included in our scheme– which forms off of the left PO rod (see [43]). (G) Stage 5: spine primordium (6-sided star- white arrow; note also spine lumen visible at this and later stages). (H) Also Stage 5: incomplete TF ring (black arrow); multi-branched spicule indicated (black arrowhead) is fated to form the interambulacral skeleton at the base of an adult spine in interambulacrum 1 (sensu [34, 35]; but see [44]). Note this view is looking down on the left side of larva. (I) Stage 6: spine primordium (white arrowhead) with base of spine (black arrowhead) extending orthogonal to that. (J) Stage 7: pre-spine (fronds present with no cross hatches; white arrowhead), and TF ring with 2nd ring <1/2 complete (white arrow). Larva viewed from left side. (K) Stage 8: adult spine with cross hatches (white arrowhead); juvenile spine (white arrow) indicated for comparison; anal view. (L)Stage 9: TF ring with 2nd ring >1/2 complete (white arrow is pointing directly to a gap in 2nd ring). (M) Stage 10: TF ring with 2nd ring complete (white arrow); anal view. (N)schematic of larva as seen in panels I-M with larger rudiment and spines becoming recognizable. (O) Stage 10. Whole larva (oriented as in N) showing, in addition to the rudiment, several of the non-rudiment juvenile skeletal characters that we do not incorporate in our staging scheme: genital plate 5 (white arrow), posterior juvenile spines articulating with genital plate 4 (white arrowhead) and right side posterior juvenile spine (black arrow; see [45]). Ocular 5 is indicated in panel F. Note that schematic does not show PO arms for simplicity. Abbreviations: Ala – anterolateral arms; M – Mouth; St – stomach; PD – postdorsal arms; PO – postoral arms, PrO – preoral arms. Scale bars B – 300 μm; C-D: 70 μm; E- 80 μm;F-M – 70 μm; O - 200 μm.

We have separated the proposed scheme into soft tissue stages (roman numerals in Table 1, Figure 1) and skeletogenic stages (arabic numerals in Table 2, Figure 2), as we have noticed heterochronic variation among and within clutches in the soft tissue stage at which skeleton is first visible. We show representative images of the defining features of the various stages in Figures 1 & 2, and provide sketches of the same in Tables 1 & 2 that are also shown in Figures 1& 2 as insets.

All of the features that we used to define the various stages are readily visible in whole mounts of live larvae with raised cover glass, and employing any compound microscope at approximately 100× total magnification. For this reason, we did not include any characters (such as the developing epineural folds and dental sacs) that typically require histological analysis for their visualization – development of a companion histological staging analysis would be a useful expansion on the staging scheme presented here. By contrast, the use of cross-polarized light (placing a polarizing filter on either side of the sample and rotating one filter relative to the other to quench the light) results in birefringence of larval and juvenile skeleton, allowing for visualization of skeletal elements deep within the rudiment of live larvae.

Several of our skeletogenic stages (Table 2, Figure 2) include more than one defining character. For example, Stage 3 is defined as having multibranched spicules and/or tube foot spicule dots present. The reason that we have combined characters in this way is because we have noted heterochronies in some larvae in the relative appearance of the two classes of characters. Therefore, in this example, we score a larva as Stage 3 if either one or more multibranched spicules or one or more tube foot spicule dots (or both) are visible in that larva. If required for specific analyses, these characters can be easily analyzed in isolation.

Although we examined juvenile skeletal structures that form both inside and outside of the juvenile rudiment, our staging scheme itself includes only juvenile skeletal structures (the interambulacral plates, spines, tube feet and the subset of ocular plates) that form inside of the rudiment itself. The reason that we chose to exclude any extra-rudiment skeleton from our staging scheme is that their ontogenic progression was heterochronic with respect to the rudiment skeleton in different larvae, even within the same culture vessel. As such, the inclusion of extra-embryonic skeletal structures in our staging scheme would have been non-informative alongside the easily visible and characterizable rudiment skeleton that we did include. Furthermore, as we demonstrate below, each of our stages as defined is non-ambiguous and well-spaced (approximately 24 hours of development at 14°C between stages), and thus the characters we include seem sufficient in order for any researcher to quickly and repeatedly characterize the stage of live individual larvae examined in a standard compound microscope.

Overview of the Proposed Scheme

A brief summary of all developmental stages is provided here, with details on each soft tissue and skeletogenic stage (as well as approximate timing at 14°C with abundant food) provided in Tables 1 & 2, respectively. The juvenile rudiment begins to form with an ectodermal invagination (the vestibule) on the left side of the larva (Stage i; Figure 1C, D), which contacts (Stage ii; Figure 1E) and then flattens against the left hydrocoel (Stage iii; Figure 1F). Soon thereafter, the rudiment shows the first signs of the adult pentameral symmetry as five bumps in the hydrocoel, pressed against the floor of the vestibule (Stage iv; Figure 1G). As these five bumps continue to penetrate the overlying vestibule, the vestibular ectoderm itself begins to reveal a pentameral pattern (Stage v; Figure 1H). During subsequent days, the five bumps (which we now refer to as the "primary podia") lengthen (Stage vi; Figure 1I), begin to curve in towards one another (i.e., towards the oral field of the rudiment; Stage vii; Figure 1J), and ultimately the tips of the primary podia touch one another (Stage viii; Figure 1K). The first skeletal elements of the juvenile are visible in soft tissue Stage vi or vii as small spicule dots usually visible near the esophagus/stomach border (Stage 1; Figure 2C). These will develop into triradiate spicules (Stage 2; Figure 2D), and then into multi-branched spicules (Stage 3; Figure 2E), the first of which are ocular plate and interambulacral plate primordia. At this same stage, skeletal spicule dots appear at the tips of each of the primary podia, which then become tri-radiate spicules (Stage 4; Figure 2F). These spicules then begin to elongate from two of the spicule arms (Stage 5; Figure 2G) to begin to form the first ring of the tube foot end plates. At this same stage, 6-sided spicules appear (Stage 5; Figure 2H), which are the primordia of the adult spines. These spicules elongate along the axis orthogonal to the original 6-sided spicule, forming the base of the incipient adult spine, at around the same time that the first tube foot skeletal rings are complete (Stage 6; Figure 2I). In the next stage (Stage 7; Figure 2J), the spine primordia continue to elongate into 6 parallel fronds lacking any cross hatches, while the second concentric tube foot skeletal ring begins to form. Stage 8 (Figure 2K) is characterized by the formation of the first cross hatch (perpendicular to and connecting two adjacent spine fronds) in at least one of the adult spines. The final two stages are defined by tube foot skeleton where the second concentric ring is >1/2 complete (Stage 9; Figure 2L) and then complete (Stage 10; Figure 2M).

DEVELOPMENTAL TIMING

By culturing individual larvae in well plates, and staging them at 0, 24 and in some cases 48 hours, we were able to make estimations of the length in hours for each of our skeletogenic stages (see Figure 3A). From these data, we plotted cumulative time through the ten skeletogenic stages in Figure 3B.We caution readers that the data that we present in Figure 3 are intended mainly for heuristic purposes; we note substantial variation among larvae (and among experiments) in the length of given stages. With that caveat in mind, Figure 3A indicates that most of our proposed skeletogenic stages are approximately 24 hours in length at 14°C.

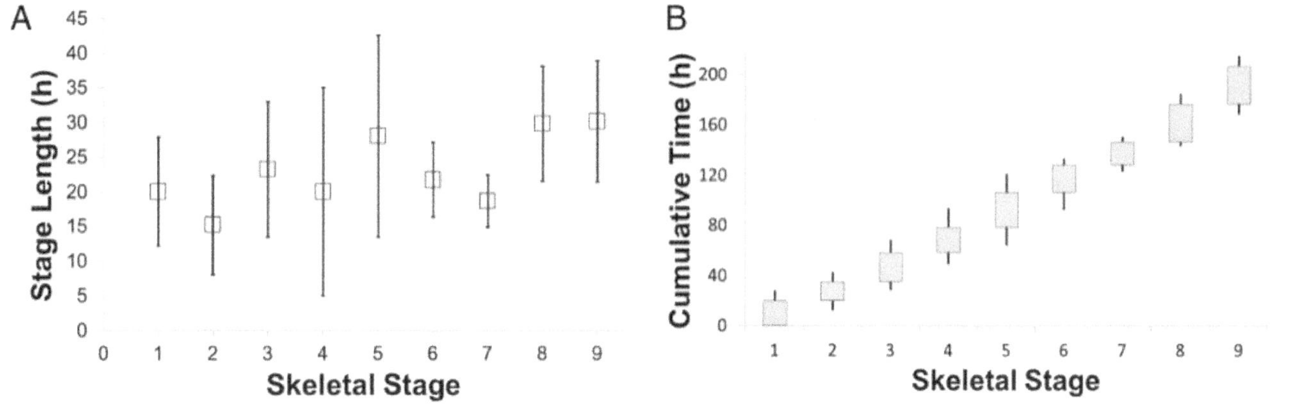

Figure 3: Temporal progression of skeletal development in *S. purpuratus* larvae. A)Mean stage length as a function of skeletal stage. Error bars are one standard error of the mean. B) Cumulative time as a function of skeletal stage, starting at the onset of Stage 1 (first appearance of any skeleton in rudiment; see Table 2). Boxes indicate the approximate cumulative time interval during which a larva is in a given stage; error bars indicate 95% confidence interval.

In order to test whether our temporary mounting methodology for viewing live larvae injured them, we compared developmental rates in mounted and unmounted larvae in a separate experiment in Guelph. On average, mounted larvae progressed 2 stages (from Stage 4 to Stage 6; see also Figure 2) during the 48 hour period of the experiment. The skeletogenic stage for mounted larvae at the end of the 48 h period was 6 ± 1.3 (Std) compared to 6 ± 1.3 (Std) for unmounted larvae. We detected no statistical difference between the two treatments using a 2-tailed independent sample *t*-test ($t_{1,19}$=-0.53; p=0.60). Converting these stage values into time (using the cumulative time data presented in Figure 3B) did not change this result (data not shown).

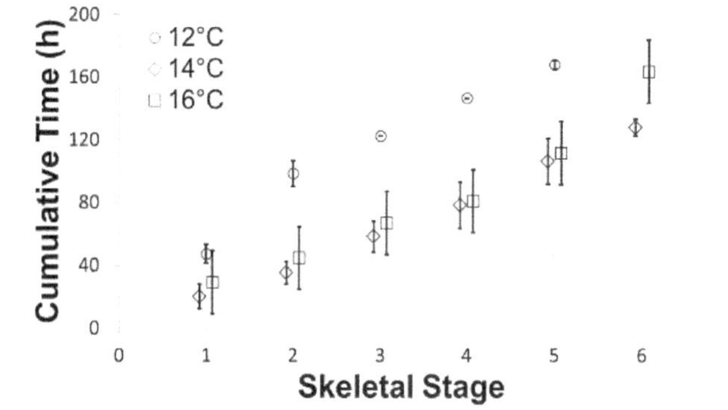

Figure 4: Comparison of cumulative time to reach specific stages at three different temperatures. The temporal progression through these stages was clearly slowest at 12°C, whereas no difference is apparent between 14°C and 16°C. Note that the data for 12°C and 16°C originate from the Seattle dataset while the data for 14°C originate from the Guelph dataset. Several other differences existed between these two experiments (see Methods). Error bars indicate 95% confidence interval.

Robustness of the Staging Scheme

To test whether our proposed staging scheme is robust under a variety of environmental conditions, we repeated our well plate experiment with larvae that had been reared in natural (as opposed to artificial) sea water, with lower food levels (but still likely *ad libitum* - [46, 47]) and at a higher temperature (16°C) – and that derived from a different source population, widely separated from the California population of urchins studied in Guelph. In this 'Seattle' experiment, we repeated our well-plate individual larval rearing methodology at two temperatures: 12°C and 16°C.Figure 4 shows a summary of the Seattle experiment, plotted alongside our main ('Guelph' experiment, 14°C) dataset. First, the range of rearing conditions and temperatures examined did not alter the relative progression of stages. Second, lower temperature rearing (12°C) resulted in slower overall progression through the skeletogenic stages. And third, our 16°C Seattle larvae progressed through their stages at a similar rate as did the Guelph 14°C larvae. Among other possible interpretations, this latter result might indicate that the higher food levels that we used in Guelph compensated for the lower temperature, or that 15°C is the optimum rearing temperature, and that 14°C and 16°C lie on either side of that optimal development peak.

In addition to this Seattle well plate experiment, we have reared multiple batches of sea urchin larvae through settlement in 5 different locations with different adult feeding and sea water conditions (see Methods). During each of these rearings, we have tracked the stages of our developing larvae and have not noticed any variations from the staging scheme that we propose here.

Spine Growth

We are aware that the skeletogenic features that we have used in our staging scheme are more qualitative than quantitative in nature. Nevertheless, there is one quantitative character that is easily scored in late stage larvae: the elongation of the growing spines. As the incipient spines increase in length, they increase the number of cross bars that form between the adjacent spine fronds in a step-wise, linear fashion (Figure 5A). Therefore, one can use the number of cross bars (=cross hatches) present in a developing spine as a numerical proxy for spine length.

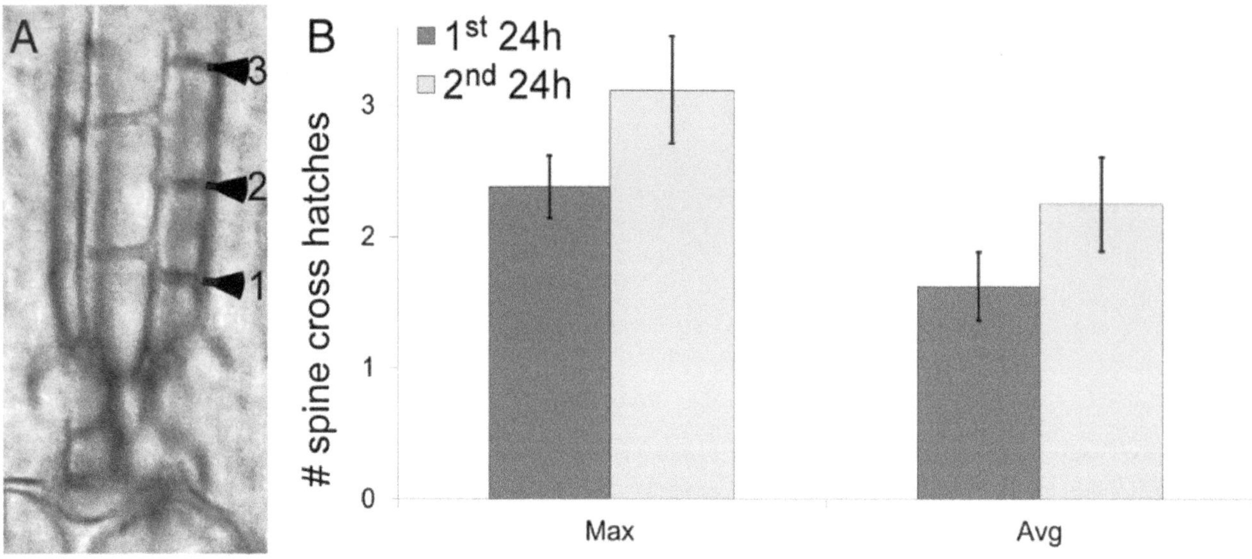

Figure 5: Spine elongation in *S. purpuratus* rudiment. A) single developing adult spine with cross hatches; this spine has 3 cross hatches on the indicated face, but notice that other faces have fewer. B) Spine elongation measured by the average (Avg) and maximum (Max) number of cross hatches in the adult spines of developing *S. purpuratus* larvae at Stage 8 or later (see Table 2). We detected no significant difference in spine elongation between the first and second 24 hours of development in well plates during the course of the experiment at 14°C. Error bars are one standard error of the mean.

Figure 5B shows the increase in the mean (Avg) and maximum (Max) number of cross hatches in our larvae during the first and second 24 hours of development in our well plate set-up at 14°C (Guelph experiment). Two conclusions are apparent from these data: 1) spines grew in our well plate experiment at a rate of about 2 cross hatches per day; and 2) there is no significant difference in the rate of spine growth when comparing the first and second 24 hours of our experiment (Max: $t_{38}=-1.7$; $p=0.10$; Avg: $t_{38}=-1.4$; $p=0.16$). If anything, the results in Figure 5B indicate a trend towards faster spine growth (Avg and Max) in the second 24 hours.

DISCUSSION

Metamorphosis is a widespread phenomenon across animals and non-animals alike [12, 18, 48,49]; in the most extreme examples, the pre-metamorphic form (e.g., larva) is so disparate from the post-metamorphic form (e.g., juvenile, adult) that biologists once classified them as being entirely unrelated (e.g., [50]). As such, metamorphosis is a period where complex ontogenetic processes occur in a relatively short time [12, 14, 18, 48, 49], a phenomenon not unlike what occurs during embryogenesis.

Many echinoderms (sea urchins, sea stars, sea cucumbers and their kin) have life histories containing notable examples of such a radical metamorphosis, where their bilaterally-symmetric larva transforms in relatively short order to a pentameral adult, a process that has fascinated biologists for over a hundred years. Despite this interest, we still lack a thorough mechanistic understanding of metamorphosis in any echinoderm, or indeed in any marine invertebrate. Thus while the purple sea urchin, *Strongylocentrotus purpuratus*, has progressed in recent years from being a useful subject for classical embryogenesis studies to revealing detailed molecular genetic mechanisms of development, a parallel advance in our understanding of its 'second embryogenesis' [51] –metamorphosis– has not materialized.

The lack of a clearly defined, sufficiently detailed, and broadly accepted staging scheme for commonly studied echinoderms (but see [31, 32]) is one explanation for the lack of parallel progress in understanding their metamorphoses. We thus offer the purple urchin staging scheme described herein as a step to overcoming this deficit, and thus facilitating a broader range of investigations into sea urchin (and ultimately echinoderm) metamorphosis. In this sense, our staging scheme was inspired by similar ones that remain widely used in*Drosophila*[52], the sea hare *Aplysia californica*[53], *Xenopus*[54] and zebrafish [55], and like ours, are focused on characters readily visible under the microscope in live individual embryos and larvae.

Comparison with Previous Echinoid Staging Schemes

Homologs of the characters that we focused on for the majority of our stages had been previously described in detail by Macbride [56–59], von Ubisch [60] and Gordon [34, 35] among others in their classic histological studies of pre-settlement juvenile development in a variety of echinoid taxa. Nevertheless, a treatment of these skeletal characters is largely absent from the purple urchin staging scheme of Smith and colleagues [37], and from the echinoid staging scheme of Chino and colleagues [38] as well. As such, both of these latter schemes compress the final 1/3 of larval development (or approximately 10 days in *S. purpuratus*) into 2–3 stages (as compared to our 10 stages; see Table 2). This seems insufficient given that this period represents the bulk of juvenile morphogenesis in the larva, and is thus when the majority of adult structures first make their appearance during ontogeny. By contrast, many of our proposed rudiment soft tissue stages are similar to those proposed by Smith and colleagues [37] and Chino and colleagues [38], as we indicate in Table 1. For this reason, and because of the detailed histological descriptions of juvenile purple urchin morphogenesis in Smith et al. [37], we see our proposed staging scheme as broadly complementary to these previous efforts.

Phenotypic Plasticity and Heterochronies

The biggest concern in proposing any staging scheme for echinoderm larvae is taking account of the substantial phenotypic plasticity in their ontogeny [61]. For this reason, we specifically avoided combining, in our proposed staging scheme, larval and juvenile characters, the developmental trajectories of which can vary substantially relative to one another depending on food and other rearing conditions [38, 40, 41, 62, 63].

Nevertheless, even when restricting our analysis to juvenile characters alone, we noted instances where different larvae exhibited heterochronies: namely, variation in the relative appearance of distinct characters. One example is the juvenile spines that develop on the right side of purple urchin larvae, outside of the rudiment proper (see [37, 45]); another is the genital plates of the test, several of which form as bifurcations and proliferations off of larval skeletal rods [see 34,35]. We left an analysis of those structures out of our staging scheme since we observed substantial heterochronic variation in their developmental trajectories relative to rudiment characters, and thus their inclusion would not be informative. We also noted much more subtle heterochronies in some of the characters that we did include; in such cases (see skeletogenic Stages 3 and 5–7), we accounted for the variation by combining more than one character into a single stage. Likewise, because we noticed heterochronic variation (even within clutches) in the soft tissue stage at which we first observed juvenile skeleton formation, we split our staging scheme into two sub-schemes: one for soft tissue characters, the other for skeletogenic characters.

To address the issue that variation in rearing conditions might likewise lead to heterochronic shifts in the ontogeny of different juvenile characters, we reared larvae under five conditions that differed in sea water chemistry (artificial versus natural), food level, temperature, method of culture mixing and source population. Despite rearing larvae with these multiple sources of variance, we did not observe any heterochronies in our staging characters. Therefore, we are confident that the broad outlines of our staging scheme should hold across *S. purpuratus*populations and in different rearing conditions.

Applicability Outside of *S. purpuratus*

One of the great advantages in studying echinoid larval development and metamorphosis is the great diversity in developmental modes, larval forms and adult habitats within a group having an excellent fossil record and robust phylogenetic hypotheses for their relationships [26, 64,65]. This suite of features makes echinoids prime candidates for a wide range of comparative studies on larval development and ecology, metamorphosis and settlement, as well as juvenile form, function and behavior.

To encourage the use of echinoids in these ways, we envision our staging scheme as a template for the development of similar schemes across echinoids. Over 120 years of research effort has been directed at descriptions of late larval and juvenile development across a wide range of echinoid species [12, 34–40, 42, 56–60, 66–89] and from these studies it is clear that there is substantial variation across echinoids in what and when juvenile skeletal elements form during late larval development. For example, pedicellariae form before settlement in a widely divergent assemblage of species (reviewed in [89]), whereas they form only after settlement in most taxa, including *S. purpuratus*. Emlet [39] undertook a comparison of the juvenile structures present shortly after settlement in a disparate group of 31 echinoid species, revealing substantial variation in early juvenile form, variation that must have been generated during late larval development. The staging scheme proposed here can be explicitly used to trace the ontogenetic basis for this variation.

Specifically, we predict that such comparative studies will uncover both heterochronies (see [90]) and heterotopies (change in the relative position of developmental events) in the formation of juvenile structures in late larvae, changes that might be hypothesized to be related to selection – or otherwise reveal constraints – on juvenile form and function. Indeed, many of the descriptive studies cited have already indicated the existence of this kind of variation; with an explicit staging scheme to use for comparisons, such as the one developed here, such variation can be easily demonstrated.

We would also advocate that our general approach used herein be applied to non-echinoid echinoderms as well, enabling more precise comparisons among and between the different classes of echinoderms. Such a broadening would help harness the power of the comparative approach for studies ranging from life history evolution to functional morphology of larvae and juveniles to the evolution of body plans in the remarkably diverse echinoderm phylum.

CONCLUSIONS

Here we present a novel, detailed staging scheme for late larval development for the purple sea urchin, *Strongylocentrotus purpuratus*. We conceived this scheme to allow sea urchin researchers to quickly and consistently identify the stage of live larvae during the period of rapid juvenile morphogenesis that occurs in the final weeks before settlement, using standard microscopy techniques. We offer a detailed comparison of our scheme to previous efforts, providing a side by side comparison of corresponding stages. This analysis reveals that our new staging scheme provides a more detailed picture of juvenile development using ontogenetically informative characters. Inspired by widely used schemes in other organisms such as*Drosophila*, zebrafish and *Xenopus*, our proposed staging scheme provides both a useful method to study late larval development in the purple urchin, and a framework for developing similar staging schemes across echinoderms. Such efforts will have a high impact on evolutionary developmental as well as larval ecology studies and facilitate research on this important deuterostome group.

METHODS

We conducted staging observations on multiple clutches of purple sea urchin (*Strongylocentrotus purpuratus*) larvae from 2010–2014, fertilized and reared through settlement in the laboratory using our modifications of standard methods; [3] details below). In order to insure consistency of our staging scheme, we reared larvae in 5 different locations in North America (SEA-Seattle, WA; FHL-Friday Harbor Labs,

WA; BML-Bodega Marine Laboratories, CA; HMS-Hopkins Marine Station, CA and UOG-Guelph, ON).

Urchins and Spawning

Adult urchins derived from two distinct USA source populations: Slip Point (Clallam Bay, WA; SEA, FHL) and The Cultured Abalone Ltd (Goleta, CA; UOG, BML, HMS). At UOG, we maintained urchins in the Hagen Aqualab, University of Guelph, in artificial seawater (Instant Ocean™; Instant Ocean) at 12°C and 34 ppt salinity. We fed the urchins rehydrated kombu kelp (*Laminaria sp.*) *ad libitum*. For the SEA and FHL rearings, we used urchins maintained in subtidal cages suspended off the floating docks at FHL, fed throughout the year *ad libitum* with drift kelp (mainly blades of *Nereocystis leutkeana*).

For the HMS and BML rearings, we used urchins maintained in the dark in flow-through natural sea water tables at HMS, fed throughout the year *ad libitum* with giant kelp (*Macrocystis pyrifera*).

We spawned adult sea urchins by gentle shaking or intra-coelomic injection with 0.5 M KCl, and fertilized spawned eggs (>90% fertilization success) with diluted sperm using standard methods [3].

Larval Culturing

For all embryo and larval culturing in Guelph, we used 0.2 μm Millipore™-filtered artificial seawater (MFASW), sterilized with UV light. We cultured embryos at 14°C until hatching (24 hours) at which point we poured off swimming embryos and set up our main cultures at 1 larva/4 ml derived from one male and one female. We used a mechanical stirring system [3] to keep larvae in suspension, and fed them a mixture of *Dunieliella tertiolecta* (12 cells/μl) and*Rhodomonas lens* (6 cells/μl). We changed greater than 95% of the water every two days by reverse filtration, and fed the larvae as above.

For our embryo and larval rearings at FHL, SEA, HMS and BML, we derived our cultures from equal-part mixtures of single-parent fertilizations [either 3 males × 1 female (3M × 1F), 3F × 1M or 2F × 2M] at an initial density of 1 embryo/ml, and maintained either using a mechanical stirring system (FHL, HMS, BML; 3]) or a shaking water bath (SEA) to keep larvae in suspension. Starting on day 3, we fed larvae a mixture of *Dunieliella teriolecta* (3 cells/μl) and*Rhodomonas spp.* (2.5 cells/μl) every 2 days following water changes as described above. We reduced the larval density to 0.1-0.25 larvae/ml on about day 16 (6 arm plutei, before rudiment invagination). At FHL and BML, we cultured our larvae at sea table temperatures, which vary from an average of 10-14°C depending on the time of year. At HMS and SEA, we maintained our larvae at constant temperatures (14°C and 16°C, respectively).

Staging Scheme

To develop the staging scheme, we identified dozens of score-able soft tissue and skeletogenic juvenile characters (see also [88]), and characterized hundreds of total larva for the presence or absence of that character. Our goal was to identify characters whose relative timing of appearance would be consistent. As such, we ended up excluding several classes of juvenile characters that we identified as "heterochronic" in their appearance from larva-to-larva or batch-to-batch, notably, those outside of the rudiment proper. In the final staging scheme, we thus settled on characters that are defined by discrete morphological features (soft and hard structures) within the developing juvenile rudiment of *S. purpuratus* larvae, visible in live specimens either under differential interference contrast (DIC) or cross-polarized light. A detailed description of these stages is presented in the Results section and in Tables 1 and 2.

Our data on approximate numbers of days from fertilization to various stages (see third column in Tables 1 and 2) are compiled from our numerous fertilizations and rearings in different seasons and locations at 14ºC and under our various culturing conditions outlined above. We therefore believe that the results are robust with respect to population differences in different geographic locations, as well as water chemistry and other factors. We note that these timing data are truly approximations, as larvae in a given batch can vary substantially from these values, both in mean stage and in the variance among stages. For example, some batches of larvae, for reasons that we and others have not identified, undergo bouts of asexual larval budding [45, 91, 92], sometimes in a majority of the larvae within a culture vessel (data not shown). The resultant larval cultures are both delayed and more variable than larvae from a more typical rearing.

Stage Length Experimental Design

To determine with more precision the lengths of individual skeletogenic stages, we conducted staging observations on individual larvae in two of our rearing locations. The majority of our staging data derive from larvae that we cultured at UOG in June 2011. We also report on our temperature comparison experiment conducted in SEA in September 2011. Throughout the Methods and Results sections, we refer to our 'Guelph' or 'Seattle' experiments, respectively.

For the Guelph staging studies, we selected 48 larvae on day 28 that had visible rudiments, and mounted them individually on a microscope slide with raised cover glass using modeling clay, in order to immobilize but not damage the larvae. We then used cross-polarized and DIC optics to stage each larva (0 hour time point) according to our staging scheme (see Results and Tables 1& 2), photographed it (see below), and gently transferred it to an individual well in a 24 well plate (Costar 3524) with 1.5 ml of MFASW and algae at the same concentrations as in their rearing conditions (see above). We maintained the well plates for 24 hours at 14°C, at which point we again mounted and photographed each larva as above, and assigned it to a stage (24 hour time point). We continued to culture 24 of these larvae for an additional 24 hours in their same wells, and photographed and staged them one final time (48 hour time point).

The Seattle protocol was similar to that described above with the following differences: 1) the experiment began on day 21, so the experimental larvae only covered skeletogenic Stages 0–6; 2) we assigned 48 larvae at random to either a 12°C or 16°C treatment to examine temperature effects on staging progression; 3) we staged each larva at 0, 24, 48 and 96 hours of the experiment; and 4) we fed larvae for the duration of the experiment at the level at which they were fed throughout development (see above), with a full water change at 48 hours.

Stage Length Calculations

For both the Guelph and the Seattle datasets, we made the following three assumptions for the calculation of skeletogenic stage lengths (in hours) for each observed larva: 1) since larvae were observed at most once per day, larvae at each observation point were assumed to be at the mid time point of the observed stage; 2) if a larva advanced one or more stages during an observation time interval (i.e. 24 or 48 hours), that larva was assumed to have progressed through all intervening stages at a constant rate (calculated as the time interval divided by the stage differential); and 3) if a larva did not progress at all during a 48 hour observation interval, the total length of that stage for that larva could not be estimated, so we scored the length of that stage for such larvae as 72 hours. We consider this maximum length of 72 hours to be quite conservative, based on our observations on cohorts of larvae that indicate that none of our skeletogenic stages are of that duration at 14-16°C (data not shown).

For example, the length of stages for a hypothetical larva that began the experiment (0 hrs) at Stage 6, progressed to Stage 7 at 24 hours, and then Stage 9 at 48 hours would be computed as follows. Based on assumption 1: the larva was at the midpoint of Stage 6 at 0 hrs, of Stage 7 at 24 hours and Stage 9 at 48 hours. Sometime during the second 24 hours, the larva entered and exited Stage 8. Based on assumption 2: the lengths of Stage 7, 8 and 9 are assumed to be equal, and thus 12 hours each (time interval=24 hours, stage differential=2 stages; therefore 24/2=12 hr). This means that Stage 7 is assumed to have begun 18 hours into the experiment, Stage 8 at 30 hours, and Stage 9 at 42 hours). And, finally, Stage 6 – which was at its midpoint at 0 hrs (see assumption 1) – is 36 hours for this larva (it is assumed that Stage 6 for this larva began 18 hours before the onset of the experiment and finished when Stage 7 is assumed to have begun: namely, 18 hours after the start of the experiment).

We calculated means, standard errors of the mean and the 95% confidence intervals for the lengths of each stage at each temperature. Note that we could not estimate the length of Stage 10 in our study, since we calculated stage lengths based on progression into the subsequent stage, and there is no Stage 11 in our scheme.

In order to confirm that our live mounting technique of larvae did not affect the developmental progression of larvae, we setup a separate experiment with a total of 24 larvae in February 2014 at UOG. 12 larvae were randomly assigned to a mounting treatment. Specifically, we mounted these 12 larvae as described above, staged them and placed them back into individual wells as described above. The other 12 larvae were placed directly into wells without mounting. After 48 hours, all 24 larvae were staged, and we compared the distribution of stages in the two sets of larvae (manipulated vs. non-manipulated).

Spine Growth

Purple urchin larvae have two types of spines: 6-sided "adult" spines and 4-sided "juvenile" spines (see [39]) for review, and for a listing of alternative names in the literature for these spine types). The latter are so named as they are juvenile-specific, and are not found in urchin adults. Due to their greater numbers, ease of scoring and regular positions within the growing rudiment, we decided to focus exclusively on the adult spines for this particular analysis. Using the same dataset described above (Guelph experiment), we counted the number of cross hatches in the first five adult spines that we could observe in each larva that had reached Stage 8. We used these data to calculate mean and maximum number of cross hatches for each larva.

We calculated growth rates of the adult spines within the first and second 24 hour period of observations to get a rough estimate of spine growth in *S. purpuratus* larvae. For this analysis we counted the regular cross bars that occur along the length of the growing adult spines (see Figure 5A) within the rudiment during the final week or so of development before settlement; since these bars occur at regular intervals along the length of the growing spine (data not shown), the number of cross bars (=cross hatches) is a convenient way to estimate spine length. For each larva in the Guelph experiment, that was at or beyond Stage 8 (see Table 2), we counted the number of complete cross hatches (i.e. with no gap) in the first five adult spines that we identified while examining larvae at 0, 24 and 48 hours. We then calculated the mean (Avg) and maximum (Max) number of cross hatches for the five counted spines in each of these larvae, and used these values to compare mean rates of spine growth (for both Avg and Max) during the first and second 24 hour periods of observation.

Note that adult spines are hexagonal around the long axis of the spine, and thus have cross bars along all six edges of the hexagon (i.e., along all six faces of the spine), and that the number of cross hatches along each of these six faces is not always consistent within a spine (e.g., see Figure 5A). We counted the maximum number of cross hatches visible on any of the six faces. For example, in a spine where the six faces had 2, 2, 3, 2, 2 and 3 cross hatches, it would receive a score of 3 cross hatches. In another example, if only one complete cross hatch appears on as few as one face, that spine would receive a score of 1 cross hatch. We analyzed these data using a one tailed *t*-test for 24 and 48 h in order to assess whether the rate of addition of cross hatches was different during the first and second 24 hours of our experiment.

Comparison to Other Staging Schemes

In Tables 1 & 2, we present comparisons between our staging scheme and the staging schemes for *S. purpuratus* and 3 other echinoids, as presented by Smith et al. [37] and Chino et al. [38] respectively. For these comparisons, we examined the images, drawings and descriptions presented by these authors, and compared them to the new soft tissue and skeletogenic stages that we present here.

Imaging

In Guelph, we imaged larvae on a Nikon Ti microscope equipped with a motorized stage. Images of larvae were taken at each time point and, if necessary, multiple views were captured of juvenile skeletons. In Seattle, we imaged only a select subset of larvae to illustrate certain stages, using a Nikon Coolpix 990 camera mounted on a Leitz Wetzlar Ortholux microscope. We captured additional images to illustrate some stages in Figure 1 using a Zeiss Axio Imager Z1 microscope at HMS. All three of these systems were equipped with DIC optics and cross-polarized light. All image are oriented with posterior to the left.

Ethics Statement

Research presented here does not require any approval and therefore complies with all necessary regulations.

ACKNOWLEDGEMENTS

We thank Friday Harbor Labs and the University of Washington, Hopkins Marine Station, Bodega Marine Laboratories, Merrill Hille, Lynn Riddiford, Chris Lowe and Brian Gaylord for facilities access; Richard Strathmann, Victoria Foe, Billie Swalla, Kevin Uhlinger and Pam Miller for urchin care; and Cory Bishop, Richard Emlet and seven anonymous reviewers for their comments on earlier versions of this study. We would also like to thank NSERC for funding (Discovery Grant #400230 to AH), a grant from the Evo-Devo-Eco Network (EDEN to AH) and the National Science Foundation (NSF 1357033 to JH and Chris Lowe).

AUTHORS' CONTRIBUTIONS

AH and JAH carried out experiments at the University of Guelph and Hopkins Marine Station (Stanford University), JAH carried out additional experiments at the University of Washington (Seattle), at Friday Harbor Laboratory, and at Bodega Marine Laboratories. Experimental planning, analysis of data and writing of the article were joint efforts by AH and JAH. Both authors read and approved the final manuscipt.

REFERENCES

1. Jasny BR, Purnell BA: The glorious sea urchin - introduction. Science. 2006, 314: 938-938. 10. 1126/ science. 314. 5801. 938.
2. Sodergren E, Consortium SUGS: The genome of the sea urchin Strongylocentrotus purpuratus. Science. 2006, 314: 941-952.
3. Strathmann MF: Reproduction and Development of Marine Invertebrates of the Northern Pacific Coast. 1987, Seattle, WA, USA: University of Washington Press
4. McEdward LR: Ecology of Marine Invertebrate Larvae. 1995, Boca Raton: CRC Press
5. Swanson RL, Byrne M, Prowse TAA, Mos B, Dworjanyn SA, Steinberg PD: Dissolved histamine: a potential habitat marker promoting settlement and metamorphosis in sea urchin larvae. Mar Biol. 2012, 159: 915-925. 10. 1007/s00227-011-1869-2.
6. Swanson RL, Marshall DJ, Steinberg PD: Larval desperation and histamine: how simple responses can lead to complex changes in larval behaviour. J Exp Biol. 2007, 210: 3228-3235. 10. 1242/jeb. 004192.
7. Bishop CD, Brandhorst BP: Development of nitric oxide synthase-defined neurons in the sea urchin larval ciliary band and evidence for a chemosensory function during metamorphosis. Dev Dyn. 2007, 236: 1535-1546. 10. 1002/dvdy. 21161.
8. Burke RD: Pheromonal control of metamorphosis in the pacific sand dollar, Dendraster excentricus. Science. 1984, 225: 442-443. 10. 1126/science. 225. 4660. 442.
9. Bishop CD, Brandhorst BP: On nitric oxide signaling, metamorphosis, and the evolution of biphasic life cycles. Evol Dev. 2003, 5: 542-550. 10. 1046/j. 1525-142X. 2003. 03059. x.
10. Cameron RA, Tosteson TR, Hensley V: The control of sea urchin metamorphosis - ionic effects. Dev Growth Differ. 1989, 31: 589-594. 10. 1111/j. 1440-169X. 1989. 00589. x.
11. Sutherby J, Giardini JL, Nguyen J, Wessel G, Leguia M, Heyland A: Histamine is a modulator of metamorphic competence in Strongylocentrotus purpuratus (echinodermata: echinoidea). BMC Dev Biol. 2012, 12: 14-10. 1186/1471-213X-12-14.
12. Hodin J: Expanding networks: Signaling components in and a hypothesis for the evolution of metamorphosis. Integr Comp Biol. 2006, 46: 719-742. 10. 1093/icb/ icl038.
13. Burke RD: Neural control of metamorphosis in Dendraster excentricus. Biol Bull. 1983, 164: 176-188. 10. 2307/1541137.
14. Heyland A, Reitzel AM, Degnan S: Emerging Patterns in the Regulation and Evolution Of Marine Invertebrate Settlement And Metamorphosis. Mechanisms of Life History Evolution: The Genetics and Physiology of Life History Traits and Trade-Offs. Edited by: Flatt T, Heyland A. 2011, London: Oxford University Press, 29-42.
15. Amador-Cano G, Carpizo-Ituarte E, Cristino-Jorge D: Role of protein kinase C, G-protein coupled receptors, and calcium flux during metamorphosis of the sea urchin Strongylocentrotus purpuratus. Biol Bull. 2006, 210: 121-131. 10. 2307/4134601.
16. Gaylord B, Hodin J, Ferner MC: Turbulent shear spurs settlement in larval sea urchins. Proc Natl Acad Sci U S A. 2013, 110: 6901-6906. 10. 1073/pnas. 1220680110.
17. Hadfield MG: Metamorphic competence is a major adaptive convergence in marine invertebrate larvae. Am Zool. 2000, 40: 1038-1038.
18. Heyland A, Moroz LL: Signaling mechanisms underlying metamorphic transitions in animals. Integr Comp Biol. 2006, 46: 743-759. 10. 1093/icb/icl023.
19. Strathmann RR: Evolution and loss of feeding larval stages of marine invertebrates. Evolution. 1978, 32: 894-906. 10. 2307/2407502.
20. Heyland A, Hodin J, Reitzel AM: Hormone signaling in evolution and development: a non-model system approach. BioEssays. 2005, 27: 64-75. 10. 1002/bies. 20136.
21. Hart M: Phylogenetic analyses of mode of larval development. Semin Cell Dev Biol. 2000, 11: 411-418. 10. 1006/scdb. 2000. 0194.
22. Strathmann RR, Eernisse DJ: What molecular phylogenies tell us about the evolution of larval forms. Am Zool. 1994, 34: 502-512.
23. McHugh D, Rouse GW: Life history evolution of marine invertebrates: new views from phylogenetic systematics.

Trends Ecol Evol. 1998, 13: 182-186. 10. 1016/S0169-5347(97)01285-8.

24. Strathmann RR: Why life histories evolve differently in the sea. Am Zool. 1990, 30: 197-207.

25. Rutherford S, Hirate Y, Swalla BJ: The hsp90 capacitor, developmental remodeling, and evolution: the robustness of gene networks and the curious evolvability of metamorphosis. Crit Rev Biochem Mol. 2007, 42: 355-372. 10. 1080/10409230701597782.

26. Smith AB: Echinoderm larvae and phylogeny. Annu Rev Ecol Syst. 1997, 28: 219-241. 10. 1146/annurev. ecolsys. 28. 1. 219.

27. Swalla BJ: Building divergent body plans with similar genetic pathways. Heredity. 2006, 97: 235-243.

28. Sumrall CD, Wray GA: Ontogeny in the fossil record: diversification of body plans and the evolution of "aberrant" symmetry in Paleozoic echinoderms. Paleobiology. 2007, 33: 149-163. 10. 1666/06053. 1.

29. Givnish TJ: How a Better Understanding of Adaptations Can Yield Better Use Of Morphology In Plant Systematics: Toward Eco-Evo-Devo. Deep Morphology: Toward a Renaissance of Morphology in Plant Systematics. Edited by: Stuessy TF, Mayer V, Horandl E. 2003, Bratislava: Regnum Vegetabile, 273-295.

30. Gilbert SF, Epel D: Ecological Developmental Biology: Integrating Epigenetics, Medicine, and Evolution. 2009, Sunderland, MA, USA: Sinauer

31. Hendler G: Development of amphioplus abditus (Verrill) (Echinodermata: Ophiuroidea). Description and discussion of ophiuroid skeletal ontogeny and homologies. Biol Bull. 1978, 154: 79-95. 10. 2307/1540776.

32. Emlet RB: Direct development of the brittle star Amphiodia occidentalis (Ophiuroidea, Amphiuridae) from the northeastern Pacific Ocean. Invertebr Biol. 2006, 125: 154-171. 10. 1111/j. 1744-7410. 2006. 00049. x.

33. Chia FS, Rice ME: Settlement and Metamorphosis of Marine Invertebrate Larvae. 1978, New York: Elsevier

34. Gordon I: The development of the calcareous test of Echinus miliaris. Proc R Soc B. 1926, 214: 259-312.

35. Gordon I: Skeletal development in arbacia, echinarachnius and leptasterias. Proc R Soc B. 1929, 217: 289-334.

36. Gosselin P, Jangoux M: From competent larva to exotrophic juvenile: a morphological study of the perimetamorphic period of Paracentrotus lividus (Echinodermata: Echinoida). Zoomorphology. 1998, 118: 31-43. 10. 1007/s004350050054.

37. Smith MM, Cruz Smith L, Cameron RA, Urry LA: The larval stages of the sea urchin, Strongylocentrotus purpuratus. J Morphol. 2008, 269: 713-733. 10. 1002/jmor. 10618.

38. Chino Y, Saito M, Yamasu K, Suyemitsu T, Ishihara K: Formation of the adult rudiment of sea urchins is influenced by thyroid hormones. Dev Biol. 1994, 161: 1-11. 10. 1006/dbio. 1994. 1001.

39. Emlet RB: Morphological evolution of newly metamorphosed sea urchins-a phylogenetic and functional analysis. Integr Comp Biol. 2010, 50: 571-588. 10. 1093/icb/icq073.

40. Heyland A, Hodin J: Heterochronic developmental shift caused by thyroid hormone in larval sand dollars and its implications for phenotypic plasticity and the evolution of nonfeeding development. Evolution. 2004, 58: 524-538. 10. 1111/j. 0014-3820. 2004. tb01676. x.

41. Strathmann RR, Fenaux L, Strathmann MF: Heterochronic developmental plasticity in larval sea urchins and its implications for evolution of nonfeeding larvae. Evolution. 1992, 46: 972-986. 10. 2307/2409750.

42. Hyman LH: The Invertebrates: Echinodermata, Volume 4. 1955, New York, NY, USA: McGraw-Hill

43. Emlet RB: Crystal axes in recent and fossil adult echinoids indicate trophic mode in larval development. Science. 1985, 230: 937-940. 10. 1126/science. 230. 4728. 937.

44. Mooi R, David B: Skeletal homologies of echinoderms. Paleontol Soc Papers. 1997, 3: 305-335.

45. Eaves AA: Potential for paired vestibules in plutei (Echinodermata, Echinoidea). Invertebr Biol. 2005, 124: 174-184. 10. 1111/j. 1744-7410. 2005. 00017. x.

46. Hart MW: Variation in suspension feeding rates among larvae of some temperate, eastern Pacific echinoderms. Invertebr Biol. 1996, 115: 30-45. 10. 2307/3226940.

47. Fox CA: Ingestion Rates Of Five Species Of Pacific Northwest Echinoderm Larvae On Natural And Artificial Particles. Masters Thesis. 2007, Georgia Southern University

48. Bishop CD, Huggett MJ, Heyland A, Hodin J, Brandhorst BP: Interspecific variation in metamorphic competence in marine invertebrates: the significance for comparative investigations into the timing of metamorphosis. Integr Comp Biol. 2006, 46: 662-682. 10. 1093/icb/icl043.

49. Bishop CD, Erezyilmaz DF, Flatt T, Georgiou CD, Hadfield MG, Heyland A, Hodin J, Jacobs MW, Maslakova SA, Pires A, Reitzel AM, Santagata S, Tanaka K, Youson JH: What is metamorphosis?. Integr Comp Biol. 2006, 46: 655-661. 10. 1093/icb/icl004.

50. Scholtz G: Zoological detective stories: the case of the facetotectan crustacean life cycle. J Biol. 2008, 7 (5): 16-10. 1186/jbiol77.

51. Muller WA: Developmental Biology. 1997, Berlin: Springer

52. Campos-Ortega JA, Hartenstein V: The Embryonic development of Drosophila melanogaster. 1985, Berlin: Springer

53. Heyland A, Vue Z, Voolstra CR, Medina M, Moroz LL: Developmental transcriptome of Aplysia californica. J Exp Zool Part B. 2011, 316B: 113-134. 10. 1002/jez. b. 21383.

54. Nieuwkoop PD, Faber J: Normal Table of Xenopus laevis (Daudin). 1994, New York: Garland Publishing Inc

55. Kimmel CB, Ballard WW, Kimmel SR, Ullmann B, Schilling TF: Stages of embryonic development of the

zebrafish. Dev Dyn. 1995, 203: 253-310. 10. 1002/aja. 1002030302.

56. MacBride EW: The development of Echinus esculentus, together with some points in the development of E. miliaris and E. acutus. Proc R Soc B. 1903, 195: 285-327.
57. MacBride EW: The artificial production of echinoderm larvae with two water vascular systems, and also of larvae devoid of a water vascular system. Proc R Soc B. 1914, 90: 259-282.
58. MacBride EW: The development of Echinocardium cordatum - part I: the external features of the development. Q J Micr Sci. 1918, 59: 487-521.
59. MacBride EW: Some further experiments in the artificial production of a double hydrocoele in the larvae of Echinus miliaris. Proc R Soc B. 1919, 196: 207-208.
60. von Ubisch L: Die entwicklung von Strongylocentrotus lividus (Echinus microtuberculatus, Arbacia pustulosa). Zeit F Wiss Zoo. 1913, 106: 409-448.
61. Miner BG: Mechanisms Underlying Feeding Structure Plasticity in Echinoderm Larvae. Mechanisms of Life History Evolution: The Genetics and Physiology of Life History Traits and Trade-Offs. Edited by: Flatt T, Heyland A. 2011, London: Oxford University Press, 221-229.
62. McEdward LR, Miner BG: Fecundity-time models of reproductive strategies in marine benthic invertebrates: fitness differences under fluctuating environmental conditions. Mar Ecol Prog Ser. 2003, 256: 111-121.
63. Reitzel AM, Miles CM, Heyland A, Cowart JD, McEdward LR: The contribution of the facultative feeding period to echinoid larval development and size at metamorphosis: a comparative approach. J Exp Mar Biol Ecol. 2005, 317: 189-201. 10. 1016/j. jembe. 2004. 11. 018.
64. Wray GA, Bely AE: Several distinct factors drive the evolution of early development in echinoderms. Development. 1994, Supplement: 97-106.
65. McEdward LR, Miner BG: Larval and life-cycle patterns in echinoderms. Can J Zool. 2001, 79: 1125-1170. 10. 1139/z00-218.
66. Mortensen T: Studies of the Development and Larval Forms of Echinoderms. 1921, Copenhagen: Bianco Lunos
67. Morris VB: Apluteal development of the sea urchin Holopneustes purpurescens Agassiz (Echinodermata, Echinoidea, Euechinoidea). Zool J Linn Soc Lond. 1995, 114: 349-364. 10. 1111/j. 1096-3642. 1995. tb00120. x.
68. Prenant M: Sur le déterminisme de la forme spiculaire chez les larves d'oursins. Soc Biol. 1926, 94: 433-435.
69. Bury H: Studies in the embryology of the echinoderms. J Cell Sci. 1889, 116: 409-449.
70. Devanesen DW: The development of the calcareous parts of the lantern of Aristotle in Echinus miliaris. Proc R Soc B. 1922, 93: 468-485. 10. 1098/rspb. 1922. 0034.
71. Emlet RB: Larval form and metamorphosis of a "primitive" sea urchin, Eucidaris thouarsi (Echinodermata: Echinoidea: Cidaroida), with implications for developmental and phylogenetic studies. Biol Bull. 1988, 174: 4-19. 10. 2307/1541754.
72. Emlet RB: Larval spicules, cilia, and symmetry as remnants of indirect development in the direct developing sea urchin Heliocidaris erythrogramma. Dev Biol. 1995, 167: 405-415. 10. 1006/dbio. 1995. 1037.
73. Fukushi T: The formation of the echinus rudiment and the development of the larval form in the sea urchin, Temnopleurus hardwickii. Bull Mar Biol Stn Asmushi. 1960, 10: 65-72.
74. Fukushi T: The external features of the development of the sea urchin, Glyptocidaris crenularis. Bull Mar Biol Stn Asmushi. 1960, 10: 57-63.
75. Garman H, Colton B: Some notes on the development of Arbacia punctulata. Lam Studies Biol Lab Johns Hopkins Univ. 1883, 2: 247-255.
76. Hinegardner RT: Growth and development of the laboratory cultured sea urchin. Bio Bull. 1969, 137: 465-475. 10. 2307/1540168.
77. Kryuchkova GA: Development of the definitive skeleton of the sea-urchin Echinocardium cordatum. Sov J Mar Biol. 1979, 5: 487-494.
78. Kryuchkova GA: Formation of the definitive skeleton in sea urchins of the genus Strongylocentrotus. Sov J Mar Biol. 1979, 5: 276-282.
79. Kryuchkova GA: Formation of the amniotic cavity and definitive skeletal development of sand dollars. Sov J Mar Biol. 1979, 5: 203-208.
80. Minsuk SB, Raff RA: Pattern formation in a pentameral animal: induction of early adult rudiment development in sea urchins. Dev Biol. 2002, 247: 335-350. 10. 1006/dbio. 2002. 0704.
81. Nunes CDP, Jangoux M: Larval growth and perimetamorphosis in the echinoid Echinocardium cordatum (Echinodermata): the spatangoid way to become a sea urchin. Zoomorphology. 2007, 126: 103-119. 10. 1007/s00435-007-0032-6.
82. Okazaki K: Normal Development to Metamorphosis. 1975, Berlin: Springer
83. Onoda K: Notes on the development of Heliocidaris crassispina with special reference to the structure of the larval body. Mem Coll Sci Kyoto Ser B. 1931, 7: 103-134.
84. Onoda K: Notes on the development of some Japanese echinoids with special reference to the structure of the larval body. Jap J Zool. 1936, 6: 637-654.
85. Theel H: On the development of Echinocyamus pusillus. Nova Acta Regiae Soc Sci Upsal Ser. 1892, 3: 1-57.
86. Thet MM, Noguchi M, Yazaki I: Larval and juvenile development of the Echinometrid sea urchin Colobocentrotus mertensii: emergence of the peculiar form of spines. Zool Sci. 2004, 21: 265-274. 10. 2108/zsj. 21. 265.
87. Vellutini BC, Migotto AE: Embryonic, larval, and juvenile development of the sea biscuit Clypeaster subdepressus (Echinodermata: Clypeasteroida). PLoS One. 2010, 5: e9654-10. 1371/journal. pone. 0009654.

88. Yajima M, Kiyomoto M: Study of larval and adult skeletogenic cells in developing sea urchin larvae. Biol Bull. 2006, 211: 183-192. 10. 2307/4134592.

89. Burke RD: Development of pedicellariae in the pluteus larva of Lytechinus pictus (Echinodermata, Echinoidea). Can J Zool. 1980, 58: 1674-1682. 10. 1139/z80-229.

90. Raff RA: Constraint, flexibility, and phylogenetic history in the evolution of direct development in sea urchins. Dev Biol. 1987, 119: 6-19. 10. 1016/0012-1606(87)90201-6.

91. Chan KYK, Grunbaum D, Arnberg M, Thorndyke M, Dupont ST: Ocean acidification induces budding in larval sea urchins. Mar Biol. 2013, 160: 2129-2135. 10. 1007/s00227-012-2103-6.

92. Vaughn D, Strathmann RR: Predators induce cloning in echinoderm larvae. Science. 2008, 319: 1503-1503. 10. 1126/science. 1151995.

CHAPTER 12

A PANCREATIC EXOCRINE-LIKE CELL REGULATORY CIRCUIT OPERATING IN THE UPPER STOMACH OF THE SEA URCHIN STRONGYLOCENTROTUS PURPURATUS LARVA

Margherita Perillo, Yue Julia Wang, Steven D. Leach and Maria Ina Arnone

Keywords: Strogylocentrotus purpuratus, Pancreas, Ptf1a, Carboxypeptidase, Pancreatic lipase, Amylase

BACKGROUND

The ability of a species to survive in a new environment is strictly related to its capacity to successfully digest and assimilate different food sources. A characteristic that is shared by all the living forms is the ability to digest large molecules available in their habitat to produce energy. Digestion relies on the function of enzymes that are able to break down proteins, lipids and sugars in the diet. The number of specialized enzyme-producing cell types involved in food digestion increases with the complexity of the animal body and the variety of the meal. Metazoans typically have specialized exocrine cell types for the secretion of digestive enzymes. For example, gland cells that produce and release zymogens in the gastric cavity are scattered along the hydra gastroepidermis [1, 2] and along the digestive filaments of corals [3]. Cells rich in granular inclusions involved in extracellular digestion are localized along the foregut and in the gut caeca of different species of flat worms [4–6]. In different regions of the *Drosophila*midgut there are enterocytes that secrete a vast array of digestive enzymes, with expression levels matching organismal requirements [7]. A specialized organ geared toward the production of digestive enzymes and hormones, the pancreas, appeared uniquely in the vertebrate lineage. The pancreas is a complex organ evolved mainly for food digestion (exocrine cells) and maintaining blood sugar levels (endocrine cells). The exocrine pancreas constitutes the majority of the mass of the pancreas and it is composed by exocrine cells grouped into acini that secrete enzymes into the intestine [8]. The pancreas originates early in development from progenitor cells located within the endodermal epithelium. Two members of the basic helix-loop-helix (bHLH) family of protein are known to be critical for the exocrine pancreas differentiation and identity, Ptf1a (pancreatic transcriptional factor 1) and Mist1 [9, 10]. Ptf1a has a double role, first in early pancreas specification, and later in acinar cell differentiation and maintenance [11–13], while Mist1 is necessary for full maturation of the acinar cell phenotype [14]. In the mouse, it has been found that in the absence of Hnf1β, *Ptf1a* expression is not induced, leading to defective specification and reduction of the ventral and dorsal pancreas [15]. Furthermore, several signaling factors

Citation: Perillo, Yue Julia Wang, Steven D. Leach and Maria Ina Arnone, A pancreatic exocrine-like cell regulatory circuit operating in the upper stomach of the sea urchin Strongylocentrotus purpuratus larva, DOI 10.1186/s12862-016-0686-0.

are involved in balancing exocrine cell proliferation and differentiation. Among all, Notch appears to prevent pancreatic exocrine development, while FGF signaling mediates growth, morphogenesis and differentiation of exocrine cells [16–18]. In both mouse and zebrafish, Notch and its target genes inhibit the activity of the Ptf1 complex in the exocrine cells, and loss of Notch function results in accelerated development of exocrine pancreas [19].

One of the most abundant and evolutionary conserved miRNA involved in pancreas development is miR-375. MiR-375 negatively regulates glucose-stimulated insulin exocytosis by targeting myothrophin mRNA, a protein involved in insulin secretion in pancreatic β-cells [20]. A loss of function approach in zebrafish revealed that miR-375 is essential for the formation of insulin producing β-cells because its knockdown mainly results in malformation of the endocrine pancreas [21]. In addition, studies on mice lacking miR-375 demonstrated that it controls β- and α-cells mass by regulating genes involved in cellular growth and differentiation [22].

Echinoderms—a group of animals that comprise sea urchins, sea stars, sea lilies, brittle stars and sea cucumbers-belong to the deuterostome clade. The relatively simple development of sea urchin embryos, coupled with the availability of molecular tools for gene perturbation, enable comparative studies on cell specification and developmental mechanisms with other deuterostomes [23]. In particular, echinoderm systems present, in the comparison with vertebrate deuterostomes, the great advantage of a simplified gene toolkit, since this group split before the duplication events occurred at the emergence of Craniata. Importantly for this study,*Strongylocentrotus purpuratus* has an indirect development that generates a bilateral free-swimming larva whose behavior is centered on feeding.

The sea urchin larval gut is a simple tripartite tract composed of a muscular esophagus that exhibits strong contractions, a large spherical stomach with sphincters at both openings, and a small tubular intestine, which exits through the anus. In the stomach of sea urchin larvae, there are cells with the morphological features of zymogen secreting cells that are able to synthesize digestive enzymes, such as β-glucanase and α-amylase [24–27]. Gastric exocrine cells have been described also in the stomach of the adult form of *S. purpuratus*. These cells are highly concentrated in the upper stomach and electron microscopy analyses reveal that they are strikingly similar to mammalian zymogen cells, such as cells of the gastric mucosa and pancreatic acinar cells [28].

Although pancreatic cell types have been extensively characterized in the main vertebrate model systems, most of the studies in non-vertebrate animals have examined gut cells at the morphological level only and molecular information is still very scarce. Therefore, a few data are currently available on the genes/genetic characterization of the pancreatic exocrine-like cell types in non-vertebrate metazoans. In this study, we use a candidate gene approach to identify and characterize the pancreatic exocrine cell type in the sea urchin embryo and larva. We defined a unique population of cells clustered in the sea urchin larva upper stomach that responds to food and expresses the homologs of the pancreatic genes *Ptf1a* and *Mist1* together with at least three digestive enzymes. Our findings are a first step to understand the evolution of pancreatic exocrine cells. We report that the components of the specification pathway and expression for zymogen genes are active in an early branching deuterostome, which suggests that the exocrine cell type may predate the evolution of the vertebrate pancreas.

METHODS

Animal Husbandry, Embryo and Larva Cultures, Feeding Experiments

Adult *S. purpuratus* were obtained from Patrick Leahy (Kerchoff Marine Laboratory, California Institute of Technology, Pasadena, CA, USA) and housed in circulating seawater aquaria at the Stazione Zoologica Anton Dohrn of Naples. Adult sea urchin maintenance, embryo and larva cultures and feeding experiments were performed as previously described [29].

RNA *Whole Mount In Situ Hybridization*

For fluorescent whole mount *in situ* hybridization (FISH), we followed the protocol outlined in Cole et al. 2009 with the modification described in [30]. Signal was developed with fluorophore-conjugated tyramide (1:400 reagent diluents, Perkin Elmer) following instructions. For all the genes, labeled probes were transcribed from linearized DNA as described in [31]. SpmiR-375 probe has been synthesized and DIG labelled from Exiqon, and the sequence is: 5'/DigN/TGACGCGAGCCGAACGAACAAA/3'DigN/. The double FISH procedure for SpCpa2L and SpmiR-375 was performed as cited with the only exception that the miRNA probe concentration was 0025 pmol/μl and the samples were hybridized 5 days at 42 °C. mir375 probe has already been used by Christodoulou et al. 2010. Primers used to amplify the riboprobes are summarized in Additional file 1. Templates of all the probes were sequenced prior to probe generation and cloned in the pGEM®-T Easy Vector (Promega, Madison, WI, USA). Sense probes were synthesized to test the specificity of the antisense probe signal. FISH was imaged with a Zeiss 510 Meta confocal microscope.

DAPT Treatment

The γ-secretase inhibitor DAPT (N-[N-(3,5-difluorophenacetyl)-L-alanyl]-S-phenylglycine t-butyl ester, Sigma–Aldrich, St. Louis, MO) [32] was dissolved in DMSO and added at 17 h to a final concentration of 8 μM, to avoid the early and toxic effects of the drug, as described in Materna et al. 2012. A corresponding volume of DMSO was added as control.

Perturbation Experiments with MO Injection

For each experiment and for each morpholino oligonucleotide (MO), 200 to 400 eggs were injected with approximately 2-4pl of oligonucleotide injection solution and each experiment was repeated three times. As a negative control, fertilized eggs were injected with the standard control morpholino (GeneTools, Philomath, OR) and compared side-by-side with knockdown embryos. The injection of the standard control morpholino did not have any effect on the development of embryos. In a pilot experiment, different concentrations of the MASOs were injected and the morphant's phenotype was observed in order to test for the right concentration and avoid secondary effects or MASO toxicity. The highest concentration that was not toxic for the embryos was used in each experiment. The injected embryos were cultured at 15 °C until 50 h and 67 h of development, when they were fixed and used to analyze the expression of target genes in FISH or qPCR experiments. A translation-blocking antisense MO against SpHnf1a was used at a final concentration in the injection solution of 200 µM (5'-CTAGTTCGTCACCCGAATGCAGCAT-3', first tested by Peterson and Davidson 2011). A translation-blocking antisense MO against SpPtf1a was newly designed and used at a final concentration of 300 µM (SpPtf1a_MO1: 5'-ATATTTCCATAGTGATCTCTGAGT-3'). A second translation-blocking antisense MO against SpPtf1a was injected at a final concentration of 150 µM and used to confirm the specific effect of SpPtf1a MO_1 (SpPtf1a_MO2: 5'-GCGCAGGTGGATTATCAAATTGTTC-3').

RNA Extraction and Quantitative Real-Time PCR (qPCR)

For qPCR analysis of temporal expression profiles, total RNA was isolated from cultures derived from three mixed batches and extracted with Eurozol (EuroClone, Celbio, Milan, Italy). For SpPtf1a morphants and relative controls and for feeding experiments, RNA was extracted using the RNAqueous kit (Ambion). For all the experiments, samples were treated with DNase I (Ambion) to remove DNA contamination as described by the manufacturer. First-strand cDNA was synthesized from total RNA using the SuperScript VILO™ cDNA Synthesis Kit (Invitrogen) following the manufacturer's instructions. Reactions were performed as described in [33] using the ViiA7 REAL TIME PCR with SYBR Green chemistry (Applied Biosystems, Foster City, CA). For all qPCR experiments, ubiquitin mRNA was used as a reference because remains relatively constant during development [34]. For knockdown and feeding experiments, three biological replicates were analyzed, ddCt values were calculated between experiment and control embryos and converted to fold differences using ubiquitin as a reference ($FD=2^{\Delta\Delta Ct}$). A threshold of 2-fold difference was chosen as a significant change. Primer sets were chosen to amplify products 100 to 200 bp in length and sequences are reported in Additional file 1. qPCR experiments in different feeding conditions and the relative statistics were performed as previously described in Perillo and Arnone 2014.

Luciferase Assay

The coding sequences of E47, Rbpl, rnPtf1a and spPtf1a were cloned into the pCDNA3.1 backbone. Ptf1a Luciferase assay was conducted as previous described [35]. Briefly, 293 t or Hela cells were plated at a density of $1x10^5$ cells/well in a 24-well plate and transfected with 512.5 ng total DNA/well using Fugene 6 transfection reagent (Promega). For individual constructs, the following concentrations were used (per well): 12.5 ng renilla (pRL-TK, Promega), 125 ng 4x ptf1a:luciferase reporter (contains 4x Ptf1a binding motif cloned from rat*Chymotrypsinogen B* promoter, gift of Dr. Masashi Kawaichi), 125 ng human E47, 125 ng Rbpl and 125 ng of either rat rnPtf1a or sea urchin spPtf1a. Different combinations of plasmids were cotransfected and pcDNA3.1 (Invitrogen) was used as filler DNA to make the total amount of DNA per well uniform across different conditions. Two days post-transfection, cells were harvested and analyzed for firefly and renilla luciferase activities using the Dual-luciferase reporter assay system (Promega) per manufactor's protocol.

Bioinformatics Analysis

To identify putative Ptf1a binding sites on sea urchin genes of interest, we used the list of Ptf1a binding sites of the mouse acinar digestive enzyme gene promoters published by Masui et. al. (2007). From this list we built base frequency tables for the E-box and TC-box separately, i.e. counts the number of times each base occurs at each position. The base frequency table has four rows (one row for each letter: A, C, G and T) and the number of columns is equal to the motif length. Then, we calculated the weight matrices for the E-box and TC-box, the matrices have the same number of rows and columns as the corresponding frequency tables, with the value at each position being the natural logarithm of the value from the frequency table divided by the number of sequences, i.e. the weight matrices contain the estimates of the log-probabilities of each base occurring at each position in true binding sites, based on the sample of known sites. Since the gap between the two boxes is about one or two helical DNA-turns, we introduced a penalty function based on a cosine function, whose period and phase shift are 10.5 and 21 base pairs separately. Thus for a given sequence, we can calculate a matching score based on the sum of the weights that correspond to the sequence, which should be equal to the log-probability of seeing that sequence given that it is a binding site, and also the gap penalty function for the distance between the boxes.

RESULTS

Characterization of Regulatory and Differentiation Signatures for Sea Urchin Pancreatic Exocrine-Like Cells

We identified sea urchin orthologs of regulatory genes with a known conserved role in pancreatic exocrine cell development. The embryonic temporal expression of the transcriptional factors (TFs) analyzed in this study agrees with data already published from other authors [36, 37], but our analysis includes also late larval stages.

The hepatocyte nuclear factor (HNF) gene family is part of the HNF homeobox class and includes HNF1α and HNF1β, two paralogs with known interchangeable functions in several contexts [38, 39]. The sea urchin genome has only one homolog named *SpHnf1*. Its expression begins in the presumptive endoderm at blastula stage and by mid-gastrula appears confined to the gut [37]. To clarify its later functions in development, we defined the spatial expression of*SpHnf1* in post-gastrula stage embryos. During gastrulation, *SpHnf1* is expressed throughout the hindgut and the midgut, but particularly enriched in a group of cells just below the cardiac sphincters (Fig. 1a). This expression fades away in the early pluteus (Fig. 1b) and it is not detectable in older stages, consistently with the low quantitative level of expression exhibited by this gene at larval stages (Fig. 1g).

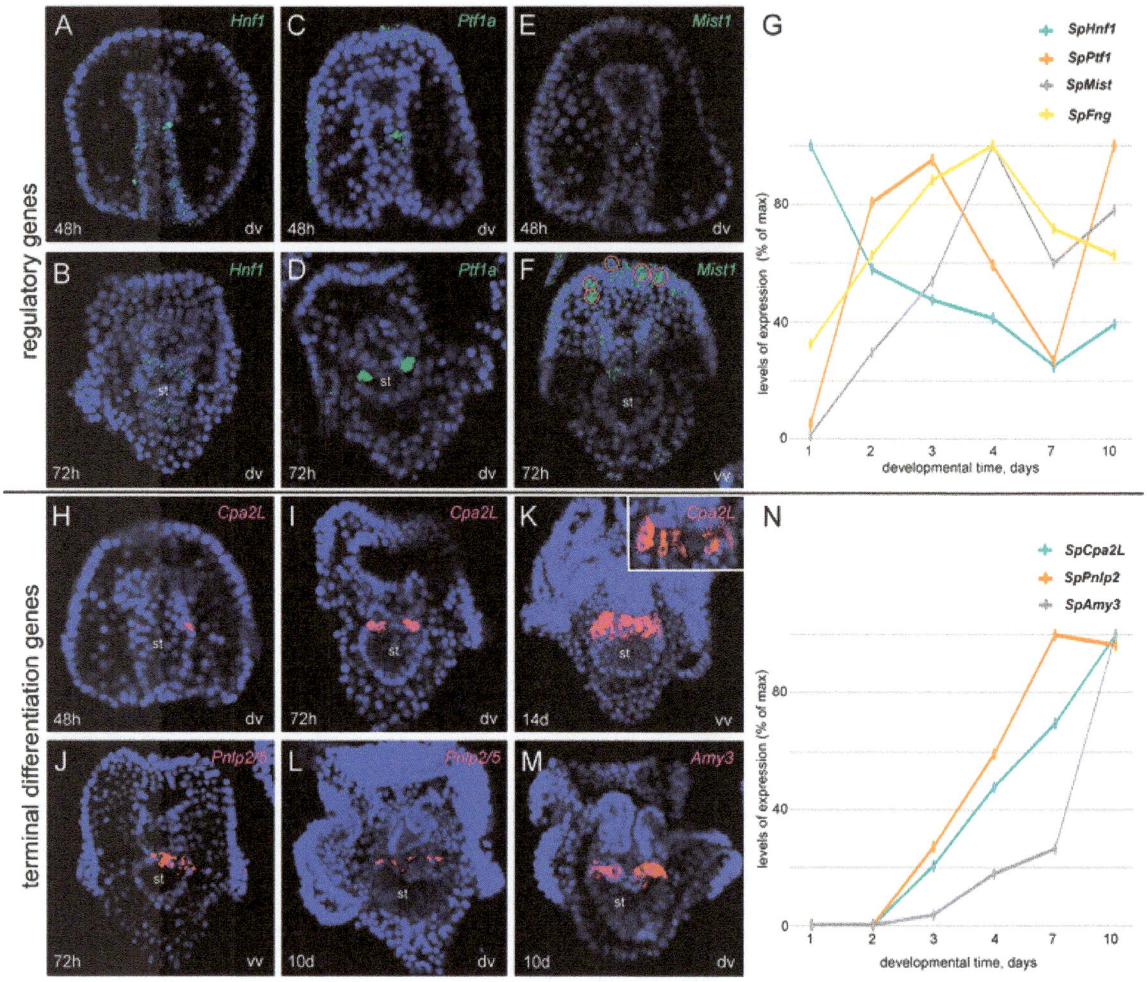

Figure 1: Expression analysis of regulatory and terminal differentiation orthologs of known pancreatic genes. mRNA localization of sea urchin regulatory (a-f, *green*) and terminal differentiation (h-m, magenta) genes. For all the figures in this paper, every picture is a full projection of merged confocal Z stacks and nuclei are stained with DAPI and depicted in blue. Red circles in f show *Mist1* RNA localization in alternating cells of the apical organ. Inset in panel k is a representative single confocal section of the upper stomach of a late larva showing that *SpCpa2L* transcripts are abundantly expressed in the entire cell.g, n Temporal expression profiles of pancreatic regulatory and terminal differentiation genes during sea urchin development. The graphs show the relative transcript abundance normalized against ubiquitin mRNA. The results are expressed as percentage of the maximum value, corresponding to the stage with the highest level of expression. Standard deviations of three technical replicas are all <0.5. For the sake of simplicity, for each panel, the species in the gene name has been omitted. Abbreviations: d, days; dv, dorsal view; h, hours post fertilization; vv, ventral view.

The sea urchin ortholog of *Ptf1a* [36] has two distinct domains of expression. At gastrula stage (48hpf, Fig. 1c), transcripts start to accumulate in cells situated right-lateral toward the dorsal side of the midgut, next to the cardiac sphincter, and some hours later it equally expands on both sides. The second domain of expression is in the ectoderm, particularly at the apical organ and in scattered cells of the ciliary band. Throughout the larval stages, the stomach cells increase in number, and we also observe an increase in *SpPtf1a* positive cells (Fig. 1d). The expression of another bHLH gene, *SpMist1* [36], starts at late gastrula stage in the upper stomach and in scattered cells of the apical organ and ciliary band (Fig. 1e, f). *SpMist1* gene expression increases during gastrulation (2 days), although at early pluteus stage (3 days)*SpPtf1a* decreases while *SpMist1* reaches its maximum expression (Fig. 1g). Similarly, the expression of both genes decreases at 7 days after fertilization and increases again after a few days.

We next identified the genes that most likely encode for enzymes used in food digestion. We compared the predicted ORFs of digestive enzyme gene families with the ESTs of an annotated sea urchin late larva cDNA library (MPMGp691, http://owww.molgen.mpg.de/ag_seaurchin/, [40]). The carboxypeptidase family was the first to be analyzed. Among 7 sea urchin orthologs, only SpCpa2L showed a significant representation in the above mentioned larval cDNA library. As shown in Fig. 1h , *SpCpa2L* transcripts are expressed at gastrula stage in two groups of cells in the upper midgut, next to the cardiac sphincter. In a temporal progression, *SpCpa2L*expression begins at the upper lateral stomach cells and later it extends ventrally from both sides until coalescing at the ventral most cells, while leaving a gap at the dorsal most part of the stomach (Fig. 1i and k). Based on the EST abundance at larval stages, *SpPnlp2* and *SpPnlp5*were selected among the lipases and blast analysis revealed that they encode the same protein, hereby named *SpPnlp2/5*. Likewise, *SpAmy3* was selected among the amylases. For both genes, transcripts are detectable one day after *SpCpa2L* starts to be expressed, and they are produced in the same above described cells of the upper stomach (Fig. 1j-m). Consistently, the levels of expression of these three genes steadily increase during development and it reaches a peak at 10 days (Fig. 1n).

Pancreatic Exocrine-Like Cells are clustered in the Sea Urchin Larva Upper Stomach

Our examination of the expression of individual pancreatic regulatory and terminal differentiation genes suggests that they are expressed within a common exocrine-like cell population. To further test this premise, we tested co-expression of these factors by double fluorescent in situ. The endodermal expression of *SpPtf1a* is initially confined to a group of ~3 cells in the midgut within the broad *SpHnf1* domain (Fig. 2a). We next asked if other homologs of pancreatic exocrine genes were co-expressed in those cells. The first digestive enzyme that appears in the *SpPtf1a* and *SpHnf1* positive domain is *SpCpa2L* (Fig. 2b). At gastrula stage, a group of cells in the dorsal-right midgut co-express *SpCpa2L* and *SpPtf1a*. Some hours later, at the pluteus stage, cells expressing these two markers appear ventrally, forming a belt-like structure (Fig. 2c; see also Fig. 2I in [41] for co-expression of these two genes at the pluteus stage). Remarkably, only a few cells of the dorsal stomach do not express these markers. Considering the high expression of *SpCpa2L* transcripts, all confined to the *Ptf1a* positive cells (*SpPtf1a+*), we used this gene expression as a marker for the pancreatic exocrine-like cells. Thus, we analyzed *SpMist*, *SpPnlp2/5* and *SpAmy3* spatial expressions relative to *SpCpa2L* at larval stages. As shown in Fig. 2c-e, all the orthologs of exocrine genes that we considered in this study are co-expressed in the *pancreatic exocrine-like* cells. Remarkably, the same cells express the evolutionary conserved pancreatic microRNA *miRNA-375*. Figure 2f shows that *miRNA-375*is strongly expressed throughout the ciliary band and in the stomach, where it is localized exclusively in the *pancreatic exocrine-like* cells. In addition, this gene is expressed in the coelomic pouches and the probe signal is stronger in the right one.

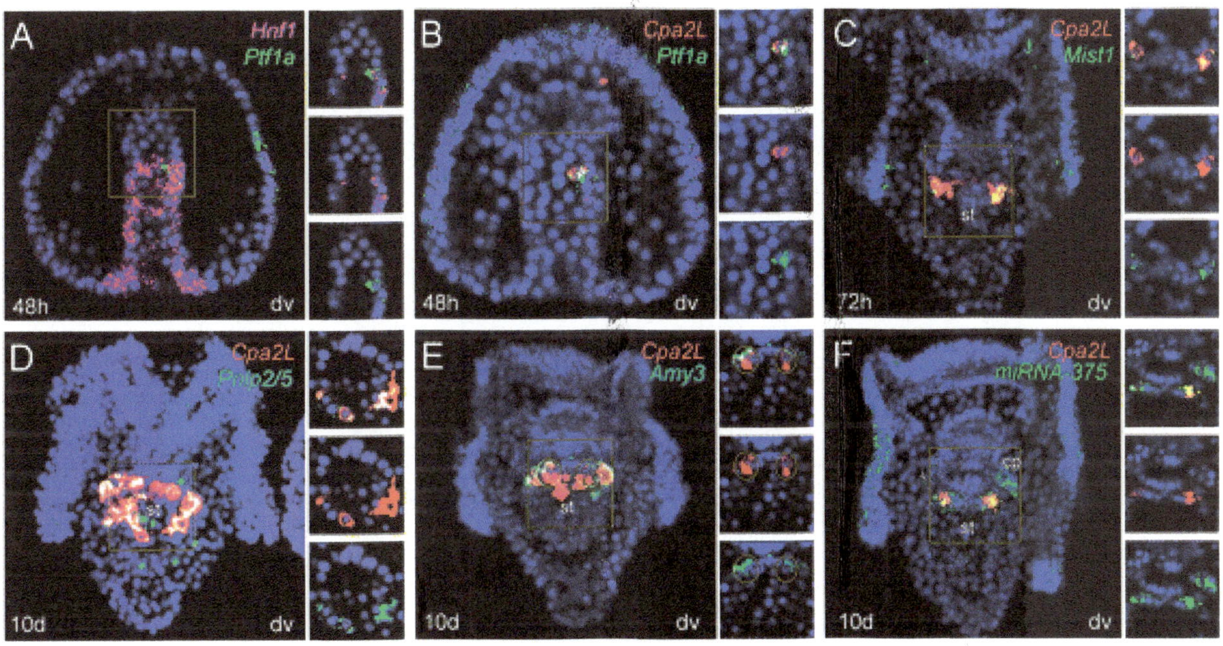

Figure 2: Co-expression analysis of markers of pancreatic exocrine cell-types. Double FISH of selected pancreatic genes in the sea urchin embryos and larva (dv). On the right of each panel, split and combined channels of single confocal sections of the gut domain expressing the two genes (the region of the embryo shown in the right insets is *underlined by a yellow square*) are provided to confirm that the two genes are indeed expressed in the same cells.

Notch Signaling Inhibits Differentiation towards Pancreatic Exocrine Cell Fate

Given its known role in pancreatic differentiation, we next tested the role of the Notch signaling pathway in sea urchin pancreatic exocrine-like cell differentiation, using the Notch inhibitor DAPT. The first striking effect of this treatment is the expansion of the *SpPtf1a*+cells in the gastrula and larva upper midgut/stomach (Fig. 3). Notably, the SpPtf1a expansion in the treated embryos never overcomes the cardiac sphincter border. Despite the strong effect on the endodermal cells, ectodermal *SpPtf1a* expression is not affected. Moreover, 35 % of the analyzed plutei showed an ectopic expression of *SpPtf1a* in a small group of cells in the anus (Fig. 3d, *red arrow*). Conversely, the expression of the early endodermal gene *SpHnf1* is not affected by the DAPT treatment at gastrula stage (Fig. 3a, b), in agreement with the previously reported data at blastula stage [42]. When Notch signaling is perturbed, the expression of *SpMist1* exhibits opposite effects than that of *SpPtf1a*. At the pluteus stage, the gene expression in the stomach is unaltered, while in the esophagus and in the apical organ *SpMist1 is turned on also in* cells that are adjacent to the wild type SpMist1+ cells(Fig. 3e and f and *arrows* in Fig. 3f).

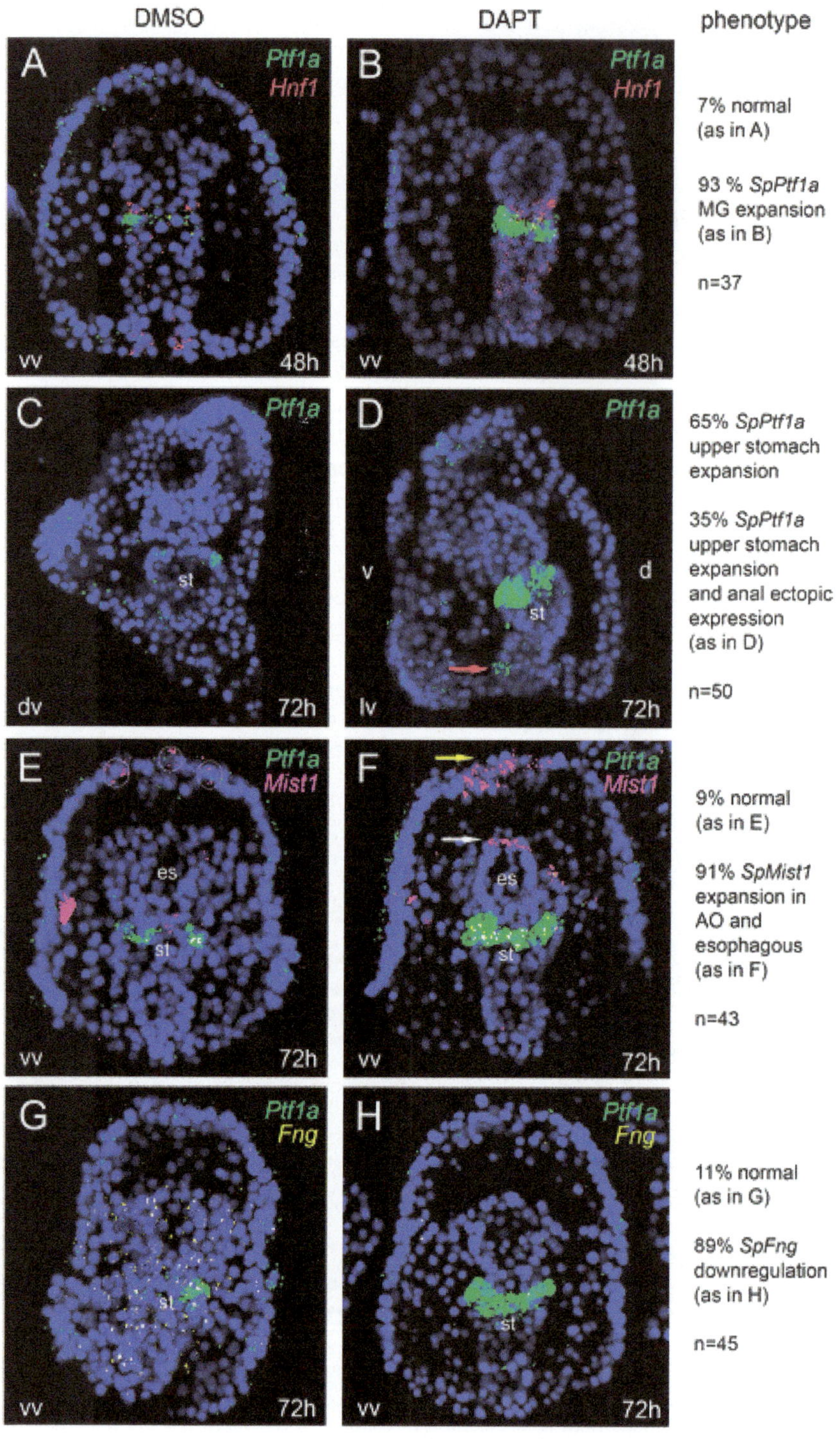

Figure 3: Spatial analysis of gene expression after Notch signaling perturbation. *Hnf1* (a, b), *Ptf1a*(c, d), *Mist1* (e, f) and *Fng* (g, h) transcript localization tested by FISH in control animals (a, c, e, g) and in animals treated with DAPT (b, d, f, h). In all the experiments, *Ptf1a* (*in green*) was used as second probe in double FISH (in a, b, e, f, g, h) since the increase of Ptf1a+cells in treated larvae confirms the phenotype. Red arrow in D indicates *Ptf1*ectopic expression in a few cells of the anus. Yellow circles in E show that *Mist1* is expressed in alternated cells of the apical organ in control larvae, while arrows in F indicate that in treated larvae *Mist1* is expressed in adjacent cells of the apical organ (*yellow arrow*) and in a few cells of the esophagus (*white arrow*). For each analyzed gene, quantification of the phenotypes is shown on the right. Abbreviations: AO, apical organ; es, esophagus; lv, lateral view; MG, midgut; st, stomach, v, ventral.

Many modulators of Notch signaling are known, among which there is the O-fucose β-1,3 N-acetylglucosaminyltransferase Fringe, whose role is to potentiate Notch activation by Delta [43]. Since a Notch signaling requiring Fringe is involved in specification of endoderm and boundary formation during sea urchin embryonic development [44], we asked if Notch inhibition affects Fringe expression also in the pluteus larva. The sea urchin ortholog of Fringe, *SpFng*, is widely expressed in the pluteus mesoderm, ectoderm and endoderm (Fig. 3g). Our analysis demonstrated that perturbation of Notch signaling causes a consistent reduction of *SpFng*expression in all the tissues where it is normally expressed (Fig. 3h), even in the pluteus.

Transcriptional Control of Sea Urchin Pancreatic Exocrine-Like Cell Development

We next asked if the TFs expressed early in development are involved in a gene regulatory module that specifies pancreatic exocrine cells. To this aim, specific morpholino oligonucleotides were injected in the sea urchin fertilized eggs to block either *SpHnf1* or *SpPtf1a* translation.*SpHnf1* MO sequence specificity has already been demonstrated by Peterson and Davidson (2001), and we used that same morpholino in our experiments. In *SpHnf1* morphants, *SpPtf1a*expression in the midgut is downregulated, while its expression in the ectoderm remains unaltered (Fig. 4a, b). Fluorescent *in situ* hybridization (FISH) or quantitative PCR (qPCR) analyses were performed on larvae to test the effects of SpPtf1a MO perturbation on putative target genes. We observed that the spatial expression of *SpCpa2L*, *SpPnlp2/5* and *SpAmy3* is strongly reduced (Fig. 4e-h). For *SpCpa2L*, which among the enzymes shows the highest transcript expression at 3 days, changes at the transcript levels were also analyzed by qPCR analysis, confirming that its expression is strongly downregulated (Fig. 4k). Conversely, in SpPtf1a morphants additional groups of *SpPtf1a*+cells appear in the upper stomach, both ventrally and dorsally (Fig. 4j) and qPCR data confirm *SpPtf1a* upregulation (Fig. 4k). Additionally, in SpPtf1 morphants there is a modest increase of *SpMist1* expression (Fig. 4k), and a slight increase of *SpFng* transcripts (Fig. 4i, j and k). Particularly, *SpFng* and *SpPtf1* seem to be expressed by adjacent and not overlapping cells (Fig. 4j, insert). To test if the effects of SpPtf1a MO were specific, we injected a second MO to block SpPtf1a translation (SpPtf1a MO_2) and we checked the expression of *SpCpa2L*, since this gene is the first TD marker of this cell type to appear and also the most abundant at pluteus stage. Also in this case, we found that the *SpCpa2L* mRNA is downregulated in the morphants (Fig. 4c and d), confirming that the two morpholinos produce comparable effects.

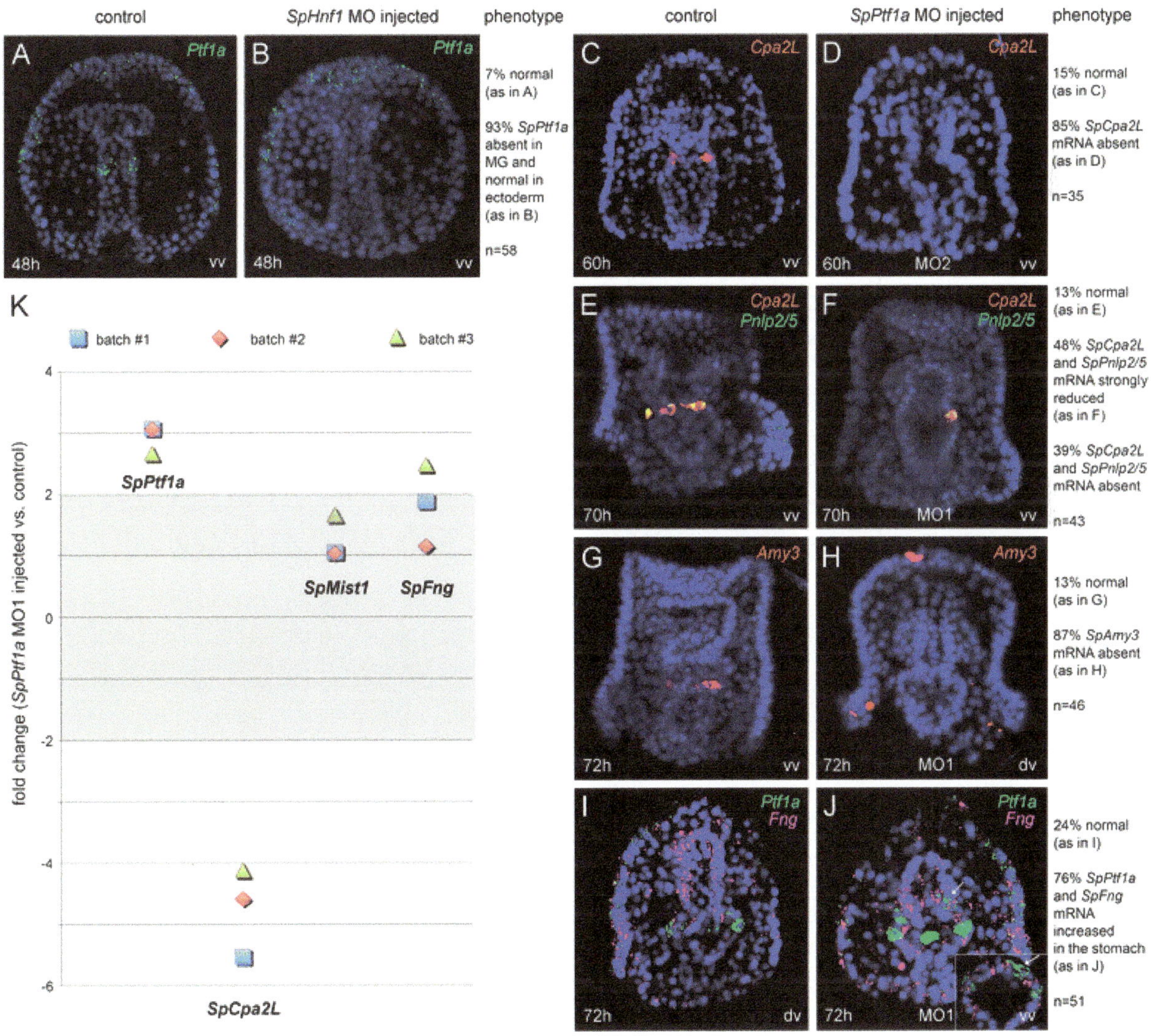

Figure 4: Analysis of gene expression after MO perturbation. *Ptf1a* mRNA localization in control embryos (a) and in embryos injected with MO directed against the translation of SpHnf1 RNA (b). *Cpa2L*, *Pnlp2/5*, *Amy3*, *Ptf1a* and *Fng* transcripts were detected by single or double FISH in control larvae (c, e, g, i) and in larvae injected with two different MOs directed against the translation of SpPtf1a RNA (d, f, h, j). Note that f, h and j show larvae injected with SpPtf1a MO1. To confirm the effects of MO1, figure d shows that SpCpa2L transcripts are absent also in larvae injected with SpPtf1a MO2. Inset in j is a representative single confocal section of the dorsal upper stomach of a late larva and the white arrow shows that Ptf1a+and Fng+cells are adjacent and not overlapping. k qPCR analysis showing the effects of SpPtf1a MO1 perturbation on transcript levels of selected pancreatic genes at 70 h. Fold changes ≥ -2 and ≤ 2 are shaded in light grey and indicate non-significant changes in gene expression.

SpPtf1a Function in Mammalian Cells

Ptf1a is part of the PTF1 heterotrimeric complex that also includes one of the common E proteins (E47) and RBPL [45]. To test the ability of the sea urchin Ptf1a protein to substitute for mammalian Ptf1a in this complex, we performed luciferase essays in which either SpPtf1a or rat Ptf1a were co-transfected with mammalian E47 and RBPL, and evaluated for an ability to transactivate a PTF1-responsive luciferase reporter containing four tandem repeats of the PTF1 binding site from the rat chymotrypsinogen promoter (Obata et al., 2001). Figure 5a and b show that, in both 293 t and HeLa cells, SpPtf1a can effectively partner with mammalian PTF1 components to activate the rat chymotrypsinogen element, albeit with lower activity than rat Ptf1a (Fig. 5b). As a correlate of this conserved function, we hypothesized that the sea urchin orthologs of mammalian PTF1a target genes would also display similar PTF1a regulatory motifs. In the promoter of genes encoding mammalian pancreatic digestive enzymes, PTF1 has been found to bind to an extended bipartite DNA sequence that consists of an E-box and a TC-box, spaced one or two helical DNA-turns apart. Masui and coauthors have reported a list of PTF1-binding sites of the acinar digestive enzyme gene promoters for mouse [45]. We used the list of binding sites from that paper and built a position weight matrix for the E-box and TC-box, taking into consideration of the distance between these two elements. This matrix was used to look for Ptf1a binding sites in the promoters of the genes that were affected in SpPtf1a morpholino perturbations. Interestingly, we identified canonical Ptf1a binding sites in sea urchin genes displaying in vivo spPtf1a dependence, including *SpFng*, *SpCpa2L*, *SpPnlp2/5* and*SpAmy3*. An alignment of the documented Ptf1a binding site in the rat *Ctrb1* gene with the putative Ptf1a binding sites in each of the *S. purpuratus* genes is shown in Fig. 5c.

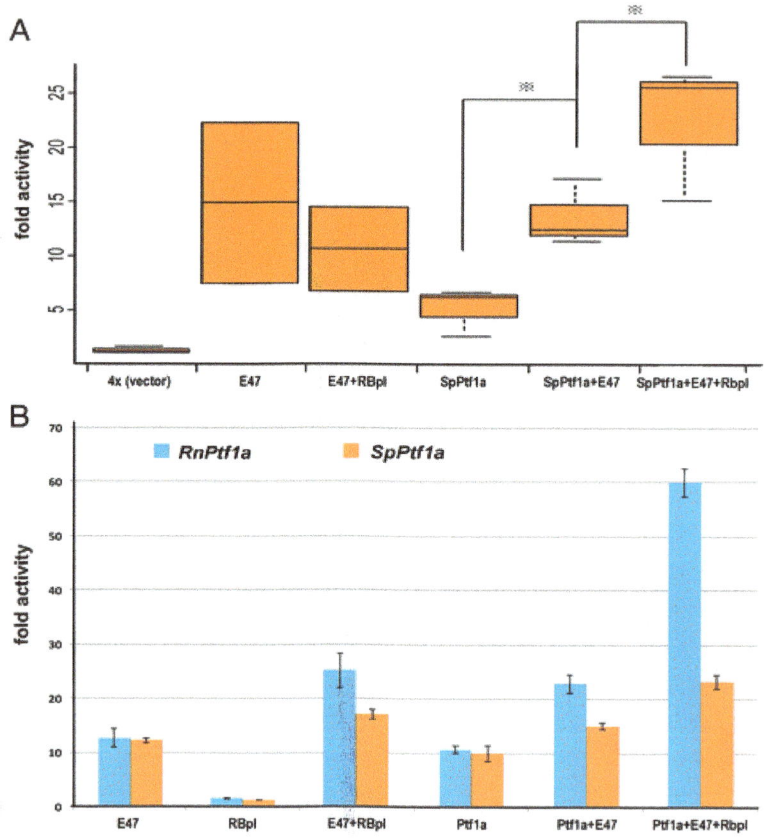

C

Ptf1a binding site:	Ebox	Gap (nt)	TCbox	Distance to TSS (nt)
on *RnCtrb1* gene:	CACCTG	5	TTCCCATG	-229
on *SpCpa2L* gene:	ATTTTG	18	TTCTCATG	-668
on *SpPnlp2/5* gene:	CACCGG	4	CACTTAGG	-306
on *SpAmy3* gene:	CATCTG	6	TTCTCACA	-7161
on *SpLfng* gene:	CAGCTG	14	ATCTTCAG	-146

Figure 5:Sea urchin Ptf1a is active and can function together with mammalian Ptf1 partners. a 293 t cells were transiently transfected with luciferase reporter containing 4x Ptf1a-responsive element (vector indicated as 4x in the figure) and/or spPtf1a, E47, Rbpl construct; renilla vector was cotransfected for normalization. Results are displayed in the box plot as fold activation over the activity of 4x reporter alone. Transfections were done in triplicate and were repeated 5 times. Luciferase assays from all transfections show similar trend, one representative result is shown. t-test was carried out for group comparison. The bottom and top of the box are the first and third quartiles, and the band inside the box is the median. The whisker represents 1.5 IQR (interquantile range, the differences between the third and first quantile). b SpPtf1a has lower activity compared with mammalian Ptf1a in the *in vitro* luciferase assay. Experiment setting is the same as (a) except that rat rnPtf1a was transfected in a second plate side-by-side for comparison. We tested SpPtf1a activity in multiple cell lines, one representative result from HeLa cells is shown. c Alignment of the closest to the transcription start site (TSS) Ptf1a binding sites found by bioinformatic analysis on *S. purpuratus* (Sp) promoters with the canonical Ptf1a binding site of the RnCtbr1 gene (ref). nt, nucleotides.

Pancreatic Exocrine-Like Gene Expression Response to Different Feeding Regimes

We next asked if pancreatic exocrine-like cells respond to different feeding regimes. To this aim,*S. purpuratus* embryos were grown and fed on algae until day 10 to reach the normal larval morphology, then food was removed from the culture media. The culture was split into two batches, of which one was fed for 24 h and the other was fasted for the same time interval. First, we observed larval morphology and found that the different feeding regimes show a clear effect on gut sizes. The stomach and intestine of larvae that were fed is bigger than the ones of the fasted larvae, suggesting that the whole gut epithelium responds to the presence of food (Fig. 6a, b). Second, we compared by qPCR the expression levels of *SpPtf1a* and the previously described digestive enzymes in feeding vs. starving conditions (Fig. 6c). The analysis shows that the expression of the digestive enzymes increases upon feeding. Differently from its putative target genes, *SpPtf1a* expression is higher in fasted larvae than in fed ones. To understand if the increased expression of those genes was due to the same pancreatic exocrine-like cells or to other cells, we tested gene expression through FISH analysis, confirming that the spatial expression of the above genes does not change in feeding or starving conditions (data not shown).

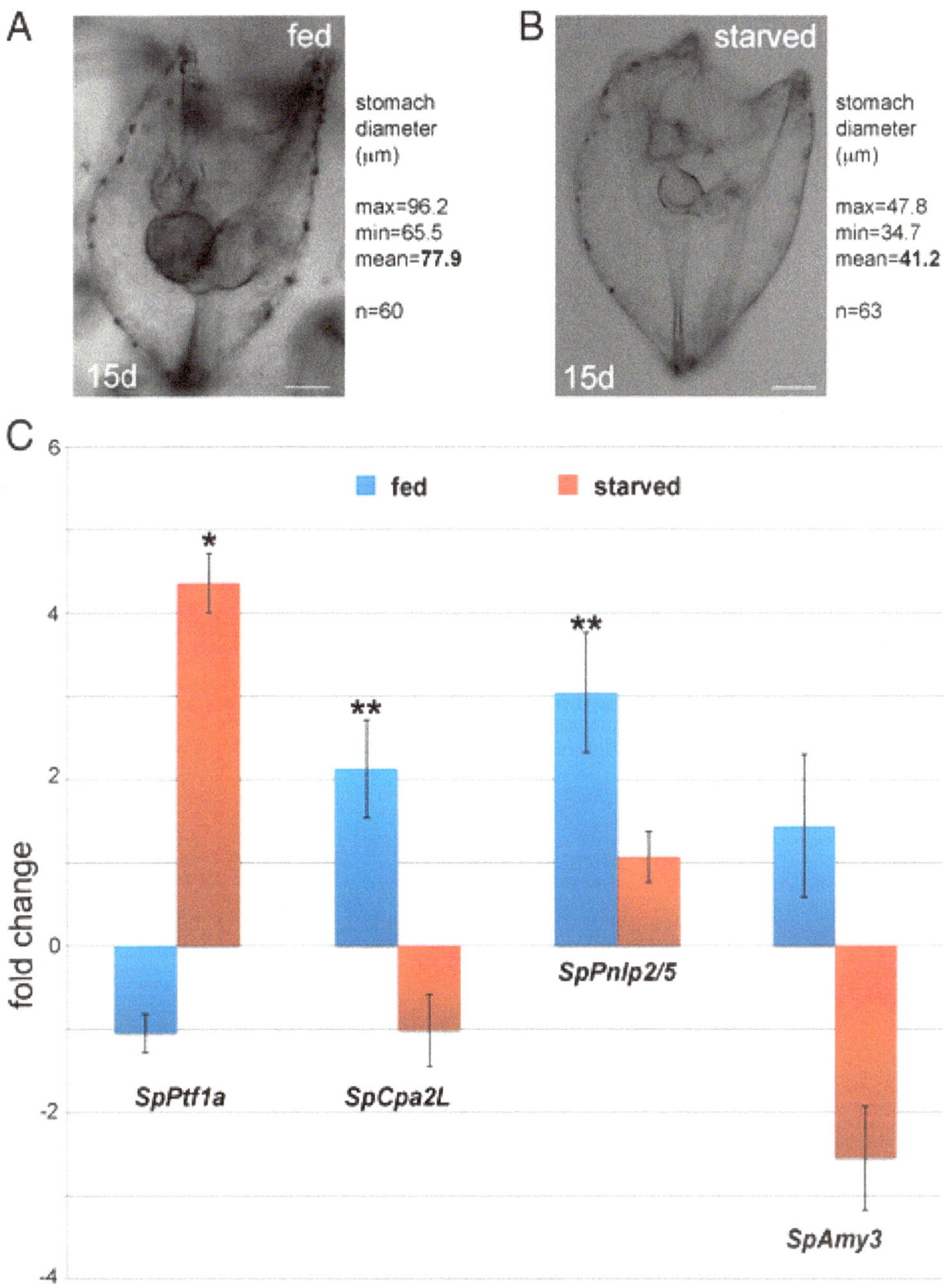

Figure 6: Effect of different feeding regimes on pancreatic exocrine cell gene expression. a and (b) are bright field images of 15-day larvae that are representative for the fed or starved conditions. Larvae are viewed from lateral right side, mouth up. Stomach diameter was measured from ventral to dorsal, along the midline of the larval body. c Quantitative PCR analysis of *SpPtf1a* and digestive enzymes transcripts in feeding and fasting conditions. Data were normalized using ubiquitin as reference gene. Three biological replicas and three technical replicas were measured. The differences between groups that resulted significant by Student's two tailed t-test are indicated as ** if $p<0.01$ (highly significant).

DISCUSSION

In the present study, we show that, in the sea urchin larva, cells specialized in food digestion are clustered in the upper stomach. These cells display a differentiation pathway similar to zymogen cells, where regulative genes (such as *Ptf1a*, *Hnf1* and *Mist1*) and terminal differentiation genes (those encoding digestive enzymes) are co-expressed. Within this upper stomach region, only a few cells on the ventral side do not express the pancreatic markers, denoting a different gene regulatory module and a different function. Figure 7a presents a schematic view of the ontogenetic process leading to the formation of pancreatic exocrine-like clustered cells in the sea urchin embryo and larva. From a temporal point of view, the increase of *SpPtf1a* expression is necessary for the following expression of at least three digestive enzymes that have been analyzed in this study. The expression of these terminal differentiation genes remains constantly high throughout the larval stage, underlying their importance for larval development. *SpPtf1a*and *SpMist1* expression oscillate, and after initial activation, decrease, suggesting that these early TFs may be necessary to activate digestive enzyme transcription but also other factors are involved in maintaining the right level of enzymes during feeding. Interestingly, the pancreatic exocrine-like cells also express the islet-specific microRNA-375. A former study showed that also miR-100, miR-125 and let-7, markers of neurosecretory endocrine cells, are localized in the same area of the sea urchin stomach [46]. These cells seem to be of the pancreatic exocrine-like type, although further analysis will be necessary to test co-expression.

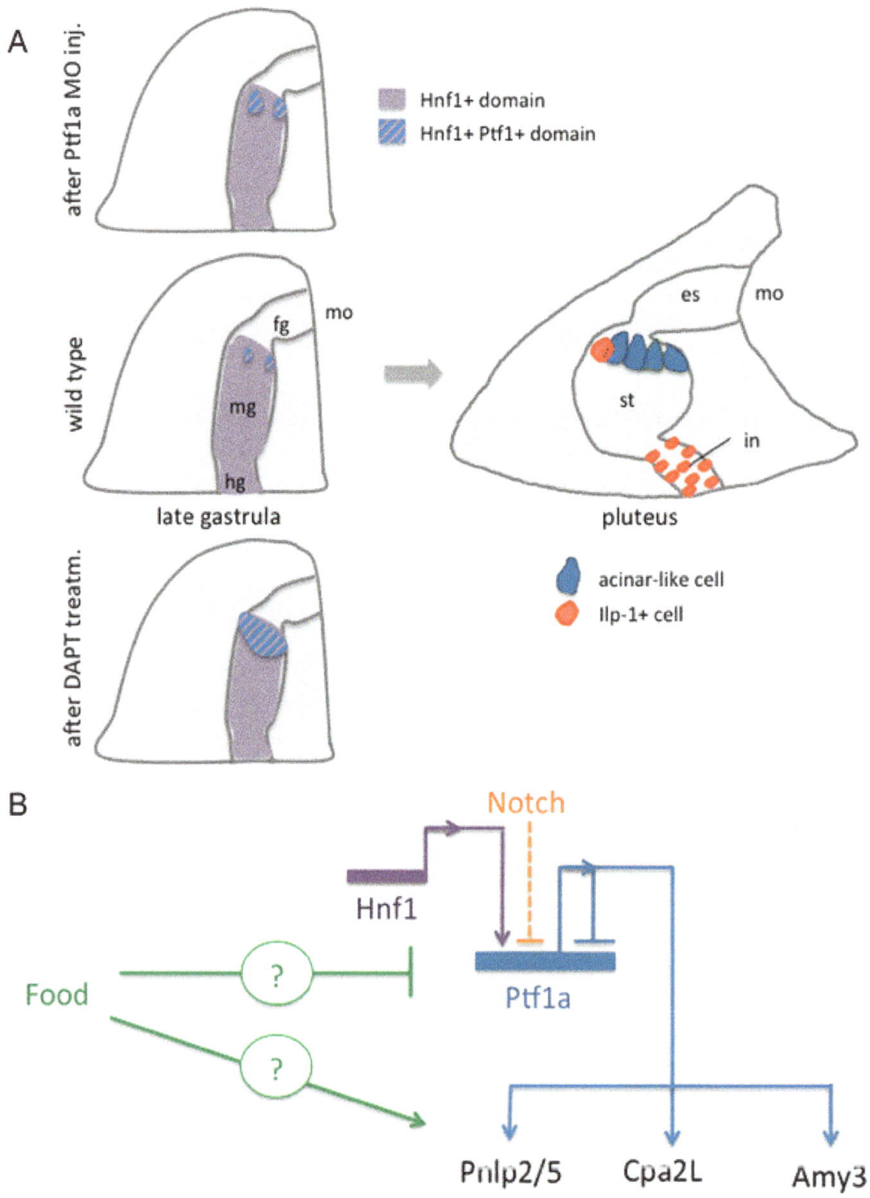

Figure 7: Proposed model of sea urchin pancreatic exocrine-like specification and differentiation. A The schemes show the development of exocrine-like cells from late gastrula (*left*) to larva (*right*). The right drawing also shows the localization of putative endocrine cells that are positive for antibody staining of a member of the insulin family (ILP-1+ cells). Note that we do not know if the ILP-1+ cells in the upper stomach are simply adjacent to the pancreatic exocrine-like cells or there are stomach cells which share both features. B Summary of regulatory interactions occurring in pancreatic exocrine-like cells unraveled by this study. Different genes are shown with different colors. The wiring among the genes is shown with solid lines. Arrows represent positive regulation, bars represent repression, and dashed lines indicate signaling events. Question marks in the food input means that the mechanism and the genes involved in the pathway are not known. Abbreviations: fg, foregut; hg, hindgut; in, intestine; mo, mouth.

In the developing pancreas, Notch signaling represses endocrine development and exocrine cell differentiation [16, 19, 47]. The early role of Notch signaling in the sea urchin embryos has been extensively studied [48–51]. In this study, we identified a putative recruitment of Notch signaling at the late gastrula stage in blocking the differentiation of stomach cells towards an exocrine-like fate in the pluteus larva. We found that Notch signaling represses *SpPtf1a*expression in those cells that are adjacent to the *SpPtf1a* + cells. Moreover, we observed an ectopic expression of *SpPtf1a* in a group of cells of the anus, suggesting that the Notch signaling could be responsible for the direct or indirect repression of *SpPtf1a* expression in these intestinal cells. Nonetheless, we observed that when Notch signaling is blocked, *SpMist1* ectopically accumulates in the cells of the apical organ that are adjacent to *SpMist1*+ cells. One possibility is that a secretory fate is repressed by a Notch lateral-inhibition mechanism in ectodermal cells that are committed to a not-secretory fate. Alternatively, this ectodermal effect may be due to an extra round of cell division. In a similar way, our observations suggest that Notch signaling prevents the exocrine fate of a group of cells of the esophagus. In some cases, we observed that when Notch signaling is blocked, pancreatic genes that are expressed in both the ectoderm and the endoderm lose expression in a specific territory, and not in the other. For instance, the transcription of *SpPtf1a* in the ectoderm and that of *SpMist1* in the stomach do not change in DAPT treated plutei, similarly to what Materna and Davidson observed for *SpShr2* and *SpDelta*in the early embryo [42]. Our data suggest that Notch signaling could have different effects not only during early development, but it can also finely regulate cell differentiation during development in the sea urchin larva. Nonetheless, to deeply analyze the fine mechanisms that underlie regulation of midgut fates, further knockdown experiments of the Notch signaling machinery will be important to enhance our understanding in the future.

The Ptf1a transcriptional complex controls maturation of pancreatic exocrine cells and many efforts has been made to identify its cofactors and targets [52, 53]. In the pancreas, *Hnf1a*activates the transcription of *Ptf1a*, which is part of a transcriptional complex that controls pancreatic exocrine cell maturation and regulates the production of digestive enzymes in acinar cells [13, 54]. Here we found that this gene regulatory module is conserved in an early-branching deuterostome. Based on the homology with mammalian binding sequences, we bioinformatically identified several putative Ptf1a binding sites on the promoters of the sea urchin digestive enzymes which transcription is strongly downregulated in Ptf1a perturbed embryos. We also showed that SpPtf1a can bind mammalian target genes to activate transcription. However, further in vivo experiments are necessary to test the functionality of these binding sites predicted *in silico*. In this study, we also demonstrated that SpPtf1a represses its own gene expression in some stomach cells. We reported that in SpPtf1a morphants, more *SpPtf1a* + cells appear as single pouches at the ventral and dorsal stomach in cells where normally the gene is never expressed. These cells appear as a distinct group of cells that alternate with cells where *SpPtf1a* is absent, like a chessboard (Fig. 4i). Conversely, blocking the Notch pathway results in a continuous ring of Ptf1a+cells (Fig. 4l). Hence, the presence of different mechanisms supports the role of a Notch lateral-inhibition model in the stomach domain. Another finding that emerged from our data is that SpPtf1a may repress the expression of *SpFng*. Due to its role, the Fringe ortholog is widely expressed in the embryos, and this could explain the borderline qPCR values in SpPtf1a morphants. However, our bioinformatic analysis found binding sites for Ptf1a in the *SpFng* promoter region. To understand if the effect of DAPT treatment on SpFng is specific for an exocrine cell fate, further experiments should define whether Ptf1a interacts with corepressor proteins, as in the case of Nkx6 in mammals [55], to block SpFng transcription and consequentially Notch activation, in a negative feedback loop manner.

A summary of the above described putative gene interactions and the proposed gene model leading to specification and differentiation of sea urchin pancreatic exocrine-like cells are schematized in Fig. 7b.

The sea urchin larva is a very plastic organism, able to adapt its metabolic and growth programs to changes in the environmental energy context, e.g. arms and ciliary band growth is influenced by different food availability [56–61]. Feeding has a central role in the life of sea urchin larvae, since they may live up to several months in the plankton, therefore they have to be considered as fully developed animals, even if their body axis will be rearranged after metamorphosis. In a previous study, we found that in the sea urchin larva, a member of the insulin family is expressed in a feeding-dependent fashion, opening new questions on whether other genes are influenced by food assumption or deprivation [29]. As expected during feeding, when more digestive enzymes are necessary, we found that the transcription of a carboxipeptidase, a pancreatic lipase and an amylase are significantly increased, whereas *SpPtf1a* expression showed an opposite behavior. It will be important to further investigate which other factors are involved in the binding of the Ptf1a complex to its targets. Previous transcriptomic analysis reported that *SpCpa2L* and *SpPnlp2/5* transcripts are strongly reduced when *SpLox* is knocked down [62] and it is known that its vertebrate ortholog, *Pdx1*, regulates the development of digestive enzyme-producing acinar cells and insulin expression [63, 64]. In the sea urchin, SpLox is expressed in a group of cells between the lower stomach and the intestine (the pyloric sphincter) [65], in a gut domain and in cells that are different from the herein characterized pancreatic exocrine-like cells. Thus, a possible scenario that could be worth testing in further experiments is that SpLox could act through a hypothetic signaling pathway to eventually regulate the expression of digestive enzymes, accordingly with food availability.

The pancreatic exocrine-like cells that we describe are localized in the same area where there are also endocrine-like cells producing a member of the insulin family, ILP1 [29]. We cannot rule out the possibility that these two cell types are just in close proximity and do not represent a unique cell-type sharing both functions. Generation of antibodies for the digestive enzymes will be necessary to test if these exocrine markers are coexpressed in the ILP+ cells, but we can at least conclude that both cell types are localized in the same area of the stomach. In the cephalocordate amphioxus, endocrine and exocrine cells do not exhibit any gut specialization [8, 66–68]. In the hagfish, an islet-like organ is independent of exocrine acini, and, in the teleost, endocrine islets are scattered among the more widespread exocrine parenchyma [69]. Thus, two possible evolutionary scenarios could be put forward. First, the pancreas and the exocrine and endocrine-like cells (closely localized) in the sea urchin gut are the result of convergent evolution, thus implying that this feature was lost in early branching chordates. Second, the clustering of the sea urchin pancreatic exocrine-like cells could be the result of an independent evolution from an ancestral pancreatic cell-type. To test these hypotheses and to infer the evolutionary origin of the pancreas, homologs of pancreatic genes should be studied in other echinoderms as well as in other non-chordate deuterostomes, such as the hemichordates.

CONCLUSIONS

Here, we report that a gene regulatory module required for vertebrate pancreatic exocrine cell development is shared by a non-chordate deuterostome. Our study, together with previous studies [65, 70], supports the idea that the sea urchin gut is a multifunctional organ and it is composed of highly differentiated cells, which in many aspects are similar to vertebrates, therefore providing an amenable system in which to study development and function of gut cells.

ACKNOWLEDGMENTS

The authors wish to thank Giovanna Benvenuto for confocal microscopy, Marco Borra for quantitative Real Time PCR and Xiaogang Zhong for bioinformatics assistance. We would also like to thank Pat Leahy (KML, Caltech, Pasadena, CA) for providing adult *S. purpuratus* and Davide Caramiello (SZN) for animal maintenance. MP would like to thank Steven Z. Swartz for stimulating discussions and paper revisions. This work was supported by MIUR (premiale PANTRAC to MIA) and NIH grant DK61215 (to SDL). M.P. has been supported by a SZN PhD fellowship and by a fellowship from POR Campania FSE 2007-2013 Project-MODO, Model Organism.

AUTHORS' CONTRIBUTIONS

MIA and MP conceived the project, SDL and YJW designed the bioinformatic analysis and the experiments in mammalian cells. MP performed the experimental work, YJW performed the luciferase essays. The authors contributed equally to the interpretation of the results and the writing of the manuscript. All authors read and approved the final manuscript.

REFERENCES

1. Chera S, de Rosa R, Miljkovic-Licina M, Dobretz K, Ghila L, Kaloulis K, Galliot B. Silencing of the hydra serine protease inhibitor Kazal1 gene mimics the human SPINK1 pancreatic phenotype. J Cell Sci. 2006;119:846–57.
2. Lentz TL. Intramitochondrial glycogen granules in digestive cells of Hydra. J Cell Biol. 1966;29:162–7.
3. Goldberg WM. Gastrodermal structure and feeding responses in the scleractinian*Mycetophyllia reesi*, a coral with novel digestive filaments. Tissue Cell. 2002;34:246–61.
4. Brennan GP, Ramasamy P. Ultrastructure of the gut caecal epithelium of *Pricea multae*(Monogenea: Polyopisthocotylea). Parasitol Res. 1996;82:312–8.
5. Jones MK, Hughes-Stamm SR, East RM, Cribb TH. Ultrastructure of the digestive tract of*Gyliauchen nahaensis* (Platyhelminthes, Digenea), an inhabitant of the hindgut of herbivorous fishes. J Morphol. 2000;246:198–211.
6. Brennan GP, Ramasamy P. Ultrastructure of the surface structures and electron immunogold labeling of peptide immunoreactivity in the nervous system of*Pseudothoracocotyla indica* (Polyopisthocotylea: Monogenea). Parasitol Res. 1996;82:638–46.
7. Lemaitre B, Miguel-Aliaga I. The digestive tract of *Drosophila melanogaster*. Annu Rev Genet. 2013;47:377–404.
8. Slack JM. Developmental biology of the pancreas. Development. 1995;121:1569–80.
9. Krapp A, Knöfler M, Ledermann B, Bürki K, Berney C, Zoerkler N, Hagenbüchle O, Wellauer PK. The bHLH protein PTF1-p48 is essential for the formation of the exocrine and the correct spatial organization of the endocrine pancreas. Genes Dev. 1998;12:3752–63.
10. Lemercier C, To RQ, Swanson BJ, Lyons GE, Konieczny SF. Mist1: a novel basic helix-loop-helix transcription factor exhibits a developmentally regulated expression pattern. Dev Biol. 1997;182:101–13.
11. Cockell M, Stevenson BJ, Strubin M, Hagenbüchle O, Wellauer PK. Identification of a cell-specific DNA-binding activity that interacts with a transcriptional activator of genes expressed in the acinar pancreas. Mol Cell Biol. 1989;9:2464–76.
12. Kawaguchi Y, Cooper B, Gannon M, Ray M, MacDonald RJ, Wright CVE. The role of the transcriptional regulator Ptf1a in converting intestinal to pancreatic progenitors. Nat Genet. 2002;32:128–34.
13. Krapp A, Knöfler M, Frutiger S, Hughes GJ, Hagenbüchle O, Wellauer PK. The p48 DNA-binding subunit of transcription factor PTF1 is a new exocrine pancreas-specific basic helix-loop-helix protein. EMBO J. 1996;15:4317–29.
14. Pin CL, Rukstalis JM, Johnson C, Konieczny SF. The bHLH transcription factor Mist1 is required to maintain exocrine pancreas cell organization and acinar cell identity. J Cell Biol. 2001;155:519–30.
15. Haumaitre C, Barbacci E, Jenny M, Ott MO, Gradwohl G, Cereghini S. Lack of TCF2/vHNF1 in mice leads to pancreas agenesis. Proc Natl Acad Sci U S A. 2005;102:1490–5.
16. Hald J, Hjorth JP, German MS, Madsen OD, Serup P, Jensen J. Activated Notch1 prevents differentiation of pancreatic acinar cells and attenuate endocrine development. Dev Biol. 2003;260:426–37.
17. Kim SK, Hebrok M. Intercellular signals regulating pancreas development and function. Genes Dev. 2001;15:111–27.
18. Norgaard GA, Jensen JN, Jensen J. FGF10 signaling maintains the pancreatic progenitor cell state revealing a novel role of Notch in organ development. Dev Biol. 2003;264:323–38.
19. Esni F, Ghosh B, Biankin AV, Lin JW, Albert MA, Yu X, MacDonald RJ, Civin CI, Real FX, Pack MA, Ball

DW, Leach SD. Notch inhibits Ptf1 function and acinar cell differentiation in developing mouse and zebrafish pancreas. Development. 2004;131:4213–24.

20. Poy MN, Eliasson L, Krutzfeldt J, Kuwajima S, Ma X, Macdonald PE, Pfeffer S, Tuschl T, Rajewsky N, Rorsman P, Stoffel M. A pancreatic islet-specific microRNA regulates insulin secretion. Nature. 2004;432:226–30.

21. Kloosterman WP, Lagendijk AK, Ketting RF, Moulton JD, Plasterk RHA. Targeted inhibition of miRNA maturation with morpholinos reveals a role for miR-375 in pancreatic islet development. PLoS Biol. 2007;5:e203.

22. Poy MN, Hausser J, Trajkovski M, Braun M, Collins S, Rorsman P, Zavolan M, Stoffel M. miR-375 maintains normal pancreatic alpha- and beta-cell mass. Proc Natl Acad Sci U S A. 2009;106:5813–8.

23. McClay DR. Evolutionary crossroads in developmental biology: sea urchins. Development. 2011;138:2639–48.

24. Burke RD. Structure of the digestive tract of the pluteus larva of *Dendraster excentricus*(Echinodermata: Echinoida). Zoomorphology. 1981;98:209–25.

25. Vacquier VD. The appearance of -1,3-glucanohydrolase activity during the differentiation of the gut of sand dollar plutei. Dev Biol. 1971;26:1–10.

26. Vacquier VD, Korn LJ, Epel D. The appearance of -amylase activity during gut differentiation in sand dollar plutei. Dev Biol. 1971;26:393–9.

27. Vacquier VD. The effects of glucose and lithium chloride on the appearance of -1,3-glucanohydrolase activity in sand dollar plutei. Dev Biol. 1971;26:11–6.

28. Holland ND, Lauritis JA: The fine structure of the gastric exocrine cells of the purple sea urchin, *Strongylocentrotus purpuratus*. Trans Amer Microscop Soc. 1968;87:201-209.

29. Perillo M, Arnone MI. Characterization of insulin-like peptides (ILPs) in the sea urchin Strongylocentrotus purpuratus: insights on the evolution of the insulin family. Gen Comp Endocrinol. 2014;205:68–79.

30. Andrikou C, Iovene E, Rizzo F, Oliveri P, Arnone MI. Myogenesis in the sea urchin embryo: the molecular fingerprint of the myoblast precursors. Evodevo. 2013;4:33.

31. Andrikou C, Pai C-Y, Su Y-H, Arnone MI: Logics and properties of a genetic regulatory program that drives embryonic muscle development in an echinoderm. Elife. 2015;4.

32. Hughes JN, Dodge N, Rathjen PD, Rathjen J. A novel role for gamma-secretase in the formation of primitive streak-like intermediates from ES cells in culture. Stem Cells. 2009;27:2941–51.

33. Rast JP, Amore G, Calestani C, Livi CB, Ransick A, Davidson EH. Recovery of developmentally defined gene sets from high-density cDNA macroarrays. Dev Biol. 2000;228:270–86.

34. Nemer M, Rondinelli E, Infante D, Infante AA. Polyubiquitin RNA characteristics and conditional induction in sea urchin embryos. Dev Biol. 1991;145:255–65.

35. Dong PDS, Provost E, Leach SD, Stainier DYR. Graded levels of Ptf1a differentially regulate endocrine and exocrine fates in the developing pancreas. Genes Dev. 2008;22:1445–50.

36. Howard-Ashby M, Materna SC, Brown CT, Chen L, Cameron RA, Davidson EH. Gene families encoding transcription factors expressed in early development of*Strongylocentrotus purpuratus*. Dev Biol. 2006;300:90–107.

37. Howard-Ashby M, Materna SC, Brown CT, Chen L, Cameron RA, Davidson EH. Identification and characterization of homeobox transcription factor genes in*Strongylocentrotus purpuratus*, and their expression in embryonic development. Dev Biol. 2006;300:74–89.

38. Bach I, Mattei MG, Cereghini S, Yaniv M. Two members of an HNF1 homeoprotein family are expressed in human liver. Nucleic Acids Res. 1991;19:3553–9.

39. Booth H, Holland P: Annotation, nomenclature and evolution of four novel homeobox genes expressed in the human germ line. Gene. 2007;387:7–14.

40. Poustka AJ, Groth D, Hennig S, Thamm S, Cameron A, Beck A, Reinhardt R, Herwig R, Panopoulou G, Lehrach H. Generation, annotation, evolutionary analysis, and database integration of 20,000 unique sea urchin EST clusters. Genome Res. 2003;13:2736–46.

41. Annunziata R, Perillo M, Andrikou C, Cole AG, Martinez P, Arnone MI. Pattern and process during sea urchin gut morphogenesis: the regulatory landscape. Genesis. 2014;52:251–68.

42. Materna SC, Davidson EH. A comprehensive analysis of Delta signaling in pre-gastrular sea urchin embryos. Dev Biol. 2012;364:77–87.

43. Moloney DJ, Panin VM, Johnston SH, Chen J, Shao L: Fringe is a glycosyltransferase that modifies Notch. Nature. 2000;406:369–375.

44. Peterson RE, McClay DR. A Fringe-modified Notch signal affects specification of mesoderm and endoderm in the sea urchin embryo. Dev Biol. 2005;282:126–37.

45. Masui T, Long Q, Beres TM, Magnuson MA, MacDonald RJ. Early pancreatic development requires the vertebrate Suppressor of Hairless (RBPJ) in the PTF1 bHLH complex. Genes Dev. 2007;21:2629–43.

46. Christodoulou F, Raible F, Tomer R, Simakov O, Trachana K, Klaus S, Snyman H, Hannon GJ, Bork P, Arendt D. Ancient animal microRNAs and the evolution of tissue identity. Nature. 2010;463:1084–8.

47. Murtaugh LC, Stanger BZ, Kwan KM, Melton DA. Notch signaling controls multiple steps of pancreatic differentiation. Proc Natl Acad Sci U S A. 2003;100:14920–5.

48. Croce JC, McClay DR. Dynamics of Delta/Notch signaling on endomesoderm segregation in the sea urchin embryo. Development. 2010;137:83–91.

49. Sherwood DR, McClay DR: Identification and localization of a sea urchin Notch homologue: insights into vegetal plate regionalization and Notch receptor regulation. Development. 1997;124:3363–3374.

50. Sherwood DR, McClay DR: LvNotch signaling plays a dual role in regulating the position of the ectoderm-endoderm boundary in the sea urchin embryo. Development. 2001;128:2221–32.

51. Materna SC, Swartz SZ, Smith J. Notch and Nodal control forkhead factor expression in the specification of multipotent progenitors in sea urchin. Development. 2013;140:1796–806.

52. Thompson N, Gésina E, Scheinert P, Bucher P, Grapin-Botton A. RNA profiling and chromatin immunoprecipitation-sequencing reveal that PTF1a stabilizes pancreas progenitor identity via the control of MNX1/HLXB9 and a network of other transcription factors. Mol Cell Biol. 2012;32:1189–99.

53. Arda HE, Benitez CM, Kim SK. Gene regulatory networks governing pancreas development. Dev Cell. 2013;25:5–13.

54. Petrucco S, Wellauer PK, Hagenbüchle O. The DNA-binding activity of transcription factor PTF1 parallels the synthesis of pancreas-specific mRNAs during mouse development. Mol Cell Biol. 1990;10:254–64.

55. Schaffer AE, Freude KK, Nelson SB, Sander M. Nkx6 transcription factors and Ptf1a function as antagonistic lineage determinants in multipotent pancreatic progenitors. Dev Cell. 2010;18:1022–9.

56. Adams DK, Sewell MA, Angerer RC, Angerer LM. Rapid adaptation to food availability by a dopamine-mediated morphogenetic response. Nat Commun. 2011;2:592.

57. Heyland A, Reitzel AM, Hodin J. Thyroid hormones determine developmental mode in sand dollars (Echinodermata: Echinoidea). Evol Dev. 2004;6:382–92.

58. Reitzel AM, Heyland A: Reduction in morphological plasticity in echinoid larvae: relationship of plasticity with maternal investment and food availability. Evol Ecol Res. 2007;9:109–121.

59. Strathmann RR, Fenaux L, Strathmann MF: Heterochronic developmental plasticity in larval sea urchins and its implications for evolution of nonfeeding larvae. Evolution. 1992;46:972–986.

60. Strathmann RR, Foley GP, Hysert AN. Loss and gain of the juvenile rudiment and metamorphic competence during starvation and feeding of bryozoan larvae. Evol Dev. 2008;10:731–6.

61. Carrier TJ, King BL, Coffman JA. Gene Expression Changes Associated With the Developmental Plasticity of Sea Urchin Larvae in Response to Food Availability. Biol Bull. 2015;228:171–80.

62. Annunziata R, Martinez P, Arnone MI. Intact cluster and chordate-like expression of ParaHox genes in a sea star. BMC Biol. 2013;11:68.

63. Ohlsson H, Karlsson K, Edlund T. IPF1, a homeodomain-containing transactivator of the insulin gene. EMBO J. 1993;12:4251–9.

64. Zhou Q, Law AC, Rajagopal J, Anderson WJ, Gray PA, Melton DA. A multipotent progenitor domain guides pancreatic organogenesis. Dev Cell. 2007;13:103–14.

65. Cole AG, Rizzo F, Martinez P, Fernandez-Serra M, Arnone MI. Two ParaHox genes, SpLox and SpCdx, interact to partition the posterior endoderm in the formation of a functional gut. Development. 2009;136:541–9.

66. Falkmer S, Dafgård E, el-Salhy M, Engström W, Grimelius L, Zetterberg A. Phylogenetical aspects on islet hormone families: a minireview with particular reference to insulin as a growth factor and to the phylogeny of PYY and NPY immunoreactive cells and nerves in the endocrine and exocrine pancreas. Peptides. 1985;6 Suppl 3:315–20.

67. Lecroisey C, Le Pétillon Y, Escriva H, Lammert E, Laudet V. Identification, evolution and expression of an insulin-like peptide in the cephalochordate *Branchiostoma lanceolatum*. PLoS One. 2015;10:e0119461.

68. Pieler T, Chen Y. Forgotten and novel aspects in pancreas development. Biol Cell. 2006;98:79–88.

69. Youson JH, Al-Mahrouki AA. Ontogenetic and phylogenetic development of the endocrine pancreas (islet organ) in fish. Gen Comp Endocrinol. 1999;116:303–35.

70. Annunziata R, Arnone MI. A dynamic regulatory network explains ParaHox gene control of gut patterning in the sea urchin. Development. 2014;141:2462–72.

CHAPTER 13

SEA URCHIN VAULT STRUCTURE, COMPOSITION, AND DIFFERENTIAL LOCALIZATION DURING DEVELOPMENT

Phoebe L Stewart, Miriam Makabi, Jennifer Lang, Carrie Dickey-Sims, Anthony J Robertson, James A Coffman and Kathy A Suprenant

BACKGROUND

The sea urchin *Strongylocentrotus purpuratus* is an important model system in developmental biology and its genome is currently being sequenced by the Human Genome Sequencing Center at the Baylor College of Medicine under the auspices of the National Human Genome Research Institute (NHGRI). Sea urchin embryos are well suited for biochemical approaches to studying the cell biology of development, as large quantities of eggs can easily be obtained, and their fertilization initiates the synchronous development of optically transparent embryos [1]. Sea urchins occupy an important phylogenetic position as basal deuterostomes, and are thus more closely related to humans than are other invertebrate model organisms such as *Drosophila* and *C. elegans*. In addition, the echinoderm lineage leading to sea urchins diverged from chordates prior to the large scale gene duplication events that occurred early in the evolution of the vertebrates, and because of this many of the genes that are found as multiple paralogues in vertebrates have only a single homolog in sea urchins. Therefore sea urchins provide a system that avoids the problem of functional redundancies between multiple paralogues that often occurs in vertebrates, and serves as a useful comparison for assessing the significance of conserved genes and regulatory linkages within the genome.

Sea urchins cells, like mammalian cells, contain abundant quantities of vaults, which are ~13MDa ribonucleoprotein particles of as yet unknown function [2]. Vaults are barrel-shaped assemblies composed of multiple copies of three proteins and small untranslated RNA molecules, called vRNA. Intriguingly vaults are up-regulated in certain human multidrug-resistant cancer cell lines, although their role in multidrug resistance remains unclear [3–6]. The high level of conservation among higher eukaryotes of both the major vault protein (MVP) sequence and the barrel-shaped vault structure suggests an important cellular role for the vault. Nucleocytoplasmic transport, sequestration of macromolecules, and protection from xenobiotics have all been proposed as possible functions for the vault [4, 7–9]. Several recent publications have supported a role for vaults as either a transporter [10, 11] or a scaffold protein [12]. Knock-out mice have been produced lacking one each of the three mammalian vault proteins: MVP; the vault poly(ADP-ribose) polymerase (VPARP); and the telomerase associated protein one (TEP1) [13–15]. These knock-out mice appear to be healthy, indicating that if the vault does perform a critical cellular function there must be a redundant or compensatory pathway in the mouse.

Citation: Phoebe L Stewart, Miriam Makabi, Jennifer Lang, Carrie Dickey-Sims, Anthony J Robertson, James A Coffman and Kathy A Suprenant, Sea urchin vault structure, composition, and differential localization during development, DOI: 10.1186/1471-213X-5-3.

Immunofluorescence studies in sea urchins indicate that MVP is present throughout the cytoplasm in cleavage-stage zygotes and in the nucleus in adult somatic cells [16]. Within the nucleus of coelomocytes, the MVP is present in particularly high concentrations in the nucleolus. This is in contrast to localization studies in most other eukaryotic cells, which show MVP as primarily cytoplasmic [17–21]. During isolation from sea urchins, vaults are found to co-purify with both microtubules and ribosomes [22–24]. The sea urchin MVP cell localization and co-purification results led to a hypothesis that vaults may play a role in nucleocytoplasmic transport of ribosomes and/or mRNA [16].

The present study provides a comparison of vaults from sea urchins and rats in terms of their molecular composition and MVP protein sequence, as well as by cryoEM imaging and three-dimensional reconstruction. In addition, images obtained by confocal microscopy are presented that indicate differential localization of MVP during sea urchin embryogenesis, suggesting that the highly conserved ribonucleoprotein vault might play a role during development.

RESULTS

Molecular Composition of Sea Urchin Vaults and Comparison with Mammalian Vaults

Vaults isolated from mammals, including rats, mice, and humans, are composed of three proteins (MVP, VPARP, and TEP1) and one or more small untranslated RNA molecules, called vRNA. The protein stoichiometry is thought to be 96 copies of MVP with 2–16 copies of VPARP and TEP1 per particle [25]. The human (99 kDa) and rat (96 kDa) homologs of MVP display 91% similarity. In humans the two high molecular weight vault proteins, VPARP and TEP1, have been sequenced and have masses of 193 and 290 kDa, respectively [18, 26–28]. vRNA accounts for ~5% of the total mass of the mammalian vault particle. Four human vRNA genes have been identified, encoding vRNAs that range from 86 to 99 bases in length [8, 28]. Rat vaults in contrast have only one species of vRNA, 141 bases in length [29], which appears as a ~37 kDa band by SDS-PAGE [25]. All of the mammalian vRNA sequences that have been sequenced thus far show ~80% sequence identity and have similar predicted secondary structures [18]. In rat vaults vRNA can be degraded by a harsh treatment with two RNases leaving the rest of the vault particle structurally intact [25].

When purified under identical conditions, the molecular composition of the sea urchin vault is more complex than that of the rat vault (Fig. 1A). For both rat and sea urchin vaults, the strongest protein band is that of MVP at ~100 kDa. We have previously demonstrated that polyclonal antiserum generated against the sea urchin ~100 kDa protein has cross-reactivity with the rat vault MVP, as well as with the *Dictyostelium* MVPα and MVPβ vault proteins [16]. SDS-PAGE indicates that the sea urchin vault has two or more high molecular weight components at ~200 kDa, similar to the VPARP and TEP1 bands for rat and mouse vaults [9,30, 31]. Searches of the *S. purpuratus* genome identified clear sequence homologs of both VPARP and TEP1 (data not shown). Purified sea urchin vaults also appear to contain vRNA, as demonstrated by the loss of a 26.5 kDa band after RNase treatment (Fig. 1B). We note that the apparent molecular weight of the sea urchin RNA molecule is smaller than that of the rat vault vRNA, 26.5 *vs.* 37 kDa.

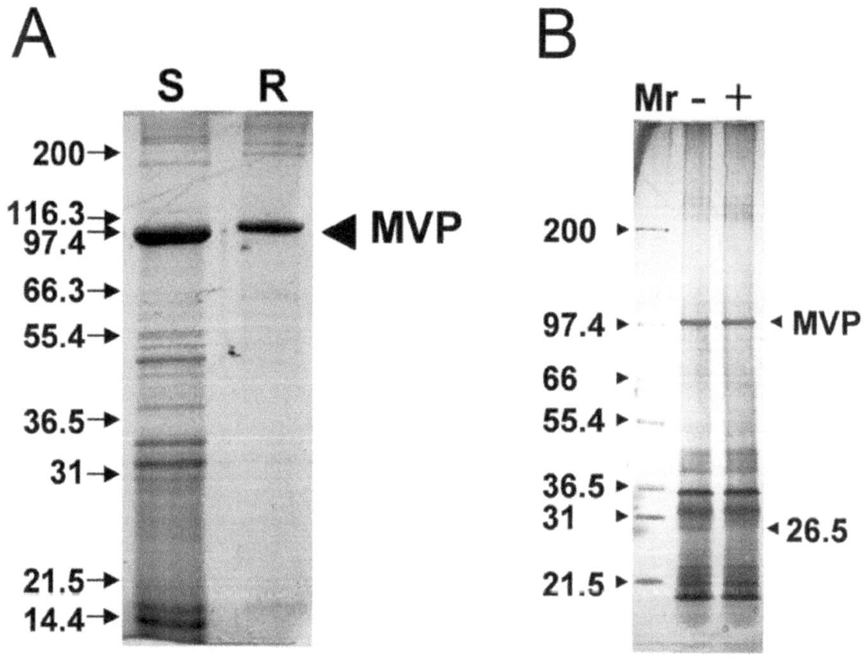

Figure 1: The molecular composition of the sea urchin vault.(A) SDS-PAGE analysis (4–16% acrylamide gradient gel) of isolated sea urchin (S) and rat (R) vaults. The band at ~100 kDa in both lanes represents the major vault protein (MVP). (B) SDS-PAGE analysis of sea urchin vaults both without (-) and with (+) RNase treatment. The 26.5 kDa band in the (-) RNase lane, which is missing in the (+) RNase lane, is thought to correspond to the sea urchin vRNA. Mr represents the molecular weight marker lane.

What is noticeably different between the sea urchin and rat vaults is the number of protein bands within the molecular weight range of 14 to 55 kDa in the sea urchin vault that are not observed in the rat vault. Since sea urchin vaults were purified in the same manner as rat vaults, we conclude that either the molecular composition of the sea urchin vault is more complex than that of the mammalian vaults, or that the sea urchin vaults have a more varied composition of molecular cargo.

Amino Acid Sequence of the Sea Urchin Major Vault Protein

To more fully characterize and identify the sea urchin MVP, we used expressed sequence tags (ESTs) and trace sequences from the *S. purpuratus* genome to assemble the coding sequence of the *SpMVP* gene (see Methods), which is apparently present as a single homolog per haploid genome. The fact that this sequence was found in several blastula-stage ESTs shows that it is expressed in the embryo. The deduced sequence of sea urchin MVP encodes a 95 kDa protein which is slightly smaller than the 96 kDa rat MVP, with 847 *vs.* 861 aa. These two MVP homologs exhibit 64% identity and 81% strong similarity (Fig. 2A). The sea urchin MVP sequence is composed of seven repeats (each 41 to 62 residues) in the N-terminal half of the protein, and a predicted coiled coil region (aa 663–762) in the C-terminal half of the protein (Fig. 2B). The rat MVP sequence has seven similar repeats of unknown function in the N-terminal half and also has a predicted coiled coil region in the C-terminal half [32, 33].

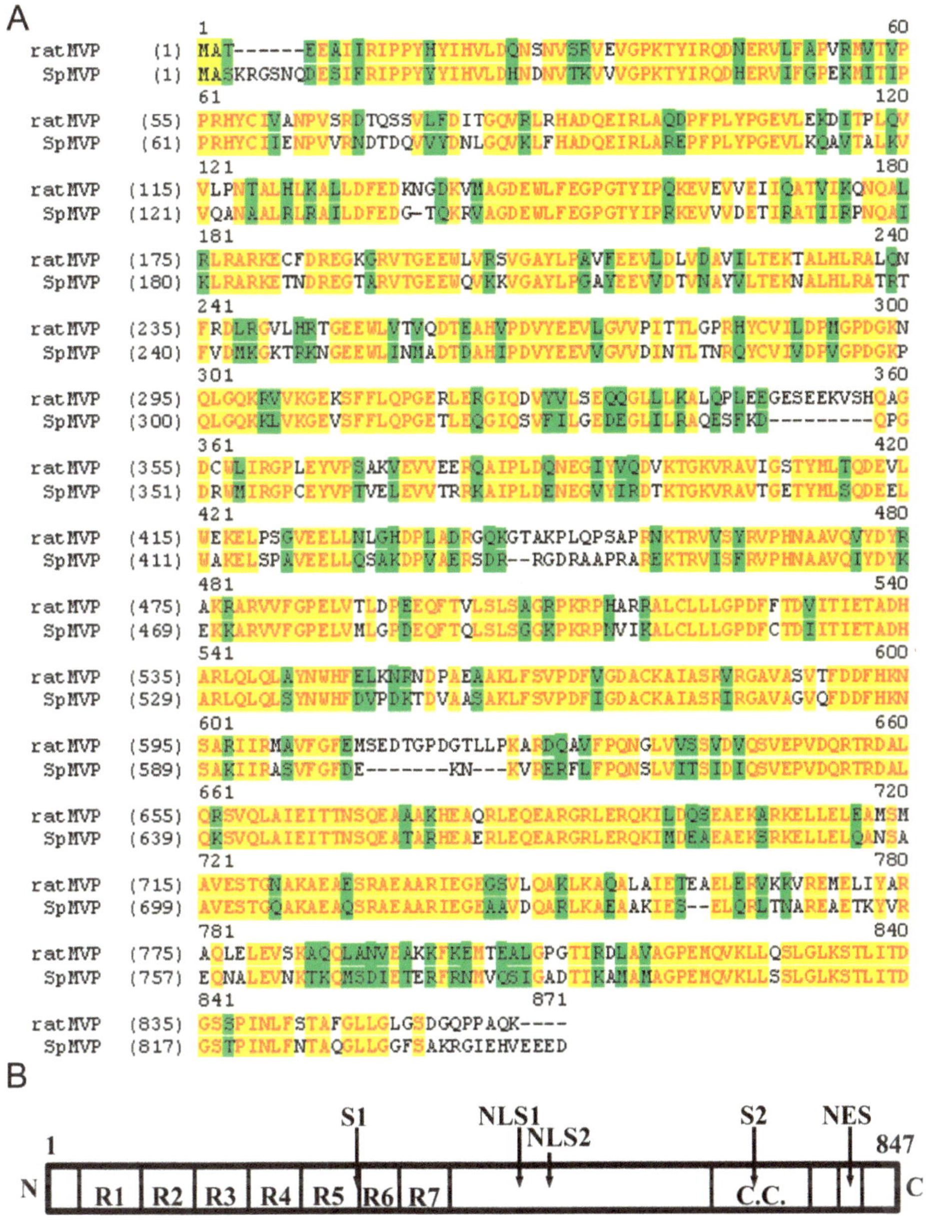

Figure 2: The sequence of the sea urchin major vault protein (SpMVP). (A) Amino acid sequence of SpMVP aligned with that of rat MVP. Identical residues are highlighted in yellow, and similar residues are highlighted in green. (B) Bar diagram of SpMVP. The sequence has seven repeats in the N-terminal half of the protein (R1-R7), each repeat consisting of 41 to 62 aa, spanning residues 32–400; and a predicted coiled coil region in the C-terminal half of the sequence (C.C.), residues 663–762. The positions of two possible nuclear localization sequences (NLS1 470-KKAR, and NLS2 498-KPKR), a putative nuclear export sequence (NES, 792–816), and two probable sumoylation sites that are conserved across six species (S1 K308-VKGE, and S2 K707-AKAE) are indicated.

In addition to comparing rat and sea urchin MVP sequences, we used pair-wise alignments to compare the amino acid sequences of several other known MVPs, including the two from*Dictyostelium* (Table 1). These comparisons indicate that the MVP sequence is highly conserved between phylogenetically distant species, and that the similarity between sea urchin and rat MVP sequences is comparable to that among distantly related vertebrates, and also to the intraspecific similarity of *Dictyostelium* MVPs.

Table 1: Differences between MVP amino acid sequences of various species

	Rat	***Xenopus***	**Zebrafish**	**Sea urchin**	***Dictyostelium*A**	***Dictyostelium*B**
Rat	0	81 (9.38 %)	120 (13.67 %)	112 (12.86 %)	151 (17.34 %)	161 (18.51 %)
Xenopus		0	91 (10.50 %)	90 (10.48 %)	124 (14.44 %)	154 (17.91 %)
Zebrafish			0	113 (12.96 %)	148 (16.91 %)	165 (18.97 %)
Sea Urchin				0	121 (14.07 %)	157 (18.15 %)
***Dictyostelium*A**					0	137 (15.91 %)
***Dictyostelium*B**						0

Pair-wise sequence alignments of MVPs from each species were used to obtain number of amino acids that are not identical, strongly similar, or weakly similar between each pair of sequences. The numbers in parentheses represent corresponding percent difference. The two*Dictyostelium* MVPs are designated A and B.

The sea urchin MVP sequence was analyzed for potential nuclear localization signals (NLS›s) and nuclear export signals (NES›s), as well as possible sumoylation sites. Although the PredictNLS server [34] did not find any putative NLS›s, visual examination of the sequence resulted identification of two basic regions with three lysine (K) or arginine (R) residues within a four amino acid stretch that are similar to the smallest consensus sequence of the monopartite type NLS [35]. Comparison of the sea urchin MVP sequence with the NES logo defined by la Cour *et al.* [36], indicates that a region of 25 residues near the C-terminus of protein (aa 792 – 816) has a close resemblance to the NES logo. Within this window we find the three most highly conserved leucine (L) residues (L804, L807 and L809); 3 additional residues matching the most favored residue type in the logo; 6 additional residues matching the second or third most favored residue type; and 3 additional residues matching the fourth, fifth of sixth most favored residue type. The Abgents Sumoylation Calculator predicts five highly probable sumoylation sites in order of probability: K308 (VKGE), K731 (LKAE), K707 (AKAE), K736 (AKIE), and K427 (AKDP) within the sea urchin MVP. With the exception of K427, the amino acid sequence surrounding the lysine acceptor residue follow the consensus sequence, ΨKXE, where Ψ represents a hydrophobic residue, K represents the target lysine, X represents any amino acid and E represents glutamic acid [37]. Sequence alignment of MVP homologs from six species showed that two of these five motifs (K308 and K707) are conserved.

Epitopes are also conserved between the rat MVP and the sea urchin MVP. Previously, we generated and purified a rabbit polyclonal antibody against the MVP that copurified with sea urchin microtubules [16]. These antibodies were shown to recognize both *Dictyostelium* and rat MVP [16]. In this report, we show that these affinity-purified antibodies also recognize the MVP in purified sea urchin vault preparations (Fig. 3). Two minor bands of ~80 kDa and 50 kDa cross-react with the anti-MVP antibodies indicating that these may be breakdown products of the 100 kDa MVP. In addition to the antibodies generated against the sea urchin MVP, anti-peptide antibodies were generated against a peptide located at the amino terminus of the rat MVP sequence [38]. This sequence is approximately 70% identical in the sea urchin and rat MVP sequence and the affinity-purified anti-peptide antibodies bind to purified sea urchin MVP (Fig. 3). This observation provides experimental support to relate the assembled coding sequence of the *SpMVP* gene to the major protein component of the purified sea urchin vault.

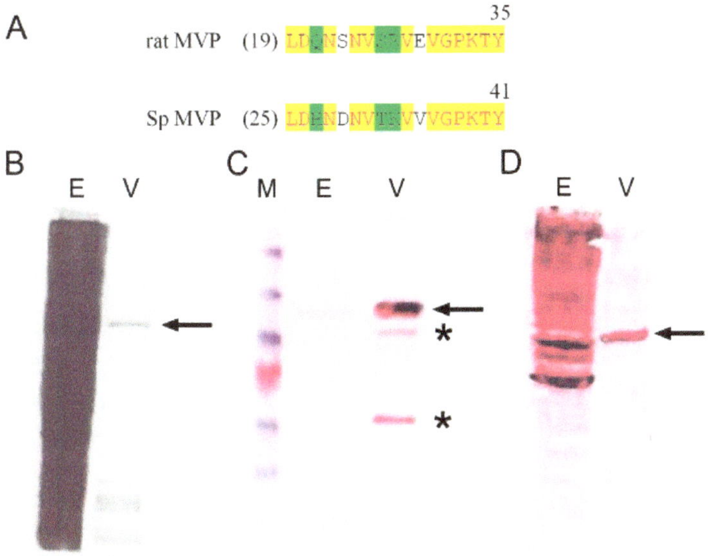

Figure 3: Western blots of isolated sea urchin vaults. (A) Peptide amino acid sequence of rat MVP (aa 19–35) aligned with Sp MVP (25–41). Anti-peptide antibodies were generated against the rat MVP sequence shown and affinity-purified as previously described [38]. Twelve of the rat MVP amino acids are conserved in the Sp MVP peptide sequence. (B) 150 µg of sea urchin egg extract proteins (lane E) and 10 µg of purified sea urchin vault proteins (lane V) were separated on this Coomassie-blue stained SDS-8% polyacrylamide mini-gel. The arrow shows the position of the 100 kDa MVP in (B, C and D). (C) Alkaline-phosphatase stained western blot showing the migration of the pre-stained protein ladder (lane M: 176.5, 113.7, 80.9, 63.8 (pink), 49.5, and 8.4 kDa polypeptides). Affinity-purified anti-sea urchin MVP antibodies [16] recognize a 100 kDa polypeptide in egg extracts (lane E) and in purified vault preparations (lane V). Asterisks indicate two bands of approximately 80 kDa and 50 kDa that may be breakdown products of the MVP. (D) Affinity-purified anti-peptide antibodies (anti-LDQN, [38]) bind to the 100 kDa MVP polypeptide in purified sea urchin vaults. This peptide antibody appears to bind non-specifically to a large number of polypeptides in the sea urchin egg extracts.

Electron Microscopy and Reconstruction of the Sea Urchin Vault

Negative-stain EM and cryoEM images of isolated sea urchin vaults reveal that they have the same overall morphology as mammalian vaults (Fig. 4). As was noted in cryomicrographs of the rat vault [9], sea urchin vaults are occasionally observed to be 'open' at the midsection. One advantage of cryoEM over negative-stain TEM is that a cryomicrograph is essentially a projection image of all of the density, external and internal, within a macromolecular assembly. In cryomicrographs, both sea urchin and rat vaults appear to have molecular contents of varying mass per particle (Fig. 4B, white arrow and [9]).

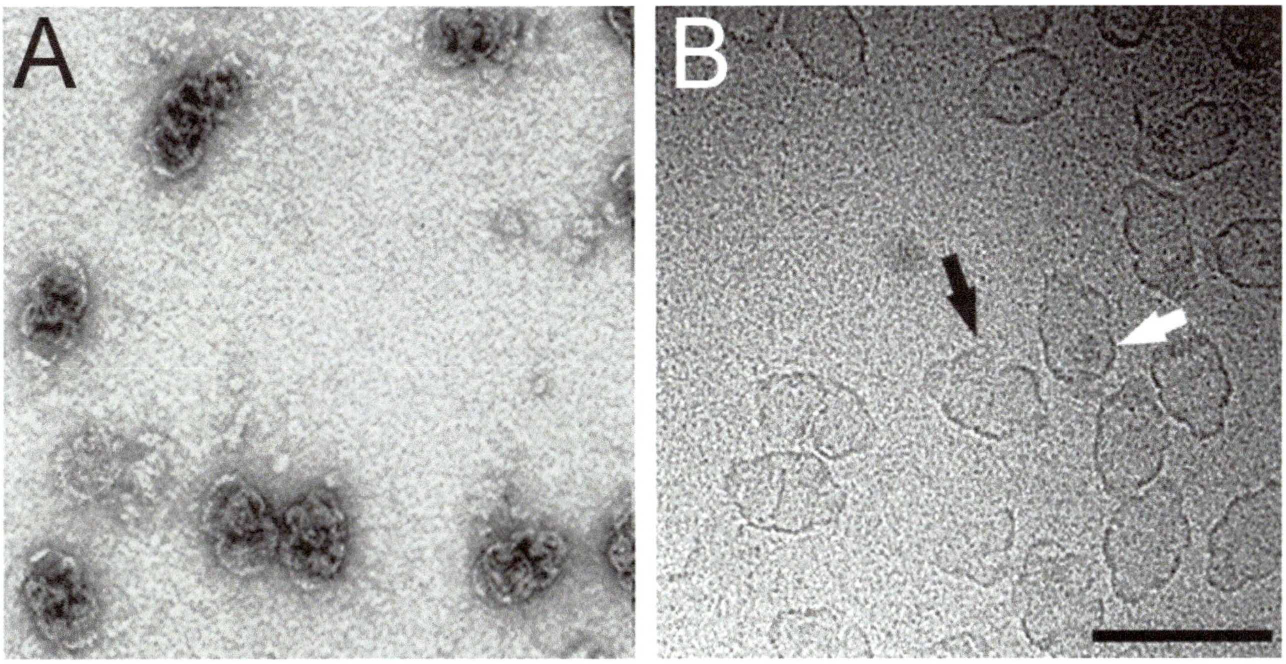

Figure 4: Negative-stain and cryoEM images of isolated sea urchin vaults. (A) Negative-stain electron micrograph. (B) Cryoelectron micrograph. The black arrow indicates a vault that is opening at the midsection, and the white arrow indicates the dark molecular contents within another particle. The scale bar represents 1,000 Å. Note that the magnification of the cryoEM image (B) is slightly higher than that of the negative-stain EM image (A).

In order to characterize the three dimensional structure of the sea urchin vault, a three-dimensional structure was calculated using cryoEM images and single particle reconstruction methods. Only particle images of well-formed, fully closed sea urchin vaults were selected for image processing. Refinement of the cryoEM data set produced a three-dimensional reconstruction at 33 Å resolution (Fig. 5). The exterior shape of the sea urchin vault reconstruction is nearly identical to the shape of the rat vault reconstruction with a central barrel section and two protruding caps [9]. When the sea urchin vault reconstruction is cropped in half the large internal cavity is revealed (Fig. 5B and 5C). As has been noted for the rat vault, the volume of the internal cavity is large enough to enclose an intact ribosome [9].

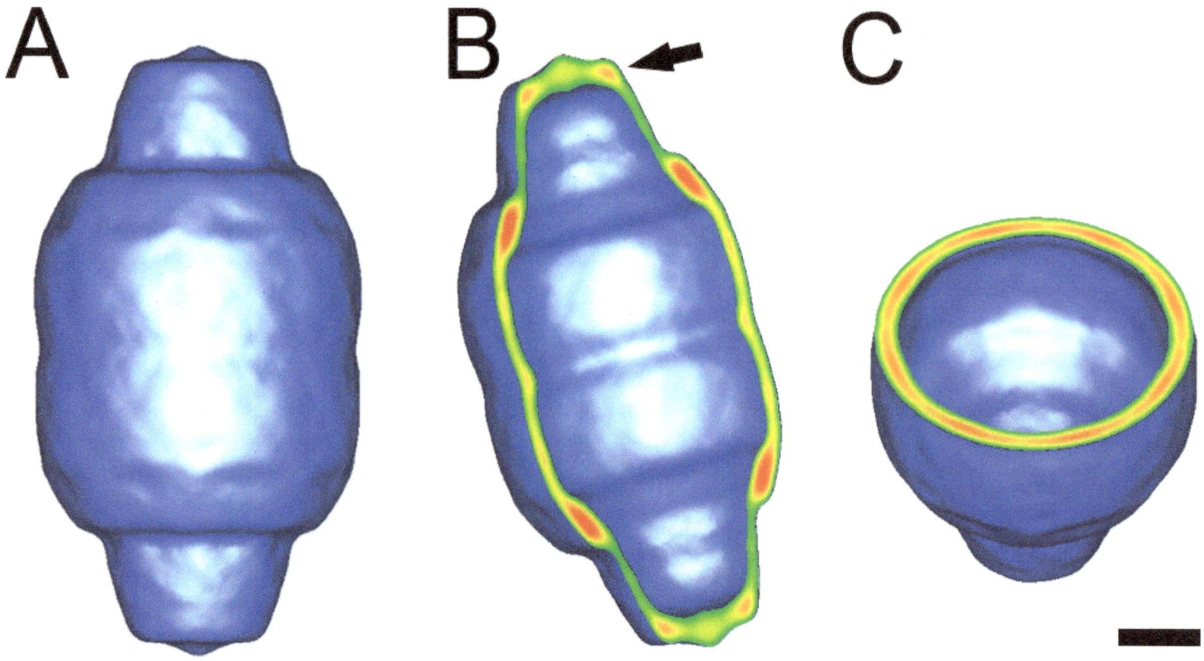

Figure 5: The sea urchin vault reconstruction at 33 Å resolution. (A) The full reconstruction, which reveals that the sea urchin vault has essentially the same exterior structure as rat and mouse vaults. (B and C) The reconstruction is shown cropped along two perpendicular axes to reveal the hollow interior. The crop planes are displayed with the strongest density in red and the weakest density in green. Note that the strongest density is in the "shoulder" region at the top and bottom of the central barrel section. The flat portion of one cap is indicated by an arrow in (B). The scale bar represents 100 Å.

CryoEM reconstructions of recombinant and tissue derived vaults often show small holes at the cap/barrel junction [9, 39]. The sea urchin vault reconstruction does not show these holes, but they are probably obscured by the low resolution of the sea urchin vault structure. Another subtle difference between the sea urchin and rat vault reconstructions is that less density is observed in the flat portion of the sea urchin vault cap (Fig. 5B, arrow). This is the same region of the vault that was identified as the vRNA binding site within the rat vault [25]. It is also a region that tends to be variable between vault reconstructions [39].

Cellular Localization of MVP during Sea Urchin Embryonic Development

Earlier immunofluorescence studies indicated that maternal and zygotic MVP is located in the cytoplasm, whereas adult MVP is predominantly nuclear [16]. The subcellular location of MVP during embryogenesis was not determined by immunofluorescence. However, cell fractionation and immunoblot analyses of embryos showed that while total amount of MVP remained constant during embryogenesis, MVP became progressively concentrated in the nuclear fractions as development proceeded from the mesenchyme blastula stage to the larval stage [16]. Here we confirm and extend this result using antibody staining and confocal microscopy to localize MVP subcellularly during embryonic development. The confocal images in Figure 6demonstrate that sea urchin MVP moves from a largely cytoplasmic distribution in the cleavage stage embryo, to a predominantly nuclear and/or perinuclear location at blastula and gastrula stages.

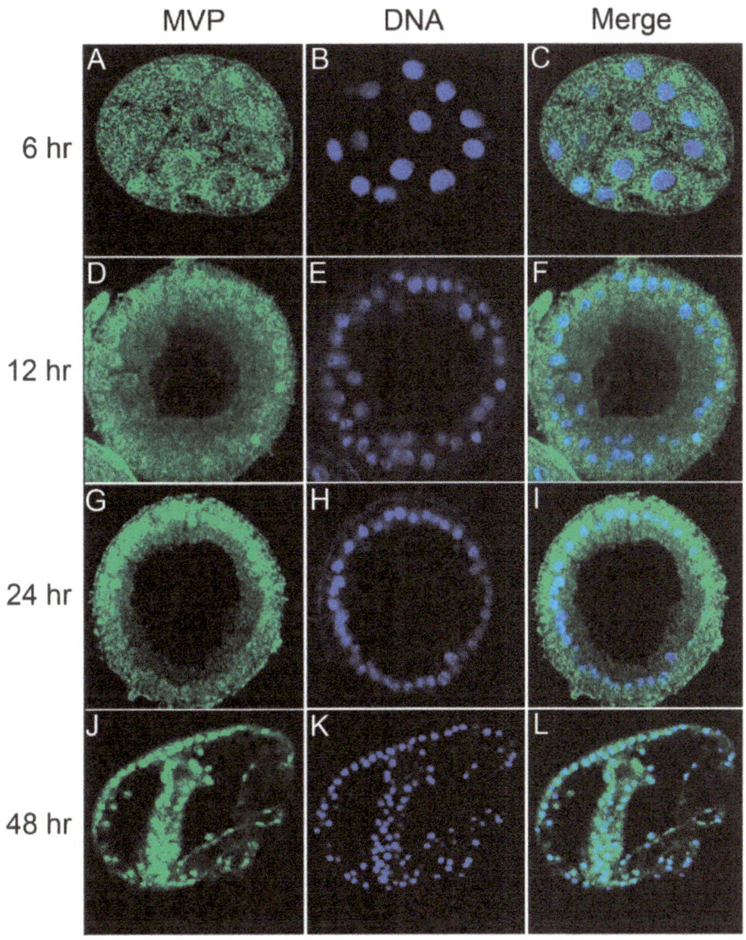

Figure 6: Confocal images of sea urchin embryos at various developmental stages. The cellular distribution of MVP is revealed by staining with an affinity-purified polyclonal antiserum against the sea urchin MVP and secondary staining with Oregon Green-conjugated anti-rabbit antibody (A, D, G, J; left column). The nuclei are stained with DAPI (B, E, H, K; middle column). By merging the MVP and DNA images coincident staining is observed as cyan (C, F, I, L; right column). Note that as development proceeds the cellular localization of MVP progressively shifts from cytoplasmic to nuclear.

DISCUSSION

In this study we used the sea urchin genome to deduce the amino acid sequence of the sea urchin MVP and characterized the structure of the sea urchin vault particle using cryoEM single particle reconstruction methods. It is known that multiple copies of MVP form the recognizable vault structure, as expression of rat MVP in insect cells leads to the assembly of vaults [40]. Given the relatively high (81%) sequence similarity observed between sea urchin and rat MVP, it is not surprising that the exterior sea urchin vault structure is quite similar to that of rat vaults [9, 39]. In addition to MVP, the rat vault is composed of the high molecular weight proteins VPARP and TEP1, as well as the small vRNA. The electrophoretic analysis of purified sea urchin vault presented here shows protein bands and one RNA band that might correspond to homologs of the mammalian vault components VPARP, TEP1, and vRNA. Moreover, sequence homologs of VPARP and TEP1 are found in the sea urchin genome.

SDS-PAGE analysis has indicated that the molecular composition of sea urchin vaults is more complex than that of rat vaults. Protein bands in the range of 14 to 55 kDa are observed that might either be additional vault components or macromolecular contents that are present in high copy numbers within the sea urchin vault. One protein in this mass range (~50 kDa) cross-reacts with MVP antisera and is likely to be a breakdown product of MVP. The identities of the additional proteins will be the subject of future investigations. It is tempting to speculate that the difference in protein composition between sea urchin and rat vaults might indicate that sea urchin vaults are transporting or sequestering a larger assortment of macromolecules.

Vaults were originally named because of their similarity to arched cathedral ceilings [41]. This name may be apt in another sense in that the vault might serve to provide a safe enclosure for its molecular contents [2]. It has been postulated that mammalian vaults might open under certain physiological conditions to allow encapsulation or release of their molecular cargo. Sea urchin vaults are occasionally observed to be partially open at the midsection in cryomicrographs. Thus sea urchin vaults, as well as mammalian vaults, might be able to regulate what is contained within the large internal cavity. Both sea urchin and rat vaults [9] appear to have molecular contents in cryomicrographs. However as the features of the contents vary from vault to vault they are inappropriately averaged or "smeared" during the image reconstruction process.

A difference map analysis between cryoEM reconstructions of intact and RNase-treated rat vaults led to the localization of the vRNA to the flat portions of the rat vault caps [25]. The reconstruction of the sea urchin vault reveals a difference in this region of the vault and appears more similar to the RNase-treated rat vault [25]. Although RNase-treatment does not harm the structure of the rat vault, it does appear to affect the sea urchin vault. Several attempts were made to collect negative-stain and cryoEM images of the RNase-treated sea urchin vaults, but no intact vault particles were observed. The difference observed in the cap density and the differing response to RNase-treatment could imply that the vRNA has a different location or a more important structural role in the sea urchin vault than in the rat vault.

The confocal images reveal a progressive enrichment of MVP in the nucleus and perinuclear cytoplasm during sea urchin embryogenesis. Our previous biochemical fractionation studies confirm that the MVP accumulates in the nucleus beginning at the mesenchyme blastula stage [16]. This important stage of development marks the transition between proliferation and differentiation. As a consequence of differentiation the cells are preparing to exit from the rapid cell cycles that followed fertilization. For example, mRNAs which were abundant in the early cleavage stages, such as cyclin mRNA, decrease dramatically at the blastula stage [42]. New post-cleavage stage mRNAs begin to accumulate at blastula stage. We suggest that vaults participate in nucleocytoplasmic transport that accompanies remodeling of nuclear architecture in preparation for terminal cell differentiation [43].

While MVP localization studies in mammalian cells reveal a largely cytoplasmic location, the published staining pattern is consistent with there also being a small amount of nuclear MVP [17, 18]. While no classical NLS sequence is found in either the sea urchin or rat MVP, there are basic sequences that might potentially fulfill the requirements of an NLS. The C-terminal region of sea urchin MVP has a region that strongly resembles the NES logo assembled from known NES's [36] and this region is fairly well conserved in MVP from six species. In addition to possible NLS and NES signals, there are several sumoylation signals within the sea urchin MVP.

Sumoylation is a dynamic and reversible process that involves the covalent attachment of the small ubiquitin-like modifier (SUMO) protein to protein substrates. Similar to ubiquination, sumoylation requires four enzymatic steps to attach the carboxyl-terminal glycine of SUMO to the ε-amino group of lysine (reviewed in [44]). However, unlike traditional ubiquination, sumoylation does not mark the protein for degradation. The wide variety of SUMO targets identified to date does not reveal a single function for this protein modification (reviewed in [45]). A variety of transcription factors and cofactors, viral proteins, proteins associated with nuclear architecture and genome surveillance, as well as signal transduction molecules appear to by sumoylated [44]. Importantly for this study, two proteins of the nuclear pore complex, RanGAP1 and Ran-binding protein 2 are both SUMO substrates. Sumoylation of RanGAP1 is important for the nucleocytoplasmic transport of this cytoplasmic nuclear import factor [46–48]. In addition, RanBP2, a docking protein at the cytoplasmic surface of the nuclear protein complex is a SUMO E3 ligase, also known as Nup358 [49–51]. These observations suggest that SUMO plays an important role in nucleocytoplasmic trafficking. In this regard, sumoylation of the major vault protein may regulate the transport of the vault particle through the nuclear pore complex.

Taken together these results suggest that the vault may play a role in delivering macromolecules to the nucleus during embryonic development in sea urchins. A trafficking role has also been proposed for vaults in human cells based on the observations that MVP interacts with the estrogen receptor in MCF-7 breast cancer cells [10] and with the tumor suppressor PTEN in HeLa cells [11]. Estrogen receptor is involved in regulating eukaryotic gene expression and can affect cellular proliferation and differentiation. Estrogen treatment is observed to increase the association of the estrogen receptor with MVP in the nuclei of MCF-7 cells, thus leading to the hypothesis that vaults may be involved in nucleocytoplasmic shuttling of the estrogen receptor and in modulating the effect of steroid hormones. PTEN negatively regulates the phosphoinositide 3-kinase pathway and cell growth. MVP has been shown by a yeast two-hybrid screen to be a dominant PTEN-binding protein. It has been postulated that perhaps the vault serves to mediate the cellular localization PTEN, thus potentially affecting cell growth.

The results presented here suggest that sea urchin vaults contain various macromolecules in addition to the two main vault-associated proteins, VPARP and TEP1. Human MVP is known to interact specifically with the tyrosine phosphatase SHP-2, an enzyme that plays an important role in intracellular signaling [12]. It has been demonstrated using MVP-deficient mouse fibroblasts that MVP helps to support cell survival. MVP is proposed to act as a scaffold protein for SHP-2 and other extracellular-regulated kinases (Erks) and thus facilitate, or somehow modulate, growth factor signaling. Perhaps the varied macromolecular contents of the sea urchin vault include proteins involved in sea urchin growth factor signaling.

CONCLUSIONS

It would seem likely that the sea urchin MVP is transported to the nucleus for a reason. Presumably cells must expend a considerable amount of energy to achieve this differential localization, as MVP is a relatively abundant protein. The MVP concentration in the unfertilized sea urchin egg is on the order of 10 μM, as determined by immunoblotting [16]. This maternal store of MVP is sufficient to assemble 10^7 intact vault particles. By comparison, the mature sea urchin egg contains approximately 4×10^8 ribosomes [52]. Although the function of the vault is not yet clearly delineated, our results on the highly conserved structure, molecular composition, and differential cellular localization of MVP suggest that vaults may be important for nucleocytoplasmic trafficking during embryonic development.

METHODS

Isolation of Sea Urchin and Rat Vaults

Vaults were purified to homogeneity from unfertilized eggs of the Pacific coast sea urchin*Strongylocentrotus purpuratus*. Animals were spawned and the eggs were collected and washed as described [53]. Washed eggs (150 ml) were homogenized with a motor driven Potter-Elvehjem-type tissue grinder in an equal volume of buffer (100 mM PIPES-KOH, pH 7.3, 1 mM $MgSO_4$, 4 mM EGTA, 2 mM DTT, 1 mM GTP, 0.2 mM PMSF, 10 μg/ml leupeptin and 1 μg/ml pepstatin A). The homogenate was centrifuged at 18 K rpm (~39,000 g) in a Beckman J20 rotor at 4°C for 45 min. The resulting supernatant fluids were decanted into a 500-ml Erlenmeyer flask at room temperature. A fresh aliquot of GTP was added such that the GTP concentration was increased by 1 mM. The DMSO was added in three aliquots with gentle mixing until a final concentration of 15% (v/v) was reached. Microtubule assembly was promoted by incubating the supernatant fluids in a 24°C water bath for 30–45 min. The microtubules were pelleted by centrifugation at 18K rpm (~39,000 g) in a Beckman J20 rotor at 24°C for 30 min. The supernatant fluids (H1S) were carefully removed and drop-frozen in liquid nitrogen and stored in a -80°C freezer. Microtubules in the pellet were purified by two subsequent cycles of assembly and disassembly as previously described [54] and used for purposes other than the current study.

Sea urchin vaults were purified from the microtubule-depleted supernatant fluids that had been previously frozen in liquid nitrogen. Approximately 150 ml of frozen supernatant (H1S) was thawed and mixed with an equal volume of 50 mM Tris-HCL, pH 7.4, 1.5 mM $MgCl_2$, 75 mM NaCl, 1 mM DTT and 1 mM PMSF. This pooled supernatant was used as the starting point for the vault purification. This extract was treated as if it was a rat liver homogenate and all subsequent purification steps were as described for rat liver vaults, including purification on sucrose density gradients and two cesium chloride gradients [9].

Rat vaults were purified from liver by sucrose-density and cesium chloride gradient centrifugation as described by the Rome laboratory [9].

Molecular Composition of Sea Urchin Vaults

Vault samples were analyzed by SDS-PAGE with the discontinuous buffer formulation of Laemmli *et al.*, [55] and stained with Coomassie Blue R-250 (Fig. 1A) or silver (Fig. 1B).

Assembly and Analysis of the Sea Urchin Major Vault Protein Sequence

Trace sequences from the *S. purpuratus* genome project [56] were used to assemble the coding sequence of the SpMVP gene. Toward this end the TBLASTN algorithm [57, 58] was performed using the rat MVP as a query sequence to search the NCBI trace archive of the sea urchin genome that is now at near 10x coverage. Due to the high level of sequence conservation it was possible to assemble the urchin sequence from the TBLASTN results using the rat MVP as a scaffold. The urchin MVP sequence was assembled in such a way that maximum similarity to the rat scaffold sequence was maintained. The fully assembled SpMVP sequence was then verified by using it to query the NCBI sea urchin EST collection (representing mRNAs expressed in the embryo and larva) by TBLASTN. Except for a small region (A728 to G785) near the C-terminal sequence, the entire SpMVP was covered by *S. purpuratus* ESTs, with the following NCBI accession numbers: BG784484, BG784394, CD305511, CD295832, CD295616, CD304000, CD310423, CD307400, CD305382, CD292179, CD305646, CD333690, CD309547, CD292477, and BG783194. The gap from A728 to G785 was covered by a sequence in the NCBI trace archives (gnl|ti|287010960). The assembled SpMVP sequence has been deposited in the NCBI database under accession number BK005641.

We note that the sea urchin MVP has a molecular weight of 95 kDa, calculated from the protein sequence. Previously the sea urchin MVP was referred to as a 107-kDa polypeptide based on its apparent molecular weight by SDS-PAGE [16]. The program COILS [59] was used to identify a likely coiled coil region in the C-terminal half of the protein (aa 663–762) [60]. The sea urchin MVP sequence was submitted to the PredictNLS server [34, 61]. No potential NLS sequences were found by this server. The two possible NLS sequences shown in Figure 2B are based on sequence observation. The sea urchin MVP sequence was compared by hand to the NES logo described by la Cour *et al.* and based on the alignment of 58 high-quality NES's [36]. The sea urchin MVP sequence was submitted to the Abgent Sumoylation Calculator [62]. Five motifs with high probability of being sumoylation sites were identified (scores 0.6889 to 0.9278). Two of these five motifs are conserved among MVP sequences of six species (*Strongy. purp., Homo sapiens, Mus musculus, Rattus norv., Danio rerio*, and *Xenopus laevis*) and are indicated in Figure 2B. The CLUSTALW alignment between sea urchin and rat MVP in Figure 2A was produced by submitting the sequences to the CLUSTALW server [63].

Negative-Stain Electron Microscopy

Isolated sea urchin vaults were applied to freshly glow-discharged formvar-coated copper grids (Electron Microscopy Sciences). Excess liquid was wicked away and the samples were briefly air dried. Samples were stained with uranyl acetate and observed using a JEOL 1200EXII transmission electron microscope.

CryoEM Imaging And Three-Dimensional Reconstruction

CryoEM grids were prepared using cryogenic plunge freezing methods described previously [9,64]. Digital cryoelectron micrographs were collected on an FEI/Philips CM120 transmission electron microscope with a LaB_6 filament, a Gatan 626 cryotransfer holder, and a Gatan slow scan CCD camera (1024 × 1024 pixels, YAG scintillator). One hundred and forty-one cryomicrographs were collected with a nominal magnification of 45,000×, and with two different defocus values (-1.0 and -0.6 μm). The pixel size in the cryomicrographs is 4.1 Å on the molecular scale as determined by calibration with a catalase crystal.

The QVIEW software package was used to extract 481 individual vault particle images with a selection box size of 200 × 200 pixels [65]. The initial translation step was performed by cross-correlating each particle image with an 180° rotated version of itself. The IMAGIC software package was used for all 3D image processing steps [66]. The published RNase-treated rat vault reconstruction was used as a search model in the first round of refinement [9]. C8 symmetry was assumed for the sea urchin vault for the first several rounds of refinement, as was used for the intact rat vault reconstruction [9]. After a few rounds of refinement the two ends of the vault appeared nearly identical, and from then on D8 symmetry was imposed, as was done for the RNase-treated rat vault reconstruction [25]. The particles images were corrected for the contrast transfer function (CTF) of the microscope using the CTF equation published by Baker *et al.* [67]. The following CTF parameters were used: Cs = 2 mm, fraction of amplitude contrast = 0.1, and kV = 120. A subset of 409 particle images that agreed best with the rest of the set were selected for the final reconstruction. The resolution was calculated using the Fourier shell correlation method with an elliptical mask applied to remove noise and disordered contents. The resolution was found to be 33 Å by the 0.5 correlation threshold criterion. The sea urchin vault reconstruction is shown filtered to 33 Å resolution and contoured so that the surface appears continuous.

All of the image processing was performed on HP/Digital alpha unix workstations. The graphics representations were produced with the AVS (Advanced Visual Systems, Inc.) software package.

Antibody Staining and Confocal Microscopy of Whole Mounted Embryos

Staged embryos were fixed in -20°C Methanol for 20 minutes on ice, then washed three times in PBS plus 0.2% Tween-20 (PBST). Embryos were blocked for 30 minutes on ice in PBST plus 5% Bovine Serum Albumin, then incubated with an appropriate dilution of an affinity purified antibody against MVP [16] overnight at 4°C. Following three washes for 5 minutes each in PBST, the embryos were incubated for 1 hour at room temperature in 1 µg/ml Oregon Green 488 goat anti-rabbit antibody. Embryos were washed three times in PBST and stained in PBST with 300 nM DAPI for 10 minutes at room temperature during the second wash. Digital confocal images were collected using a Leica TCS confocal microscope.

ACKNOWLEDGEMENTS

This work was supported by grants from the National Institutes of Health (COBRE award 1P20RR15563) with matching support from the State of Kansas and the University of Kansas (K.A.S.); the National Science Foundation (MCB-9722353 to P.L.S. and MCB-9982377 to K.A.S.); and institutional funding by the Stowers Institute for Medical Research (J.A.C.). The authors thank Dr. Kristi Neufeld for her advice on predicting NES and NLS.

AUTHORS' CONTRIBUTIONS

PLS directed the cryoEM and three-dimensional image processing as well as helped to draft the manuscript. MM performed the cryomicroscopy and image processing. JL provided technical assistance. CDS performed the antibody staining and confocal imaging of whole-mounted embryos. AJR assembled the SpMVP sequence and performed the sequence alignments used to calculate sequence similarities and differences. JAC directed the research pertaining to vault localization in the embryo, and drafted Figure 6 and the relevant parts of the manuscript. KAS conceived of the study, participated in its design and coordination, purified the sea urchin vaults, and helped to draft the manuscript. All authors read and approved the final manuscript.

REFERENCES

1. Suprenant KA, Rebhun LI: Assembly of unfertilized sea urchin egg tubulin at physiological temperatures. J Biol Chem. 1983, 258 (7): 4518-4525.
2. Suprenant KA: Vault ribonucleoprotein particles: sarcophagi, gondolas, or safety deposit boxes?. Biochemistry. 2002, 41 (49): 14447-14454. 10.1021/bi026747e.
3. Scheffer GL, Wijngaard PL, Flens MJ, Izquierdo MA, Slovak ML, Pinedo HM, Meijer CJ, Clevers HC, Scheper RJ: The drug resistance-related protein LRP is the human major vault protein. Nat Med. 1995, 1 (6): 578-582. 10.1038/nm0695-578.
4. Izquierdo MA, Scheffer GL, Flens MJ, Giaccone G, Broxterman HJ, Meijer CJ, van der Valk P, Scheper RJ: Broad distribution of the multidrug resistance-related vault lung resistance protein in normal human tissues and tumors. Am J Pathol. 1996, 148 (3): 877-887.
5. Ikeda K, Oka M, Yamada Y, Soda H, Fukuda M, Kinoshita A, Tsukamoto K, Noguchi Y, Isomoto H, Takeshima F, Murase K, Kamihira S, Tomonaga M, Kohno S: Adult T-cell leukemia cells over-express the multidrug-resistance-protein (MRP) and lung-resistance-protein (LRP) genes. Int J Cancer. 1999, 82 (4): 599-604. 10.1002/(SICI)1097-0215(19990812)82:4<599::AID-IJC21>3.0.CO;2-R.
6. Siva AC, Raval-Fernandes S, Stephen AG, LaFemina MJ, Scheper RJ, Kickhoefer VA, Rome LH: Up-regulation of vaults may be necessary but not sufficient for multidrug resistance. Int J Cancer. 2001, 92 (2): 195-202. 10.1002/1097-0215(200102)9999:9999<::AID-IJC1168>3.0.CO;2-7.
7. Chugani DC, Rome LH, Kedersha NL: Evidence that vault ribonucleoprotein particles localize to the nuclear pore complex. J Cell Sci. 1993, 106 (Pt 1): 23-29.
8. Kickhoefer VA, Rajavel KS, Scheffer GL, Dalton WS, Scheper RJ, Rome LH: Vaults are up-regulated in multidrug-resistant cancer cell lines. J Biol Chem. 1998, 273 (15): 8971-8974. 10.1074/jbc.273.15.8971.
9. Kong LB, Siva AC, Rome LH, Stewart PL: Structure of the vault, a ubiquitous cellular component. Structure Fold Des. 1999, 7 (4): 371-379. 10.1016/S0969-2126(99)80050-1.
10. Abbondanza C, Rossi V, Roscigno A, Gallo L, Belsito A, Piluso G, Medici N, Nigro V, Molinari AM, Moncharmont B, Puca GA: Interaction of vault particles with estrogen receptor in the MCF-7 breast cancer cell. J Cell Biol. 1998, 141 (6): 1301-1310. 10.1083/jcb.141.6.1301.
11. Yu Z, Fotouhi-Ardakani N, Wu L, Maoui M, Wang S, Banville D, Shen SH: PTEN associates with the vault particles in HeLa cells. J Biol Chem. 2002, 277 (43): 40247-40252. 10.1074/jbc.M207608200.
12. Kolli S, Zito CI, Mossink MH, Wiemer EA, Bennett AM: The major vault protein is a novel substrate for the tyrosine phosphatase SHP-2 and scaffold protein in epidermal growth factor signaling. J Biol Chem. 2004, 279 (28): 29374-29385. 10.1074/jbc.M313955200.

13. Liu Y, Snow BE, Hande MP, Baerlocher G, Kickhoefer VA, Yeung D, Wakeham A, Itie A, Siderovski DP, Lansdorp PM, Robinson MO, Harrington L: Telomerase-associated protein TEP1 is not essential for telomerase activity or telomere length maintenance in vivo. Mol Cell Biol. 2000, 20 (21): 8178-8184. 10.1128/MCB.20.21.8178-8184.2000.

14. Mossink MH, van Zon A, Franzel-Luiten E, Schoester M, Kickhoefer VA, Scheffer GL, Scheper RJ, Sonneveld P, Wiemer EA: Disruption of the murine major vault protein (MVP/LRP) gene does not induce hypersensitivity to cytostatics. Cancer Res. 2002, 62 (24): 7298-7304.

15. Liu Y, Snow BE, Kickhoefer VA, Erdmann N, Zhou W, Wakeham A, Gomez M, Rome LH, Harrington L: Vault poly(ADP-ribose) polymerase is associated with mammalian telomerase and is dispensable for telomerase function and vault structure in vivo. Mol Cell Biol. 2004, 24 (12): 5314-5323. 10.1128/MCB.24.12.5314-5323.2004.

16. Hamill DR, Suprenant KA: Characterization of the sea urchin major vault protein: a possible role for vault ribonucleoprotein particles in nucleocytoplasmic transport. Dev Biol. 1997, 190 (1): 117-128. 10.1006/dbio.1997.8676.

17. Kedersha NL, Rome LH: Vaults: large cytoplasmic RNP's that associate with cytoskeletal elements. Mol Biol Rep. 1990, 14 (2-3): 121-122. 10.1007/BF00360441.

18. Kickhoefer VA, Siva AC, Kedersha NL, Inman EM, Ruland C, Streuli M, Rome LH: The 193-kD vault protein, VPARP, is a novel poly(ADP-ribose) polymerase. J Cell Biol. 1999, 146 (5): 917-928. 10.1083/jcb.146.5.917.

19. Herrmann C, Golkaramnay E, Inman E, Rome L, Volknandt W: Recombinant major vault protein is targeted to neuritic tips of PC12 cells. J Cell Biol. 1999, 144 (6): 1163-1172. 10.1083/jcb.144.6.1163.

20. Berger W, Spiegl-Kreinecker S, Buchroithner J, Elbling L, Pirker C, Fischer J, Micksche M: Overexpression of the human major vault protein in astrocytic brain tumor cells. Int J Cancer. 2001, 94 (3): 377-382. 10.1002/ijc.1486.

21. Schroeijers AB, Reurs AW, Scheffer GL, Stam AG, de Jong MC, Rustemeyer T, Wiemer EA, de Gruijl TD, Scheper RJ: Up-regulation of drug resistance-related vaults during dendritic cell development. J Immunol. 2002, 168 (4): 1572-1578.

22. Suprenant KA, Tempero LB, Hammer LE: Association of ribosomes with in vitro assembled microtubules. Cell Motil Cytoskeleton. 1989, 14 (3): 401-415.

23. Suprenant KA: Microtubules, ribosomes, and RNA: evidence for cytoplasmic localization and translational regulation. Cell Motil Cytoskeleton. 1993, 25 (1): 1-9.

24. Hamill D, Davis J, Drawbridge J, Suprenant KA: Polyribosome targeting to microtubules: enrichment of specific mRNAs in a reconstituted microtubule preparation from sea urchin embryos. J Cell Biol. 1994, 127 (4): 973-984. 10.1083/jcb.127.4.973.

25. Kong LB, Siva AC, Kickhoefer VA, Rome LH, Stewart PL: RNA location and modeling of a WD40 repeat domain within the vault. RNA. 2000, 6 (6): 890-900. 10.1017/S1355838200000157.

26. Nakayama J, Saito M, Nakamura H, Matsuura A, Ishikawa F: TLP1: a gene encoding a protein component of mammalian telomerase is a novel member of WD repeats family. Cell. 1997, 88 (6): 875-884. 10.1016/S0092-8674(00)81933-9.

27. Harrington L, McPhail T, Mar V, Zhou W, Oulton R, Bass MB, Arruda I, Robinson MO: A mammalian telomerase-associated protein. Science. 1997, 275 (5302): 973-977. 10.1126/science.275.5302.973.

28. Kickhoefer VA, Stephen AG, Harrington L, Robinson MO, Rome LH: Vaults and telomerase share a common subunit, TEP1. J Biol Chem. 1999, 274 (46): 32712-32717. 10.1074/jbc.274.46.32712.

29. Kickhoefer VA, Searles RP, Kedersha NL, Garber ME, Johnson DL, Rome LH: Vault ribonucleoprotein particles from rat and bullfrog contain a related small RNA that is transcribed by RNA polymerase III. J Biol Chem. 1993, 268 (11): 7868-7873.

30. Kedersha NL, Heuser JE, Chugani DC, Rome LH: Vaults. III. Vault ribonucleoprotein particles open into flower-like structures with octagonal symmetry. J Cell Biol. 1991, 112 (2): 225-235. 10.1083/jcb.112.2.225.

31. Kickhoefer VA, Liu Y, Kong LB, Snow BE, Stewart PL, Harrington L, Rome LH: The Telomerase/vault-associated protein TEP1 is required for vault RNA stability and its association with the vault particle. J Cell Biol. 2001, 152 (1): 157-164. 10.1083/jcb.152.1.157.

32. Herrmann C, Kellner R, Volknandt W: Major vault protein of electric ray is a phosphoprotein. Neurochem Res. 1998, 23 (1): 39-46. 10.1023/A:1022445302710.

33. van Zon A, Mossink MH, Schoester M, Scheffer GL, Scheper RJ, Sonneveld P, Wiemer EA: Structural domains of vault proteins: a role for the coiled coil domain in vault assembly. Biochem Biophys Res Commun. 2002, 291 (3): 535-541. 10.1006/bbrc.2002.6472.

34. Cokol M, Nair R, Rost B: Finding nuclear localization signals. EMBO Rep. 2000, 1 (5): 411-415. 10.1093/embo-reports/kvd092.

35. Tinland B, Koukolikova-Nicola Z, Hall MN, Hohn B: The T-DNA-linked VirD2 protein contains two distinct functional nuclear localization signals. Proc Natl Acad Sci USA. 1992, 89 (16): 7442-7446.

36. la Cour T, Gupta R, Rapacki K, Skriver K, Poulsen FM, Brunak S: NESbase version 1.0: a database of nuclear export signals. Nucleic Acids Res. 2003, 31 (1): 393-396. 10.1093/nar/gkg101.

37. Rodriguez MS, Dargemont C, Hay RT: SUMO-1 conjugation in vivo requires both a consensus modification motif and nuclear targeting. J Biol Chem. 2001, 276 (16): 12654-12659. 10.1074/jbc.M009476200.

38. Eichenmüller B, Kedersha N, Solovyeva E, Everley P, Lang J, Himes RH, Suprenant KA: Vaults bind directly to microtubules via their caps and not their barrels. Cell Motil Cytoskeleton. 2003, 56 (4): 225-236. 10.1002/cm.10147.

39. Mikyas Y, Makabi M, Raval-Fernandes S, Harrington L, Kickhoefer VA, Rome LH, Stewart PL: Cryoelectron microscopy imaging of recombinant and tissue derived vaults: localization of the MVP N termini and VPARP. J Mol Biol. 2004, 344 (1): 91-105. 10.1016/j.jmb.2004.09.021.

40. Stephen AG, Raval-Fernandes S, Huynh T, Torres M, Kickhoefer VA, Rome LH: Assembly of vault-like particles in insect cells expressing only the major vault protein. J Biol Chem. 2001, 276 (26): 23217-23220. 10.1074/jbc.C100226200.

41. Kedersha NL, Rome LH: Isolation and characterization of a novel ribonucleoprotein particle: large structures contain a single species of small RNA. J Cell Biol. 1986, 103 (3): 699-709. 10.1083/jcb.103.3.699.

42. Kelso-Winemiller L, Yoon J, Peeler MT, Winkler MM: Sea urchin maternal mRNA classes with distinct development regulation. Dev Genet. 1993, 14 (5): 397-406.

43. Angerer LM, Angerer RC: Patterning the sea urchin embryo: gene regulatory networks, signaling pathways, and cellular interactions. Curr Top Dev Biol. 2003, 53: 159-198.

44. Seeler JS, Dejean A: Nuclear and unclear functions of SUMO. Nat Rev Mol Cell Biol. 2003, 4 (9): 690-699. 10.1038/nrm1200.

45. Muller S, Ledl A, Schmidt D: SUMO: a regulator of gene expression and genome integrity. Oncogene. 2004, 23 (11): 1998-2008. 10.1038/sj.onc.1207415.

46. Matunis MJ, Coutavas E, Blobel G: A novel ubiquitin-like modification modulates the partitioning of the Ran-GTPase-activating protein RanGAP1 between the cytosol and the nuclear pore complex. J Cell Biol. 1996, 135 (6 Pt 1): 1457-1470. 10.1083/jcb.135.6.1457.

47. Mahajan R, Delphin C, Guan T, Gerace L, Melchior F: A small ubiquitin-related polypeptide involved in targeting RanGAP1 to nuclear pore complex protein RanBP2. Cell. 1997, 88 (1): 97-107. 10.1016/S0092-8674(00)81862-0.

48. Mahajan R, Gerace L, Melchior F: Molecular characterization of the SUMO-1 modification of RanGAP1 and its role in nuclear envelope association. J Cell Biol. 1998, 140 (2): 259-270. 10.1083/jcb.140.2.259.

49. Pichler A, Gast A, Seeler JS, Dejean A, Melchior F: The nucleoporin RanBP2 has SUMO1 E3 ligase activity. Cell. 2002, 108 (1): 109-120. 10.1016/S0092-8674(01)00633-X.

50. Kirsh O, Seeler JS, Pichler A, Gast A, Muller S, Miska E, Mathieu M, Harel-Bellan A, Kouzarides T, Melchior F, Dejean A: The SUMO E3 ligase RanBP2 promotes modification of the HDAC4 deacetylase. EMBO J. 2002, 21 (11): 2682-2691. 10.1093/emboj/21.11.2682.

51. Miyauchi Y, Yogosawa S, Honda R, Nishida T, Yasuda H: Sumoylation of Mdm2 by protein inhibitor of activated STAT (PIAS) and RanBP2 enzymes. J Biol Chem. 2002, 277 (51): 50131-50136. 10.1074/jbc.M208319200.

52. Griffith JK, Griffith BB, Humphreys T: Regulation of ribosomal RNA synthesis in sea urchin embryos and oocytes. Dev Biol. 1981, 87 (2): 220-228. 10.1016/0012-1606(81)90145-7.

53. Suprenant KA, Foltz Daggett MA: Sea urchin microtubules. Curr Top Dev Biol. 1995, 31: 65-99.

54. Suprenant KA, Marsh JC: Temperature and pH govern the self-assembly of microtubules from unfertilized sea-urchin egg extracts. J Cell Sci. 1987, 87 ((Pt 1)): 71-84.

55. Laemmli UK: Cleavage of structural proteins during the assembly of the head of bacteriophage T4. Nature. 1970, 227 (259): 680-685.

56. Sea Urchin Genome Project [http://www.hgsc.bcm.tmc.edu/projects/seaurchin/].

57. Karlin S, Altschul SF: Methods for assessing the statistical significance of molecular sequence features by using general scoring schemes. Proc Natl Acad Sci USA. 1990, 87 (6): 2264-2268.

58. Karlin S, Altschul SF: Applications and statistics for multiple high-scoring segments in molecular sequences. Proc Natl Acad Sci USA. 1993, 90 (12): 5873-5877.

59. COILS - Prediction of Coiled Coil Regions in Proteins [http://www.ch.embnet.org/software/COILS_form.html].

60. Lupas A, Van Dyke M, Stock J: Predicting coiled coils from protein sequences. Science. 1991, 252 (5010): 1162-1164.

61. PredictNLS Server [http://cubic.bioc.columbia.edu/services/predictNLS/] .

62. Abgent SUMOplot Sumoylation Calculator [http://www.abgent.com/default.php?page=sumoplot].

63. CLUSTALW Server [http://npsa-pbil.ibcp.fr/cgi-bin/npsa_automat.pl?page=/NPSA/npsa_clustalw.html].

64. Adrian M, Dubochet J, Lepault J, McDowall AW: Cryo-electron microscopy of viruses. Nature. 1984, 308 (5954): 32-36. 10.1038/308032a0.

65. Shah AK, Stewart PL: QVIEW: software for rapid selection of particles from digital electron micrographs. J Struct Biol. 1998, 123 (1): 17-21. 10.1006/jsbi.1998.4011.

66. van Heel M, Harauz G, Orlova EV, Schmidt R, Schatz M: A new generation of the IMAGIC image processing system. J Struct Biol. 1996, 116 (1): 17-24. 10.1006/jsbi.1996.0004.

67. Baker TS, Olson NH, Fuller SD: Adding the third dimension to virus life cycles: three-dimensional reconstruction of icosahedral viruses from cryo-electron micrographs. Microbiol Mol Biol Rev. 1999, 63 (4): 862-922.

CHAPTER 14

PURIFICATION AND BIOCHEMICAL CHARACTERIZATION OF A CELLULASE FROM THE DIGESTIVE ORGANS OF THE SHORT-SPINED SEA URCHIN STRONGYLOCENTROTUS INTERMEDIUS

Syuto Hasegawa, Kazuhiro Ura, Hiroyuki Tanaka, Takao Ojima, Yasuaki Takagi

Keywords: Cellulase, Digestive organ, Enzymatic properties, Protein isolation, Sea urchin

INTRODUCTION

Cellulose, a major cell-wall component in plants, is a linear polysaccharide consisting of β-1,4-linked d-glucose residues that hydrolyzes the β-1,4-glycosyl linkages of the cellulose chain to produce cello-oligosaccharides, such as cellotriose and cellobiose [1]. This enzyme family is grouped into endolytic (endo-1,4-β-glucanase, EC 3.2.1.4) and exolytic (cellobiohydrolase, EC 3.2.1.91) enzymes. Complete depolymerization of cellulose to glucose requires another sort of cello-oligosaccharide-degrading enzyme, 1,4-β-glucosidase (EC 3.2.1.21). Thus far, cellulolytic enzymes have been isolated and characterized from bacteria [2], fungi [2], plants [3], molds [4], microbes [5], arthropods [6–11], nematodes [12], mollusks [13–19] and sea urchin [20].

The sea urchin is a well-known marine organism distributed widely from the rocky intertidal areas to the abyssal depths in all of the world's oceans. The gonads of sea urchins are a high value seafood product, and considerable research is being undertaken worldwide on the development of sea urchin aquaculture. More recently, considerable attention has been directed toward the development of artificial diets that enhance the quantity and quality of roe produced from sea urchin aquaculture [21, 22]. To develop such artificial diets and aquaculture systems, it is important to understand the digestive physiology of sea urchins. Although numerous edible sea urchins are kelp feeders, a surprisingly limited amount of information exists on the biochemical characterization of cellulases in the digestive system of sea urchins, although several reports have been published on the detection of cellulase activity in the digestive organs of sea urchins [13, 23–29]. In the respective studies, cellulase activity was analyzed using crude extracts from the digestive organs and, therefore, the exact biochemical characterization of the cellulase has remained unclear. Nishida et al. [20] reported the isolation, biochemical characterization and cDNA cloning of cellulase from the

Citation: Syuto Hasegawa, Kazuhiro Ura, Hiroyuki Tanaka, Takao Ojima, Yasuaki Takagi, Purification and biochemical characterization of a cellulase from the digestive organs of the short-spined sea urchin Strongylocentrotus intermedius, DOI: 10.1007/s12562-012-0528-y.

digestive organs of the Northern sea urchin *Strongylocentrotus nudus*. However, these authors focused on the biochemical characterization of the cellulase at a temperature that is not the considered to be the growth temperature of the Northern sea urchin. Thus, the isolation and biochemical characterization of digestive enzymes at the growth temperature of sea urchin is an important step toward understanding the digestive physiology of this organism and the development of efficient aquaculture systems. To the best of our knowledge, there has been not study on the biochemical characterization of cellulases from sea urchin at its growth temperature.

In the study reported here, we isolated and biochemically characterized a cellulase from the digestive organs of short-spined sea urchin, *Strongylocentrotus intermedius*. We also compared the biochemical characterization of the cellulases from the short-spined sea urchin and Northern sea urchin and discussed digestive physiology in terms of the biochemical characterizations of the cellulases at physiological growth temperatures.

MATERIALS AND METHODS

Materials

Digestive organs from the short-spined sea urchin were obtained at the local fish market in Hakodate, Hokkaido Prefecture, Japan. The pooled tissues (1.5 kg) were thoroughly rinsed in sea water to remove the intestinal contents and then dehydrated by the addition of 1.5 l acetone at −20 °C. Precipitates were collected by centrifugation at 10,000 *g* for 15 min, rinsed five times in 0.5 l cold acetone, then rinsed again in 0.5 l acetone at room temperature and finally collected by centrifugation at 10,000 *g* for 15 min. The acetone-precipitated tissues were dried at room temperature, and the acetone-dried powder was stored at −20 °C until use.

Purification of Cellulase from the Short-Spined Sea Urchin

Crude enzyme was extracted from 2.5 g acetone-dried powder from the sea urchin digestive tracts with 100 ml sodium phosphate (10 mM; pH 7.0) at 0 °C for 30 min with occasional stirring. The enzyme extract was centrifuged at 10,000 *g* for 15 min and the supernatant then dialyzed against 10 mM sodium phosphate (pH 7.0) at 4 °C for 24 h. The supernatant was then applied to a TOYOPEARL-DEAE 650M column (2.0 × 26 cm) (Toyo Soda Mfg Co., Tokyo, Japan). After unadsorbed materials were washed out with 10 mM sodium phosphate (pH 7.0), proteins were eluted by a linear NaCl gradient from 0 to 300 mM (total volume 700 ml) in 10 mM sodium phosphate (pH 7.0) at a 90 ml/h flow rate. The eluate was collected in 8-ml fractions and carboxymethyl cellulose (CMCase) activity was determined. CMCase activity was detected in the fractions eluted at approximately 0.2 M NaCl (pooled as the DE-1 fraction). The DE-1 fraction was subjected to the same chromatography procedure after dialysis in 10 mM sodium phosphate (pH 7.0). The active fractions (pooled as the DE-2 fraction) were concentrated to 1.2 ml with a Vivaspin 20 ml centrifugal concentrator (Sartorius, Goettingen, Germany). The DE-2 fraction was subjected to ion exchange chromatography using a Mono Q 5/50 GL column (GE Healthcare UK, Buckinghamshire, UK) after dialysis in 10 mM sodium phosphate (pH 7.0). The retained proteins were eluted with a linear NaCl gradient from 0 to 300 mM at a 60 ml/h flow rate using a fast protein liquid chromatography system (GE Healthcare Biosciences). The active fractions were concentrated to 1.0 ml with a Vivaspin 5 ml centrifugal concentrator (Sartorius) and then subjected to gel filtration chromatography using a Superdex 75 column (GE Healthcare Biosciences) at a 60 ml/h flow rate. A 59-kDa protein with high CMCase activity was eluted as a single major peak.

Cellulase Activity Assay

A standard assay for cellulase activity was performed with a reaction mixture containing 0.52 % carboxymethyl cellulose (CMC; medium viscosity, Bio Medics, Orange, CA) in 10-mM sodium phosphate (pH 7.0) at 30 °C. Reduced sugar produced by the reaction was determined using the method described by Park and Johnson [30]. One unit of cellulase is defined as the amount of enzyme that produces reduced sugar equivalent to 1 μmol glucose during 1 min. The temperature dependence of the cellulase was determined over a range of 5 to 60 °C and at pH 7.0; pH dependence was determined at 30 °C in the following reaction buffers: 10 mM HCl-glycine (pH 3–4), 10 mM CH_3COOH–CH_3COONa (pH 4–5), 10 mM NaH_2PO_4–Na_2HPO_4 (pH 5–7), 10 mM Tris–HCl (pH 7–9) and 10 mM NaOH–glycine (pH 9–10). The thermal stability was characterized by incubating the isolated enzyme at different temperatures (5–60 °C) and pH 7.0 for 30 min followed by assaying the activity as described below.

Effects of Metal Ions on the Cellulase Activity

The effects of several metal ions and EDTA on the activity of the purified cellulase (1.5 units/ml) activity was determined in a reaction mixture containing 1 mM (final concentration) of $AgNO_3$, $CaCl_2$, $CdCl_2$, $CoCl_2$, $CuSO_4$, $FeCl_2$, $MgCl_2$, $MnCl_2$, $ZnCl_2$ and EDTA in 10 mM sodium phosphate (pH 7.0) at 30 °C for 60 min, followed by an assay of the activity as described below.

Sodium Dodecyl Sulfate-Polyacrylamide Gel Electrophoresis

Sodium dodecyl sulfate-polyacrylamide gel electrophoresis (SDS-PAGE) was performed with 3 % stacking and 10 % separating gels as described by Laemmli et al. [31]. Samples were treated with equal volumes of sample buffer containing 4 % SDS, 14 % glycerol and 10 % 2-mercaptoethanol at 100 °C for 2 min or equal volumes of sample buffer containing 4 % SDS and 20 % glycerol. Gels were stained with 0.1 % Coomassie brilliant blue R-250 in ethanol, acetic acid and deionized water (DW) (40:10:50), then destained in ethanol, acetic acid, glycerol and DW (200:50:25:725). Molecular weights were estimated using Broad Range Standard Markers (New England Biolabs, Ipswich, MA).

Detection of Cellulase Activity in Polyacrylamide Gels

Cellulase activity in the polyacrylamide gels was detected using a modification of the method reported by Beguin [32]. Briefly, enzyme samples were treated with equal volumes of sample buffer containing 0.25 % SDS and 10 % glycerol. After SDS-PAGE, SDS was removed by washing the gel twice for 30 min in 10 mM sodium phosphate (pH 7.0) and 25 % isopropanol, followed by another two washes for 10 min in 10 mM sodium phosphate (pH 7.5). Agar gels (2 %; thickness 5 mm) containing 0.1 % CMC in 10 mM Tris–HCl (pH 7.5) were prepared. Excess liquid on the polyacrylamide gels was removed by wiping the gel surface, and the gels were then laid on top of the agar gels and incubated at 37 °C for 3 h. After incubation, the agar gels were stained with 0.1 % Congo red solution and then destained with 1 M NaCl. All procedures were performed at 4 °C except for the incubation.

Thin-Layer Chromatography

Thin-layer chromatography (TLC) for analysis of degradation products of cellulose and cello-oligosaccharides was performed on a TLC-60 plate (Merck, Darmstadt, Germany) and a developing solvent consisting of *n*-butanol/acetic acid/water (2:1:1, v/v/v). The sugar developed on the plate was detected by heating at 120 °C for 15 min after being sprayed with 10 % (v/v) sulfuric acid ethanol. Cellulose powders were purchased from Advantec (San Diego, CA) and Merck. The 0.1-ml reaction mixture (0.5 mg of phosphoric acid-swollen cellulose and crystalline cellulose, 5 mM sodium phosphate, pH 7.0, 0.26 U of purified cellulase) was incubated at 30 °C for up to 24 h. The supernatant of the reaction mixture was subjected to TLC. The reaction mixture of the 0.05 ml containing 0.05 mg of G2 (cellobiose)–G6 (cellohexaose), 5 mM sodium phosphate (pH 7.0), and 0.13 U of purified cellulase was incubated at 30 °C for 20 h, then subjected to TLC.

Protein Concentration

Protein concentrations were determined with bovine serum albumin fraction V (Wako Pure Chemicals, Osaka, Japan) as a standard protein using either the biuret method [33] or the method described elsewhere by Lowry et al. [34].

Amino acid sequencing

The N-terminal amino acid sequence of the intact protein was determined by electrophoretically blotting a sample onto a polyvinyldifluoridine membrane (Advantec) and performing the sequencing in an ABI 492HT protein sequencer (Applied Biosystems, Foster city, CA, USA) following the manufacturer's instructions.

RESULTS

Cellulase Isolation

A cellulase was isolated from the acetone-dried powder of the digestive organs of the short-spined sea urchin. Isolation was performed using a combination of ion-exchange chromatography and gel filtration together with an assay for CMCase activity (Fig. 1). The isolation procedures resulted in a target protein that was purified to a 1,156-fold concentration with a 5.4 % final yield and 28.9 U/mg specific activity (Table 1). The purified short-spined sea urchin cellulase showed a single band with an apparent molecular weight of 59 kDa, as determined by SDS-PAGE under reducing conditions (Fig. 2a), and was stained by Congo red (Fig. 2b).

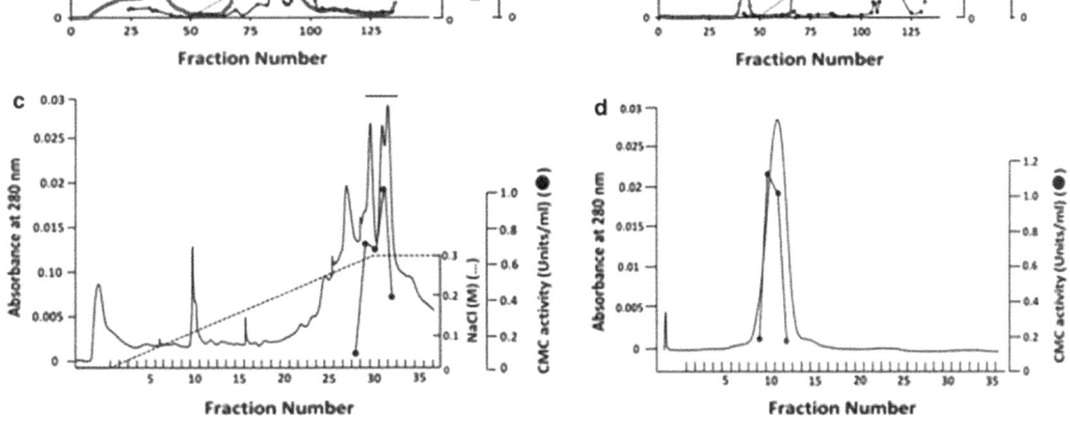

Figure 1: Purification of short-spined sea urchin (*Strogylocentrotus intermedius*) cellulase. a First TOYOPEARL–DEAE 650M chromatography. b Second TOYOPEARL-DEAE 650M chromatography. c MonoQ column chromatography. d Superdex 75 gel filtration. The chromatography conditions are described in the Materials and methods. The fractions indicated by *solid bars* were pooled and designated as the DE-1 fraction (a; fractions eluted at approximately 0.2 M NaCl), DE-2 fraction (b; active fractions obtained after DE-1 fraction was subjected to the same chromatography procedure after dialysis in 10 mM sodium phosphate), MonoQ fraction (c) and Superdex 75 fraction (d).

Table 1: Purification of cellulase from the short-spined sea urchin (*Strongylocentrotus intermedius*)

Purification step[a]	Total protein (mg)	Specific activity (U/mg)	Total activity (U)	Purification yield	
				Fold	%
Crude extract	1,677.8	0.025	42.12	1	100
Fist DEAE fraction	24.9	0.743	18.53	29.7	43.9
Second DEAE fraction	5.08	2.27	11.53	90.8	27.3
Mono Q	0.20	16.9	3.39	676	8.0
Superdex 75	0.08	28.9	2.31	1,156	5.4

[a]The chromatography conditions are described in the Materials and methods

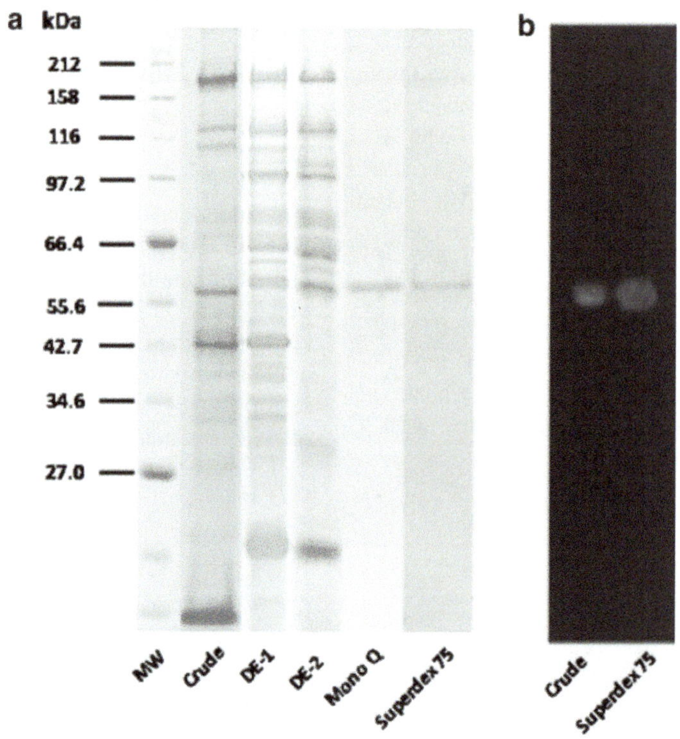

Figure 2: a Sodium dodecyl sulfate-polyacrylamide gel electrophoresis (SDS-PAGE) of short-spined sea urchin cellulase from various purification steps. b Detection of cellulase activity in polyacrylamide gels. Gels were stained by Congo red. *Crude* Crude extract, *DE-1* DE-1 fraction, *DE-2* DE-2 fraction, *MonoQ* MonoQ fraction, *Superdex 75* Superdex 75 fraction,*MW* molecular-weight marker proteins.

Enzymatic Properties of Short-Spined Sea Urchin Cellulase

The basic properties of short-spined sea urchin cellulase were investigated and compared with those of cellulase from the Northern sea urchin [20]. As shown in Fig. 3, the optimal temperature and pH of short-spined cellulase was 30 °C and pH 8.0, respectively. These properties differ from those of the Northern sea urchin cellulase, which are 35 °C and pH 6.3, respectively (Fig. 3a, b). The thermal stability was characterized as the temperature at which activity decreased to 50 % with treatment for 30 min at pH 7.0; this temperature was found to be 32 and 37 °C for short-spined sea urchin and Northern sea urchin cellulase, respectively (Fig. 3c). Table 2 shows the effect of metal ions and EDTA on the activity of the short-spined sea urchin cellulase. The activity of cellulase was strongly inhibited by the addition of Ag^{+} and Cu^{2+}. The TLC analysis showed that short-spined sea urchin cellulase has a higher degradation activity toward phosphoric acid-swollen cellulose than toward crystalline cellulose (Fig. 4a, b). The short-spined sea urchin cellulase degraded the phosphoric acid-swollen cellulose to produce cellotetraose, cellotriose and cellobiose as major products (Fig. 4a). It also degraded the crystalline cellulose to produce cellotetraose, cellootriose and cellobiose; however, the degradation activity on this substrate was very low (Fig. 4b). Figure 4c shows the degradation of the cello-oligosaccharides. The short-spined sea urchin cellulase readily degraded both cellohexaose and cellopentaose to cellotetraose, cellotriose, cellobiose and glucose, and slowly degraded cellotetrose to cellotriose, cellobiose and glucose. However, it scarcely degraded cellotriose and cellobiose.

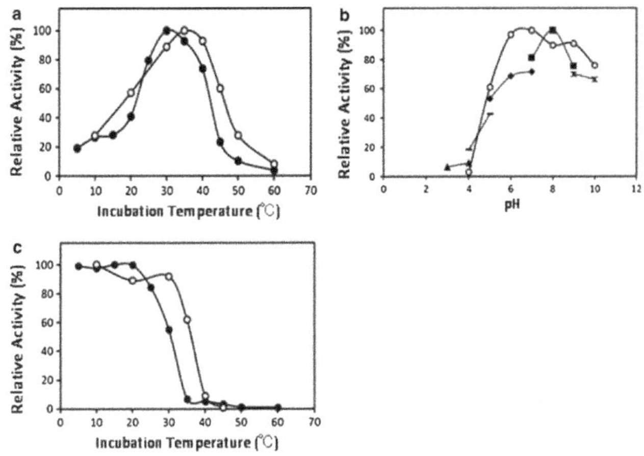

Figure 3: Temperature, pH dependence and thermal stability activities of short-spined sea urchin and Northern sea urchin cellulases. a Activities of the cellulases were measured at various temperatures in a reaction mixture containing 0.52 % carboxymethyl cellulose (CMC) and 10 mM sodium phosphate (pH 7.0). b Activities were measured in the presence of 0.52 % CMC at 30 °C in various reaction buffers at varying pH, including 10 mM HCl-glycine (pH 3–4) (*filled triangles*), 10 mM CH_3COOH–CH_3COONa (pH 4–5) (*solid lines*), 10 mM NaH_2PO_4–Na_2HPO_4 (pH 5–7) (*filled diamonds*), 10 mM Tris–HCl (pH 7–9) (*filled squares*), 10 mM NaOH–glycine (pH 9–10) (*x symbols*) . **c** Thermal stability of the purified enzyme was measured by pre-incubation at various temperatures at a constant pH, and the activity was assayed in a reaction mixture containing 0.52 % CMC, 10 mM sodium phosphate (pH 7.0). *Filled circles* Short-spined sea urchin cellulase; *open circles* Northern sea urchin cellulase (Nishida et al. [20]).

Table 2: N-terminal amino acid sequence of the cellulases

Organisms	N-terminal amino acid sequence of the cellulase
Strongylocentrotus intermedius (short-spined sea urchin)	SDYXDVLHKSILFFE
Strogylocentrotus nudus (Northern sea urchin)	YDYXDVIHKSILFFE

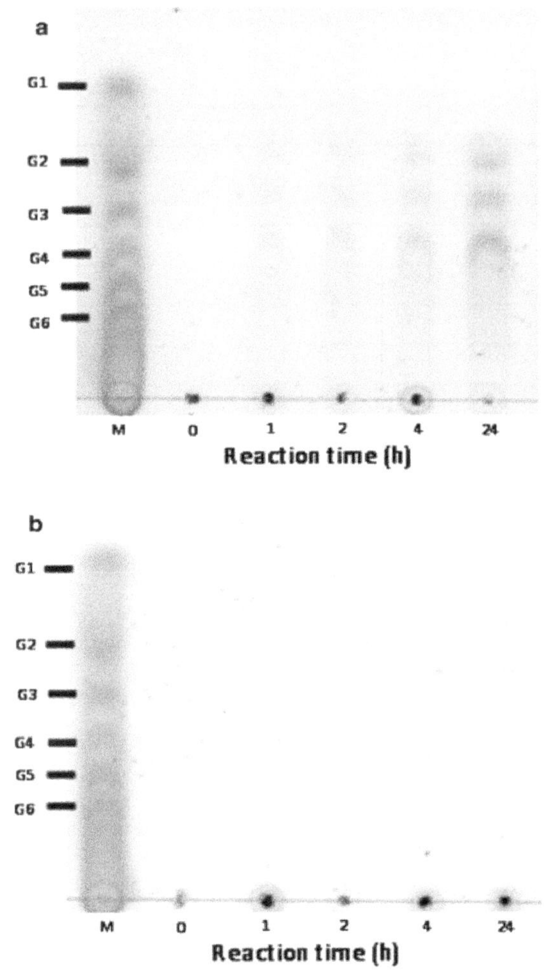

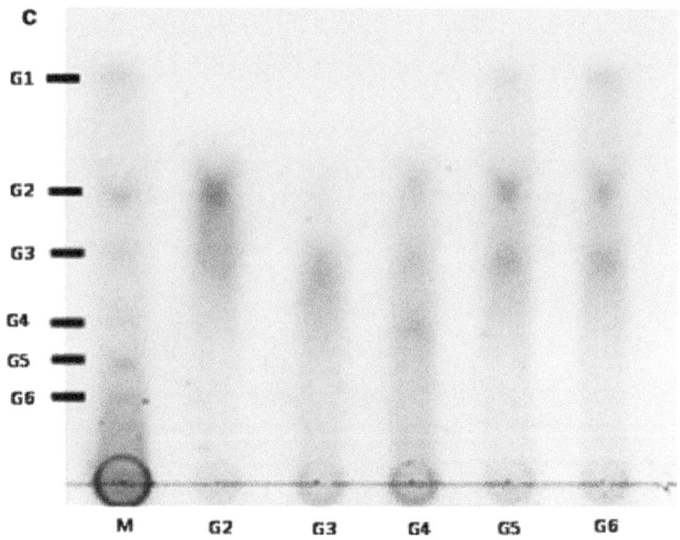

Figure 4: Degradation of phosphoric acid-swollen cellulose, crystalline cellulose and cello-oligosaccharides by the short-spined sea urchin cellulase. The reaction mixture of 0.1 ml containing 0.5 mg of phosphoric acid-swollen cellulose (a) or crystalline cellulose (b), 5 mM sodium phosphate (pH 7.0) and 0.26 U of purified cellulase was incubated at 30 °C for up to 24 h. The supernatant of the reaction mixture was subjected to thin-layer chromatography (TLC). The reaction mixture of 0.05 ml containing 0.05 mg of G2–G6, 5 mM sodium phosphate (pH 7.0) and 0.13 U of purified cellulase was incubated at 30 °C for 20 h, then subjected to TLC (**c**). *G1* Glucose, *G2* cellobiose, *G3* cellotriose, *G4*cellotetraose, *G5* cellopentaose, *G6* cellohexaose, *M* marker sugars.

N-Terminal Amino-Acid Sequence

Using a protein sequencer, we determined the 15 residue N-terminal amino acid sequence of short-spined sea urchin cellulase to be SDYXDVLHKSILFFE. This sequence showed 86 % identity with the Northern sea urchin cellulase sequence at amino acid positions 18–32 (Table 3).

Table 3: Effects of metal ions and EDTA on the activity of cellulase of *Strongylocentrotus intermedius*

Compound	Relative activity (%)
None	100
$AgNO_3$	5.5
$CaCl_2$	97.0
$CdCl_2$	100.5
$CoCl_2$	95.2
$CuSO_4$	43.5
$FeCl_2$	83.6
MgCl	96.1
MnCl	106.7
$ZnCl_2$	101.7
EDTA	93.9

EDTA, Ethylenediaminetetraacetic acid

DISCUSSION

To date, cellulases have been isolated and characterized from various organisms. However, biochemical characterization of the cellulases in marine organisms has been limited. In sea squirts and sea urchins, cellulase primary structures have been analyzed by means of genomic and expressed sequence tag analyses [35, 36]. In more recent studies, cellulases have been isolated from abalone and sea urchin for biochemical characterization and primary structure determination [18, 20]. In our study, short-spined sea urchin cellulase was purified using a combination of ion-exchange chromatography and gel filtration together with CMCase activity assays. The molecular weights of cellulases have been reported from various organisms, such as plants (50–70 kDa [3]), molds (42 kDa [4]), arthropods (40–47 kDa [6, 10, 11]), nematodes (49 kDa [12]), rotifer (62 kDa [37]), mollusks (44–70 kDa [15–18]) and sea urchin (54 kDa [20]), and the range has been found to be quite wide. The molecular weight of purified short-spined sea urchin cellulase was estimated to be 59 kDa, a value which falls well within the range of previously reported cellulases from marine invertebrates, including abalone [18] and Northern sea urchin [20]. Furthermore, the N-terminal amino-acid sequence of purified short-spined sea urchin cellulase showed 86 % identity with that of Northern sea urchin [20], providing good proof that the purified protein is actually the cellulase of short-spined sea urchin.

In this study, we investigated the effects of metal ions and EDTA on the activity of short-spined sea urchin cellulase. To the best of our knowledge, this is the first such report in marine organisms. The activity of the enzyme was strongly inhibited with the addition of Ag^{+} and Cu^{2+}. To date, there are very few published studies on the effect of metal ions on the activity of cellulase. In *Bacillus sphaericus*, the activity of cellulase was inhibited by adding Fe^{3+} and $Hg^{2+}$and stimulated by adding Co^{2+} and K^{+} [38]. Working with various *Bacillus* strains, Mawadza et al. [39] isolated and characterized two types of cellulases (HR66 and CH43). The activity of HR66 and CH43 was stimulated by adding Co^{2+} and inhibited by adding Hg^{2+}. The activity of CH43 was also inhibited by adding Ag^{+}, but that of HR66 was stimulated by adding Ag^{+}. Other metal ions (K^{+}, Mg^{2+}, Cu^{2+}, Ca^{2+}, Ni^{2+}, Zn^{2+}, Fe^{3+}) had no effects on the activity of either HR66 or CH43. Based on their results, the authors suggested that the effects of metal ions on the activity of cellulase was different depending on the type of cellulase and organisms. We found that the short-spined sea urchin cellulase showed a higher degradation activity toward phosphoric acid-swollen cellulose than to crystalline cellulose. Moreover, the degradation pattern was similar to that of the Northern sea urchin [20]. The short-spined sea urchin cellulase scarcely degraded cellotriose and cellobiose. β-Glucosidase activity was reported to be present in crude extracts from digestive organs of the sea urchin [40, 41], leading to the suggestion that in the sea urchin cellulose is first degraded to cellotriose and cellobiose by the cellulase, followed by the possible degradation of these cello-oligosaccharides to glucose by β-glucosidase. However, no data have been published on the isolation and biochemical characterization of β-glucosidase in sea urchin. In addition, in another study a cellulase was purified which raised a specific antibody against the enzyme from abalone; this antibody reacted with three protein bands by western blotting analysis, suggesting that there are several types of cellulase in marine organisms (Ojima, unpublished data). Further studies are being contemplated to determine the degradation mechanism of cellulose in sea urchin.

In echinoderms, less attention has been paid to the relationship between digestive enzymes and digestive physiology. Almost every study performed to date has focused on the biochemical characterization of digestive enzymes at temperatures not considered to be the growth temperature of sea urchin. Consequently, enzyme activity has been assayed at temperatures ranging from 25 to 35 °C in various studies [28, 42–45], but few studies have focused on assaying digestive enzyme activity at the growth temperature of echinoderms. Klinger et al. [46] showed that there was no difference in digestive enzyme activity between polar and subtropical sea urchins at 15 and 25 °C after sea urchins had been reared at 16 or 23 °C. Thus far, there are several reports on the detection of cellulase activity in the digestive organs of sea urchin [13,23–29], and the activity of digestive enzymes from sea urchins has been analyzed using crude extracts from the digestive organs. However, the physiological function of digestive enzymes in the digestive system of sea urchins remains unclear. The isolation and biochemical characterization of the digestive enzymes is an important step towards understanding the digestive physiology of the sea urchin. In echinoderms, the isolation and biochemical characterization of a cellulase from the digestive organs of Northern sea urchin was first reported by Nishida et al. [20]. However, this report also focused on the biochemical characterization of a cellulase.

In our study, we isolated a cellulase from the digestive tract of the short-spined sea urchin and found that the optimum temperature (approximately 30 °C) for short-spined sea urchin cellulase activity was lower than that of Northern sea urchin cellulase (approximately 35 °C) [20]. Moreover, the enzyme activity remained at 20–40 % between 5 and 20 °C, which is the growth temperature for short-spined sea urchin. In our previous study, we isolated and biochemically characterized the proteolytic enzyme subtilase from the digestive tracts of the short-spined sea urchin, which is a major proteolytic enzyme in sea urchin, and determined that the optimum temperature is approximately 50 °C and that enzyme activity remains at 40–50 % between 10 and 20 °C [47]. The thermal stability of the short-spined sea urchin cellulase was also characterized, and although the enzyme activity remained at a high level between 5 and 20 °C, it decreased abruptly above 20 °C and was 50 % at 32 °C. In contrast, in the Northern sea urchin, the enzyme activity remained at a high level between 10 and 30 °C and was 50 % at 37 °C. Gonadal growth and growth rate are significantly suppressed in the short-spined sea urchin when rearing occurs at a water temperature of >21 °C versus <20 °C [48]. Also, the value of assimilation efficiency in the short-spined sea urchin was found to be 55 % at approximately 5 °C after kelp *Laminaria japonica* was supplied as food [49, 50]. In addition, gonadal weight in the short-spined sea urchin increased at approximately 5 °C after the kelp *L. japonica* had been supplied as food (T. Kayaba, personal communication), but it did not change in the Northern sea urchin under the same feeding conditions (K. Ura, unpublished data). In *S. droebachiens* and *Tripneustes esculentus*, both residents of low-temperature water zones, the value of assimilation efficiency is 50–60 % in winter [51, 52]. Other sea urchins (*S. droebachiens*) which live in water zones characterized by the same low temperature as the habitat of the short-spined sea urchin are known to undergo a decrease in survival rate at water temperatures of >19 °C, with the lethal temperature being approximately 24 °C [53]. In comparison, in the Northern sea urchin, which can live at higher water temperatures than the short-spined sea urchin, the value of assimilation efficiency is approximately 50 % at 10–25 °C; however, assimilation efficiency is approximately 10 % at 5 °C after kelp is supplied as food [54]. From these results, we conclude that the activity of the short-spined sea urchin cellulose is weaker at high water temperatures than the Northern sea urchin cellulose.

The pH of the food canal of the sea urchin in an unfed state food is around pH 7.0 [55]. In the short-spined sea urchin, the optimal pH of purified cellulase was found to be pH 8.0; in contrast, that of the Northern sea urchin is pH 7.0, although enzyme activity was high at around pH 6.0–9.0. In our previous study, we isolated and biochemically characterized the proteolytic enzyme subtilase from the digestive tracts of the short-spined sea urchin and found that the optimum pH for subtilase activity is approximately pH 9.0. These data suggest that the digestive enzymes of sea urchin function under weak alkaline conditions. However, to date there have been no published studies on the changes in the pH of the food canal of sea urchin after feeding. Further work is needed to determine the physiological conditions of the food canal and the role of each of the enzymes in the food canal of sea urchin.

In Northern sea urchin, although cellulase activity has been detected in the esophagus, stomach, and intestine (T. Ojima, unpublished data), the localization of the cellulase in the digestive organs is still unclear. In our previous study, we isolated subtilase and raised a specific antibody against the enzyme [47]. Moreover, we found that subtilase is localized in granules in the anterior and posterior stomach and the anterior intestine [56]. From these data, it is likely that the localization of subtilase and cellulase may differ. Further work is needed to determine the localization of cellulase in the digestive organs of sea urchin.

In conclusion, we isolated and biochemically characterized a cellulase from the digestive organs of the short-spined sea urchin. We demonstrated that cellulase from the short-spined sea urchin possesses a physiological function at low temperatures, which is the growth temperature of this organism. Further work is in progress to confirm the localization of cellulase in the digestive organs in order to gain an understanding of the digestive system of sea urchin.

ACKNOWLEDGMENTS

This study was supported by a Grant-in-Aid for Scientific Research from the Ministry of Education, Science and Culture, Japan.

REFERENCES

1. Bhat MK, Bhat S (1997) Cellulose degrading enzymes and their potential industrial applications. Biotech Adv 15:583–620
2. Tomme P, Warren RAJ, Gilkes NR (1995) Cellulose hydrolysis by bacteria and fungi. Adv Microbiol Physiol 37:1–81
3. Brummel DA, Lashbrook CC, Bennett AB (1994) Plant endo-1,4-β-d-glucanase. ACS Symp Ser 566:100–129
4. Blume JE, Ennis HL (1991) A *Dictyostelium discoideum* cellulase is a member of a spore germination-specific gene family. J Biol Chem 266:15432–15437
5. Moriya S, Ohkuma M, Kudo T (1998) Phylogenetic position of symbiotic protest*Dinemympha exlis* in the hindgut of the termite *Reticulitermes speratus* inferred from the protein phylogeny of elongation factor 1α. Gene 210:221–227
6. Watanabe H, Nakamura M, Tokuda G, Yamaoka I, Scrivener AM, Noda H (1997) Site of secretion and properties of endogenous endo-β-1,4-glucanase components from*Reticuliterms speratus* (Kolbe), a Japanese subterranean termite. Insect Biochem Mol Biol 27:305–313
7. Watanabe H, Noda H, Tokuda G, Lo N (1998) A cellulase gene of termite origin. Nature 394:330–331
8. Tokuda G, Watanabe H, Matsumoto T, Noda H (1997) Cellulose digestion in the wood-eating higher termite, *Nasutitermes takasagonesis* (Shiraki): distribution of cellulases and properties of endo-β-1,4-glucanase. Zool Sci 14:83–93
9. Tokuda G, Lo N, Watanabe M, Slaytor M, Matsumoto T, Noda H (1999) Metazoan cellulase genes from termites: intron/exon structure and sites of expression. Biochem Biophys Acta 1447:146–159
10. Xue XM, Anderson AJ, Richardson NA, Anderson AJ, Xue GP, Mather PB (1999) Characterization of cellulase activity in the digestive system of the redclaw cray-fish (*Cherax quadricarinatus*). Aquaculture 180:373–386
11. Sugimura M, Watanabe H, Lo N, Saito H (2003) Purification characterization, cDNA cloning and nucleotide sequence of a cellulase from the yellow-spotted longicorn beetle,*Psacothea hilaris*. Eur J Biochem 270:3455–3460
12. Smant G, Stokkermans JPWG, Yan Y, De Boer JM, Baum TJ, Wang X, Hussey RS, Gommers FJ, Henrissat B, Davis EL, Helder J, Schots A, Bakker J (1998) Endogenous cellulases in animals: isolation of β-1,4-endoglucanases genes from two species of plant-parasitic cyst nematodes. Proc Natl Acad Sci USA 95:4906–4911
13. Yokoe Y, Yasumasu I (1964) The distribution of cellulase in invertebrates. Comp Biochem Physiol 13:323–338
14. Marshall JJ (1973) Purification of β-1,4-glucan hydrolase (cellulase) from the snail, *Helix pomatia*. Comp Biochem Physiol 44B:981–988
15. Marshall JJ, Grand RJA (1976) Characterization of a β-1,4-glucan hydrolase from the snail, *Helix pomatia*. Comp Biochem Physiol 53B:231–237
16. Anzai H, Nisizawa K, Matsuda K (1984) Purification and characterization of a cellulase from *Dolabella auricularia*. J Biochem 96:1381–1390
17. Xu B, Hellman U, Ersson B, Jason J-C (2000) Purification, characterization and amino-acid sequence analysis of a thermostable, low molecular mass endo-β-1,4-glucanase from blue mussel, *Mytilus edulis*. Eur J Biochem 267:4970–4977
18. Suzuki K, Ojima T, Nishita K (2003) Purification and cDNA cloning of a cellulase from abalone *Haliotis discus hannai*. Eur J Biochem 270:771–778
19. Wang J, Ding M, Li Y, Chen Q, Xu G, Zhao F (2003) Isolation of a multifunctional enzogeneous cellulase gene from mollusk, *Ampullaria crossean*. Acta Biochem Biophys Sin 35:941–946
20. Nishida Y, Suzuki K, Kumagai Y, Tanaka H, Inoue A, Ojima T (2007) Isolation and primary structure of a cellulase from the Japanese sea urchin *Strongylocentrotus nudus*. Biochimie 89:1002–1011
21. Kelly MS (2004) Echinoderms. In: Heinzeller T, Nebeisick JH (eds) Munchen. Taylor and Francis, London, pp 283–289
22. Lawrence JM (2007) Biology and ecology. In: Lawrence JM (ed) Edible sea urchins. Elsevier Science, Amsterdam, pp 1–9
23. Hultin E, Wanntorp I (1966) Viscosimetric determination of cellulase activity in the intestine of the sea urchin: reaction mechanism and equilibrium constant for

cellulase stabilization with calcium. Acta Chem Scand 20:2667–2677

24. Elyakova LA (1972) Distribution of cellulases and chitinases in marine invertebrates. Comp Biochem Physiol 43B:67–70
25. Elyakova LA, Shevchenko NM, Avaeva SM (1981) A comparative study of carbohydrase activities in marine invertebrates. Comp Biochem Physiol 69B:905–908
26. Suzuki M, Nikuchi R, Onishi T (1984) The polysaccharide degradation activity in digestive tract of sea urchin *Strongylocentrotus nudus*. Bull Jpn Soc Sci Fish 50:1255–1260
27. Yamaguchi K, Araki I, Aoki T, Tseng C-H, Kitamikodo M (1989) Algal cell wall-degrading enzymes from viscera of marine animals. Nippon Suisan Gakkaishi 55:105–110
28. Obrietan K, Drinkwine M, Williams DC (1991) Amylase, cellulase and protease activities in surface and gut tissues of *Dendraster excentricus*, *Pisaster ochraceus* and*Strongylocentrotus droebachiensis* (Echinodermata). Mar Biol 109:53–57
29. Gomez-Pinchetti JL, Garcia-Reina G (1993) Enzymes from marine phycophages that degrade cell walls of seaweeds. Mar Biol 116:553–558
30. Park J, Johnson MJ (1949) A submicrodetermination of glucose. Biol Chem 181:149–151
31. Laemmli UK (1970) Cleavage of structural proteins during the assembly of the head of bacteriophage T4. Nature 227:680–685
32. Beguin P (1983) Detection of cellulase activity in polyacrylamide gels using Congo red-stained agar replicas. Anal Biochem 131:333–336
33. Gornall AG, Bardawill CJ, David MM (1949) Determination of serum proteins by means of the biuret reaction. J Biol Chem 177:751–766
34. Lowry OH, Rosebrough NJ, Farr AL, Randall RJ (1951) Protein measurement with the Folin phenol reagent. J Biol Chem 193:265–275
35. Zhu XD, Mahairas G, Illies M, Cameron RA, Davidson EH, Ettensohn CA (2001) A large-scale analysis of mRNA expressed by primary mesenchyme cells of the sea urchin embryo. Development 128:2615–2627
36. Dehal P, Satou Y, Campbell RK, Chapman J, Degnan B et al (2002) The draft genome of*Ciona intestinalis*: insights into chordate and vertebrate origins. Science 298:2157–2167
37. Chun CZ, Hur SB, Kim YT (1997) Purification and characterization of an endoglucanase from the marine rotifer, *Brachionus plicatilis*. Mol Biol Int 43:241–249
38. Singht J, Batra N, Sobti RC (2004) Purification and characterization of alkaline cellulase produced by a novel isolate, *Bacillus sphaericus* JS1. J Microbiol Biotechnol 31:51–56
39. Mawadza C, Hatti-Kaul R, Zvauya R, Mattiasson B (2000) Purification and characterization of cellulase produced by two *Bacillus* strains. J Biotechnol 83:177–187
40. Molodtsov NV, Vafina MG (1972) The distribution of β-*N*-acetylglucosaminidase in marine invertebrates. Comp Biochem Physiol 41B:113–120
41. Molodtsov NV, Vafina MG, Kim A, Sundukova EV, Artyukov AA, Bliknov YG (1974) Glycosidases of marine invertebrates from Poset Bay, Sea of Japan. Comp Biochem Physiol 48B:463–470
42. Klinger TS (1984) Activities and kinetics of digestive α- and β-glucosidase and β-galactosidase of five species of echinoids (Echinodermata). Comp Biochem Physiol 78A:597–600
43. Holzman TF, Russo SF, Williams DC (1985) Effects of feeding and starvation on proteolytic and tryptic activities in pyloric caecal tissues and duct fluids of the sea star *Pisaster ochraceus*. Mar Biol 90:55–59
44. Farrand AL, Williams DC (1988) Isolation, purification and partial characterization of four digestive proteases from the purple seastar *Pisaster ochraceus*. Mar Biol 97:231–236
45. Nakatani N, Kobayashi I (1996) Enzymatic properties of a-amylase from sea urchin,*Strongylocantrotus nudus*. Comp Biochem Physiol 113B:383–386
46. Klinger TS, Hsierh HL, Pangallo RA, Chen CP, Lawrence JM (1986) The effect of temperature on feeding, digestion, and absorption of *Lyechinus variegates* (Lamarck) (Echinodermata: Echinoidea). Physiol Zool 59:332–336
47. Ura K, Tanaka E, Todo T, Gotoh T, Shimizu M, Ojima T, Takagi Y (2009) Purification of subtilase from short-spined sea urchin *Strongylocentrotus intermedius*. Bull Fish Sci Hokkaido Univ 58:21–28
48. Chang Y-Q, Lawrence JM, Cao X-B, Lawrence AL (2005) Food consumption, absorption, assimilation and growth of the sea urchin *Strongylocentrotus intermedius* fed a prepared food and the alga *Laminaria japonica*. J World Aquac Soc 36:68–75
49. Fuji A (1962) Studies on the biology of the sea urchin, V. Food consumption of*Strogylocentrotus intermedius*. Jpn J Ecol 12:181–186
50. Fuji A (1967) Ecological studies on the growth and food consumption of Japanese common littoral sea urchin, *Strogylocentrotus intermedius* (A.Agassiz). Mem Fac Fish Hokkaido Univ 12:83–160
51. Miller RJ, Mann KH (1973) Ecological energetics of the seaweed zone in marine bay on the Atlantic coast of Canada. III. Energy transformations by sea urchins. Mar Biol 18:99–114
52. Larson BR, Vadas RL, Keser M (1980) Feeding and nutritional ecology of the sea urchin*Strongylocentrotus droebachiensis* in Maine USA. Mar Biol 59:49–62
53. Pearce CM, Williams SW, Yuan F, Castell JD, Robinson SMC (2005) Effect of temperature on somatic growth and survivorship of early post-settled green sea urchin,*Strongylocentrotus droebachiensis* (Muller). Aquac Res 36:600–609
54. Agatsuma Y (2007) Ecology of *Strogylocentrotus nudus*. In: Lawrence JM (ed) Edible sea urchins. Elsevier Science, Amsterdam, pp 443–457

55. Stott FC (1995) The food canal of the sea-urchin *Echinus esculentus* L. and its functions. Proc Zool Soc Lond 125:63–86

56. Ura K, Hasegawa S, Tanaka E, Gotoh T, Ojima T, Takagi Y (2012) Immunohistochemical analysis of the proteolytic enzyme subtilase in the digestive organs of the sea urchin Strongylocentrotus intermedius. Anat Rec 295:73–77

CHAPTER 15

PARACENTRIN 1, A SYNTHETIC ANTIMICROBIAL PEPTIDE FROM THE SEA-URCHIN PARACENTROTUS LIVIDUS,INTERFERES WITH STAPHYLOCOCCAL AND PSEUDOMONAS AERUGINOSA BIOFILM FORMATION

Domenico Schillaci, Maria Grazia Cusimano, Angelo Spinello, Giampaolo Barone, Debora Russo, Maria Vitale, Daniela Parrinello and Vincenzo Arizza

Keywords: AMP (Antimicrobial peptides); Biofilm; Staphylococci; Pseudomonas aeruginosa; Paracentrotus lividus

INTRODUCTION

Many natural antimicrobial peptides show a high specificity for prokaryotes and a low toxicity for eukaryotic cells, and for their mode of action the development of resistance by pathogenic bacteria is considered unlikely (Hancock and Rozek [2002]). At present, there has been an increase of interest in these molecules as potential new antimicrobials (Bax et al. [2000]; Mor [2000]).

In human medicine, chronic and persistent forms of some infectious diseases depend on the ability of pathogenic bacteria to develop bacterial communities called biofilms. Opportunistic pathogens, such as staphylococcal strains and *Pseudomonas aeruginosa* show a great ability to produce biofilms that preventing infected wounds to heal, render the treatment extremely challenging. In veterinary medicine, biofilms are believed to be involved in many diseases such as pneumonia, liver abscesses, enteritis, wound infections and mastitis, which is one of the most common diseases in dairy cattle (Clutterbuck et al. [2007]). *Staphylococcus aureus,* a major pathogen of mastitis has good in vitro antimicrobial susceptibility, but the therapy used to treat the affected animals is often disappointing and results in chronic infections due to the growth of bacteria as biofilms (Melchior et al. [2006]). The biofilm of *P. aeruginosa* is a severe complicating factor in bovine mastitis, which is often associated with contaminated udder washing water or contaminated intramammary dry-cow preparations (Melchior et al. [2009]).

The treatment of biofilm-associated infections is complicated because microbial biofilms are typically highly resistant to conventional antibiotics (Gilbert et al. [2002]). The discovery of anti-infective agents active not only against planktonic microorganisms but also against biofilms represents an important goal for an effective control of infections (Projan and Youngman [2002]).

Citation: Schillaci et al.: Paracentrin 1, a synthetic antimicrobial peptide from the sea-urchin Paracentrotus lividus, interferes with staphylococcal and Pseudomonas aeruginosa biofilm formation. AMB Express 2014 4:78. doi:10.1186/s13568-014-0078-z.

The antimicrobial defence system of marine invertebrates is an interesting source of new anti-infective agents (Arizza [2013]). We focused, particularly, on the effector cells of the echinoderm immune system, the coelomocytes. In a recent study, the antimicrobial activity of a 5-kDa peptide fraction from coelomocyte cytosol (5-CC) of the *Paracentrotus lividus,* the sea-urchin from Mediterranean sea, was demonstrated in relation to a group of important human pathogens. The anti-biofilm activity of 5-CC was shown in *S. epidermidis* 1457, a clinical strain isolated from an infected central venous catheter, against reference staphylococcal biofilms and against *Candida albicans* and *Candida tropicalis* (Schillaci et al. [2010]; Schillaci et al. [2012]). We showed the presence of three principal peptides, in the 5-CC content, whose molecular weights were respectively 1251.7, 2088.1, and 2292.2. These peptides are the (9–19), (12–31), (24–41) fragments of a β-thymosin of *P. lividus*. We focused particularly on the smallest peptide, that we called paracentrin 1, 11 amino acids in length, because showed the common chemical-physical characteristics of an antimicrobial peptide (Wang and Wang [2004]).

The present study was aimed at evaluating the antibacterial and anti-biofilm activity of a chemically synthesized paracentrin 1 (SP1) against a group of staphylococcal reference strains and isolates and against *P. aeruginos* a ATCC 15442.

MATERIALS AND METHODS

Synthetic Peptide

The SP1 was purchased from CASLO, Lyngby, Denmark utilizing the following sequence EVASFDKSKLK derived from ESI-MS analysis. The peptide was synthesized using Fmoc solid phase technology and the peptide content and purity was determined by high performance liquid chromatography (HPLC) and mass spectrometry (MS) analysis.

Net molecular charge was calculated by using the following formula (1) and the pI was calculated by using a web tool (http://isoelectric.ovh.org/).

$$z = \sum_i N_i \frac{10^{pKa_i}}{10^{pH}} - \sum_j N_j \frac{10^{pH}}{10^{pH} + 10^{pKa_j}} \quad (1)$$

Hydrophobic mean value was calculated by using the Liu-Deber hydrophobicity scale ([Liu and Deber 1998]). Secondary structure was evaluated utilizing the algorithm of Wang et al. ([2009]), present in antimicrobial peptide calculator web site (http://aps.unmc.edu/AP/prediction/prediction_main.php). The helical wheel projection was performed using the Gautier et al. ([2008]) algorithm present in the HeliQuest site (http://heliquest.ipmc.cnrs.fr) with a window size of 11 residues.

Microbial Strains

The staphylococcal reference strains used were: *Staphylococcus aureus* ATCC 29213, *Staphylococcus aureus* ATCC 25923, *Staphylococcus aureus* ATCC 6538, and *Staphylococcus epidermidis* RP62A, known for its ability to form a biofilm (Schumacher-Perdreau et al. [1994]). Four staphylococcal isolates from the Istituto Zooprofilattico Sperimentale, Sicily (IZS) bacterial collection including strain 657 isolated from a milk sample from an individual sheep affected by mastitis, strain 688 and 700 and strain 702 isolated from bulk milk samples from different sheep flocks. The isolates were selected on blood agar plates and on Mannitol Salt Agar (Difco, Sparks, MD). The colonies were typed by API Staph strip (bio-Mérieux) and tube coagulase test was performed (Canicatti and Roch [1989]).

Pseudomonas aeruginosa ATCC 15442, the reference strain in official tests for antibacterial evaluation in vitro (UNI EN European Standard), was also used in this study.

Minimum Inhibitory Concentrations (MIC)

MICs were determined by a micro-method described previously (Schillaci et al. [2005]). Briefly, a series of solutions of SP1 was prepared with a range of concentrations from 25 to 0.07 mg/ml(obtained by two-fold serial dilution). The serial dilutions were made in Tryptic Soy Broth (TSB) (Merck) in a 96-wells plate; to each well was added 10 μl of a bacterial suspension from a 24 h culture containing ~1 × 10^6 CFU/ml. The plate was incubated at 37°C for 24 h; after this time, the MICs were determined by a microplate reader (ELX 800, Bio-Tek Instruments), and defined as the lowest concentration of compound whose O.D., read at 570 nm, was comparable with the negative control wells (broth only, without inoculum).

Evaluation of Biofilm Formation

All the bacterial reference strains were tested for their ability to form biofilms. Briefly, bacteria were grown in Tryptic Soy Broth (TSB, Sigma) containing 2% glucose overnight at 37°C in a shaking bath and then diluted 1:200 to a suspension with optical density (OD) of about 0.040 at 570 nm corresponding to ~10^6 CFU/ml. Polystyrene 24-well tissue culture plates were filled with 1 ml of diluted suspension and incubated for 24-hours at 37°C. Then, the wells were washed three times with 1 ml of sterile phosphate-buffered saline (PBS) and stained with 1 ml of safranin 0.1% v/v for 1 min. The excess stain was removed by placing the plates under running tap water. Plates were dried overnight in inverted position at 37°C. Safranin stained adherent bacteria in each well were re-dissolved to homogeneity in 1 ml of 30% v/v glacial acetic acid, and the OD was read at 492 nm. Each assay was performed in triplicate and repeated at least twice.

Biofilm Prevention Assay

Procedure described above was used to evaluate the activity of SP1 in preventing biofilm formation. Polystyrene 24-well tissue culture plates were filled with 1 ml of diluted bacterial suspension (OD of about 0.040 at 570 nm), obtained and diluted as previously seen, and sub-MIC concentrations, ranging from 6.2 to 0.7 mg/ml, of SP1 were directly added to the bacterial suspension at time zero and incubated at 37°C for 24 hours. After that time the wells were washed and stained with safranin as seen in biofilm forming assay. By comparing the average optical density of the growth in control wells with that in the sample wells, the following formula was used to calculate the percentages of inhibition for each concentration of the sample:

$$\text{Inhibition}\ (\%) = \frac{\text{OD}_{570}\ \text{growth control} - \text{OD}_{570}\ \text{sample}}{\text{OD}_{570}\ \text{growth control}} \times 100 \quad (2)$$

Each assay was performed in triplicate and assays were repeated at least twice.

Scanning Electron Microscopy

The effects of SP1 on formation of bacterial biofilm were morphologically evaluated by scanning electron microscopy (SEM). Glass slides in the bottom of a polystyrene 24-well tissue culture plates, were filled with 1 ml of a *S. epidermidis* RP62A suspension, obtained and diluted as previously seen, and a sub-MIC concentration of 3.1 mg/ml of SP1 was directly added at time zero and incubated at 37°C for 24 hours. After that time, the glass slides were gently washed four times with PBS to remove non-adherent bacteria and fixed with 2.5% glutaraldehyde-2% paraformaldehyde in 0.1 M cacodilate buffer (pH 7.4) for 30 min at 4°C. The bacterial preparation were washed with phosphate saline buffer PBS and post-fixed in osmium tetroxide 1% for 30 min at 4°C, followed by an ethanol dehydration series: 15 minutes in 50:50 ethanol: H_2O, 15 minutes in 75:25 ethanol: H_2O, 15 minutes in 95:5 ethanol: H_2O, and 30 minutes in 100% ethanol than a critical point drying procedure was followed, and the preparations were mounted on aluminium stubs, and gold coated in a sputter coater. Imaging was conducted with a LEO 420 scanning electron microscope as previously reported (Arizza et al. [2011]).

Haemolytic Assay

The assay was performed as described (Arizza et al. [2013]). In brief, freshly collected rabbit erythrocytes (RE) with heparin, kindly provided by the "Zooprophylaxis Institute of Sicily" (Palermo, Italy), were washed (400 g for 10 min at 4°C) to remove the buffy coat, and the erythrocytes obtained were washed three times with phosphate-buffered saline (PBS: 6 mM KH_2PO_4; 30 mM Na_2HPO_4; 0.11 M NaCl; pH 7.4) and suspended in 10 ml PBS to obtain an 80 × 10^6 cells/ml suspension. Aliquots of 200 µL of RE suspension were mixed with 200 µl of SP1 concentrations (1.5, 3.2, 6.2 and 50 mg/ml) prepared in PBS, two-fold sequentially diluted v/v with a PBS. After an 1 h incubation at 37°C, the reaction mixture was centrifuged at 800 g for 15 min at 4°C to remove debris and residual erythrocytes. The O.D. of haemoglobin release was measured spectrophotometrically at a 541 nm wavelength. Spontaneous haemoglobin release (0%) was estimated incubating the RE with PBS while the complete haemolysis (100%) was assessed incubating the erythrocytes in a solution of 0.1% Triton-X 100 in distilled water and the haemolysis percentage was calculated according to the following equation:

$$\text{Hemolysis}\ (\%) = \frac{\text{OD}_{541}\text{Hb release in the reaction mixture} - \text{OD}_{541}\text{spontaneous Hb release}}{\text{OD}_{541}\ \text{complete Hb release}} \times 100 \quad (3)$$

Molecular Dynamics Simulations

The molecular folding of the peptide in aqueous solution was investigated *in silico* by molecular dynamics (MD) simulations, following recently reported procedures (Lauria et al. [2014]; Lentini et al. [2014]). In details, a 400 ns of MD simulation was carried out at 300 K, in the explicit water solvent and in the presence of 150 mM Na^+ and Cl^- counterions, using the Amber99SB-ILDN force field (Lindorff-Larsen et al. [2010]) implemented in the GROMACS 4.6.5 software package (Pronk et al. [2013]).

RESULTS

Molecular Dynamics of SP1

SP1 is an 11-residue-long cationic peptide mainly enriched by residues such as lysine, with a pI of 10.72 and a net charge of +1 at pH 7.0.

To investigate SP1 in physiological conditions, we performed a MD simulation using the model represented in Figure 1. The *in silico* study provided a comprehensive picture of the dynamic equilibrium existing among the possible backbone conformations of the small oligopeptide. In fact, the root mean square deviations (RMSD) in Figure 2 show that the peptide conformation continuously and consistently changes during the 400 ns of simulation. Essentially all possible backbone conformations are assumed. However, interestingly, between 150 and 250 ns the structure is more stable and preserved, before becoming again flexible and random. To analyse this time region quantitatively, we have reported the Ramachandran plot in Figure 3 of snapshots sampled every 10 ns. This plot shows that in this time range the different residues assume preferential local conformations. In particular, residues 2, 3, 4 and 8 are in the beta sheet conformation, residues 5, 6 and 9 are in the left-handed helix conformation, residue 7 is in the alpha helix conformation, while residue 10 is in the beta sheet conformation and only sporadically in the alpha helix conformation. The corresponding structure is depicted in Figure4. The results obtained allow us to confirm that polar and apolar residues are not segregated on opposing surfaces through the long axis of the oligopeptide. In this peculiar conformation, the one that most frequently occurs in the investigated simulation time, the peptide possesses a hydrophobic non-amphipathic core constituted by V^2, A^3, S^4, F^5 with a hydrophobic region placed between E^1 and D^6. These features confer to SP1 properties similar to the AMPs with hydrophobic core.

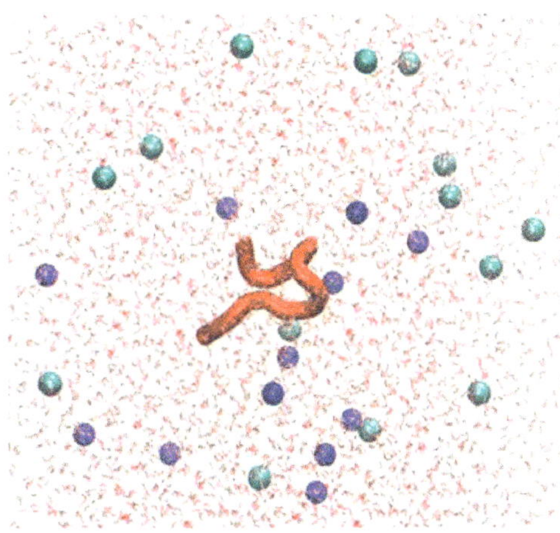

Figure 1: Example of MD simulation box, showing the SP1 molecule, the solution counterions, Na$^+$(in blue) and Cl$^-$(in cyan), and the explicit water molecules (in red).

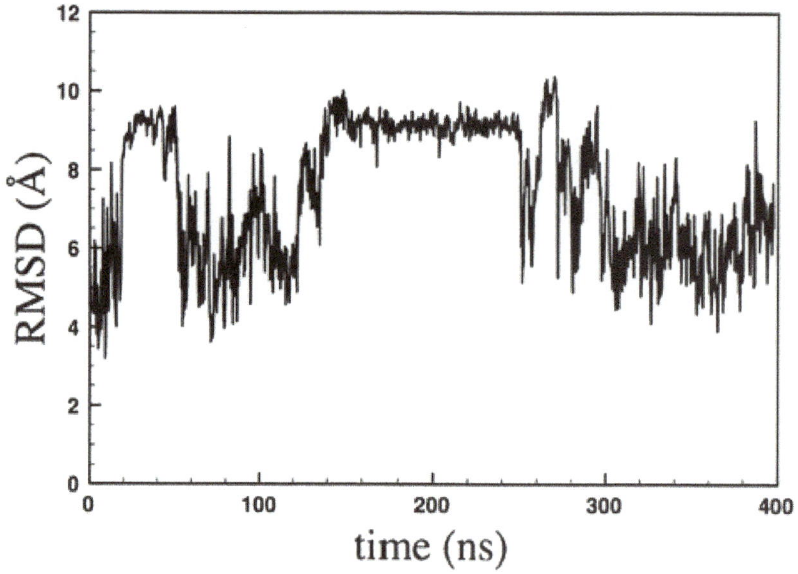

Figure 2: Plot of the RMSD obtained for SP1 up to 400 ns of MD simulation.

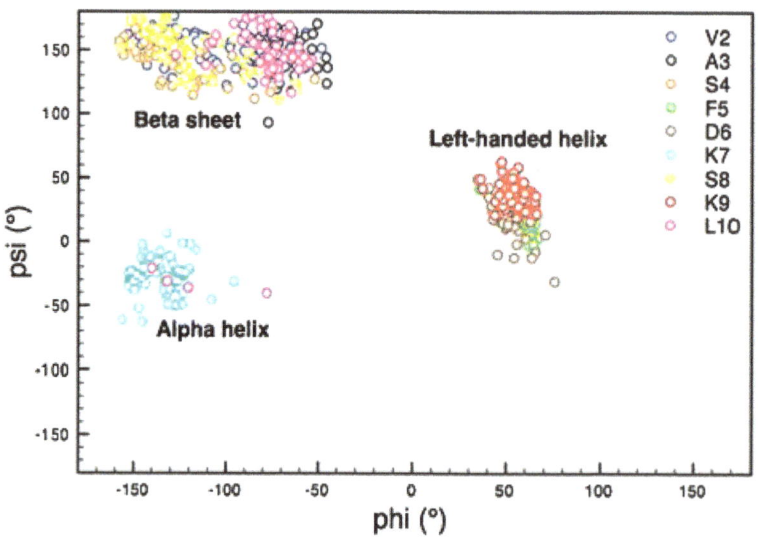

Figure 3: Ramachandran plot showing the values of psi and phi angles assumed by residues 2–10 of SP1 between 150 and 250 ns of the MD simulation.

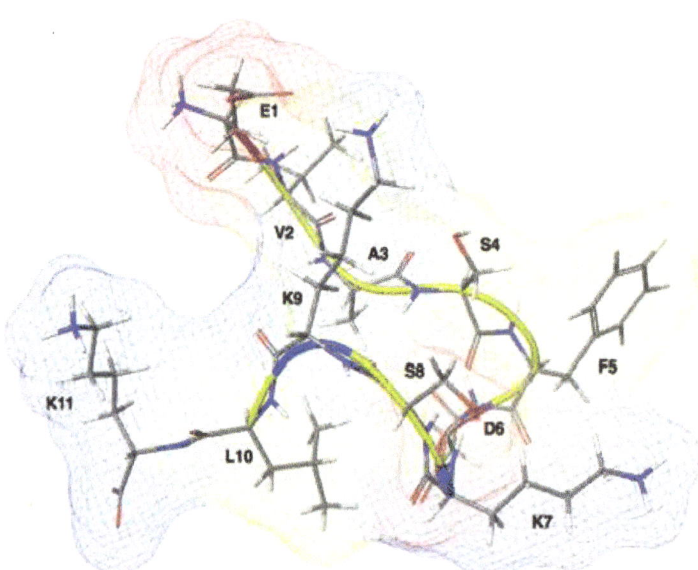

Figure 4: Molecular representation of SP1 at about 200 ns of MD showing the non amphipathic structure of the peptide. The potential surface is superimposed. Color code: acidic residues in red, basic residues in blue, and hydrophobic residues in yellow.

Antibacterial Activity of SP1

SP1 was tested at concentrations ranging from 25 to 0.07 mg/ml against a group of Gram positive and Gram negative reference strains. The antibacterial activity of SP1, expressed as minimum inhibitory concentrations (MICs) against planktonic cells of staphylococcal reference strains and isolates from veterinary origin and *P. aeruginosa* is listed in Table 1. All bacterial strains showed susceptibility to the antibacterial action of SP1. In particular the strain of *S. epidermidis* RP62A resulted to be the most susceptible, indeed, the MIC value was 6.2 mg/ml, while others attested to a value of 12.5 mg/ml. SP1 was active against all isolates of animal origin strains and in particular showed the highest activity against the strain #657 with a value of 6.2 mg/ml (Table 1).

Table 1: MIC values of SP1 tested against bacterial strains

Bacterial strains	MIC (mg/ml)
S. aureus ATCC 25923	12.5
S. aureus ATCC 29213	12.5
S. aureus ATCC 6538	12.5
S. epidermidis RP62A	6.2
S. aureus 100	12.5
S. aureus 657	6.2
S. aureus 700	12.5
S. aureus 702	12.5
P. aeruginosa ATCC15442	12.5

MIC values for each strain are the means of three different assays performed in duplicate

Interference with Biofilm Formation

The interference with biofilm formation of SP1 against staphylococcal reference strains as *S. aureus* 25923, *S. aureus* 29213, *S. aureus* 6538, *S. epidermidis* RP62A and *P. aeruginosa*15442 was observed. The inhibition was very evident to the highest concentrations of SP1 at 6.2 mg/ml, when the values, for all strains, reached approximately 80%. At the lowest concentrations of SP1 the degree of inhibition is reduced by following a dose dependence. *P. aeruginosa* strain 15442 was the most sensitive inhibitory activity of SP1, in fact, at a concentration of 3.1 mg/ml, a value of about 73% inhibition was observed, while for the other strains the inhibition values were of around 50% (Figure 5).

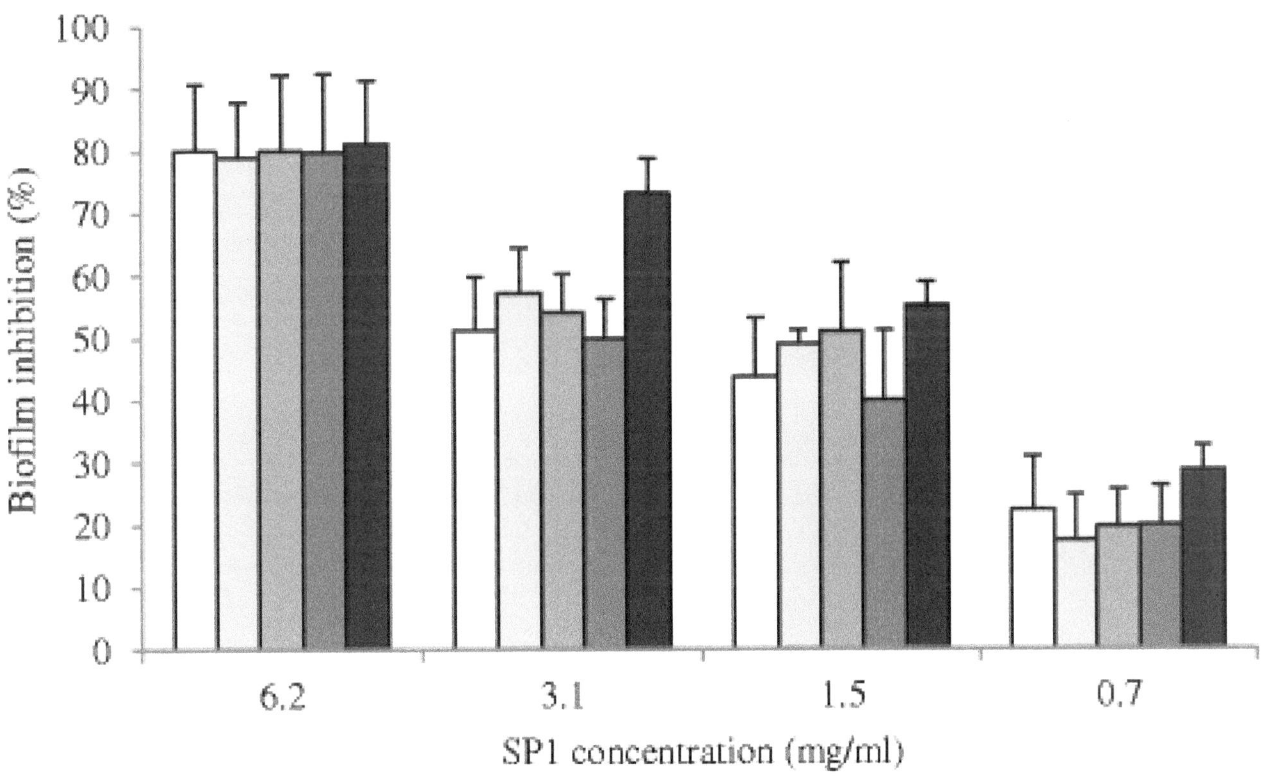

Figure 5: Interference with biofilm formation of SP1 against reference staphylococcal and *P. aeruginosa* strains. The percentage of inhibition were evaluated comparing the samples with not-treated 24 h old biofilms and staining with safranin. Bacterial strains ( ) *S. aureus* 25923, ( ) *S. aureus* 29213, ( ) *S. aureus* 6538, ( ) *S. epidermidis*RP62A, ( ) *P. aeruginosa* 15442. Data for each strain are the mean of three distinct experiments ±S.D.

SEM analysis was carried out to visualize the detailed architecture of biofilm and the antibiofilm effects of SP1 on bacterial biofilm of *S. epidermidis* RP62A (Figure 6). The control biofilm after 24 h was composed by multilayered conglomerated bacterial cells clusters that produced a dense biofilm matrix (Figure 6A). Sub MIC concentration (3.1 mg/ml) of SP1 was able to inhibit the formation of a multistratified structure, in fact in the sample treated with SP1 we found few monostratified bacterial cell clusters respect to the growth control (Figure 6B).

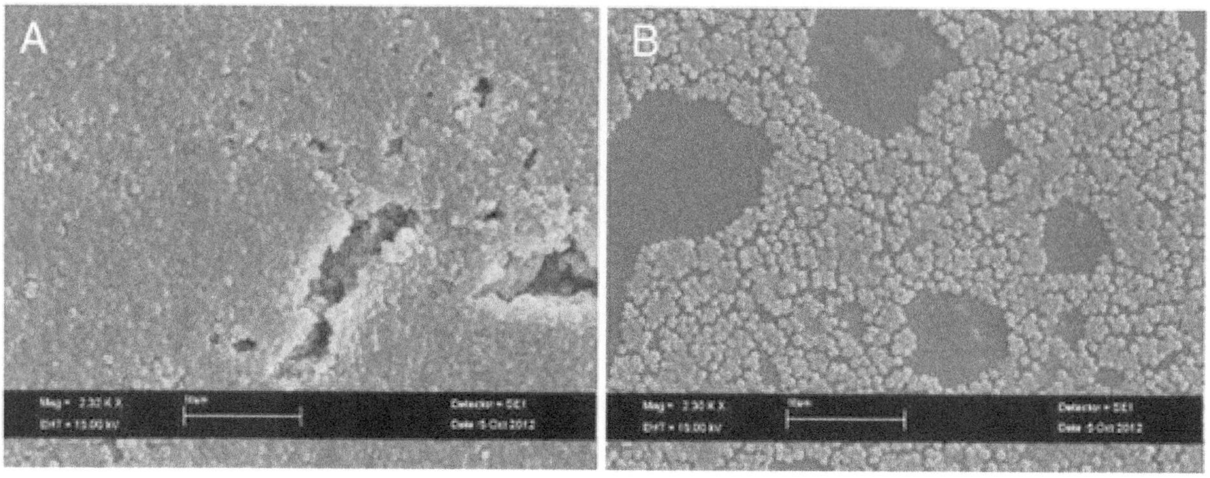

Figure 6: SEM micrography showing the effect of sub-MIC concentration of SP1 on *S. epidermidis* RP62A biofilm formation. A) Growth control (not treated with SP1);B) sample treated with a sub-MIC concentration (3.1 mg/ml) of SP1.

Haemolytic Assay

The haemolytic activity of antimicrobial peptides against mammalian erythrocytes is often used as a preliminary evaluation of their selective toxicity and of the interactions of the cationic peptides with negatively charged membranes (Fischer et al. [2003]; Wu et al. [2010]) The performed hemolytic experiment to evaluate the interaction and the potential toxicity of the peptides SP1 did not show a measurable toxic effect against RBC (~1%) at MIC concentrations. A slight haemolytic activity, about 11.8%, was evident only at the highest concentration (50 mg/ml) (Figure 7).

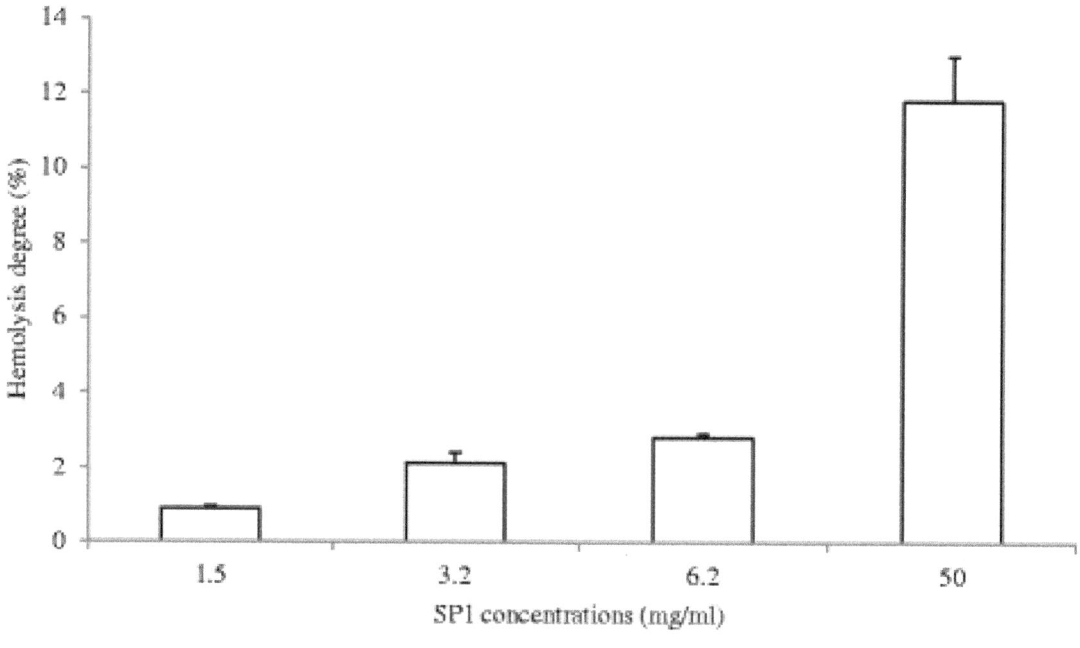

Figure 7: Hemolytic activity of SP1 peptides from *Paracentrotus lividus* hemocytes against rabbit blood cells at different concentrations: 1.5 mg/ml; 3.2 mg/ml; 6.2 mg/ml; 50 mg/ml. Data are the mean value of three separate experiments and expressed as percentage of hemolysis ±SD.

DISCUSSION

The echinoderms are considered a good source for AMPs and a variety of peptides with antimicrobial properties have been isolated from them. Antimicrobial activity has been reported in gonads of the asteroid *Marthasterias glacialis* (Stabili and Pagliara [1994]), and*Paracentrotus lividus* contains, in the low molecular weight fraction (<5 kDa) of acid precipitate of their coelomocytes, peptides with antimicrobial activity against staphylococcal biofilms. In addition, we recently reported that immune mediators cells in the echinoderm *Holothuria tubulosa* is a source of novel AMPs with anti-staphylococcal biofim activity (Schillaci et al. [2013]). Moreover, two cystein-rich AMPs, named centrocyns, have been characterized in the green sea-urchin *Strongylocentrotus droebachiensis* (Li et al. [2010]).

The synthetic fragment SP1 of β-thymosin extracted from coelomocytes of the *P. lividus* shares structural characteristics of many antimicrobial peptides: it is a cationic peptide and possesses a significant proportion (~40%) of hydrophobic or non polar residues (Hancock and Lehrer [1998]; Zasloff

[2002]). Although the α-helix is the most common secondary structure for the antimicrobial activity of AMPs (Mor and Nicolas [1994]; Skerlavaj et al. [1996]; Storici et al. [1994]; Tossi et al. [1994]) the *in silico* study suggests that the conformation assumed by SP1 (Figure 4) might be responsible for the observed antimicrobial activity. The helix structure of echinoderm AMPs shares a characteristic amphipathic structure with alternating hydrophobic and polar residues along the primary structure. According to the Shai-Matsuzaki-Huang model, the peptides bind to the membrane surface first and then with their amphipathic structure they enter into the membrane, breaking up the lipid chains and forming transient pore; this process can cause a collapse of the membrane at a critical peptide concentration (Hancock and Chapple [1999]; Huang [2000]; Matsuzaki [1998]; Shai [2002]). The SP1 peptide has a structure very different from other echinoderm AMPs (Schillaci et al. [2013]; Schillaci et al. [2012]). Moreover, such structure, unexpectedly, does not have an amphipathic nature with a hydrophobic face opposite to a hydrophilic one, because the polar charged residues and those hydrophobic are not arranged uniformly. The peptide contains a sequence of four hydrophobic/non polar residues which contribute to constitute a hydrophobic core, flanked at both ends by cationic and polar residues that can solubilize the peptides in aqueous solution providing a binding site for bacterial membranes. This particular conformation is different from other antimicrobial peptides, designed as transmembrane mimetic models and that spontaneously become inserted into the cell membranes (Chan et al. [2004]; [Liu and Deber 1998]; Stark et al. [2002]).

We found that the SP1 showed a broad antimicrobial activity against important pathogens such as *S. aureus* and *P. aeruginosa* but it acts at high concentration (12.5 or 6.2 mg/ml) against planktonic forms of these two microorganisms. Such weak activity is comparable to that reported for a different Echinoderm, *Holothuria tubulosa* (Schillaci et al. [2013]) and to that of some described innate human defence protein like lactoferrin in vitro (de Andrade et al. [2014]). MD results suggest that the observed weak activity could be due to the low stability of SP1, in fact, as showed in RMSD plot (Figure 2), the molecule is stable only in a limited time period and in this period it can present the active conformation of Figure 1. However, SP1 shows an interesting additional antimicrobial effect interfering with biofilm formation *in vitro* of the above cited pathogens, at lower concentrations than MIC evaluated against the planktonic forms. The prevention of biofilm formation – rather than its elimination – is the best strategy to contrast the growth, as a sessile community, of many pathogens.

We do not know the antimicrobial mechanisms of SP1, but we could speculate as for other cationic non-amphipathic microbial peptides that the positive charges carried by the peptide are essential for the membrane binding through electrostatic interaction between residues with anionic phospholipids. The peptide-membrane interaction could be responsible for membrane aggregation by peptides bridging simultaneously two membranes or for negative curvature of the membrane asymmetry that can form tubes (Lamaziere et al. [2007]). The synthetic SP1 could act like Dermaseptin S9, a non-amphipathic antimicrobial peptides produced by the skin of the South American hylid frog, *Phyllomedusa sauvagei*, that contains, centrally located, a hydrophobic core that can insert the peptide in interior membrane (Lequin et al. [2006]). Dermaseptin S9 exerts a microbicidal activity by perturbating both the membrane interface and the hydrophobic core of the bacterial membrane (Auvynet et al. [2008]).

In a preliminary evaluation of selective toxicity, it was gratifying to note the general lack of hemolysis of rabbit erythrocytes by SP1, even at peptide concentrations up to 50 mg/ml. Such selectivity of cationic non amphipathic antimicrobial peptides for bacterial membranes may be explicable, in part, by the differences in the compositions of eukaryotic and prokaryotic membranes (Matsuzaki [1999]; Zasloff [2002]). The outer leaflet of mammalian cells is predominantly composed of the zwitterionic phospholipids phosphatidylcholine and sphingomyelin (Verkleij et al. [1973]), along with cholesterol (Turner and Rouser [1970]), while bacterial membranes contain mainly anionic phospholipids and no cholesterol (Brock [1974]). In addition, the outer surfaces of Gram-negative bacteria contain lipopolysaccharides, while those of Gram-positive bacteria contain teichoic acid, which in both cases add to the negative charge of the bacterial surface (Brock [1974]). The cationic nature of native antimicrobial peptides clearly contributes to their preferential recognition by the negatively charged outer surfaces of bacterial membranes (Oren and Shai [1998]; Shai [1999]).

The extensive clinical use of classical antibiotics has led to the growing emergence of many medically relevant resistant strains of pathogens (Patel [2005]). Therefore, the development of a new class of antimicrobials with a different mechanism of action than conventional antibiotics has become critical. The cationic antimicrobial peptides could represent such a new class (Andreu and Rivas [1998]; Hancock [1997]; Sitaram and Nagaraj [2002]). The development of resistance to membrane active peptides whose sole target is the cytoplasmic membrane is not expected because this would require substantial changes in the lipid composition of cell membranes of microorganisms (Hancock [2001]).

The anti-adhesion property showed by SP1 against *S. aureus* and *P. aeruginosa* strains is very interesting considering that the two opportunistic pathogens are able to form biofilms in open wounds, such as chronic diabetic foot ulcers, or infected wounds in clinical and veterinary medicine. New antimicrobial agents that are effective against staphylococci and *P. aeruginosa*to treat infected wounds are needed, and a potential topical application of SP1 could be supposed. Moreover, the two species are also involved in food spoilage and biofilm formation of food transmitted pathogens and an application of SP1 in the food processing is possible and closer, at this stage of our study, than application in clinical or veterinary health.

A chemotherapeutic approach combining conventional antibiotics and novel anti-biofilm agents could be a new strategy for the treatment of biofilm-associated infections like mastitis in veterinary field or the topical treatment of infected wounds in clinical and veterinary setting.

Finally, the tested synthetic peptide is a good starting point to design new synthetic derivatives with modified chemical-physical properties, with the aim to improve their antimicrobial activity against pathogens and their pharmaceutical potential (Brogden and Brogden [2011]; Huang et al. [2010]).

ACKNOWLEDGEMENTS

Financial support from Fondi di Ateneo 2006, 2007 Università degli Studi di Palermo (Italy) and Ministero della Salute IZS is gratefully acknowledged.

REFERENCES

1. Andreu D, Rivas L: Animal antimicrobial peptides: an overview. *Biopolymers* 1998,47(6):415-433. 10.1002/(SICI)1097-0282(1998)47:6<415::AID-BIP2>3.0.CO;2-D
2. Arizza V: Marine biodiversity as source of new drugs. *Ital J Zool* 2013, 80(3):317-318. doi:10.1080/11 250003.2013.830370
3. Arizza V, Parrinello D, Cammarata M, Vazzana M, Vizzini A, Giaramita FT, Parrinello N: A lytic mechanism based on soluble phospholypases A2 (sPLA2) and beta-galactoside specific lectins is exerted by *Ciona intestinalis* (ascidian) unilocular refractile hemocytes against K562 cell line and mammalian erythrocytes. *Fish Shellfish Immun* 2011, 30(4–5):1014-1023. doi:10.1016/j.fsi.2011.01.022
4. Arizza V, Vazzana M, Schillaci D, Russo D, Giaramita FT, Parrinello N: Gender differences in the immune system activities of sea urchin *Paracentrotus lividus* . *Comp Biochem Physiol A Mol Integr Physiol* 2013, 164(3):447-455. doi:10.1016/j.cbpa.2012.11.021
5. Auvynet C, El Amri C, Lacombe C, Bruston F, Bourdais J, Nicolas P, Rosenstein Y:Structural requirements for antimicrobial versus chemoattractant activities for dermaseptin S9. *FEBS J* 2008, 275(16):4134-4151. doi:10.1111/j.1742-4658.2008.06554.x
6. Bax R, Mullan N, Verhoef J: The millennium bugs - the need for and development of new antibacterials. *Int J Antimicrob Ag* 2000, 16(1):51-59. doi:10.1016/S0924-8579(00)00189-8
7. Brock TD: *Biology of Microorganisms*. Prentice-Hall, Inc, Englewood Cliffs N.J; 1974.
8. Brogden NK, Brogden KA: Will new generations of modified antimicrobial peptides improve their potential as pharmaceuticals? *Int J Antimicrob Ag* 2011,38(3):217-225. doi:10.1016/j.ijantimicag.2011.05.004
9. Canicatti C, Roch P: Studies on Holothuria polii (Echinodermata) Antibacterial Proteins.1. Evidence for and Activity of a Coelomocyte Lysozyme. *Experientia* 1989, 45(8):756-759. doi:10.1007/Bf01974579
10. Chan C, Burrows LL, Deber CM: Helix induction in antimicrobial peptides by alginate in biofilms. *J Biol Chem* 2004, 279(37):38749-38754. doi:10.1074/jbc. M406044200
11. Clutterbuck AL, Woods EJ, Knottenbelt DC, Clegg PD, Cochrane CA, Percival SL: Biofilms and their relevance to veterinary medicine. *Vet Microbiol* 2007, 121(1–2):1-17. doi:10.1016/j.vetmic.2006.12.029
12. de Andrade FB, de Oliveira JC, Yoshie MT, Guimarães BM, Gonçalves RB, Schwarcz WD:Antimicrobial activity and synergism of lactoferrin and lysozyme against cariogenic microorganisms. *Braz Dent J* 2014, 25: 165-169. 10.1590/0103-6440201302257
13. Fischer D, Li Y, Ahlemeyer B, Krieglstein J, Kissel T: In vitro cytotoxicity testing of polycations: influence of polymer structure on cell viability and hemolysis. *Biomaterials* 2003, 24(7):1121-1131. doi:10.1016/ S0142-9612(02)00445-3
14. Gautier R, Douguet D, Antonny B, Drin G: HELIQUEST: a web server to screen sequences with specific α-helical properties. *Bioinformatics* 2008, 24(18):2101-2102. doi:10.1093/bioinformatics/btn392
15. Gilbert P, Allison DG, McBain AJ: Biofilms in vitro and in vivo: do singular mechanisms imply cross-resistance? *J Appl Microbiol* 2002, 92: 98s-110s. doi:10.1046/j.1365-2672.92.5s1.5.x
16. Hancock REW: Peptide antibiotics. *Lancet* 1997, 349(9049):418-422. doi:10.1016/S0140-6736(97)80051-7
17. Hancock RE: Cationic peptides: effectors in innate immunity and novel antimicrobials. *Lancet Infect Dis* 2001, 1(3):156-164. doi:10.1016/S1473-3099(01)00092-5
18. Hancock RE, Chapple DS: Peptide antibiotics. *Antimicrob Agents Chemother* 1999,43(6):1317-1323.
19. Hancock RE, Lehrer R: Cationic peptides: a new source of antibiotics. *Trends Biotechnol* 1998, 16(2):82-88. 10.1016/S0167-7799(97)01156-6
20. Hancock RE, Rozek A: Role of membranes in the activities of antimicrobial cationic peptides. *FEMS Microbiol Lett* 2002, 206(2):143-149. 10.1111/j.1574-6968.2002.tb11000.x
21. Huang HW: Action of antimicrobial peptides: Two-state model. *Biochemistry* 2000, 39(29):8347-8352. doi:10.1021/Bi0009461
22. Huang YB, Huang JF, Chen YX: Alpha-helical cationic antimicrobial peptides: relationships of structure and function. *Protein Cell* 2010, 1(2):143-152. doi:10.1007/ s13238-010-0004-3
23. Lamaziere A, Burlina F, Wolf C, Chassaing G, Trugnan G, Ayala-Sanmartin J: Non-metabolic membrane tubulation and permeability induced by bioactive peptides. *Plos One* 2007, 2(2):e201. doi:10.1371/journal.pone.0000201
24. Lauria A, Bonsignore R, Terenzi A, Spinello A, Giannici F, Longo A, Almerico AM, Barone G:Nickel(II), copper(II) and zinc(II) metallo-intercalators: structural details of the DNA-binding by a combined experimental and computational investigation. *Dalton Trans* 2014, 43: 6108-6119. http://dx.doi.org/10.1039/ c3dt53066c 10.1039/c3dt53066c
25. Lentini L, Melfi R, Di Leonardo A, Spinello A, Barone G, Pace A, Palumbo Piccionello A, Pibiri I: Towards a rationale for the PTC124 (Ataluren) promoted readthrough of premature stop codons: a computational approach and GFP-reporter cell-based assay. *Mol Pharm* 2014, 11: 653-664. http://dx.doi.org/10.1021/ mp400230s 10.1021/mp400230s
26. Lequin O, Ladram A, Chabbert L, Bruston F, Convert O, Vanhoye D, Chassaing G, Nicolas P, Amiche

M: Dermaseptin S9, an alpha-helical antimicrobial peptide with a hydrophobic core and cationic termini. *Biochemistry* 2006, 45(2):468-480. doi:10.1021/bi051711i

27. Li C, Haug T, Moe MK, Styrvold OB, Stensvåg K: Centrocins: Isolation and characterization of novel dimeric antimicrobial peptides from the green sea urchin, Strongylocentrotus droebachiensis. *Dev Comp Immunol* 2010, 34: 959-968. doi: 10.1016/j.dci.2010.04.004

28. Lindorff-Larsen K, Piana S, Palmo K, Maragakis P, Klepeis JL, Dror RO, Shaw DE:Improved side-chain torsion potentials for the Amber ff99SB protein force field. *Proteins* 2010, 78: 1950-1958. http://dx.doi.org/10.1002/prot.22711

29. Liu L-P, Deber CM: Guidelines for membrane protein engineering derived from de novo designed model peptides. *Pept Sci* 1998a, 47(1):41-62. doi:10.1002/(SICI)1097-0282(1998)47:1<41::AID-BIP6>3.0.CO;2-X

30. Liu LP, Deber CM: Guidelines for membrane protein engineering derived from de novo designed model peptides. *Biopolymers* 1998b, 47(1):41-62. doi:10.1002/(SICI)1097-0282(1998)47:1<41::AID-BIP6>3.0.CO;2-X

31. Matsuzaki K: Magainins as paradigm for the mode of action of pore forming polypeptides. *Biochim Biophys Acta Rev Biomembr* 1998, 1376(3):391-400. doi:10.1016/S0304-4157(98)00014-8

32. Matsuzaki K: Why and how are peptide–lipid interactions utilized for self-defense? Magainins and tachyplesins as archetypes. *Biochim Biophys Acta Rev Biomembr* 1999, 1462(1–2):1-10. http://dx.doi.org/10.1016/S0005-2736(99)00197-2

33. Melchior MB, Vaarkamp H, Fink-Gremmels J: Biofilms: a role in recurrent mastitis infections? *Vet J* 2006, 171(3):398-407. doi:10.1016/j.tvjl.2005.01.006

34. Melchior MB, van Osch MHJ, Graat RM, van Duijkeren E, Mevius DJ, Nielen M, Gaastra W, Fink-Gremmels J: Biofilm formation and genotyping of Staphylococcus aureus bovine mastitis isolates: Evidence for lack of penicillin-resistance in Agr-type II strains. *Vet Microbiol* 2009, 137(1–2):83-89. doi:10.1016/j.vetmic.2008.12.004

35. Mor A: Peptide-based antibiotics: A potential answer to raging antimicrobial resistance. *Drug Develop Res* 2000, 50(3–4):440-447. doi:10.1002/1098-2299(200007/08)50:3/4<440::Aid-Ddr27>3.0.Co;2–4

36. Mor A, Nicolas P: The Nh2-terminal alpha-helical domain 1–18 of Dermaseptin is responsible for antimicrobial activity. *J Biol Chem* 1994, 269(3):1934-1939.

37. Oren Z, Shai Y: Mode of action of linear amphipathic α-helical antimicrobial peptides. *Pept Sci* 1998, 47(6):451-463. doi:10.1002/(SICI)1097-0282(1998)47:6<451::AID-BIP4>3.0.CO;2-F

38. Patel R: Biofilms and antimicrobial resistance. *Clin Orthop Relat Res* 2005, 437:41-47. 10.1097/01.blo.0000175714.68624.74

39. Projan SJ, Youngman PJ: Antimicrobials: new solutions badly needed. *Curr Opin Microbiol* 2002, 5(5):463-465. 10.1016/S1369-5274(02)00364-8

40. Pronk S, Pall S, Schulz R, Larsson P, Bjelkmar P, Apostolov R, Shirts MR, Smith JC, Kasson PM, van der Spoel D, Hess B, Lindahl E: GROMACS 4.5: a high-throughput and highly parallel open source molecular simulation toolkit. *Bioinformatics*2013, 29: 845-854. http://dx.doi.org/10.1093/bioinformatics/btt055

41. Schillaci D, Petruso S, Sciortino V: 3,4,5,3 `,5 `-Pentabromo-2-(2 `-hydroxybenzoyl)pyrrole: a potential lead compound as anti-Gram-positive and anti-biofilm agent. *Int J Antimicrob Ag* 2005, 25(4):338-340. doi:10.1016/j.ijantimicag.2004.11.014

42. Schillaci D, Arizza V, Parrinello N, Di Stefano V, Fanara S, Muccilli V, Cunsolo V, Haagensen JJ, Molin S: Antimicrobial and antistaphylococcal biofilm activity from the sea urchin *Paracentrotus lividus* . *J Appl Microbiol* 2010, 108(1):17-24. doi:JAM4394 [pii] 10.1111/j.1365-2672.2009.04394.x

43. Schillaci D, Vitale M, Cusimano MG, Arizza V: Fragments of beta-thymosin from the sea urchin Paracentrotus lividus as potential antimicrobial peptides against staphylococcal biofilms. In *Thymosins in Health and Disease Ii.* Edited by: Goldstein AL, Garaci E. Blackwell Science Publ, Oxford; 2012:79-85.

44. Schillaci D, Cusimano MG, Cunsolo V, Saletti R, Russo D, Vazzana M, Vitale M, Arizza V:Immune mediators of sea-cucumber Holothuria tubulosa (Echinodermata) as source of novel antimicrobial and anti-staphylococcal biofilm agents. *AMB Express* 2013, 3(1):35. doi:10.1186/2191-0855-3-35

45. Schumacher-Perdreau F, Heilmann C, Peters G, Götz F, Pulverer G: Comparative analysis of a biofilm-forming Staphylococcus epidermidis strain and its adhesion-positive, accumulation-negative mutant M7. *FEMS Microbiol Lett* 1994,117: 71-78. 10.1111/j.1574-6968.1994.tb06744.x

46. Shai Y: Mechanism of the binding, insertion and destabilization of phospholipid bilayer membranes by α-helical antimicrobial and cell non-selective membrane-lytic peptides. *Biochim Biophys Acta Rev Biomembr* 1999,1462(1–2):55-70. http://dx.doi.org/10.1016/S0005-2736(99)00200-X

47. Shai Y: Mode of action of membrane active antimicrobial peptides. *Biopolymers*2002, 66(4):236-248. doi:10.1002/Bip.10260

48. Sitaram N, Nagaraj R: The therapeutic potential of host-defense antimicrobial peptides. *Curr Drug Targets* 2002, 3(3):259-267. doi:10.2174/1389450023347786

49. Skerlavaj B, Gennaro R, Bagella L, Merluzzi L, Risso A, Zanetti M: Biological characterization of two novel cathelicidin-derived peptides and identification of structural requirements for their antimicrobial and cell lytic activities. *J Biol Chem* 1996, 271(45):28375-28381. 10.1074/jbc.271.45.28375

50. Stabili L, Pagliara P: Antibacterial protection in *Marthasterias glacialis* eggs: characterization of

lysozyme-like activity. *Comp Biochem Physiol B Biochem Mol Biol* 1994, 109(4):709-713. 10.1016/0305-0491(94)90134-1

51. Stark M, Liu LP, Deber CM: Cationic hydrophobic peptides with antimicrobial activity. *Antimicrob Agents* 2002, 46(11):3585-3590. 10.1128/AAC.46.11.3585-3590.2002

52. Storici P, Scocchi M, Tossi A, Gennaro R, Zanetti M: Chemical synthesis and biological-activity of a novel antibacterial peptide deduced from a pig myeloid cDNA. *FEBS Lett* 1994, 337(3):303-307. doi:10.1016/0014-5793(94)80214-9

53. Tossi A, Scocchi M, Skerlavaj B, Gennaro R: Identification and characterization of a primary antibacterial domain in Cap18, a lipopolysaccharide-binding protein from rabbit leukocytes. *FEBS Lett* 1994, 339(1–2):108-112. doi:10.1016/0014-5793(94)80395-1

54. Turner JD, Rouser G: Precise quantitative determination of human blood lipids by thin-layer and triethylaminoethylcellulose column chromatography. I Erythrocyte lipids. *Anal Biochem* 1970, 38(2):423-436. 10.1016/0003-2697(70)90467-7

55. Verkleij AJ, Zwaal RFA, Roelofsen B, Comfurius P, Kastelijn D, van Deenen LLM: The asymmetric distribution of phospholipids in the human red cell membrane. A combined study using phospholipases and freeze-etch electron microscopy.*Biochim Biophys Acta Rev Biomembr* 1973, 323(2):178-193. http://dx.doi.org/10.1016/0005-2736(73)90143-0

56. Wang Z, Wang G: APD: the Antimicrobial Peptide Database. *Nucleic Acids Res*2004, 32(Database issue):D590-D592. doi:10.1093/nar/gkh025

57. Wang G, Li X, Wang Z: APD2: the updated antimicrobial peptide database and its application in peptide design. *Nucleic Acids Res* 2009, 37(Database issue):D933-D937. doi:10.1093/nar/gkn823

58. Wu G, Wu H, Fan X, Zhao R, Li X, Wang S, Ma Y, Shen Z, Xi T: Selective toxicity of antimicrobial peptide S-thanatin on bacteria. *Peptides* 2010, 31: 1669-1673. doi:10.1016/j.peptides.2010.06.009

59. Zasloff M: Antimicrobial peptides of multicellular organisms. *Nature* 2002,415(6870):389-395. doi:10.1038/415389a

CHAPTER 16

ANTI-INFECTIOUS AND ANTI-INFLAMMATORY EFFECTS OF PEPTIDE FRAGMENTS SEQUENTIALLY DERIVED FROM THE ANTIMICROBIAL PEPTIDE CENTROCIN 1 ISOLATED FROM THE GREEN SEA URCHIN, STRONGYLOCENTROTUS DROEBACHIENSIS

Camilla Björn, Joakim Håkansson, Emma Myhrman, Veronika Sjöstrand, Tor Haug, Kerstin Lindgren, Hans-Matti Blencke, Klara Stensvåg and Margit Mahlapuu

Keywords: Antibacterial peptide, Wound infection, Antibiotic resistance, MRSA

INTRODUCTION

Infectious diseases are among the leading causes of death worldwide. According to the World Health Organization, acute infectious diseases account for 25% of deaths globally and for 45% in developing countries, killing about 15 million people per year (WHO 2004). Several orders of magnitude higher is the number of people who survive the initial infection but suffer from conditions where microbes cause a chronic inflammation (Schmidt 2004). The treatment of infections has recently been complicated due to emerging increase of bacterial resistance towards conventional antibiotics. The occurrence of MRSA in hospital environments has been reported as a significant problem since its recognition over 50 years ago (Jevons 1961). In recent years, the emergence of community-acquired strains has rapidly grown, as in some settings, community-acquired MRSA represents greater than 50% of the total *S. aureus*infections (Giordano et al. 2007). In addition, enterococci, leading to surgical wound and urinary tract infections, are becoming intrinsically resistant to many antibiotics. Also, multi-drug resistant strains of *Pseudomonas aeruginosa* are associated with severe adverse clinical outcomes (Jones et al. 2005; Moet et al. 2007). Resistance development is not confined to antibiotics only. Silver products have been considered to carry a low risk of resistance development, however burn wound studies have demonstrated that bacteria, in particular *P. aeruginosa*, may become resistant to silver sulfadiazine and silver nitrate (Vermeulen et al.2007).

AMPs are receiving increased attention as potential therapeutic candidates in infectious disease treatment. AMPs are gene-encoded, ribosomally synthesised peptides which are widespread in nature, and appear to be important

Citation: Camilla Björn, Joakim Håkansson, Emma Myhrman, Veronika Sjöstrand, Tor Haug, Kerstin Lindgren, Hans-Matti Blencke, Klara Stensvåg, Margit Mahlapuu, Anti-infectious and anti-inflammatory effects of peptide fragments sequentially derived from the antimicrobial peptide centrocin 1 isolated from the green sea urchin, Strongylocentrotus droebachiensis, DOI: 10.1186/2191-0855-2-67.

defence molecules in all eukaryotic phyla. Most naturally occurring AMPs carry a net positive charge (*i.e.* cationic) and are composed of 12–50 residues where close to 50% of them are hydrophobic (Hancock and Diamond 2000; Teixeira et al. 2012). In membrane-like environments AMPs tend to form amphipathic structures, *i.e.* structures with separate hydrophobic and hydrophilic domains. The positive charge of AMPs presumably enables interactions with the negatively charged outer leaflet of bacterial membranes and/or bacterial cell wall components, whereas their amphipathic character causes membrane permeabilization (Devine and Hancock 2002; Teixeira et al. 2012). Notably, although AMPs are widely distributed in nature and bacteria have been exposed to these molecules for millions of years, widespread resistance has not been reported (Fjell et al. 2012). Many AMPs have also shown to possess favourable activities other than bactericidal effect, such as anti-inflammatory properties (Auvynet and Rosenstein 2009; Bowdish et al. 2005; Scott and Hancock 2000). Based on these properties, AMPs have emerged as novel promising agents for both topical and systemic treatment of infectious diseases (Schuerholz et al. 2012; Yount and Yeaman 2012).

Invertebrates are proving to be a rich source of AMPs, possibly because of their more pronounced reliance on innate immune functions in their defence against microbial pathogens compared to vertebrates. Many of the peptides isolated from marine species are structurally unique, display significant biological activities and are therefore expected to be useful drug leads (Sperstad et al. 2011). In a previous study, we detected broad-spectrum antibacterial activity in coelomocyte (blood cell) extracts of the green sea urchin, *S. droebachiensis* (Haug et al. 2002). Subsequently, an AMP named centrocin 1 was isolated and characterized. Centrocin 1 is a heterodimer formed by monomers of 30 (heavy chain) and 12 amino acid residues (light chain) connected by a single disulphide bridge (Li et al. 2010). Synthesis and subsequent antimicrobial testing of the different monomers showed that the cationic heavy chain is equally active as the original dimeric peptide. On the other hand, the neutral light chain showed no antimicrobial activity (Li et al. 2010). Helical wheel projection of the heavy chain indicates that this sequence has an amphipathic structure, a feature common for many membrane-active AMPs. Interestingly, centrocin 1 heavy chain contains a brominated tryptophan on position two (Li et al. 2010) with unknown function.

In the present study, centrocin 1 heavy chain (CEN1 HC-Br) and derivatives thereof, were synthesised and screened for *in vitro* bactericidal activity against selected microbial strains in four different assay media. The peptides were also evaluated for *in vitro* anti-inflammatory properties as measured by inhibition of secretion of tumor necrosis factor alpha (TNF-α) and interleukin-6 (IL-6) from lipopolysaccharide (LPS) stimulated macrophages derived from the monocytic cell line THP-1, and for their cytotoxicity to the same cells. The debrominated analogue of centrocin 1 heavy chain, CEN1 HC, was evaluated for its ability to reduce bacterial counts in two different infected wound models, an *in vivo* excision wound model in rats, and an*ex vivo* pig skin model. In addition, resistance development against CEN1 HC was assessed in the bacterial strains *S. aureus* and MRSA.

MATERIALS AND METHODS

Peptides and Antibiotics

The heavy chain of centrocin 1 (CEN1 HC-Br), the C-terminal truncated peptide (CEN1 HC-Br (1–20)) and its debrominated equivalent (CEN1 HC (1–20)) were purchased from BIOMOL International LP (Exeter, UK). The non-halogenated heavy chain CEN1 HC and an analogue where the free cysteine in position 25 is exchanged with serine, CEN1 HC (Ser), were purchased from Biopeptide Co., Inc. (San Diego, CA). The peptides (Table 1) were all synthesised using Fmoc solid phase technology, and the peptide content and purity were determined by high performance liquid chromatography (HPLC) and mass spectrometry (MS) analysis. The brominated peptides were all synthesised with 5-Bromo-D/L-tryptophan in position 2. Mupirocin (Bactroban, 2% ointment, GlaxoSmithKline, Brentford, UK) and fusidic acid (Fucidin, 2% cream, LEO Pharma, Malmö, Sweden) were used as comparators in assessment of*in vitro* microbicidal effect. In the *in vitro* resistance assay, mupirocin (Applichem, Darmstadt, Germany) in powder form was used as comparator.

Table 1: Amino acid sequence of centrocin 1 heavy chain (CEN1 HC-Br) and its synthetic fragments and analogues

Peptide	**Sequence**	**Net charge at pH = 7.0**	**Ratio hydrophobic/ total residues**
CEN1 HC-Br	GW(Br)FKKTFHKVSHAVKSGIHAGQRGCSALGF	+ 5.2	53%
CEN1 HC	GWFKKTFHKVSHAVKSGIHAGQRGCSALGF	+ 5.2	53%
CEN1 HC-Br (1–20)	GW(Br)FKKTFHKVSHAVKSGIHA	+ 4.3	60%
CEN1 HC (1–20)	GWFKKTFHKVSHAVKSGIHA	+ 4.3	60%
CEN1 HC (Ser)	GWFKKTFHKVSHAVKSGIHAGQRGSSALGF	+ 5.3	50%

In vitro Microbicidal Effect

All the synthesised peptides were tested for their *in vitro* microbicidal activity against *S. aureus*(American Type Culture Collection (ATCC) 12600) and *P. aeruginosa* (ATCC 15442) using a minimal microbicidal concentration (MMC) assay as previously described (Haversen et al.2010). Bacteria were cultured in 3.7% brain heart infusion (BHI) broth (Difco, BD Diagnostics, Franklin Lakes, NJ) on a shaker at 250 rpm overnight at 37°C. The culture was diluted 1:10 in fresh 3.7% BHI and incubated for two additional hours to reach log-phase growth. The bacteria were pelleted by centrifugation at 900 × g and suspended in 0.037% BHI to a concentration of 10^7 CFU/

mL, estimated by measuring optical density at 600 nm (OD_{600}). To investigate the influence of physiological conditions on microbicidal effect of the peptides, four different assay media were used: 0.037% BHI, 0.037% BHI supplemented with 85 or 150 mM sodium chloride (NaCl), and 50% heat inactivated simulated wound fluid (h.i. SWF) composed of a 50:50 mixture of 0.1% peptone (Oxoid, Basingstoke, UK) in 150 mM NaCl and fetal bovine serum (FBS; PAA Laboratories GmbH, Pasching, Austria), diluted two times in ultra pure water. The peptides were serially diluted by twofold steps from 200 to 1.56 mg/L in the assay media used, and 100 µL of the peptide solutions were mixed with 5 µL bacterial suspensions in a 96 well plate (Nunc, Roskilde, Denmark) and incubated at 37°C for two hours. An aliquot of 5 µL from each well was aspirated and added as drops onto blood agar plates (Columbia agar; Oxoid, Basingstoke, UK) supplemented with 5% defibrinated horse blood (Swedish National Veterinary Institute (SVA), Uppsala, Sweden), and the plates were incubated overnight at 37°C. All samples were processed as duplicates. The minimal test compound concentration causing ≥99% reduction of bacteria was defined as the MMC_{99}. The concentration of the bacterial suspension used in the assay was confirmed by viable count estimation on blood agar plates (Oxoid).

CEN1 HC was further screened for microbicidal activity against a panel of microbial strains using the MMC_{99} assay, as described above, except using only two different assay media; 0.037% BHI and 50% h.i. SWF. The following microbial strains were tested for sensitivity: *S. aureus* (ATCC 12600), MRSA (ATCC 33591), *Staphylococcus epidermidis* (ATCC 12228),*Streptococcus pyogenes* (ATCC 12344), *Propionibacterium acnes* (ATCC 6919) *Escherichia coli* (ATCC 11775), *P. aeruginosa* (ATCC 15442), *Klebsiella pneumoniae* (Culture collection, University of Gothenburg (CCUG) 59413, clinical isolate resistant to penicillins, cephalosporins, aztreonam and carbapenems, with the reference strain ATCC 13883), *Acinetobacter baumannii*(CCUG 58437, clinical isolate resistant to tobramycin, trimsulfa, ciprofloxacin, cefotaxim, ceftazidim, meropenem, pipera/tazobactam, with the reference strain ATCC 19606) and the yeast strain *Candida albicans* (ATCC 64549). Mupirocin (GlaxoSmithKline) and fusidic acid (LEO Pharma) were used as comparative control antibiotics.

In vitro Anti-Inflammatory Effect

The human monocytic cell line THP-1 (ATCC TIB-202) was cultured in RPMI 1640 medium (PAA Laboratories GmbH) supplemented with 10% FBS (PAA Laboratories GmbH), 1 mM sodium pyruvate (Sigma-Aldrich, St. Louis, MO), and 20 mM 4-(2-hydroxyethyl)-1-piperazineethanesulfonic acid (HEPES; PAA Laboratories GmbH). The cell density was adjusted to 10^6 cells/mL and 100 µL of the cell suspension was added to each well of 96 well plates (Sarstedt, Nümbrecht, Germany). The cells were treated with 10 ng/mL phorbol 12-myristate 13-acetate (PMA; Sigma-Aldrich) to differentiate the monocytes into macrophage-like cells. After 48 h, the cells were stimulated by adding 0.1 ng/mL LPS into the medium specified above, except that 10% FBS was replaced with 5% h.i. FBS. The LPS concentration used was selected based on *in vitro* titration experiments to give a close to maximum release of cytokines into the cell culture medium over a period of six hours (data not shown). The indicated concentrations of the peptides were added 30 min after addition of LPS. After six hours of incubation, the plates were centrifuged for six minutes at 400 × g and the supernatants were collected and kept frozen at −20°C until analysed for TNF-α and IL-6 production. The cytokine levels were analysed using enzyme-linked immunosorbent assay (ELISA; R&D Systems, Minneapolis, MN).

In vitro Cytotoxic Effect on THP-1 Cells Assessed by TACS MTT Assay

THP-1 cells were cultured, differentiated, stimulated and the peptides were added as described above. Triton X-100 (ICN Biomedicals Inc., OH) was used as a positive control, and added at the same time as the peptides. After six hours of incubation, TACS MTT (3-(4, 5-Dimethylthiazol-2-yl)-2, 5-diphenyltetrazolium bromide) assay (R&D Systems) was performed according to the manufacturer's instructions. In short, 10 µL MTT reagent was added to each well and the plates were incubated for two hours at 37°C. Subsequently, 100 µL detergent reagent (100 mg/L) was added and the plates were incubated in dark at room temperature overnight. The plates were shaken for 10 min at 300 rpm and the absorbance at 570 nm with the reference wavelength 650 nm was measured.

In Vitro Resistance Assay

S. aureus (ATCC 12600) and MRSA (ATCC 33591) were cultured according to the protocol described above and diluted to 10^8 CFU/mL in 0.037% BHI as estimated by measuring OD_{600}. The multistep resistance assay was performed as previously described (Kosowska-Shick et al.2006). Peptide dilutions were prepared in twofold dilution steps in the concentration range from 0.75 mg/L to 96 mg/L in a volume of 1 mL using 0.37% BHI medium. The same peptide concentration range was used during all passages and new peptide dilutions were prepared each day from stock solutions. Mupirocin (Applichem), the active substance in the conventional topical antibiotic Bactroban, was used as a comparator. The antibiotic control was handled in the same way as the peptides except that the concentration range of mupirocin was 0.003 mg/L to 3 mg/L.

An aliquot of 10 µL (10^8 CFU/mL) of the bacterial suspension was added to the peptide dilutions, to the positive antibiotic control, and to a negative control incubated without peptide, and the suspensions were incubated at 37°C for 24 h. Turbidity of the overnight cultures was determined by OD_{600} measurements. The OD_{600} of bacteria/peptide suspensions was calculated by subtracting the OD_{600} of the peptide mixed with medium from the OD_{600} measured on the overnight cultures. To create an optimal selection pressure, the overnight culture having the highest peptide concentration without marked reduction of bacterial viability (defined as OD_{600}value of >85% compared to the value measured for the negative control sample) was further passed.

The bacteria/peptide suspension to be further passed was diluted in 0.37% BHI to 10^8 CFU/mL. A volume of 10 µL (10^8 CFU/mL) from the selected inoculum was transferred to vials with 1 mL fresh peptide solutions diluted in 0.37% BHI. The suspensions were incubated at 37°C for 24 h. The same procedure was performed with the antibiotic control. The remaining bacterial suspensions were centrifuged at 900 × g for ten minutes, then resuspended in a 50:50 mixture of 3.7% BHI and 80% glycerol and immediately frozen at −80°C and kept frozen until analysis. The 24-h incubation and inoculation procedure described above was repeated for 14 days. MMC_{99} assay was performed, as described above, on the overnight cultures after 0, 1, 4, 7 and 14 passages to evaluate resistance induction.

In Vivo Excision Wound Model in Rats

The ability of CEN1 HC and CEN1 HC (Ser) to reduce bacterial counts was estimated in an excision wound model in rat designed based on previously published studies (Gisby and Bryant2000; McRipley and Whitney 1976; Rittenhouse et al. 2006; Saymen et al. 1972). Briefly, female Sprague–Dawley rats (200–250 g, Charles River Laboratories, Sulzfeldt, Germany) were housed at the Laboratory of Experimental Biomedicine, Gothenburg, Sweden. They were kept in a 12-h light–dark cycle and were cared for in accordance with regulations for the protection of laboratory animals. All animal experiments were performed after prior approval from the local Ethics Committee for Animal Studies at the Administrative Court of Appeals in Gothenburg, Sweden. The rats were acclimatised in cages for a minimum of five days before surgery, with four rats in each cage. They had free access to water and pellets (Lab For, Lantmännen, Sweden). The rats were anaesthetised during the whole experiment and the anaesthesia was induced by an intra-peritoneal injection of a mixture of fentanyl (272 µg/kg; B. Braun Melsungen AG, Melsungen, Germany) and medetomidine hydrochloride (545 µg/kg; Domitor, Orion Pharma Animal Health AB, Sollentuna, Sweden). The backs of the rats were shaved and swabbed with 70% chlorhexidine alcohol (Fresenius Kabi AB, Uppsala, Sweden). Six 10 mm × 10 mm full thickness wounds were made, separated by the distance of 5 mm. Bacteria were prepared according to the same protocol as described above and diluted to 2×10^9 CFU/mL. The wounds were seeded with 20 µL of either MRSA (ATCC 33591) or *P. aeruginosa* (ATCC 10145) for the treatment with CEN1 HC, and with 20 µL of MRSA (ATCC 33591) for the treatment with CEN1 HC (Ser). Two hours after infection, 100 µL of peptide in H_2O (0.1–2 mg/mL) or placebo (H_2O) was added to each wound. The treatment was randomised for all wounds. All animals were euthanized two hours after treatment by an overdose of pentobarbital sodium (Pentobarbital vet, APL, Stockholm, Sweden). To estimate bacterial counts, the whole wound area was dissected and transferred to a micro-centrifuge tube and placed on ice. A volume of 500 µL of Kligman buffer (0.1% Triton X-100 in 0.075 M phosphate buffer, pH 7.9) was added and the tube was vortexed for two minutes, followed by shaking for ten minutes at 1400 rpm. Each suspension was diluted in four tenfold serial steps in diluted Kligman buffer (0.05% Triton X-100 in 0.0375 M phosphate buffer, pH 7.9). Aliquots of 50 µL from each of the four dilutions were seeded on blood agar plates (Oxoid) and incubated at 37°C overnight. Plates containing 30–300 CFU were counted. All personnel involved in the surgery, harvesting the bacteria and counting the plates were blinded to the different treatments.

Ex Vivo Model Using Pig Skin

The ability of CEN1 HC and CEN1 HC (Ser) to reduce bacterial counts was estimated in an *ex vivo* wound model using pig skin as previously described (Schmidtchen et al. 2009). The pigs used in the study were a mixed breed of Yorkshire, Hampshire and Swedish Pigham. The pig was shaved after euthanization and skin from the back was removed, packed in plastic foil and frozen at −20°C for storage. Before the experiment, the skin was taken out from the freezer and the subcutaneous fat was removed with a scalpel. The skin was put in a petri-dish with Kleenex paper tissues (Kimberly Clark GmbH, Koblenz, Germany) in the bottom, wetted with sterile H_2O, and cleaned with 70% ethanol. Punch biopsies were made, approximately 0.5–1 mm in thickness and 3 mm in diameter. The top cylinder of a cut 1.5 mL micro-centrifuge tube (diameter approximately 9 mm) was glued around each punch wound with ethyl cyanoacrylate glue (Loctite super glue gel, Henkel Norden, Stockholm, Sweden). The area inside the cylinder was washed two times with 250 µL of sterile H_2O.

Bacteria were cultured according to the same protocol as described above and diluted to a concentration of 10^7 CFU/mL. A volume of 100 µL of the *S. aureus* (ATCC 12600) suspension was applied to the wound area. The lid of the petri-dish was applied to create a moist chamber, and the skin was incubated for two hours at 37°C. A volume of 100 µL peptide solutions in H_2O (0.1–2 mg/mL) or placebo (H_2O) was added to the infected area and the sample was incubated for another four hours at 37°C. The liquid in the cylinder was removed and the remaining bacteria were harvested by adding 210 µL of Kligman buffer (0.1% Triton X-100 in 0.075 M phosphate buffer, pH 7.9) and the skin and wound inside the cylinder was scratched with a plastic loop using moderate force. The suspension was transferred to a micro-centrifuge tube, the procedure was repeated once and the two fractions of liquid from the infected area were pooled. The suspensions were diluted in four tenfold serial steps in diluted Kligman buffer (0.05% Triton X-100 in 0.0375 M phosphate buffer, pH 7.9) and an aliquot of 50 µL from each dilution was seeded on horse blood agar plates (Oxoid). The plates were incubated at 37°C for 24 h and plates containing 30–300 CFUs were counted. All personnel involved in the surgery, harvesting the bacteria and counting the plates were blinded to the different treatments.

Statistical Analysis

The data is presented as % of the control group ± SEM (%). The results were analysed with Student's *t*-test, with a value of $p < 0.05$ considered statistically significant. Grubb's test was performed to statistically identify outliers (Grubbs 1969).

RESULTS

In vitro Microbicidal Effect of Peptides Derived from Centrocin 1

An MMC_{99} assay was used to examine the ability of centrocin 1 heavy chain (CEN1 HC-Br), its debrominated analogue (CEN1 HC), and the C-terminal truncated variants of both peptides, *i.e.*CEN1 HC-Br (1–20) and CEN1 HC (1–20) to kill bacteria *in vitro*. In order to characterize the functional significance of a free cysteine residue in the sequence of centrocin 1 heavy chain, the cysteine to serine substituted variant CEN1 HC (Ser) was also included. The sequences of the peptides are shown in Table 1. The concentration that killed ≥99% bacteria was defined as MMC_{99}. The experiments were performed using 0.037% BHI, which is the most commonly used media in this type of assessment. To mimic the physiological conditions in wounds, the data obtained in 0.037% BHI were complemented with experiments using 0.037% BHI supplemented with monovalent cations (85 mM or 150 mM NaCl), as well as using 50% of h.i. SWF.

All peptides evaluated exerted similar microbicidal activities in 0.037% BHI against *S. aureus*and *P. aeruginosa* (MMC_{99} = 12.5–25.0 mg/L; Table 2a and b). For CEN1 HC-Br and CEN1 HC, addition of monovalent cations (85 mM or 150 mM NaCl) to 0.037% BHI resulted in somewhat reduced microbicidal effect against *S. aureus* (MMC_{99} = 25.0–50.0 mg/L), while no reduction in effect was observed against *P. aeruginosa,* compared to 0.037% BHI only. For CEN1 HC-Br (1–20), CEN1 HC (1–20) and CEN1 HC (Ser), physiological-like sodium chloride concentrations completely abolished the bactericidal effect against *S. aureus* ($MMC_{99} \geq 200$ mg/L) and reduced the effect against *P. aeruginosa* (MMC_{99} = 12.5–100.0 mg/L). When the MMC_{99} assay was performed in 50% h.i. SWF, the MMC_{99} values for CEN1 HC-Br and CEN1 HC were elevated to 25.0–50.0 mg/L for *S. aureus* and 100.0 mg/L for *P. aeruginosa*. No bactericidal effect was observed against *S. aureus* or *P. aeruginosa* with CEN1 HC-Br (1–20), CEN1 HC (1–20), and CEN1 HC (Ser) at concentrations up to 200 mg/L, when 50% h.i. SWF was used as assay medium (Table 2a and b).

Table 2: Microbicidal effect of CEN1 HC-Br, CEN1 HC, CEN1 HC-Br (1–20), CEN1 HC (1–20) and CEN1 HC (Ser) against *S. aureus* (a) and *P. aeruginosa* (b)

(a)				
	***S. aureus*, medium and MMC_{99} (mg/L)**			
Test substance	**0.037% BHI**	**0.037% BHI + 85 mM NaCl**	**0.037% BHI + 150 mM NaCl**	**50% SWF**
CEN1 HC-Br	25	25	50	50
CEN1 HC	12.5	25	50	25
CEN1 HC-Br (1–20)	12.5	>200	>200	>200
CEN1 HC (1–20)	12.5	>200	>200	>200
CEN1 HC (Ser)	25	>200	>200	>200
(b)				
	***P. aeruginosa* , medium and MMC $_{99}$ (mg/L)**			
Test substance	**0.037% BHI**	**0.037% BHI + 85 mM NaCl**	**0.037% BHI + 150 mM NaCl**	**50% SWF**
CEN1 HC-Br	25	12.5	25	100
CEN1 HC	12.5	12.5	12.5	100
CEN1 HC-Br (1–20)	12.5	12.5	25	>200
CEN1 HC (1–20)	12.5	25	100	>200
CEN1 HC (Ser)	12.5	25	50	>200

The substances were tested using a MMC_{99} assay in concentrations up to 200 mg/L in four different assay media. Peptides were added to bacterial cells in duplicate (n = 2). Data are presented as maximum values from at least two independent experiments. h.i. = heat inactivated.

The microbicidal effect of CEN1 HC was further evaluated against a panel of Gram-positive bacteria (*S. aureus*, MRSA, *S. pyogenes, S. epidermidis*, and *P. acnes*) and Gram-negative bacteria (*E. coli, P. aeruginosa, K. pneumoniae*, and *A. baumannii*) as well as the yeast *C. albicans*. The results demonstrated that 3.1–6.3 mg/L or 6.3–100 mg/L of CEN1 HC was required to kill ≥99% of all the bacterial strains in 0.037% BHI broth and 50% h.i. SWF, respectively, indicating a broad range of microbicidal activities against both Gram-positive and Gram-negative bacteria assessed in both media. Interestingly, the ability to kill the yeast strain*C. albicans* appears to be impaired in 50% h.i. SWF ($MMC_{99} \geq 200$ mg/L). The MMC_{99} values for CEN1 HC compared to the conventional antibiotics mupirocin (GlaxoSmithKline) and fusidic acid (LEO Pharma) are shown in Table 3.

Table 3: Microbicidal effect of CEN1 HC against Gram-positive and Gram-negative bacterial strains and a yeast strain

	CEN1 HC MMC_{99}(mg/L)		**fusidic acid MMC_{99}(mg/L)**		**mupirocin MMC_{99}(mg/L)**	
Microorganism, strain	**0.037% BHI**	**50% h.i. SWF**	**0.037% BHI**	**50% h.i. SWF**	**0.037% BHI**	**50% h.i. SWF**
S. aureus ATCC 12600	6.3	12.5	6.3	6.3	3.1	3.1
MRSA ATCC 33591	6.3	25	6.3	6.3	3.1	3.1
S. pyogenes ATCC 12344	3.1	50	>200	>200	6.3	3.1

*S. epidermidis*ATCC 12228	3.1	12.5	6.3	6.3	3.1	3.1
P. acnes ATCC 6919	6.3	12.5	nd	nd	>200	>200
E. coli ATCC 11775	6.3	25	>200	>200	>200	>200
*P. aeruginosa*ATCC 15442	6.3	100	>200	>200	>200	>200
*K. pneumoniae*CCUG 59413	3.1	12.5	>200	>200	>200	>200
*A. baumannii*CCUG 58437	3.1	6.3	>200	>200	>200	>200
C. albicans ATCC 64549	6.3	>200	>200	>200	>200	>200

The substances were tested using a $MMC_{99\ SS}$assay in concentrations up to 200 mg/L in 0.037% BHI or 50% h.i. SWF. Peptides were added to bacterial cells in duplicate (n=2). Data are presented as maximum values from at least two independent experiments. h.i.=heat inactivated, nd=not determined.

In vitro Anti-Inflammatory Effects of Peptides Derived Fromcentrocin 1

The anti-inflammatory effect of CEN1 HC-Br, CEN1 HC, CEN1 HC (1–20) and CEN1 HC (Ser) was studied in macrophages derived from the human monocytic cell line THP-1. All peptides tested demonstrated a significant reduction in the release of the pro-inflammatory cytokine TNF-α in LPS-stimulated cells. However, the half maximal inhibitory concentration (IC_{50}) value was markedly higher for the truncated variant CEN1 HC (1–20), compared to the other peptides (Figure 1a-d, Table 4). In line with these results, the peptides CEN1 HC-Br, CEN1 HC, CEN1 HC (1–20) reduced the production of another inflammation marker, IL-6, in LPS-stimulated cells, with the effect of CEN1 HC-Br and CEN1 HC being more pronounced compared to the truncated variant CEN1 HC (1–20) (Figure 2a-c, Table 4). In these experiments, the peptides were added to the cell culture medium 30 min after addition of LPS. Based on previous findings, this time interval should be sufficient for LPS to bind to the cell receptor and thus, to exclude that the peptide would neutralize the effect of LPS on cytokine production only by scavenging this agent (Elass-Rochard et al. 1998; Haversen et al. 2000). To rule out the possibility that the reduction in cytokine production by these peptides was related to peptide-induced decline in cell viability, the cell survival was measured after six hours of LPS-stimulation. The amount of viable cells after the treatment with CEN1 HC, CEN1 HC (1–20), and CEN1 HC (Ser) was approximately 90% of the cells treated with LPS only, indicating that addition of the peptides at concentrations up to 125 mg/L, did not show any significant cytotoxic effect. CEN1 HC-Br demonstrated higher cytotoxicity compared to the other peptides as the concentration of 100 mg/L reduced the amount of viable cells by approximately 50% (data not shown).

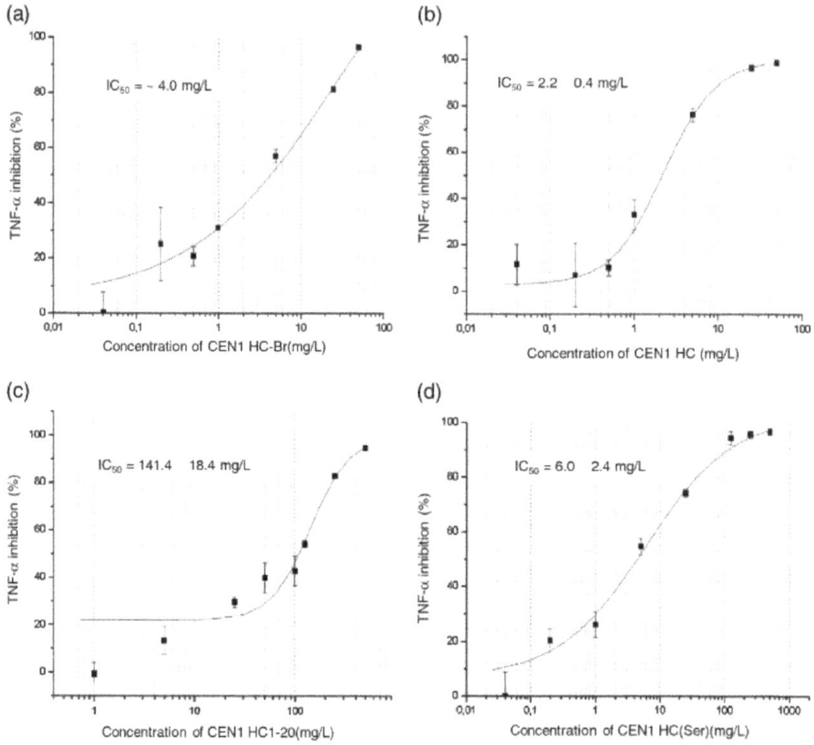

Figure 1: Effects of CEN1 HC-Br (a), CEN1 HC (b), CEN1 HC (1–20) (c) and CEN1 HC (Ser) (d) on TNF-α secretion from THP-1 cells. The peptides were added to PMA treated THP-1 cells in triplicate 30 min after the addition of LPS (0.1 ng/mL). Peptide concentrations ranging from 0.04 to 50 mg/L of CEN1 HC and CEN1 HC-Br, 0.04 to 500 mg/L of CEN1 HC (Ser) and 1 to 500 mg/L of CEN1 HC (1–20) were tested. Cytokine levels were measured in the cell supernatants by ELISA after six hours of stimulation. Data are presented as mean ± SEM with stimulated cytokine levels without peptide added set to 0%. Data was fitted to curves using a 4-parameter fit in the Origin software. The IC_{50} values were automatically calculated by the software. Due to an unsatisfactory curve fitting, $IC_{50}$for CEN1 HC-Br was estimated to be approximately 4 mg/L.

Table 4: Inhibitory effects of CEN1 HC-Br, CEN1 HC, CEN1 HC (1–20) and CEN1 HC (Ser) on TNF-α and IL-6 secretion from THP-1 cells

	IC_{50}(mg/L)	
Test substance	**TNF-α**	**IL-6**
CEN1 HC-Br	~4.0	0.0040±0.053
CEN1 HC	2.2±0.4	0.33±0.14
CEN1 HC (1–20)	141.4±18.4	9.8±8.6
CEN1 HC (Ser)	6.0±2.4	nd

Data are presented as half maximal inhibitory concentration (IC_{50}) values±SEM, as calculated by Origin software from dose–response curves using a 4-parameter fit. Due to an unsatisfactory curve fitting, IC_{50} for CEN1 HC-Br on TNF-α was estimated to approximately 4 mg/L. nd=not determined.

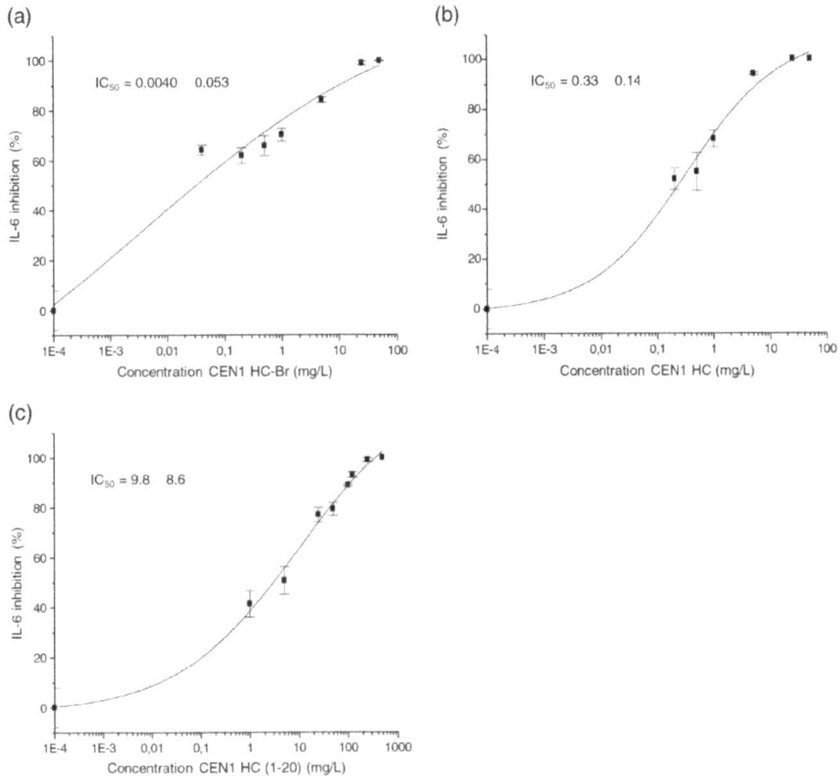

Figure 2: Effects of CEN1 HC-Br (a), CEN1 HC (b) and CEN1 HC (1–20) (c) on IL-6 secretion from THP-1 cells. The peptides were added to PMA treated THP-1 cells in triplicate 30 min after the addition of LPS (0.1 ng/mL). Peptide concentrations ranging from 0.04 to 50 mg/L of CEN1 HC and CEN1 HC-Br, and 1 to 500 mg/L CEN1 HC (1–20) were tested. Cytokine levels were measured in the cell supernatants by ELISA after six hours of stimulation. Data are presented as mean±SEM with stimulated cytokine levels without peptide added set to 0%. Data was fitted to curves using a 4-parameter fit in the Origin software. The IC_{50} values were automatically calculated by the software.

In Vitro Resistance Development towards CEN1 HC

The potential for resistance development against CEN1 HC was evaluated in two strains of Gram-positive bacteria, *S. aureus* and MRSA, by multistep dilution assay, and compared to the antibiotic mupirocin. For both strains the concentration of mupirocin killing ≥99% of the bacteria was significantly increased during 14 days of cultivation (from 3 to 24 mg/L for *S. aureus* and from 3 to >384 mg/L in MRSA). On the other hand, the concentration of CEN1 HC killing ≥99% of the bacteria was increased only twofold comparing values at day 0 and 14 (from 6 to 12 mg/L for both *S. aureus* and MRSA), which is within the limits of assay variation.

Antimicrobial Effect of CEN1 HC and CEN1 HC (Ser) in Infected Wound Models

In order to study a potential topical antimicrobial effect of CEN1 HC and CEN1 HC (Ser), an *in vivo* rat excision wound model and an *ex vivo* pig skin model were used. In the rat excision wound model full thickness wounds were infected with MRSA or *P. aeruginosa*, subsequently treated with peptide CEN1 HC or placebo (H_2O), and bacterial counts were estimated two hours after the treatment. No animal death or local reactions in connection to the application of the test article were observed during the experimental period. CEN1 HC potently reduced the level of both bacterial strains in this model (Figures 3, 4). The effect of CEN1 HC (Ser) was evaluated only in wounds infected with MRSA, and this peptide failed to significantly reduce bacterial counts in this model (data not shown).

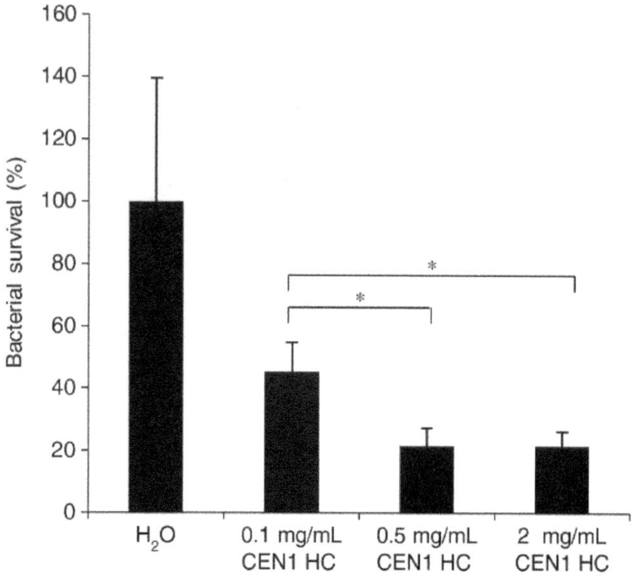

Figure 3: *In vivo* bacterial clearance of MRSA from excision wounds in rats after treatment with CEN1 HC. Wounds infected with MRSA (4×10^7 CFU) were treated with CEN1 HC, in concentrations of 0.1, 0.5, and 2 mg/mL, and sampling/bacterial counting was performed two hours later. Results are presented as mean relative bacterial survival (%) ± SEM compared to the non-treated control group (n = 14 for all the groups, except for the group treated with 0.5 mg/mL CEN1 HC, where n = 15). One value in each group, except for the group treated with 0.5 mg/mL of CEN1 HC, were considered outliers ($p<0.01$, Grubbs' test) and were discarded. * = $p<0.05$.

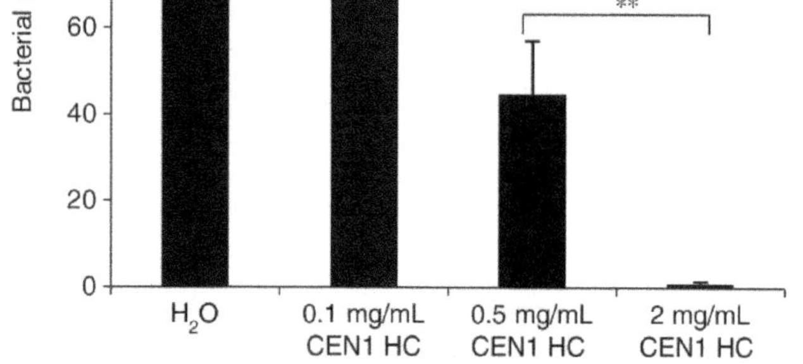

Figure 4: *In vivo* bacterial clearance of *P. aeruginosa* from excision wounds in rats after treatment with CEN1 HC. Wounds infected with *P. aeruginosa* (4×10^7 CFU) were treated with CEN1 HC, in concentrations of 0.1, 0.5, and 2 mg/mL, and sampling/bacterial counting was performed two hours later. Results are presented as mean of relative bacterial survival (%) ± SEM compared to the non-treated control group (n = 15 for all the groups). ** = $p<0.01$, *** = $p<0.001$.

In the pig skin *ex vivo* study, punch biopsies were made on the skin from the back of pigs. The wounded areas were seeded with *S. aureus*, treated with CEN1 HC, CEN1 HC (Ser) or placebo (H_2O), and bacterial counts were estimated four hours after treatment. As demonstrated in Figure 5, CEN1 HC reduced the level of bacteria in this model in a dose dependent manner. Treatment with CEN1 HC at the concentration of 0.5 and 2 mg/mL reduced bacterial counts with 98 and 99%, respectively, compared to the placebo treatment. Treatment with CEN1 HC (Ser) also significantly reduced the amount of bacteria. However, the effect was less pronounced compared to CEN1 HC (2 mg/mL of CEN1 HC (Ser) reduced bacterial counts with 67%, Figure 5).

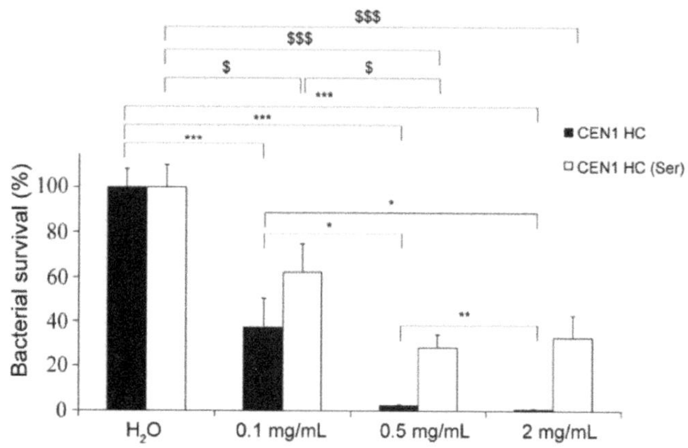

Figure 5: *Ex vivo* bacterial clearance of *S. aureus* from pig skin wounds after treatment with CEN1 HC and CEN1 HC (Ser). Wounds infected with *S. aureus* (10^6 CFU) were treated with CEN1 HC, in concentrations of 0.1, 0.5 and 2 mg/mL, and sampling/bacterial counting was performed four hours after treatment. Results are presented as mean of relative bacterial survival (%) ± SEM compared to the non-treated control group (n = 10 for all groups except for the groups treated with 0.5 and 2 mg/mL of CEN1 HC (Ser), where n = 9). One outlier ($p<0.01$, Grubbs' test) was discarded from each of the groups treated with 0.5 and 2 mg/mL CEN1 HC (Ser). * = $p<0.05$, ** = $p<0.01$, *** = $p<0.001$ comparing CEN1 HC to the control group, $ = $p<0.05$, $$$ = $p<0.001$ comparing CEN1 HC (Ser) to the control group.

DISCUSSION

The treatment of infectious diseases poses major challenges for healthcare systems worldwide due to rapid increase in bacterial strains resistant to conventional antibiotics. AMPs have emerged as novel promising agents for both topical and systemic treatment of infections. The AMPs are considered less prone to resistance development than conventional antibiotics due to the rapid microbicidal effect combined with broad mode of action (Fjell et al. 2012).

Drug development based on AMPs is complicated by the fact that the antimicrobial activities of numerous naturally occurring AMPs are antagonised by physiological salt concentrations, which might reflect the increasing ionic strength under high salt conditions weakening the electrostatic charge interactions of AMPs with bacterial membranes (Tam et al. 2002). Thus, attempts are directed towards developing therapeutic AMPs where salt-resistant properties are achieved through chemical modifications of the peptide structure (Fjell et al. 2012; Park et al. 2004; Tam et al. 2002). Peptides isolated from marine organisms, that are adapted to salt-rich environments, are likely to be less sensitive to ionic strength, and are therefore highly interesting from a drug development perspective.

In this study, the peptide previously isolated from coelomocyte (blood cell) extracts of the green sea urchin, *S. droebachiensis*, centrocin 1 heavy chain (CEN1 HC-Br) (Li et al. 2010) and the variants thereof, were evaluated for the potential as novel anti-infectious agents by measuring their antibacterial and anti-inflammatory effect *in vitro*. CEN1 HC-Br contains a brominated tryptophan in position two in the amino acid sequence (Li et al. 2010) with unknown function. AMPs containing brominated tryptophan have previously been characterized from several marine organisms such as hagfish intestinal antimicrobial peptides (HFIAPs), isolated from Atlantic hagfish or styelin D, isolated from tunicate *Styela clava* (Taylor et al. 2000; Uzzell et al.2003). It has been speculated that the unusual amino acid bromotryptophan makes these AMPs less susceptible to proteolytic degradation and may increase the biological activity of the peptide (Li et al. 2010). To elucidate the functional significance of the bromination of the tryptophan residue, the debrominated variant of the centrocin 1 heavy chain, *i.e.* CEN1 HC, was included in the assessments. To map the peptide region necessary of the biological activity, the C-terminal truncated variants of both peptides, CEN1 HC-Br (1–20) and CEN1 HC (1–20), were also tested. A free cysteine residue in the peptide sequence may potentially complicate the product development process due to the possible heterogeneity of the product caused by formation of disulphide bonds between molecules. Therefore, we investigated if the cysteine residue in CEN1 HC (position 25) could be replaced by a similar amino acid. The most commonly used replacement for cysteine is serine, which in terms of geometry and volume occupancy is a highly isosteric analogue of cysteine. A cysteine to serine modified variant of CEN1 HC, CEN1 HC (Ser), was compared for microbicidal and anti-inflammatory properties to its non-substituted equivalent.

When the CEN1 HC-Br derived peptides were evaluated for anti-bacterial activity against *S. aureus* and *P. aeruginosa*, all peptides showed pronounced microbicidal effect in assay medium containing low salt concentrations and no serum (0.037% BHI). The peptides CEN1 HC-Br and CEN1 HC were generally salt and wound fluid tolerant and significant microbicidal activity was observed in the presence of 150 mM NaCl and in 50% h.i. SWF. Notably, physiological sodium chloride concentrations as well as h.i SWF severely decreased or fully eliminated the bactericidal effect of CEN1 HC-Br (1–20), CEN1 HC (1–20) and CEN1 HC (Ser). The microbicidal effect of the peptides and their sensitivity to physiological-like conditions varied dependently on the bacterial strain tested, possibly due to differences between the cell wall/membrane properties of the Gram-positive (*S. aureus*) and Gram-negative (*P. aeruginosa*) bacteria. Interestingly, debromination of tryptophan in CEN1 HC-Br did not impair the bactericidal effect in neither condition of low salt nor under physiological-like conditions.

The anti-inflammatory properties of CEN1 HC-Br, CEN1 HC, CEN1 HC (1–20) and CEN1 HC (Ser) were evaluated by measuring secretion of the most commonly used inflammation markers TNF-α and IL-6 in human monocyte derived macrophages stimulated with LPS. In this assay, CEN1 HC-Br, CEN1 HC and the Ser-substituted peptide CEN1 HC (Ser) exerted a more potent anti-inflammatory effect compared to the C-terminal truncated variant CEN1 HC (1–20). No obvious difference in efficacy was seen comparing CEN1 HC-Br and CEN1 HC, while the first peptide had higher cytotoxicity compared to the debrominated equivalent.

Based on these experiments, we conclude that debromination of the tryptophan residue of CEN1 HC-Br did not result in any reduced effect under the experimental conditions tested. Importantly, CEN1 HC had a more favourable safety profile compared to the original peptide, with no lytic activity observed against mammalian cells at the concentrations tested, thus showing a clear dissociation of antibacterial and anti-eukaryotic cell activities. C-terminal truncation of the full-length peptide variants resulted in reduction in salt-tolerability of the microbicidal effect as well as in reduced anti-inflammatory properties. Also, the peptide with the free cysteine residue, CEN1 HC had improved salt and serum tolerability profile in the MMC_{99}assay compared to the serine-modified equivalent. Based on these observations, it was concluded that from the panel of peptide variants tested, CEN1 HC showed the best efficacy and safety profile. Thus, CEN1 HC was selected to be further evaluated for its *in vitro* and *in vivo*antimicrobial properties.

CEN1 HC was shown to have broad microbicidal effect against microorganisms appearing in topical and parenteral infections including Gram-positive bacteria (*S. aureus*, *S. pyogenes*, *S. epidermidis*, and *P. acnes*), Gram-negative bacteria (*E. coli*, *P. aeruginosa*, *K. pneumonia*, and*A. baumannii*) and the yeast *C. albicans*. CEN1 HC also showed pronounced effect against the pathogen MRSA. Importantly, during the cultivation of 14 days, the bacterial strains tested (*S. aureus* and MRSA) failed to develop significant resistance towards CEN1 HC, indicating that these strains could not circumvent the action of this AMP, while significant resistance profile was observed toward the conventional antibiotic mupirocin used as comparator.

CEN1 HC demonstrated marked effect on reducing bacterial counts in infected wound models in two different animal species. In a rat model of infected full thickness excision wounds, treatment by CEN1 HC significantly reduced the bacterial counts of MRSA as well as *P. aeruginosa*, the most common pathogens in topical infections in man. The antibacterial effect against *S. aureus*was confirmed in an infected wound model in pig skin. In line with the *in vitro* tests, CEN1 HC (Ser) had markedly lower efficacy in these two animal models, compared to the cysteine-containing equivalent.

In summary, we demonstrate that CEN1 HC, a chemically synthesised 30 amino acid peptide sequentially derived from the previously described AMP centrocin 1, has a broad spectrum microbicidal effect, resistant to physiological salt concentration and serum containing wound fluid, combined with anti-inflammatory action. Importantly, *S. aureus* and MRSA failed to develop resistance against this peptide. CEN1 HC significantly reduced bacterial counts of *S. aureus*, MRSA and *P. aeruginosa* in animal models of infected wounds. Based on this study, CEN1 HC appears to be a promising agent in the topical treatment of infections, and further studies are warranted to evaluate the applicability of this AMP in clinical settings.

ACKNOWLEDGEMENTS

This work was supported by grants from the University of Tromsø, grants from the Marine Biotechnology in Tromsø (MABIT) research program (grant number BS0023 and

BS0034) and by grants from the Pergamum AB Research and Development Fund, Gothenburg, Sweden. We gratefully acknowledge Dr. Inger Mattsby-Baltzer (Department of Infectious Medicine, Sahlgrenska Academy, University of Gothenburg, Gothenburg, Sweden) for the assistance in setting up the culturing system for *P. acnes*.

COMPETING INTERESTS

Some of the authors of this paper have links to the Pergamum company. The links are as follows: Camilla Björn, Joakim Håkansson, Emma Myhrman, Veronika Sjöstrand, Kerstin Lindgren, and Margit Mahlapuu – fulltime employee at the time of the investigation. Pergamum contributed financially (including salaries), by providing laboratory space and payment for the study costs (animals and housing, drugs, and similar). All other authors declare that they have no competing interests.

REFERENCES

1. Auvynet C, Rosenstein Y: Multifunctional host defense peptides: antimicrobial peptides, the small yet big players in innate and adaptive immunity. *FEBS J* 2009,276: 6497–6508. 10.1111/j.1742-4658.2009.07360.x
2. Bowdish DM, Davidson DJ, Scott MG, Hancock RE: Immunomodulatory activities of small host defense peptides. *Antimicrob Agents Chemother* 2005, 49: 1727–1732. 10.1128/AAC.49.5.1727-1732.2005
3. Devine DA, Hancock RE: Cationic peptides: distribution and mechanisms of resistance. *Curr Pharm Des* 2002, 8: 703–714. 10.2174/1381612023395501
4. Elass-Rochard E, Legrand D, Salmon V, Roseanu A, Trif M, Tobias PS, Mazurier J, Spik G:Lactoferrin inhibits the endotoxin interaction with CD14 by competition with the lipopolysaccharide-binding protein. *Infect Immun* 1998, 66: 486–491.
5. Fjell CD, Hiss JA, Hancock RE, Schneider G: Designing antimicrobial peptides: form follows function. *Nat Rev Drug Discov* 2012, 11: 37–51. 10.1038/nrd3591
6. Giordano P, Weber K, Gesin G, Kubert J: Skin and skin structure infections: treatment with newer generation fluoroquinolones. *Ther Clin Risk Manag* 2007, 3: 309–317. 10.2147/tcrm.2007.3.2.309
7. Gisby J, Bryant J: Efficacy of a new cream formulation of mupirocin: comparison with oral and topical agents in experimental skin infections. *Antimicrob Agents Chemother* 2000, 44: 255–260. 10.1128/AAC.44.2.255-260.2000
8. Grubbs FE: Procedures for detecting outlying observations in samples.*Technometrics* 1969, 11: 1–21. 10.1080/00401706.1969.10490657
9. Hancock RE, Diamond G: The role of cationic antimicrobial peptides in innate host defences. *Trends Microbiol* 2000, 8: 402–410. 10.1016/S0966-842X(00)01823-0
10. Haug T, Kjuul AK, Styrvold OB, Sandsdalen E, Olsen OM, Stensvag K: Antibacterial activity in Strongylocentrotus droebachiensis (Echinoidea), Cucumaria frondosa (Holothuroidea), and Asterias rubens (Asteroidea). *J Invertebr Pathol* 2002, 81: 94–102. 10.1016/S0022-2011(02)00153-2
11. Haversen L, Kondori N, Baltzer L, Hanson LA, Dolphin GT, Duner K, Mattsby-Baltzer I:Structure-microbicidal activity relationship of synthetic fragments derived from the antibacterial alpha-helix of human lactoferrin. *Antimicrob Agents Chemother* 2010,54: 418–425. 10.1128/AAC.00908-09
12. Haversen LA, Engberg I, Baltzer L, Dolphin G, Hanson LA, Mattsby-Baltzer I: Human lactoferrin and peptides derived from a surface-exposed helical region reduce experimental Escherichia coli urinary tract infection in mice. *Infect Immun* 2000,68: 5816–5823. 10.1128/IAI.68.10.5816-5823.2000
13. Jevons MP: "Celbenin" - resistant Staphylococci. *Br Med J* 1961, 14: 124–125.
14. Jones RN, Biedenbach DJ, Sader HS, Fritsche TR, Toleman MA, Walsh TR: Emerging epidemic of metallo-beta-lactamase-mediated resistances. *Diagn Microbiol Infect Dis*2005, 51: 77–84. 10.1016/j.diagmicrobio.2004.07.003
15. Kosowska-Shick K, Clark C, Credito K, McGhee P, Dewasse B, Bogdanovich T, Appelbaum PC:Single- and multistep resistance selection studies on the activity of retapamulin compared to other agents against Staphylococcus aureus and Streptococcus pyogenes. *Antimicrob Agents Chemother* 2006, 50: 765–769. 10.1128/AAC.50.2.765-769.2006
16. Li C, Haug T, Moe MK, Styrvold OB, Stensvag K: Centrocins: isolation and characterization of novel dimeric antimicrobial peptides from the green sea urchin, Strongylocentrotus droebachiensis. *Dev Comp Immunol* 2010, 34: 959–968. 10.1016/j.dci.2010.04.004
17. McRipley RJ, Whitney RR: Characterization and quantitation of experimental surgical-wound infections used to evaluate topical antibacterial agents. *Antimicrob Agents Chemother* 1976, 10: 38–44. 10.1128/AAC.10.1.38
18. Moet GJ, Jones RN, Biedenbach DJ, Stilwell MG, Fritsche TR: Contemporary causes of skin and soft tissue infections in North America, Latin America, and Europe: report from the SENTRY Antimicrobial Surveillance Program (1998–2004). *Diagn Microbiol Infect Dis* 2007, 57: 7–13. 10.1016/j.diagmicrobio.2006.05.009
19. Park IY, Cho JH, Kim KS, Kim YB, Kim MS, Kim SC: Helix stability confers salt resistance upon helical antimicrobial peptides. *J Biol Chem* 2004, 279: 13896–13901. 10.1074/jbc.M311418200
20. Rittenhouse S, Singley C, Hoover J, Page R, Payne D: Use of the surgical wound infection model to determine the efficacious dosing regimen of retapamulin, a novel topical antibiotic. *Antimicrob Agents Chemother* 2006, 50: 3886–3888. 10.1128/AAC.00183-06
21. Saymen DG, Nathan P, Holder IA, Hill EO, Macmillan BG: Infected surface wound: an experimental model and a method for the quantitation of bacteria in infected tissues. *Appl Microbiol* 1972, 23: 509–514.

22. Schmidt FR: The challenge of multidrug resistance: actual strategies in the development of novel antibacterials. *Appl Microbiol Biotechnol* 2004, 63: 335–343. 10.1007/s00253-003-1344-1

23. Schmidtchen A, Pasupuleti M, Morgelin M, Davoudi M, Alenfall J, Chalupka A, Malmsten M:Boosting antimicrobial peptides by hydrophobic oligopeptide end tags. *J Biol Chem*2009, 284: 17584–17594. 10.1074/jbc.M109.011650

24. Schuerholz T, Brandenburg K, Marx G: Antimicrobial peptides and their potential application in inflammation and sepsis. *Crit Care* 2012, 16: 207. 10.1186/cc11220

25. Scott MG, Hancock RE: Cationic antimicrobial peptides and their multifunctional role in the immune system. *Crit Rev Immunol* 2000, 20: 407–431.

26. Sperstad SV, Haug T, Blencke HM, Styrvold OB, Li C, Stensvag K: Antimicrobial peptides from marine invertebrates: challenges and perspectives in marine antimicrobial peptide discovery. *Biotechnol Adv*2011,29:519–530.10.1016/j.biotechadv.2011.05.021

27. Tam JP, Lu YA, Yang JL: Correlations of cationic charges with salt sensitivity and microbial specificity of cystine-stabilized beta -strand antimicrobial peptides. *J Biol Chem* 2002, 277: 50450–50456. 10.1074/jbc.M208429200

28. Taylor SW, Craig AG, Fischer WH, Park M, Lehrer RI: Styelin D, an extensively modified antimicrobial peptide from ascidian hemocytes. *J Biol Chem* 2000, 275: 38417–38426. 10.1074/jbc.M006762200

29. Teixeira V, Feio MJ, Bastos M: Role of lipids in the interaction of antimicrobial peptides with membranes. *Prog Lipid Res* 2012, 51: 149–177. 10.1016/j.plipres.2011.12.005

30. Uzzell T, Stolzenberg ED, Shinnar AE, Zasloff M: Hagfish intestinal antimicrobial peptides are ancient cathelicidins. *Peptides* 2003, 24: 1655–1667. 10.1016/j.peptides.2003.08.024

31. Vermeulen H, van Hattem JM, Storm-Versloot MN, Ubbink DT: Topical silver for treating infected wounds. *Cochrane Database Syst Rev* 2007, CD005486. 10.1002/14651858.CD005486.pub2

32. WHO: *The World Health Report 2004 – changing history. Annex Table 2: Deaths by cause, sex and mortality stratum in WHO regions, estimates for 2002.* World Health Organization, Geneva, Switzerland; 2004:120.

33. Yount NY, Yeaman MR: Emerging themes and therapeutic prospects for anti-infective peptides. *Annu Rev Pharmacol Toxicol* 2012, 52: 337–360. 10.1146/annurev-pharmtox-010611-134535

CHAPTER 17

GROWTH AND REPRODUCTIVE BIOLOGY OF THE SEA STARASTROPECTEN ARANCIACUS (ECHINODERMATA, ASTEROIDEA) ON THE CONTINENTAL SHELF OF THE CATALAN SEA (NORTHWESTERN MEDITERRANEAN)

Marc Baeta, Eve Galimany and Montserrat Ramón

Keywords: Asteroidea, Starfish, Mediterranean and echinoderm

BACKGROUND

Sea stars play an essential ecological role as high-level predators in the structure and function of intertidal and subtidal benthic communities, where they promote heterogeneity and diversity [23]. Changes in the abundance of these predators can have far-reaching consequences on the population characteristics of many of the invertebrates on which they prey [11]. Despite the ecological importance of sea stars, research has been mainly limited to certain species and to areas where they have an economic impact because they feed on important shellfish or on reef-building corals [28].

Astropecten (Fam. Astropectinidae) is one of the most species-rich genera among sea stars and its members are distributed worldwide, from polar to tropical seas and from intertidal areas to the deep sea [49]. They are voracious predators, feeding mainly on gastropods and bivalves. They have an intra-oral feeding mechanism, swallowing prey intact [11]. In general they exhibit a patchy spatial distribution as a consequence of a combination of biotic (e.g., prey and congener abundance) and abiotic factors (e.g., depth and sediment properties) [36].Astropecten spp. have a pair of gonads and a pair of pyloric caeca (digestive organs) located in each arm [44]. Gonad and pyloric cecum indices are widely used as indicators of reproductive effort and nutritional condition. Resources stored in pyloric cecum seem to play an important role in the seasonal production of gonads in some asteroid species [29].

There is a vast literature on the genus Astropecten worldwide, mainly focusing on their feeding behavior, diet characteristics [3, 11] and early stages of development [22, 27]. However, aspects of their biology such as growth and reproductive biology have been less analyzed than those of other sea star genera [44]. Growth in field populations has been studied

Citation: Marc Baeta, Eve Galimany, Montserrat Ramón, Growth and reproductive biology of the sea starAstropecten aranciacus (Echinodermata, Asteroidea) on the continental shelf of the Catalan Sea (northwestern Mediterranean), DOI: 10.1186/s10152-016-0453-z.

in both tropical [43] and temperate waters [15, 35]. Ventura [44] observed a growth decrease coinciding with the seasonal upwelling season in Brasil. Freeman et al. [15] observed that A. irregularis growth in North Wales was highly seasonal, with maximum growth occurring between mid-summer and mid to late autumn. Ventura [44] Astropecten spp. reproduction has been analyzed in temperate [11, 15, 18] and in tropical waters [31, 47, 48]. Reproduction is usually seasonal in temperate waters and spawning takes place with the increase in water temperatures [15]; however, in tropical waters sympatric Astropecten species seem to react differently to changes in sea water temperature and then, spawning occurs at different seasons [44].

Astropecten aranciacus (Linnaeus 1758) inhabits the Mediterranean Sea and the eastern Atlantic from Portugal to Angola, including the Canary, Madeira and Cape Verde Islands. It is one of the largest sea stars in the Mediterranean Sea, with a body diameter of up to 60 cm. It lives on sandy, muddy and detrital bottoms, as well as on seagrass beds, including Zostera,Cymodocea and Posidonia, between 1 and 200 m depth [42]. A. aranciacus is the dominant species on the continental shelf of the Maresme coast (northwestern Mediterranean Sea), where it co-occurs with other species of the same genus (A. platyacanthus, A. irregularis pentacanthus, A. bispinosus, A. spinulosus and A. jonstoni) between 5 and 30 m depth [2]. Baeta and Ramón [3] observed that Astropecten sea stars have niche segregation, showing a partitioning of the resources and occupying different microhabitats to avoid food competition. At depths between 50 and 200 m in the same area, A. aranciacus co-occurs with A. irregularis pentacanthus and other non-congeneric species (Luidia ciliaris, Tethyaster subermis,Peltaster placenta, Anseropoda placenta and Hacelia attenuata) [2].

Astropecten aranciacus has a defined behavioral pattern: it remains buried in the sand during the daytime, emerges just before sunset and looks for prey all night, and buries itself again just a few hours after dawn [37]. In the study area A. aranciacus feeds mainly on the bivalvesGlycymeris glycymeris and Callista chione, together with the gastropod Cyclope neriteabetween 5 and 30 m depth [3]; and almost exclusively on the bivalve Timoclea ovata between 50 and 200 m depth [2]. Recently, a major decrease in its population has been detected in the Mediterranean Sea [50].

This paper examines for the first time growth and reproduction aspects of A. aranciacus. For this purpose, we have analyzed modal progression of size classes, histological sections of gonads and variations of the pyloric caeca and gonadal indices. In addition, the influence of exogenous factors such as photoperiod, sea water temperature and chlorophyll (Chl a) concentration on the reproductive biology of the species was also analyzed.

METHODS

Study Site

The study was carried out on the continental shelf of the Catalan Sea (northwestern Mediterranean), between the Tordera River delta (northeast) and the mouth of the Tiana seasonal stream (southwest) (Fig. 1). This area is called the Maresme coast, and it is a reflective, sandy and straight, wave-dominated plain which displays a NE–SW trend. The continental shelf has a mean width of 21 km and a gradient of between 0.01° and 10°, with the shelf break occurring at 130 ± 20 m [14]. Its seabed consists of coarse to medium sandy bottoms nearshore and mud and/or sandy bottoms offshore [13].

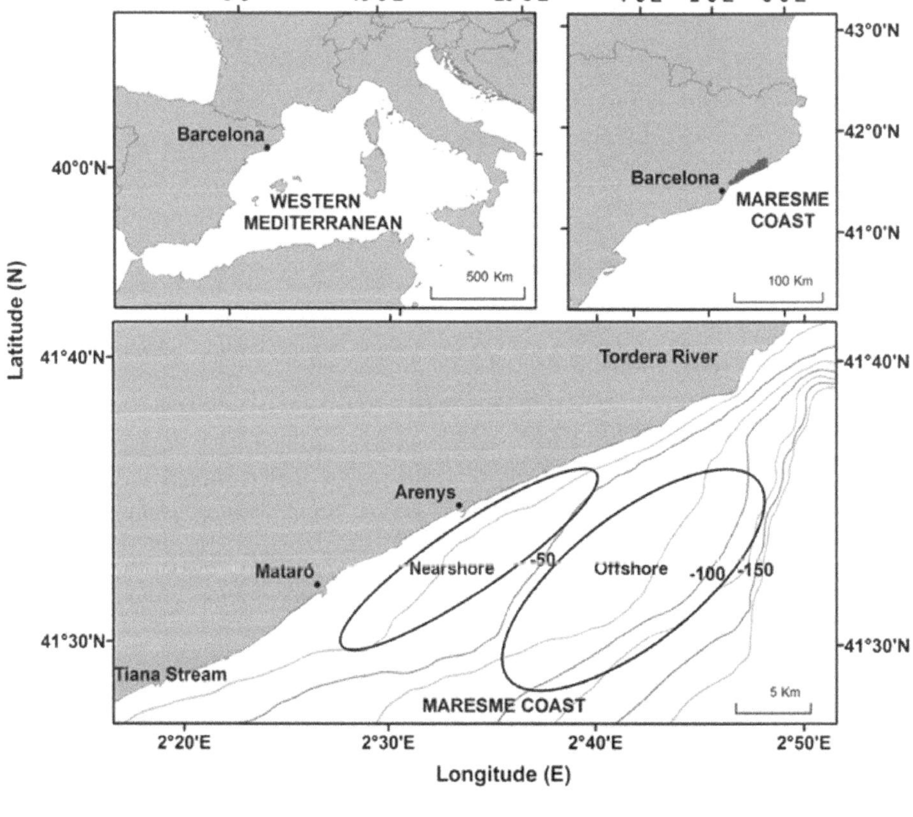

Figuer 1: Study area showing the location of the sampling areas on the Maresme coast (circles) and the 50, 100 and 150 m isobaths.

Sampling

Monthly samples were obtained on the Maresme coast between November 2009 and October 2012. Sampling was performed by fishermen on board two vessels from the Arenys de Mar harbor fishing fleet, which use commercial fishing gears according to their fishing modality (shellfish and trawlers) and fishing grounds. The impossibility of sampling the whole study area with the same gear due to local fishing regulations forced us to subdivide the study site into two nearby areas (Fig. 1): nearshore, where the shellfish fleet operate between 5 and 30 m depth; and offshore, where trawlers are allowed to operate below 50 m depth. The strip between 30 and 50 m depth is not covered by these fishing boats in their daily activities. The nearshore area was sampled during daylight onboard the artisanal fishing boat Esparta (10 m in length; 100 hp), which belonged to the smooth clam fishery fleet and used clam dredges. The clam dredge consisted of a metal frame (mouth length, 70 cm; mouth height, 53 cm), a toothed lower bar and a rectangular metal grid box (mesh size, 29 × 29 mm). Each tow lasted approximately 30 min at a speed of 1–1.2 km, and the mean towed area was around 800 m². The offshore area (between 50 and 150 m depth) was sampled during daylight onboard the small trawl fishing boat San Benito (10.75 m in length; 105 CV), which targeted black bellied angler (Lophius budegassa) and royal sea cucumber (Parastichopus regalis). The trawl net was cone-shaped, with a mouth 4 m wide × 0.5 m high and a quadrangular mesh size of 40 × 40 mm. Each tow lasted approximately 2 h at a speed of 1.8–2 km. All the A. aranciacus caught, regardless of the gear used, were transported in a cool box (4 °C) to the laboratory, where they were weighed and measured. The length of the sea stars was defined as R, major radius length or distance from the center of the oral disk to the tip of the longest arm, and was measured using calipers to the nearest 0.1 mm.

Growth

Length–weight relationship was estimated by combining biometrical data from individuals collected monthly between November 2009 and October 2012 in both bathymetrical ranges. The relationship between the length and weight of one organism is usually expressed by the equation $W = aL^{b}$, where W is body weight (g), L is length (cm), a is the intercept and b is the allometric coefficient. Values of b significantly different from 3 indicated that growth in weight is relatively faster than growth in length (positive allometry $b > 3$) or lower (negative allometry $b < 3$).

Nearshore and offshore samples were analyzed separately to study growth due to the different gear selectivity. Sea star length (R) was categorized into 20-mm classes and grouped every 2 months to obtain bimonthly length-frequency distributions. Growth was estimated from the modal class progression analysis in the length-frequency distributions, using the Bhattacharya method [5], previously used to examine growth in other asteroids [7, 30] and included in the FISAT II software [16]. The parameters of the von Bertalanffy growth function ($L_t = L_\infty(1-e^{-k(t-t0)})$) were estimated by nonlinear regression using the FiSAT II software package, where, from left to right, L_t is the predicted length at time t (mm), L_∞ is the asymptotic length (mm), k is the growth constant (year^{-1}), t is the age (year) of the A. aranciacus, and t_0 is the age at which $L_t = 0$ [21].

Reproduction

All nearshore and offshore individuals collected monthly, from May 2011 to September 2012, were used for reproduction analyses. Each sea star was damp-dried and weighed to the nearest 0.001 g. The oral surface was opened by separating the plates along the ambulacral grooves to expose the viscera. The number of immature individuals was annotated, and all gonads from adults were examined (between 18 and 34 gonads per month). Gonads and pyloric caeca of the five arms were removed, damp-dried and weighed to 0.001 g accuracy. Gonads were fixed in 10 % formalin for 24 h, cleaned with deionized water and preserved with 70 % ethanol. The gonad tissue was dehydrated in graded ethanol, and then infiltrated and embedded with glycol-methacrylate resin. Gonads were cut into 4-μm sections and stained in Lee's methylene blue-basic fuchsin. Microscopic examination of the sections allowed sex determination and assignation of each individual to a gonad maturation state. Development was divided into five stages based on an adaptation of scales used for other asteroid species, where each stage is determined according to the frequency and characteristics of gametes, presence of phagocytes and cell remains, as well as thickness and shape of the gonadal wall [4, 9, 10, 32, 38].

The gonadal index (GI) and the pyloric caeca index (PCI) were calculated as [17]:

$$\frac{\text{Wet organ weight}}{\text{Total body wet weight}} \times 100$$

Size at first maturation (size at which gonads of 50 % of the individuals are mature, R50 %) was calculated by plotting the percentage of individuals with mature gonads versus size [12]. To analyze the influence of some exogenous factors on the reproductive biology of this species, surface water temperature was obtained from the Barcelona buoy of "Puertos del Estado" (41.32°N/02.20°E), photoperiod data from the "Instituto Geográfico Nacional, Ministerio de Fomento" and the monthly Chl a concentration was obtained from MODIS satellite imagery, using a resolution of 4 km with 0.1 × 0.1 degrees areas.

Statistical Analysis

Sex ratio was tested using a Chi-square (X^2) test. Differences in the mean gonadosomatic index values between sexes were assessed for each month using a Mann–Whitney U test. A Kruskal–Wallis nonparametric ANOVA was used to check for differences in pooled GI and PCI data among months. Relations among PCI, temperature, photoperiod and Chl a (independent variables) with the reproductive cycle represented by GI (dependent variable) were assessed through a step-by-step multiple linear regression analysis (GLM) using STATA v1.2 statistical Software (Stata Corporation, College Station, TX, USA). Statistical significance was accepted at $p < 0.05$ [4].

RESULTS

Sea stars collected from both nearshore and offshore areas had different size compositions (Fig. 2). A. aranciacus major radius length (R) ranged between 11 and 214 mm in nearshore samples, with a mean size of 81.40 mm (SD = 39.44), whereas R ranged between 47 and 280 mm in offshore individuals, with a mean size of 161.33 mm (SD = 44.47). The length–weight relationship can be expressed as $W = 0.0025L^{2.22}$ ($R^2 = 0.95$; $F_{(1.487)} = 10{,}179$, N = 1365). Concerning shape variation, the length weight relationship revealed that weight is negatively allometric with length ($p < 0.001$, t = 35.54), i.e., A. aranciacus becomes relatively more slender as it grows.

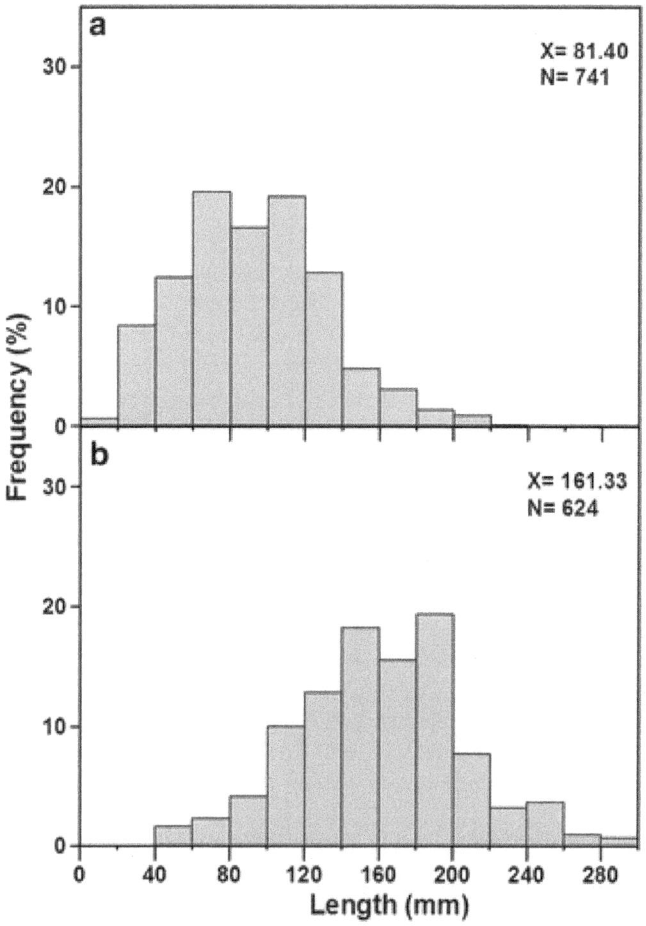

Figure 2:Astropecten aranciacus length-frequency distributions of individuals collected from November 2009 to October 2012**.** a Nearshore (5–30 m depth) and b offshore (50–150 m depth). X is the mean size (mm) and N is the number of sea stars analyzed.

Growth

The A. aranciacus bimonthly length-frequency distributions at both bathymetrical ranges are shown in Fig. 3. Recruitment nearshore was recorded twice in the whole period of study, in February 2010 and in April 2012. No recruitment or individuals smaller than 100 mm were found in the offshore samples. The progression of the identified cohorts over the 3 years of study can be seen in Fig. 4. In the nearshore samples (Fig. 4a) a cohort (C-1) was observed from April 2012 (R = 47.56 mm) to August 2012 (R = 69.15 mm). Another cohort (C-2) was observed from February 2010 (R = 34.40 mm) to October 2012 (R = 123.62 mm), with a size increase of 89.22 mm. A third cohort (C3) was detected from December 2009 (R = 113.02 mm) to June 2010 (R = 115.51 mm). In the offshore samples (Fig. 4b), we also observed three cohorts. The C-4 cohort was observed from December 2010 (R = 103.04 mm) to October 2012 (R = 144.65 mm); another cohort (C-5) was found from December 2009 (R = 125.58 mm) to August 2012 (R = 171.08 mm); and the last one (C-6) was occasionally detected from February 2011 (R = 206.5 mm) to April 2012 (R = 217.64 mm). Growth rates could be obtained following the mean length of the long lasting cohorts C-2, C-4 and C-5 over time. The nearshore C-2 cohort showed higher growth rates during summer and autumn (18–24 mm from June to October) in the 3 years studied. The offshore cohort C-4 showed higher growth between February–August 2011 (22.8 mm) and February–October 2012 (10.65 mm). The offshore cohort C-5 showed higher growth between April and August 2010 (14.44 mm). The estimated von Bertalanffy growth parameters for nearshore individuals were L_∞ = 136.75 mm, k = 0.44 $year^{-1}$ and t_0 = 0.0017 (N = 741; R^2 = 0.73), whereas for offshore individuals they were L_∞ = 205 mm, k = 0. 32 $year^{-1}$ and t_0 = 0.0011 (N = 624; R^2 = 0.82).

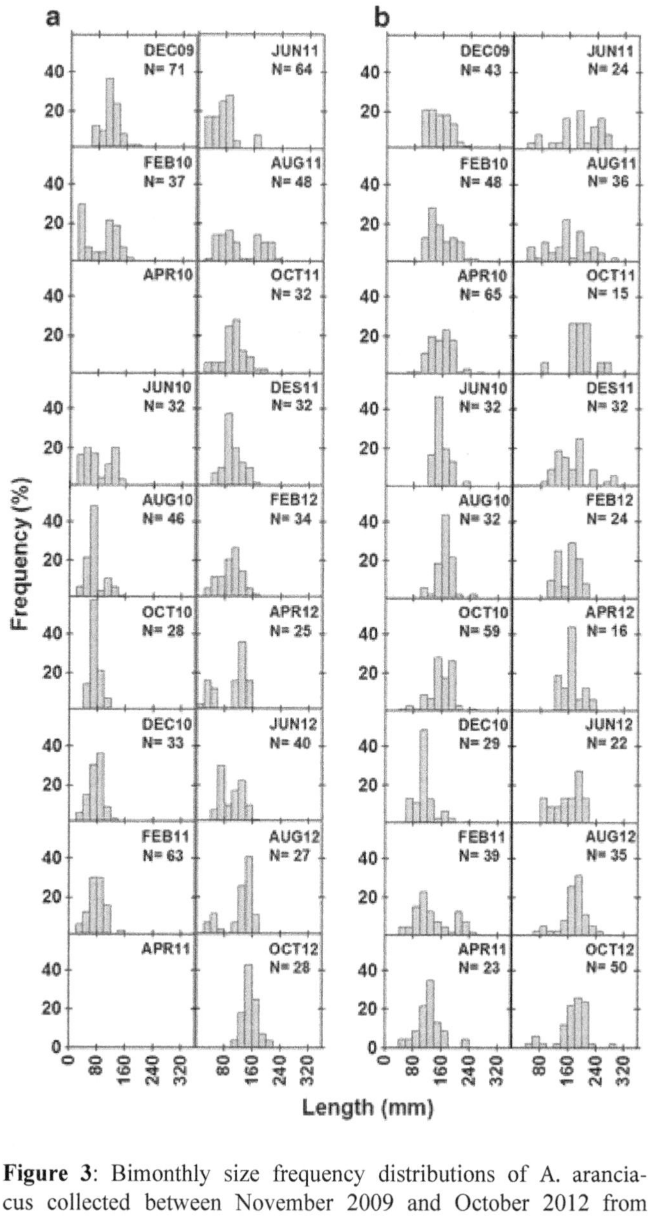

Figure 3: Bimonthly size frequency distributions of A. aranciacus collected between November 2009 and October 2012 from Maresme coast. a Nearshore (5–30 m depth) and b offshore (50–150 m depth). Data for April 2010 and April 2011 nearshore are not presented because the sample size was <10 sea stars.

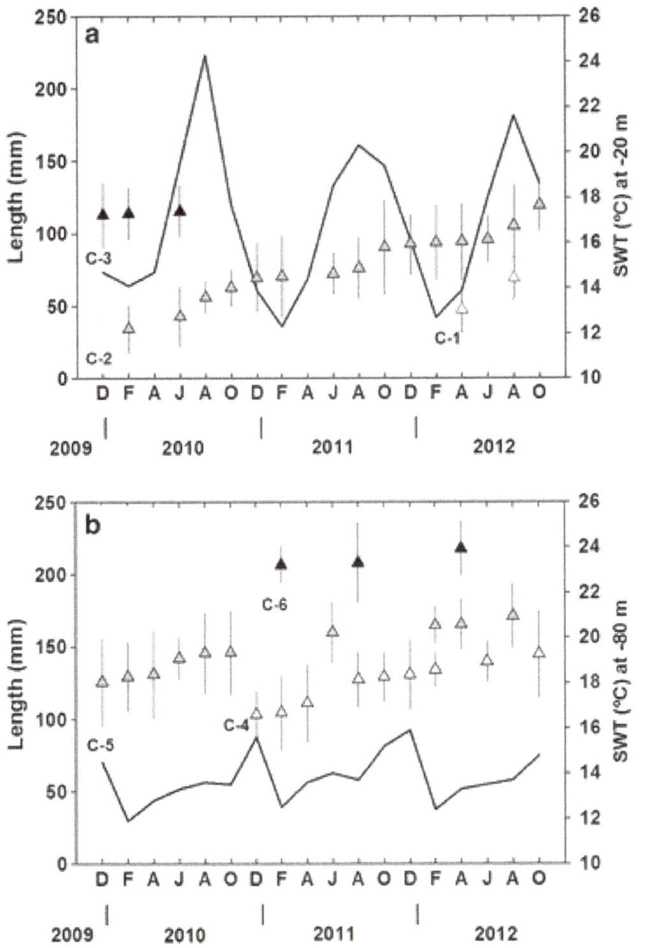

Figure 4: Astropecten aranciacus growth between November 2009 and October 2012 estimated from bimonthly progression of mean length of each cohort (±SD) in length-frequency distributions in both bathymetrical ranges: a 5–30 m and b 50–150 m. Each symbolcorresponds to different sea star cohorts. In a white triangles cohort C-1, gray trianglesC-2 and black triangles C-3; in b white triangles cohort C-4, gray triangles C-5 and black triangles C-6. The black line sea water temperature at −20 m (a) and at −80 m (b).

Gonads Morphology and Development

The reproductive system of A. aranciacus comprises two separate gonads in each arm (ten gonads in total). Each gonad was arranged as a series of tufts, located in and attached to the proximal part of each arm, near the disk on the inner face of the lateral wall, below the pyloric caeca, suspended in the celom along the entire length of the arm. Mature gonads were pale cream in males and deep yellow in females. The gonadal wall consisted of two-sac structures (the inner and the outer sacs) separated by the genital celom, which is typical for asteroids.

The five stages that characterize the gonadal development of A. aranciacus are the following:

Ovogenesis (Fig. 5). I-Recovery Stage Ovaries' lumen was partially filled with amorphous material, cell debris and phagocytes. The gonadal wall was thick, and the two-sac structures were evident; at advanced recovery stage the surface of the germinal epithelium was lined with oogonia and sometimes with early vitellogenic oogonia (Fig. 5a). II-Growing Stage Early in the growing stage (Fig. 5b), the ovarian surface epithelium was thick and mainly covered by previtellogenic oocytes, although some early vitellogenic oocytes could also be found. Late in the growing stage (Fig. 5c), the ovarian surface epithelium appeared thinner, oocytes accumulated in the lumen and vitellogenic oocytes were dominant. Oocytes showed an eccentric nucleus (Fig. 5c). The cytoplasm of the oocytes appeared blue together with oogonia, whereas the nuclei were stained in pink. III-Mature Stage Ovaries were densely packed and stacked with fully grown oocytes in the lumen, and the surface epithelium appeared thin. Oogonia and previtellogenic oocytes were uncommon and restricted to the germinal layer. The gonadal wall was the thinnest observed in the whole reproductive cycle (Fig. 5d). IV-Partly Spawned StageOvaries were poorly packed with fully grown oocytes in the lumen with large empty spaces; previtellogenic oocytes were scarce and not exclusively restricted to the germinal layer. Relict oocyte debris and phagocytes were present in the lumen. The ovary wall remained thin (Fig. 5e). Sometimes atresia was observed. V-Spent Stage Ovaries were filled with amorphous material with relict oocytes, the gonadal wall became thicker, and the two-sac structure was more evident. Relict oocyte debris and phagocytes were present in the lumen (Fig. 5f).

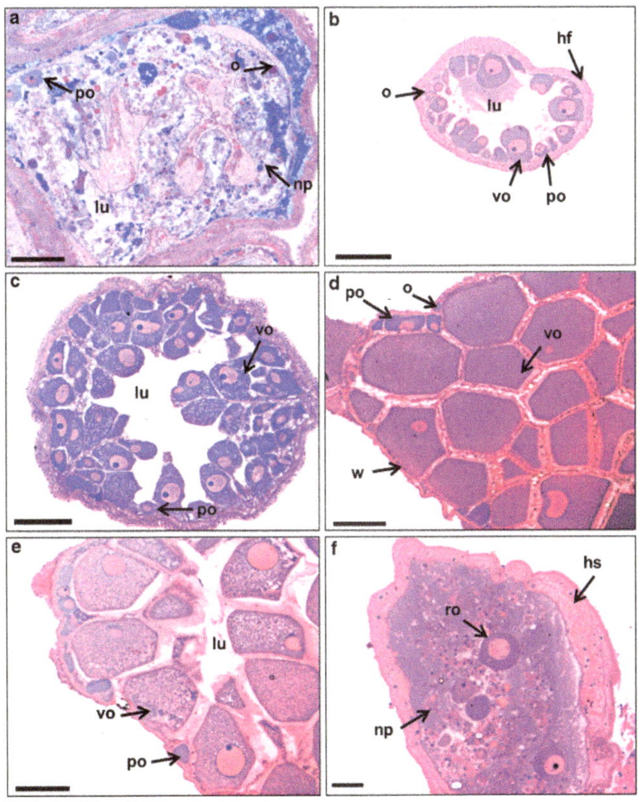

Figure 5: Photomicrographs of A. aranciacus ovaries showing different gonadal stages: a Stage I, Recovery; b, c Stage II, Growing; d Stage III, Mature; e Stage IV, Partly Spawned; f Stage V, Spent. hf hemal invaginations, hs hemal sinus, lu lumen, o oogonia, po previtellogenic oocyte, vo vitellogenic oocyte, ro relict oocyte, np nutritional phagocytes. Scale bars500 µm.

Spermatogenesis (Fig. 6). I-Recovery Stage Testes were small and remained contracted; the lumen was almost empty and partially filled with nutritive phagocytes; the gonadal wall was thick; and the two gonadal sacs were visible. Spermatogenic columns (consisting in spermatogonia) started to develop and grow (Fig. 6a). II-Growing Stage The testes had an intense spermatogenesis activity; they started to widen and became larger. Early in the growing stage, spermatogenic columns were projected into the central lumen and filled almost the totality of the volume; the gonadal wall began to shrink. As the growing stage progressed, testes enlarged and the number of spermatozoa increased from the border of the columns to the central lumen; at the same time the spermatogenic columns became larger as the gonadal wall shrank (Fig. 6b, c). III-Mature Stage Testes were densely packed and compressed with spermatozoa in the lumen, surrounded by very thin spermatogenic epithelium of spermatogenic columns. The wall of the testes was the thinnest observed in the whole reproductive cycle (Fig. 6d). IV-Partly Spawned Stage Testes looked similar to those in the mature stage but spermatozoa were poorly packed; there were spaces between patchy concentrations of spermatozoa in the lumen. Sometimes a gap between the peripheral spermatogenic layer and the central mass of spermatozoa was visible. The gonadal wall became thicker, and inner and outer sacs were visible again (Fig. 6e). V-Spent Stage Testes shrank and became smaller and contracted; the lumen was almost empty, with some nutritive phagocytes and isolated spermatozoa. The wall was thick and the two gonadal sacs became evident (Fig. 6h).

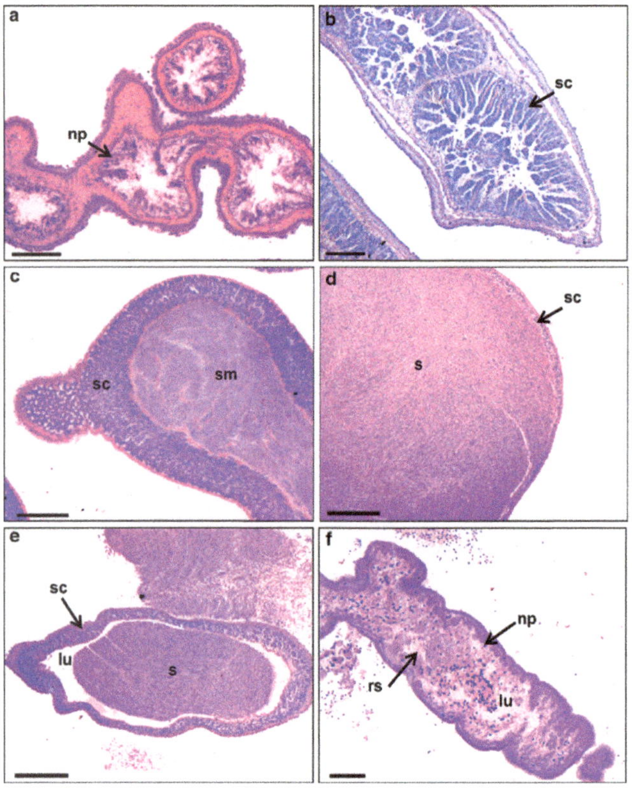

Figure 6: Photomicrographs of A. aranciacus testes showing different gonadal stages: a Stage I, Recovery; b, c Stage II, Growing; d Stage III, Mature; e Stage IV, Partly Spawned; f Stage V, Spent. hf hemal invaginations, hs hemal sinus, lu lumen, sc spermatogenic columns, spspermatogonia, sm spermatids, s spermatozoa, rs residual spermatozoa, np nutritional phagocytes. Scale bars 50 μm.

Reproductive Cycle

Results showed that females in Stage I (Recovery) were present throughout the study. Ovaries in Stage II (Growing) were also found all year round except in spring (May 2011 and from March to May 2012) (Fig. 7a). The proportion of females with ovaries in Stage II began to increase in June for both study years, and the maximum frequency of occurrence (90 %) was recorded in September 2011. Accordingly, females with mature gonads (Stage III) were observed from October 2011 to May 2012, and most mature females (>40 %) were recorded between February and April. Partially spawned gonads (Stage IV) were observed from December to June, with the highest frequency being found in May 2012. The majority of the spent ovaries (Stage V) were found between March and July for both 2011 and 2012.

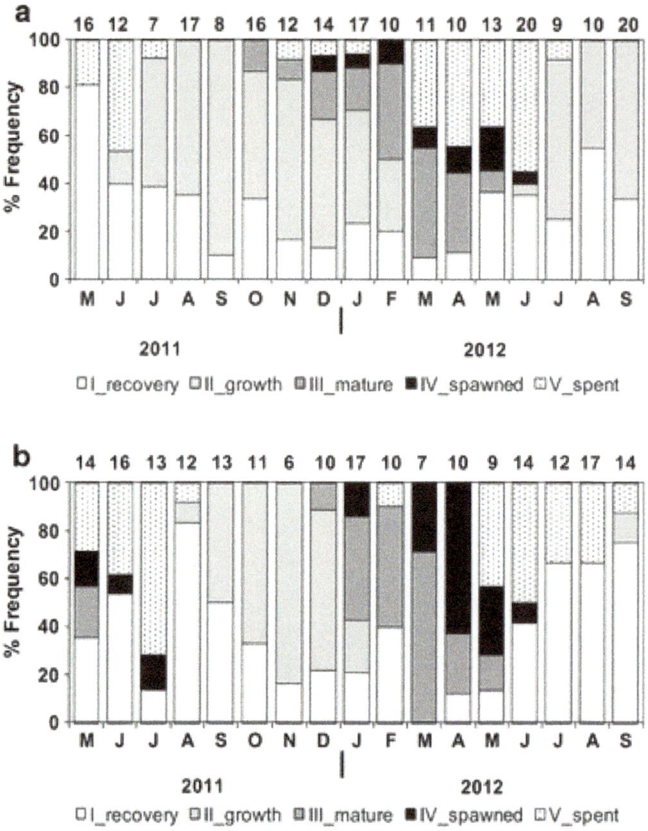

Figure 7: Annual gametogenesis cycle of A. aranciacus between May 2011 and September 2012. aFemales; b males. Numbers on the top of each column indicate the number of individuals analyzed monthly per sex.

Similarly to females, males in Stage I (Recovery) were present throughout the study except for March 2012. Spermatogenesis (Stage II) started between August and September and lasted until January, but was dominant between October and December (Fig. 7b). In accordance, mature testes (Stage III) appeared afterward and were observed in some individuals from December to May. Partially spawned (Stage IV) individuals were found between January and July, with a higher frequency in April 2012. Though a few spent testes (Stage V) were observed in February 2012, most of the individuals in this reproductive stage were observed from March to August.

Sex Ratio and Size at First Maturity

Sea stars did not show sexual dimorphism. A total of 312 individuals were collected from nearshore (87 females, 89 males and 136 immature) and 253 individuals from offshore (135 females, 116 males and 2 immature). The Chi-square (χ^2) test did not show significant differences between sexes; the ratio between males and females was almost 1:1 throughout the year in both bathymetrical ranges: nearshore ($\chi^2 = 10.186$ df = 15; $p > 0.05$) and offshore ($\chi^2 = 19.856$; df = 16, $p > 0.05$). The size of the smallest sea star with gonads was R = 56 mm for females and R = 55 mm for males. The size at first maturity (R50 %) was estimated to beR = 112 mm considering both sexes together.

Organ Indices

No significant differences between sexes were found when the monthly gonadal and pyloric caeca indices were compared (Mann–Whitney, $p > 0.05$); values for both sexes were then pooled for further analyses. The gonadal index increased from 0.18 ± 0.24 in Stage I (Recovery) to a maximal value of 5.6 ± 4.42 in Stage III (Mature) (Fig. 8). Once spawning occurred, the GI values declined progressively from Stage IV (Partly spawned) (3.52 ± 2.44) to Stage V (Spent) (0.43 ± 0.44). GI values for A. aranciacus showed significant differences between months (H = 110.07; df = 16, $p < 0.001$) (Fig. 9). Monthly values revealed an increase from October 2011 to March 2012, when GI reached its maximum (3.62 %). After that peak it decreased until July 2012 (0.14 %). Significant differences were also observed among months for pyloric caeca index (H = 58.76; df = 16, $p < 0.001$). The highest PCI values were recorded between March and May 2012 (>4.5 %), whereas the lowest were recorded in the summer months (June, July, August 2011, and July, August, September 2012). The surface sea water temperature (SSWT) at the study site had maximum values in summer months (July–September) and minimum values in winter months (January to March) (Fig. 10). Chl a monthly values recorded during our study period were lower between July and September, and two peaks were recorded: one in December 2011 (0.46 mg m^{-3}) and one in March 2012 (0.93 mg m^{-3}). The multiple regression analysis of GI with respect to the PCI and environmental variables was performed in two steps, resulting in the following model: GI = 0.90 + 0.09PCI − 0.03SSWT. The model showed that the total explained variance over the GI in terms of the selected variables was 84.58 % ($R^2 = 0.8458$, R = 0.88, $p < 0.05$), indicating that the PCI and surface sea water temperature (SSWT) are correlated with GI. The photoperiod and Chl a did not show any correlation with GI; however, the spawning season coincided with the time period of highest levels of Chl a.

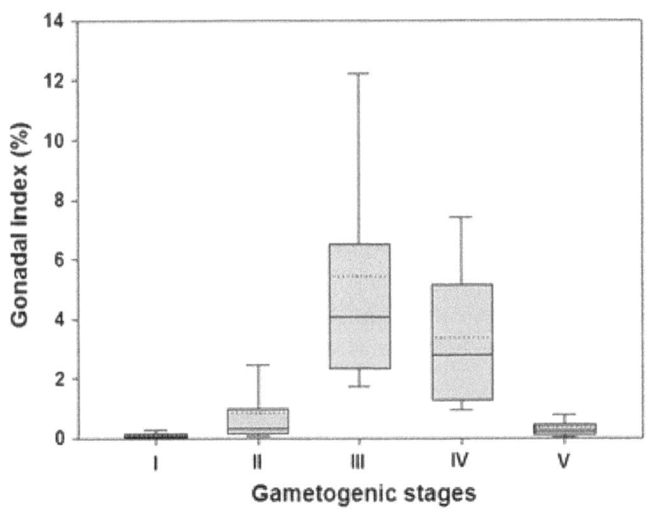

Figure 8: Astropecten aranciacus variations in the gonadal index of different gametogenic stages. Stage I, Recovery; Stage II, Growing; Stage III, Mature; Stage IV, Partly Spawned; Stage V, Spent. The boundary of the box closest to zero the 25th percentile; the continuous line within the box marks the median; the discontinuous line within the box marks the mean; and the boundary of the box farthest from zero the 75th percentile. Whiskers (error bars) above and below the box the 95th and 5th percentiles.

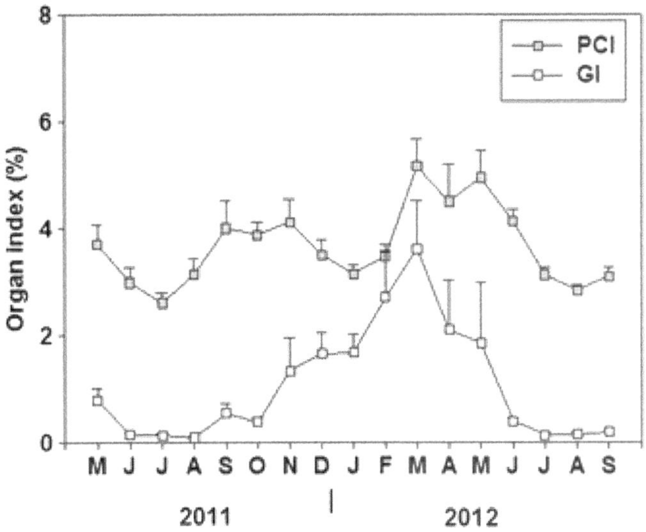

Figure 9:Seasonal changes in gonadal index (GI, mean, +SE) and pyloric caeca index (PCI, mean, +SE).

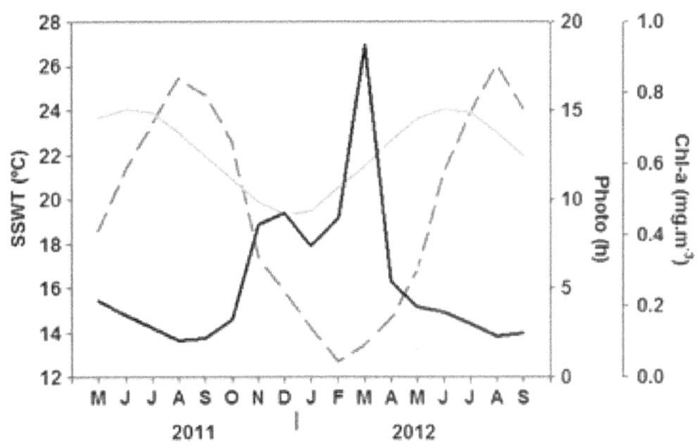

Figure 10:Monthly mean of sea surface water temperature (SSWT; gray long dash); photoperiod (Photo; gray line), expressed in number of daylight hours every 24 h; mean values of chlorophyll a (Chl a) concentration from May 2011 to September 2012 (black line).

DISCUSSION

Astropecten aranciacus individuals smaller than 80 mm were abundant in shallow waters (5–30 m depth), whereas large individuals (>130 mm) were observed mostly offshore (50–150 m depth). The use of different gears to sample both bathymetric ranges in this study may have caused bias collecting small sizes offshore, but no bias for larger sizes in shallow waters could occur. Therefore, any bigger individuals of A. aranciacus inhabiting depths between 5 and 30 m would have been collected in our sampling. Burla et al. [8] studied the activity pattern of A. aranciacus by scuba diving in Sardinia (Italy) and found only small specimens. These authors suggested that adults and juveniles may dwell in different parts of the ecological zone occupied by the species, the fully grown specimens invading the deeper waters. Our study is the first to sample most of the bathymetrical distribution range of A.

aranciacus, allowing us to confirm a bathymetrical segregation related to size in this species. Ventura and Fernandes [45] reported size-depth trends for Astropecten brasilensis in Cabo Frio, Brazil, where they found larger individuals with depth. Segregation by size were also reported for the asteroid Protoreaster nodosus, whose large specimens were found between 0 and 33 m depth, whereas small specimens were exclusively found in shallower habitats (<5 m) [6]. Young sea stars might prefer shallow areas where food and shelters to avoid predation are usually more abundant [30].

The major radius length–weight relationship showed a negative allometry in A. aranciacus, thereby they invest more energy in body enlargement than in fattening, as previously observed in other Astropecten species [19, 46] and sea stars genera [6, 7]. Absolute growth has been extensively analyzed in sea stars (see references in Lawrence and Lane [29], but few studies have established growth rates in natural populations of the genus Astropecten [15, 35, 40, 43] (Table 1). The studied A. aranciacus population exhibited a seasonal growth pattern, slightly different for nearshore and offshore cohorts. Individuals living offshore had a longer growth period but their growth rate was slower, as is expected in older specimens. Seasonal variation in growth had been described for other Astropecten species, which usually showed a decrease in growth rate associated with gonad maturation [15, 35, 43]. According to our study, the growth rate for A. aranciacus seems to be related to sea water temperature as higher growth rates were recorded nearshore when sea water temperature was also higher. Moreover, sea water temperature increased slightly from February to December offshore, coinciding with the period of highest growth rate for those cohorts. Ventura [44] reported that the decrease in growth rate of A. brasiliensis and A. cingulatus coincided with the seasonal upwelling period in Brazil, suggesting that it could be related to sea water temperature decline. Other factors such as the amount and quality of available food can also affect growth, as reported by Guillou and Guillaumin [20] in Asterias rubens.

Table 1: Estimated parameters of the von Bertalanffy growth equation for Astropecten species at different study areas

Species	Depth (m)	L_{∞}(mm)	K (year^{-1})	t_0	Study area	References
A. latespinus	0.1–5	42.02	0.04	0.0325	Fukuoka, Japan (Sea of Japan)	Nojima [35]
A. cingulatus	45–60	81	1.2	–	Maçambaba Beach, Brasil (SW Atlantic)	Ventura [43]
A. brasilensis	45–60	140	0.75	–	Maçambaba Beach, Brasil (SW Atlantic)	Ventura [43]
A. marginatus	6–15	65	0.50	–	Guarana, Brasil (SW Atlantic)	Tavares et al. [40]
A. aranciacus	5–30	136.75	0.44	0.0017	Maresme Coast, Spain (NW Mediterranean)	Present study
A. aranciacus	50–150	205	0.32	0.0011	Maresme Coast, Spain (NW Mediterranean)	Present study

The A. aranciacus population studied herein seems to have variable reproductive and recruitment success. Sea stars as other echinoderms show usually a high degree of spatial and temporal variability in settlement and recruitment [24]. Nil or low recruitment over a period of time is not surprising in A. aranciacus, i.e., Hörstadius [22] reported that they have a long pelagic larval phase (80 days) under laboratory conditions. During this period Astropectenlarvae were subject to hydrological processes which may displace them long distances [15, 41]. Regarding the main hydrological process in the study area, the prevailing currents are controlled by the cyclonic eddy of the western Mediterranean derived from the Liguro–Provencal current [34]. Although local larvae could be transported away to the southwest, recruitment was already observed 2 of the 3 years studied on the nearshore area, indicating that some larvae settled. The lack of recruitment offshore may be related to the absence of favorable environmental conditions for larvae settlement and/or to predation on juveniles. During different samplings, offshore, we observed the presence of Luidia ciliaris and Tethyaster subermis, potential predators [23] that could play an important role in the recruitment failure of A. aranciacus.

Results showed that both A. aranciacus sexes had a synchronous gametogenesis but females reached the maturity stage slightly earlier than males, in accordance with previous results obtained in Asterias amurensis by Byrne et al. [9]. The studied A. aranciacus population did not differ significantly in the male–female proportions, coinciding with previous results in other sexually reproducing asteroids (since sex-ratios can be highly biased in the case of asexual asteroids) [4, 26]. The development of the A. aranciacus gonads reflected that mature and partially spawned individuals remained for an extended period of time. The GI value was highest in March, indicating that during that month the gonads of A. aranciacus were in maximum development. Spawning likely started in March in part of the population and extended until June–July, in accordance with the decrease in GI and the presence of post-spawning individuals evidenced by the histological analysis of the gonads.

In general the reproductive cycles of sea stars are correlated with a combination of endogenous (i.e., hormones) and exogenous factors (i.e., temperature, photoperiod, food availability, lunar cycle, tidal flux and light intensity) [33]. Sea star oocytes at the peak of the breeding season are fully grown and arrested at prophase I of meiosis [39]. Hormonal stimulation (methyladenine, 1-MA) from the adhering follicle cells reenter the oocytes in the meiotic cycle before being spawned into seawater [25]. However, the direct influence of exogenous factors remains still poorly understood [33]. In Astropecten species, breeding season commonly coincides with the increase in water temperatures in temperate areas [15]. For instance, these authors reported that the spawning periods for A. irregularis coincided with the rise from 8 to 15 °C in seawater temperatures (North Wales). Our results showed a similar pattern, because the spawning season of A. aranciacus likely began in spring, when the sea water temperatures started to increase in the Mediterranean. We also observed that the beginning of A. aranciacus spawning coincided with the highest Chl a concentration in the sea water, which was caused by the spring phytoplankton bloom typical of the Mediterranean Sea. The coincidence between the highest phytoplankton concentration and the spawning has been previously described in other echinoderms, but it is rare in sea stars [33]. The majority of the species of the genusAstropecten have planktonic larvae (bipinnaria). Bipinnaria larva nourish essentially on detritus, bacteria, phytoplankton (mainly small diatoms and small flagellates) and small zooplankton [1]. The synchronism between the spawning season of A. aranciacus, and the phytoplankton bloom could be a strategy to guarantee higher survival of early developmental stages.

An inverse correlation between GI and PCI has usually been described in most sea stars, suggesting the transfer of nutrients from the pyloric caeca to the gonads during gametogenesis [29]. However, this interaction has not been observed in species inhabiting areas with stable environmental conditions [4]. This relationship was not observed in the studied A. aranciacuspopulation despite the fact that it inhabits a seasonally changing environment, in accordance with the results obtained by Ventura et al. [48] in A. cingulatus. Individuals of A. aranciacuscollected in the offshore zone of the study area had large amount of preys in the stomach throughout the year. Most of these prey items were the venerid Timoclea ovata (>55 %, abundance in 224 sea stars analyzed), and each sea star stomach contains a mean of 49 T. ovata items of a mean size of 5.77 mm (SD = 5.52, N = 9682) [2]. Therefore, food does not seem to be a limiting factor in the Maresme coast, suggesting that sea stars do not need to store large amounts of nutrients in the pyloric cecum, because nutrients can be easily obtained from the environment throughout all year. This suggests for A. aranciacus an "income breeding" reproductive strategy rather than "capital breeding", i.e., reproduction is financed using current energetic income and not from stored energetic capital. This strategy may explain why an inverse correlation between GI and PCI was not observed in this population.

Our findings provide a better understanding of the biology of A. aranciacus in the NW Mediterranean Sea and are also relevant for the future conservation of this sea star, since its populations seem to decrease in some places of the Mediterranean Sea.

AUTHORS' CONTRIBUTIONS

All co-authors carried out the adquisition, analysis and interpretation of data, drafting and critical revision of the Manuscript.

ACKNOWLEDGEMENTS

This study is part of the doctoral dissertation of MB. We are grateful to the fishermen 'Confraria de Pescadors d'Arenys', in particular the clam dredge fishermen (Miguel Angel Ruiz Navas, Jose Luís López Ramírez, Mauricio Pulido and their sailors), and Jaume Germà from the bottom trawl fleet. We are also grateful to all sampling participants: Lucía Martínez, Marta Coca, Eduard Albella and Nil García. We deeply thank Anabel Colmenero for her advice and tips on histology techniques and for her constructive comments and suggestions on reproductive biology. We are very grateful to Dr. Francesc Peters for providing us with the chlorophyll data. The authors also thank Puertos del Estado (Instituto Geográfico Nacional, Ministerio de Fomento) for the environmental data and Raquel Ubach (Autonomous University of Barcelona) for the help with the map.

REFERENCES

1. Ayukai T. Resource availability to the larvae of the crown-of-thorns starfish in Great Barrier Reef waters: an overview. In: Engelhardt U, Lassig B, editor. The possible causes and consequences of outbreaks of crown-of-thorns starfish. Great Barrier Reef Marine Authority Workshop Series; 1993. p. 21–35.
2. Baeta M. Ecology of some keystone invertebrates inhabiting shallow soft bottom communities of the Maresme coast (NW Mediterranean Sea). Dissertation, University of Barcelona; 2015.
3. Baeta M, Ramón M. Feeding ecology of three species of Astropecten (Asteroidea) coexisting on shallow sandy bottoms of the northwestern Mediterranean Sea. Mar Biol. 2013;160:2781–95.
4. Benítez-Villalobos F, Martínez-García M. Reproductive biology of the starfish Pharia pyramidatus (Echinodermata: Asteroidea) from the Mexican tropical Pacific. J Mar Biol Assoc UK. 2012;92:1409–18.
5. Bhattacharya PK. Efficient estimation of a shift parameter from grouped data. Ann Math Stat. 1967;38:1770–87.
6. Bos AR, Gumanao GS, Alipoyo JCE, Cardona LT. Population dynamics, reproduction and growth of the Indo-Pacific horned sea star, Protoreaster nodosus (Echinodermata; Asteroidea). Mar Biol. 2008;156:55–63. doi:10.1007/s00227-008-1064-2.
7. Bos AR, Gumanao GS, Van Katwijk MM, Mueller B, Saceda MM, Tejada RP. Ontogenetic habitat shift, population growth, and burrowing behavior of the Indo-Pacific beach star,Archaster typicus (Echinodermata; Asteroidea). Mar Biol. 2011;158:639–48.
8. Burla H, Ferlin V, Pabst B, Ribi G. Notes on the ecology of Astropecten aranciacus. Mar Biol. 1972;14:235–41. doi:10.1007/BF00348285.
9. Byrne M, Morrice MG, Wolf B. Introduction of the northern Pacific asteroid Asterias amurensis to Tasmania: reproduction and current distribution. Mar Biol. 1997;127:673–85. doi:10.1007/s002270050058.
10. Carvalho A, Ventura C. The reproductive cycle of Asterina stellifera (Möbius) (Echinodermata: Asteroidea) in the Cabo Frio region, southeastern Brazil. Mar Biol. 2002;141:947–54. doi:10.1007/s00227-002-0881-y.
11. Christensen AM. Feeding biology of the sea star Astropecten irregularis Pennat. Ophelia. 1970;8:1–134.

12. Conand C. Sexual cycle of three commercially important holothurian species (Echinodermata) from the lagoon of New Caledonia. Bull Mar Sci. 1981;31:523–43.
13. Díaz JI, Maldonado A. Transgressive sand bodies on the Maresme continental-shelf, western Mediterranean-sea. Mar Geol. 1990;91:53–72.
14. Ercilla G, Estrada F, Casas D, Durán R, Nuez M, Alonso B, Farrán M. The El Masnou infralittoral sedimentary environment (Barcelona province, NW Mediterranean Sea): morphology and Holocene seismic stratigraphy. Sci Mar. 2010;74:179–96.
15. Freeman SM, Richardson CA, Seed R. Seasonal abundance, spatial distribution, spawning and growth of Astropecten irregularis (Echinodermata: Asteroidea). Estuar Coast Mar Sci. 2001;53:39–49.
16. Gayanilo FCJ, Sparre P, Pauly D. FAO-ICLARM Stock Assessment Tools II (FiSAT II). Revison version. User's guide. Rome: FAO; 2005.
17. Giese AC. On the biochemical constitution of some echinoderm. In: Boolootian RA, editor. Physiology of echinodermata. New York: Interscience; 1966. p. 757–96.
18. Grant A, Tyler PA. An analysis of the reproductive pattern in the sea star Astropecten irregularis Pennant from the Bristol Channel. Int J Invertebr Reprod Dev. 1986;93:345–61.
19. Guilherme P, Rosa L. Seasonal variation in body size and diet of the sea star Astropecten marginatus (Paxillosida, Astropectinidae) off coast of Paraná, Southern Brazil. Rev Biol Trop. 2013;62:59–68.
20. Guillou M, Guillaumin A. Variations in the growth rate of Asterias rubens (L.) from west and south Brittany (France). In: Keegan BF, O'Connor BDS, editor. Proceeding of the 5th international echinoderms conference. Rotterdam: A.A. Balkema; 1984. p. 513–21.
21. Gulland JA. Fish stock assessment: a manual of basic methods. Chichester: Wiley; 1983.
22. Hörstadius S. Über die Entwicklung von Astropecten aranciacus L. Mar Ecol PSZNI. 1938;17:221–312.
23. Jangoux M, Lawrence JM. Echinoderm nutrition. Rotterdam: A.A. Balkema; 1982.
24. Jangoux M, Lawrence J. Echinoderm studies 6. Rotterdam: A.A. Balkema; 2001.
25. Kanatani H. Maturation-inducing substance in starfishes. Int Rev Cytol. 1973;35:253–98.PubMed
26. Kim YS. Histological observations of the annual change in the gonad of the starfish,Asterias amurensis. Fac Fish Hokkaido Univ. 1968;19:97–108.
27. Komatsu M, Nojima S. Development of the seastar, Astropeten gisselbrechti Doderlein. Pac Sci. 1985;39:274–82.
28. Lawrence JM. Starfish: biology and ecology of the Asteroidea. Baltimore: The Johns Hopkins University Press; 2013.
29. Lawrence J, Lane J. The utilization of nutrients by postmetamorphic echinoderms. In: Jangoux M, Lawrence J, editors. Echinoderm nutrition. Rotterdam: Balkema Press; 1982. p. 331–71.
30. Manzur T, Barahona M, Navarrete SA. Ontogenetic changes in habitat use and diet of the seastar Heliaster helianthus on the Coast of central Chile. J Mar Biol Assoc U K. 2010;90:537–46.
31. Marion KR, Watts SA, McClintock JB, Schimer G, Hopkins TS. Seasonal gonad maturation in Astropecten articulatus from the Northern Gulf of Mexico. In: Mooi R, Telford M, editors. Echinoderms. San Francisco: A.A. Balkema; 1998. p. 279–84.
32. McClary DJ, Mladenov PV. Reproductive pattern in the brooding and broadcasting sea starPteraster militaris. Mar Biol. 1989;103:531–40. doi:10.1007/bf00399585.
33. Mercier A, Hamel J. Endogenous and exogenous control of gametogenesis and spawning in echinoderms. Adv Mar Biol. 2009;55:1–291.
34. Millot C. Circulation in the western Mediterranean Sea. J Mar Syst. 1999;20:423–42.
35. Nojima S. Ecological studies of the sea star Astropecten latespinosus Meissner II. Growth rate and differences in growth pattern of immature and mature sea stars. Publ Amakusa Mar Biol Lab Kyushu Univ. 1982;62:85–94.
36. Ribi G, Burla H, Ochsner P. Beobachtungen über vorkommen, abundanzen und körpergrößen von mediterranen arten der gattung Astropecten. Helgol Wiss Meeres. 1976;28:304–17.
37. Ribi G, Scharer R, Ochsner P. Stomach contents and size-frequency distributions of two coexisting sea star species, Astropecten aranciacus and A. bispinosus, with reference to competition. Mar Biol. 1977;43:181–5.
38. Rubilar T, Pastor de Ward CT, Díaz de Vivar ME. Sexual and asexual reproduction ofAllostichaster capensis (Echinodermata: Asteroidea) in Golfo Nuevo. Mar Biol. 2005;146:1083–90. doi:10.1007/s00227-004-1530-4.
39. Santella L, Limatola L, Thai Chun J. Chapter: 13. Actin cytoskeleton and fertilization in starfish eggs. In: Sawada H, Inoue N, Iwano M, editors. Sexual reproduction in animals and plants. Berlin: Springer; 2014. p. 141–55.
40. Tavares Y, Silvéiro G, Borges P, Braz B, Borzone C. Dinâmica populacional de Astropecten marginatus (echinodermata:asteroidea) no sul do Brasil XII Congreso Latino-Americano de Ciencias do Mar Florianópolis; 2007.
41. Thorson G. Reproduction and larval development of Danish marine bottom invertebrates. Medd Kom Dan Fisk Havunders Plankton. 1946;4:1–523.
42. Tortonese E. Echinodermata. Italia: Calderini Bologna; 1965.
43. Ventura C. Gametogenesis and growth of two a Astropecten species from a midshelf upwelling region, Cabo Frio, Brazil. In: Candia Carnevali MD, Bonasuro F, editors. Research proceedings of the 5th European conference on echinoderms. Milan: A.A. Balkema; 1999. p. 167.
44. Ventura C. Astropecten. In: Lawrence JM, editor. Starfish: biology and ecology of the Asteroidea. Baltimore: The Johns Hopkins University Press; 2013. p. 101–8.

45. Ventura C, Fernandes F. Bathymetric distribution and population size structure of paxillosid seastars (Echinodermata) in the Cabo Frio upwelling ecosystem of Brazil. Bull Mar Sci. 1995;56:268–82.

46. Ventura C, Junqueira A, Fernandes F. The relation between body size and number of prey in starfish (Echinodermata: Asteroidea). In: David B, Gille A, Féral J, Roux M, editors. Echinoderms through time. Rotterdam: A.A. Balkema; 1994. p. 375–80.

47. Ventura C, Falcao A, Santos J, Fiori C. Reproductive cycle and feeding periodicity in the starfish Astropecten brasiliensis in the Cabo Frio upwelling ecosystem (Brazil). Int J Invertebr Reprod Dev. 1997;31:135–41.

48. Ventura C, Santos J, Falcao A, Fiori C. Reproduction and food intake in Astropecten cingulatus in the upwelling environment of Cabo Frio. In: Mooi R, Telford M, editor. Echinoderms: San Francisco proceedings of the ninth international echinoderm conference. San Francisco, California, USA: A.A. Balkema; 1998. p. 313–18.

49. Zulliger DE, Lessios HA. Phylogenetic relationships in the genus Astropecten Gray (Paxillosida: Astropectinidae) on a global scale: molecular evidence for morphological convergence, species-complexes and possible cryptic speciation. Zootaxa. 2010;2504:1–19.

50. Zulliger D, Tanner S, Ruch M, Ribi G. Genetic structure of the high dispersal Atlanto-Mediterranean sea star Astropecten aranciacus revealed by mitochondrial DNA sequences and microsatellite loci. Mar Biol. 2009;156:597–610. doi:10.1007/s00227-008-1111-z.

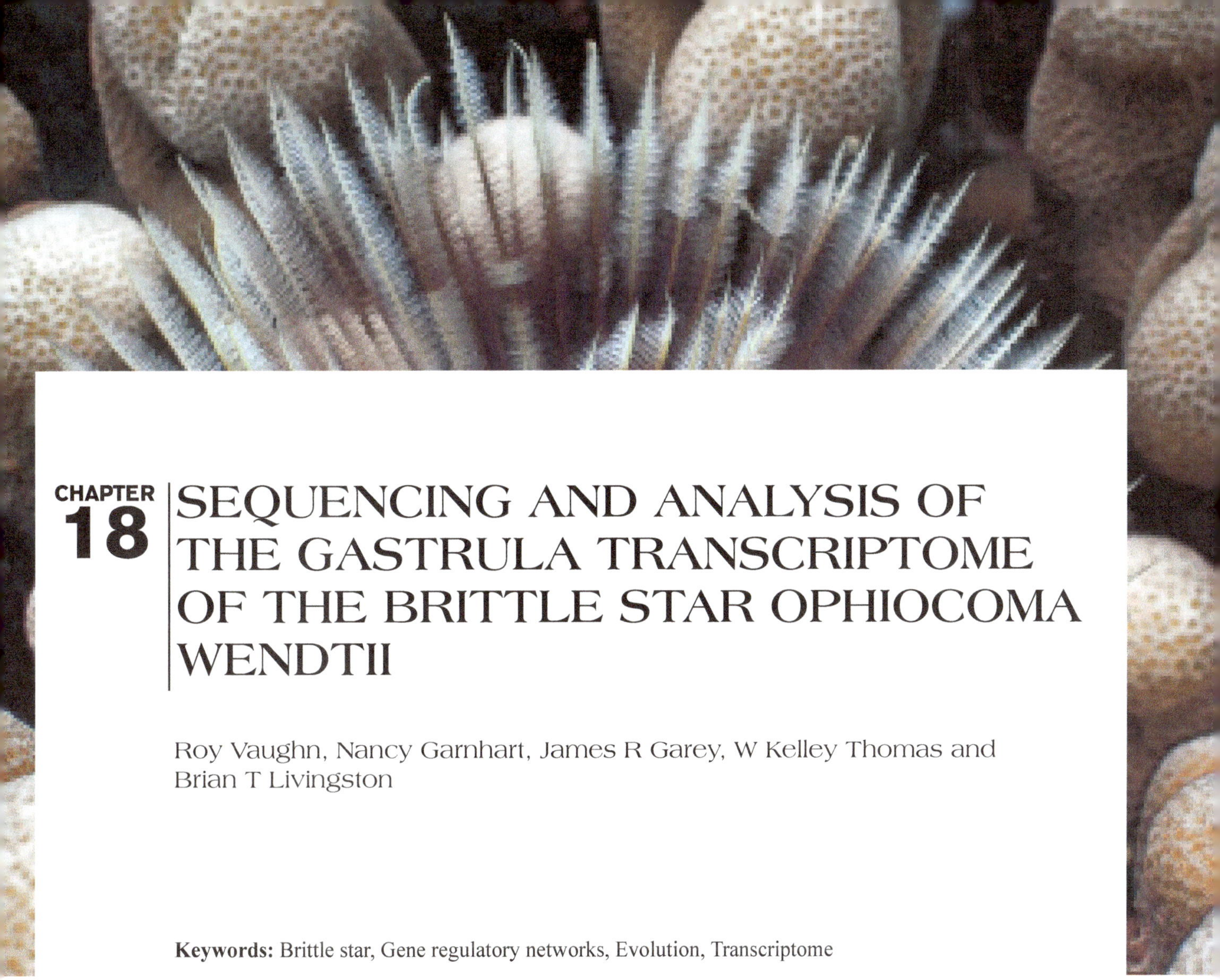

CHAPTER 18

SEQUENCING AND ANALYSIS OF THE GASTRULA TRANSCRIPTOME OF THE BRITTLE STAR OPHIOCOMA WENDTII

Roy Vaughn, Nancy Garnhart, James R Garey, W Kelley Thomas and Brian T Livingston

Keywords: Brittle star, Gene regulatory networks, Evolution, Transcriptome

BACKGROUND

Sea urchins (Class Echinoidea) have been used as model organisms in developmental biology for more than a century. Over the last two decades intensive work has led to a fairly detailed understanding of the gene regulatory network (GRN) controlling the differentiation of the embryonic germ layers during development in the species *Strongylocentrotus purpuratus*[1–6]. An initial draft of the *S. purpuratus* genome was completed in 2006 [7] and is now in its third revision [6]. Several expression databases for various embryonic stages have also been constructed using expressed sequence tags (ESTs) [8–11], microarrays [12], and NanoString RNA counting [13]. Here we begin to examine the conservation and divergence in the gene regulatory networks expressed at the gastrula stage in a member of a different echinoderm class, the Ophiuroidea. Our results indicate that, although there are differences that reflect changes in the developmental processes, there is a shared 'genetic toolkit' central to the echinoderm gastrula, a key stage in embryonic development.

Citation: Roy Vaughn, Nancy Garnhart, James R Garey, W Kelley Thomas, Brian T Livingston, Sequencing and analysis of the gastrula transcriptome of the brittle star Ophiocoma wendtii, DOI: 10.1186/2041-9139-3-19.

The echinoderms consist of five living classes: Asteroidea (starfish), Echinoidea (sea urchins and sand dollars), Ophiuroidea (brittle stars) Holothuroidea (sea cucumbers), and Crinoidea (sea lilies and feather stars). The crinoids appear first in the fossil record and are clearly the most basal anatomically. The other four classes appear to have all diverged within a very short geological period around 500 million years ago [14], and the exact phylogenetic relationship of the brittle stars to the other classes remains uncertain due to conflicts between molecular, morphological, and embryological evidence [15, 16] (Figure 1). The embryos of all echinoderm classes share some features, including holoblastic cleavage and similar cell movements during gastrulation. However, there are notable differences, such as the formation of micromeres in sea urchins but not brittle stars (Figure 2), the absence of pigment cells in brittle stars, and the formation of an embryonic skeleton in sea urchins and brittle star embryos, but not in the other groups. What is currently unclear is how these similarities and differences in development are reflected in the pattern of gene transcription. Davidson and Erwin [17] have suggested that key gene regulatory subcircuits central to the formation of major morphological features ('kernels') are very highly conserved by stabilizing natural selection, both because they are critical to the formation of a complete viable body and because their internal linkages and feedback loops make their component genes mutually dependent. A refinement of this idea is that some of the component transcription factors may be exchanged for others as long as the overall input/output logic and reliability of the circuit and its resulting function are maintained [18]. This suggests that many of the regulatory kernels shown to be important in sea urchin gastrulation would be conserved in the other echinoderm groups.

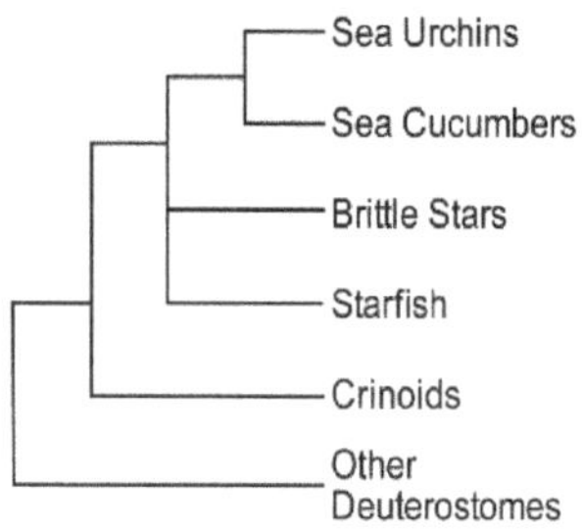

Figure 1: Phylogeny of echinoderms. All evidence indicates that crinoids are the most basal. The other four groups all diverged within a very short geological timeframe around 500 million years ago. Urchins and sea cucumbers are generally considered to form a clade of the most derived. It remains unclear whether the brittle stars group more closely with this clade or with starfish, due to conflicts between molecular, morphological, and embryological evidence [14–16].

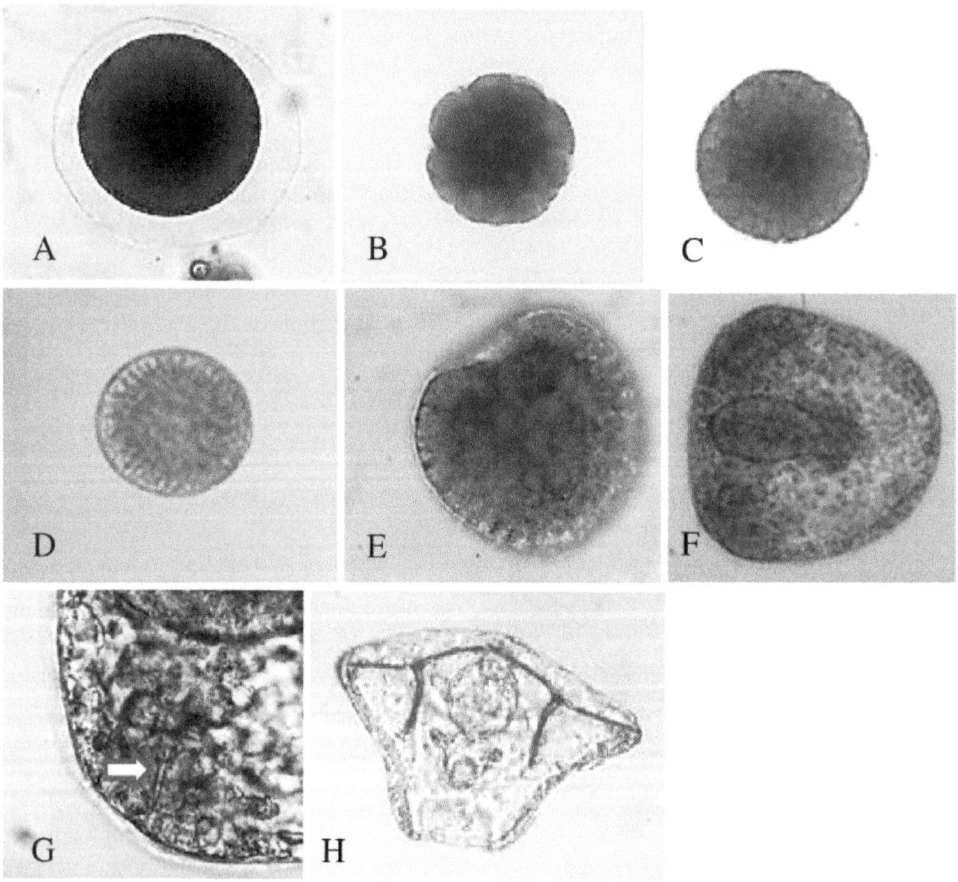

Figure 2: *Ophiocoma wendtii* embryonic development. Stages (A) egg, (B) 16 cell (5 h), (C) hatched blastula (16 to 18 h), (D) mesenchyme blastula (24 h to 26 h), (E) early gastrula (28 to 30 h), (F) gastrula (38 to 40 h), (G) ventrolateral cluster with skeletal spicule (arrow) at 40 h, (H) pluteus (80 h).

The set of genes that control skeleton formation in echinoderms may represent such a circuit under evolutionary constraints. All echinoderms form skeletons as adults; however, only sea urchins and brittle stars form extensive embryonic skeletal spicules. It has recently been shown that most of the same regulatory genes that underlie skeletogenesis in the sea urchin embryo are also expressed in the construction of the adult skeleton in both sea urchins and starfish [19]. The embryonic skeletons of sea urchins and brittle stars are thus thought to be derived characters resulting from early activation of an adult gene regulatory network in the embryo.

The process of embryonic skeletogenesis has been extensively studied in sea urchins [20]. Asymmetric fourth and fifth cleavages produce four small micromeres and four larger micromeres at the vegetal pole. Descendants of the larger micromeres ingress into the blastocoel just prior to gastrulation and become the primary mesenchyme (PMC), which soon produces the embryonic skeleton. Micromeres are a derived character unique to crown group sea urchins (euechinoids) [21]. Brittle stars and the more basal sea urchin groups form very similar embryonic skeletons from apparently homologous mesenchymal cells without prior unequal cleavages [22], Figure 2.

We have sequenced and characterized the 40-h gastrula transcriptome of the brittle star*Ophiocoma wendtii*. The gastrula stage was chosen because it represents the point in development at which the three primary germ layers diverge, with ingression of mesenchymal cells and invagination of the gut. At this point in sea urchins the gene regulatory networks that specify the germ layers are established and the genes that define the differentiated states of the tissues have begun to be activated. The early gastrula therefore expresses the greatest number and diversity of developmentally important genes. We report here that the brittle star gastrula expresses genes of all functional classes and appears to share many key developmental regulatory components with other echinoderms. Some regulatory genes, as well as genes expressed in differentiated tissues in the sea urchin gastrula, were not expressed in the brittle star gastrula.

METHODS

Animals and Embryos

Brittle stars (*O. wendtii*) were collected from reefs and rubble piles in the shallow waters of Florida Bay near the Keys Marine Laboratory, Long Key, Florida, between April and October. Animals were sorted by sex, with gravid females identified by the presence of swollen purple gonads visible through the bursal slits. Sperm was obtained from two to three males by injection of 1 to 3 mL of 0.5 to 1.0 M KCl. Shedding of eggs from females was induced by a combination of heat and light shock. Animals were placed in containers in the dark with aeration at 30 to 32°C. Periodically the animals were exposed to bright light. Developing embryos were cultured at 25 to 27°C in filtered sea water. RNA was isolated using Trizol (Life Technologies, Carlsbad, CA, USA) following the manufacturer's protocol.

Characterization of Transcriptome Sequences

Sequencing and assembly of contiguous sequences was carried out as described by Meyer *et. al.*[23]. The comparisons of gastrula transcriptomes were followed by an all-by-all BLAST [24] approach where each comparison was databased. These results were then queried for the identification of orthologous genes using a reciprocal best BLAST (RBB) strategy, and for the identification of gene families following the method of Lerat, *et al.*[25] as implemented previously [26, 27]. Gene families and singletons were then annotated using Homology Inspector (HomIn) software, a Java program that stores and queries a set of gene families using the database tool db4o for Java version 7.12 [28]. HomIn links gene families with annotation information including Kyoto Encyclopedia of Genes and Genomes (KEGG) Orthology Database (KO) categories [29], Clusters of Orthologous Groups of proteins (COGs) [30, 31], Gene Ontology (GO) categories [32], or any other available annotation.

Search for GRN Components

Glean3 predicted protein sequences for genes involved in the *S. purpuratus* developmental gene regulatory network were retrieved from SpBase [6] using the official gene name. These were used as queries to search the brittle star gastrula transcriptome sequences using TBLASTN at default settings. The best hit for each query was then used to search back against both sea urchin protein sequences and GenBank reference proteins using BLASTX. Sea urchin genes which had RBB hits to brittle star with e-values of 1e-9 or better in both directions were designated as present in the brittle star gastrula transcriptome. These sequences can be found in Genbank using accession numbers JX60067 to JX60106.

Database and Analyses

Results from the automated BLAST searches were saved to a Microsoft Access (Microsoft Corporation, Redmond, WA, USA) database. This database and Microsoft Excel (Microsoft Corporation, Redmond, WA, USA) were used for the analyses involving presence/absence of expression, functional classes, and numbers of matches to other databases. Rarefaction curves were generated using EcoSim software [33]. For the functional class analysis, KEGG ortholog clusters were used if they included genes from at least one animal taxon. When a KEGG cluster participates in more than one pathway within a functional class, it was counted only once within the larger functional class. For example, K00128 aldehyde dehydrogenase (NAD+) is part of five different pathways within the class of carbohydrate metabolism and two pathways in lipid metabolism, among many others, but was counted only once within each class in Figure 3B, and once in the total number of distinct KEGG animal clusters in Figure 3A.

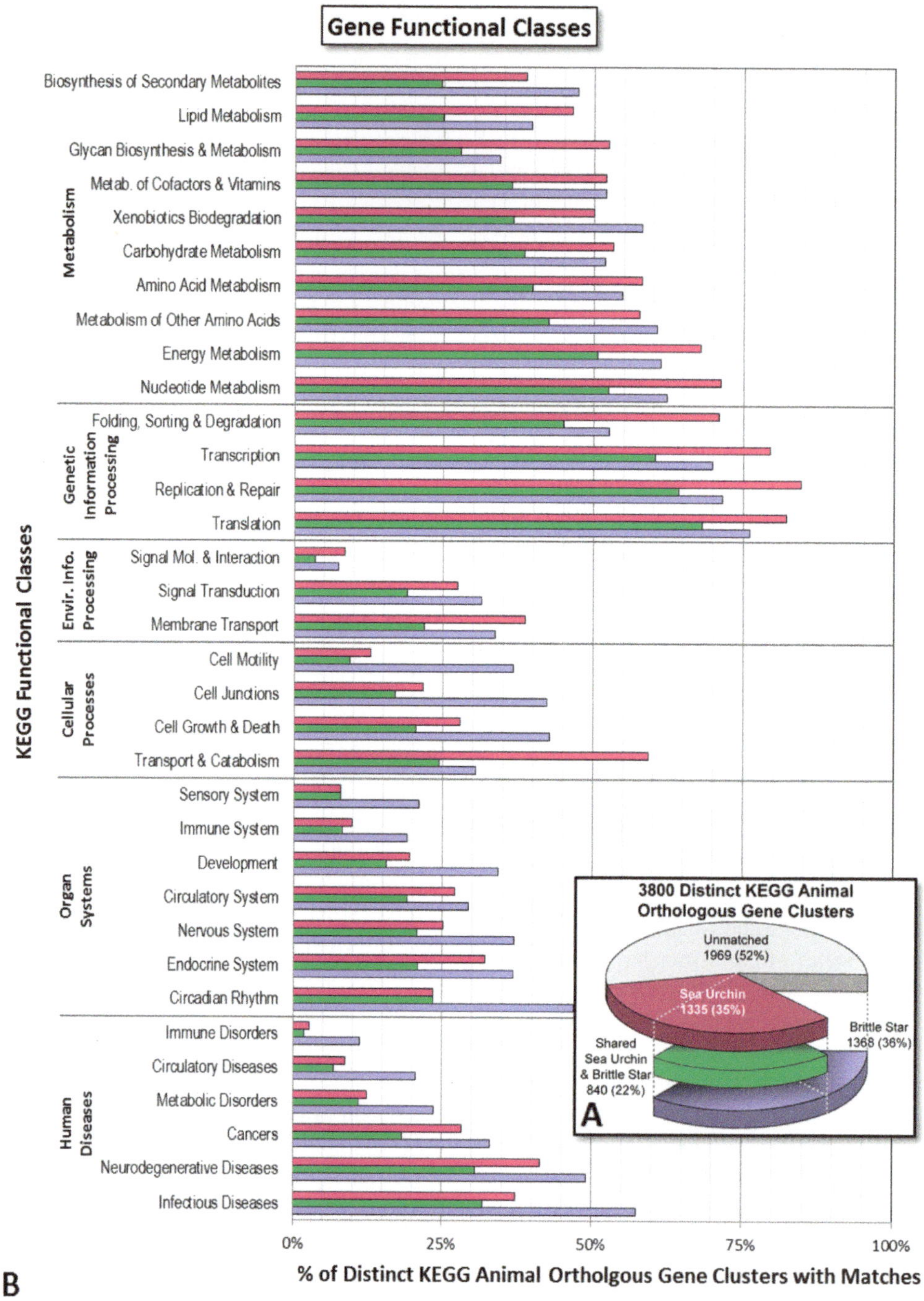

Figure 3: Gene functional classes found in brittle star gastrula transcriptome. (A)*Ophiocoma wendtii* sequences were compared to the KEGG Orthology database by reciprocal best BLAST. Of 3,800 distinct KEGG animal gene clusters, 36% had significant matches to brittle star (blue), and 35% had matches to sea urchin (purple). Green shows the overlap between these two sets, indicating the KEGG clusters that match to both organisms (22%). (B) When sorted into functional classes, an average of 43%, 39%, and 28% of the KEGG clusters within each class had matches to brittle star, to sea urchin, or to both, respectively, with a majority of classes having similar representation in both organisms.

RESULTS AND DISCUSSION

Embryonic Development of *O. Wendtii*

The key stages of *O. wendtii* development are shown in Figure 2. The egg is pigmented, and pigment granules are retained during cleavage stages but disappear in the blastula. Cleavage is radial and holoblastic and is equal throughout cleavage, such that the micromeres characteristic of the sea urchin fourth cleavage division are not produced. A hollow blastula is formed and cells ingress into the blastula to initiate gastrulation. The number of ingressing cells seems much larger than is typical in sea urchins, but we have not quantified the number or traced the lineage of individual cells. Archenteron formation occurs through invagination and convergent extension. A second group of mesenchyme cells forms at the tip of the archenteron and gives rise to the coelomic pouches, but no pigment cells appear. The skeletogenic mesenchyme cells gather in ventrolateral clusters as in sea urchins and begin to form the mineralized skeleton. The timing of development to hatching blastula is similar to sea urchins. However, following the invagination of endoderm, brittle star development proceeds at a slower rate relative to sea urchins. There is an initial invagination at 26 to 30 h post-fertilization, but this persists for several hours before overt endomesoderm development proceeds. Also, unlike that seen in sea urchins, the elongation of the skeletal rods is delayed relative to the extension of the archenteron, such that the archenteron has extended one-third to halfway across the blastocoel before skeletal elements appear. When the gut is fully formed the skeleton is still composed of relatively small tri-radiate spicules. These then elongate such that the pluteus larva is very similar to that of sea urchins. The stage at which we isolated RNA for sequencing analysis is similar to Figure 2F. We chose that point when skeletal elements were first visible.

Sequencing and Assembly

Pyrosequencing was performed on mRNA from gastrula stage brittle star embryos. After cleaning and trimming, there were 354,586 sequencing reads with a total of 75,031,136 bp(Figure 4A). Sequencing read lengths ranged from 16 to 439 bp, with approximately three-fourths of the reads being between 200 and 300 bp. Less than one percent were longer than 300 bp. Reads of 15 bp or shorter after trimming were not used for contig assembly. A total of 14,261 contigs were assembled, with a combined length of 5,488,581 bp (Figure 4B). Median length increased by 23% over that of the unassembled reads (282 vs. 229), while average length increased by 81% (384 vs. 212). Roughly two-thirds had lengths between 100 and 400 bp. The average number of reads per contig was 16.3, with a median of 5, a mode of 2, and a maximum of 8,989. Coverage or depth ranged from 1x to 8549.4x, with an average of 7.1, median of 3.5, and standard deviation of 44.6 (Figure 4C).

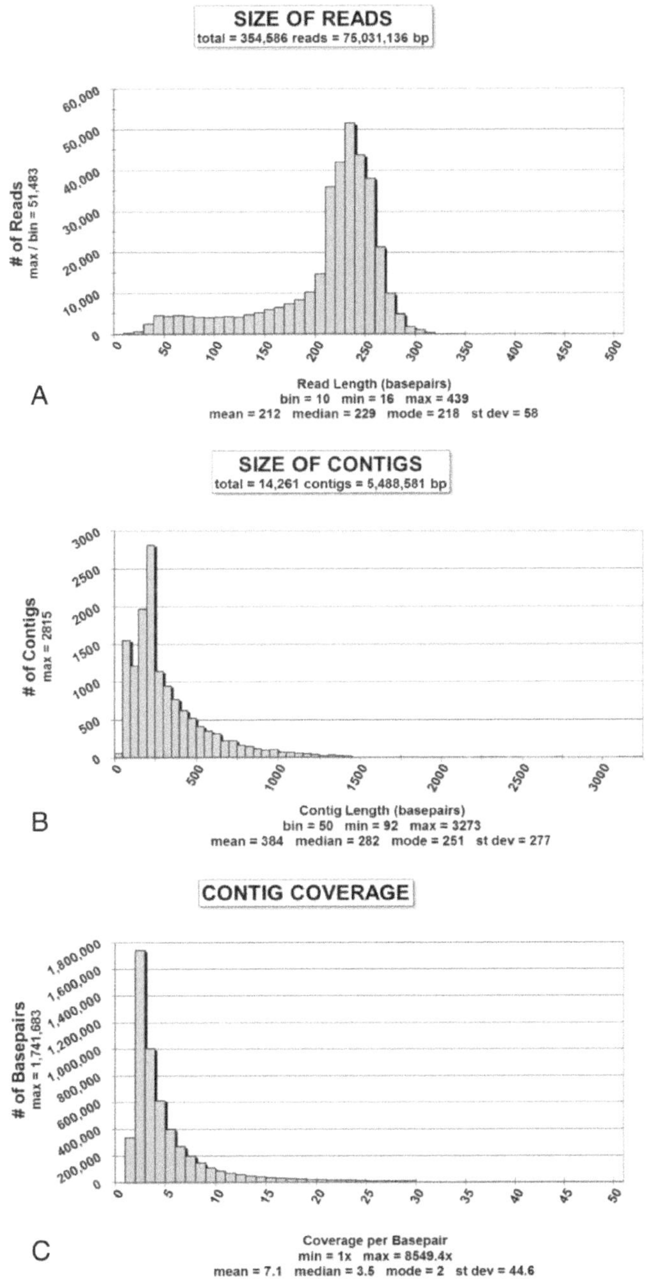

Figure 4: Pyrosequencing of brittle star transcriptome. (A) After cleaning and trimming, 354,586 reads totaled 75,031,136 bp. Approximately 3/4 had lengths between 200 and 300 bp. Less than one percent were longer than 300 bp. (B) A total of 14,261 contigs were assembled, with a combined length of 5,488,581 bp. Median length increased by 23% over that of the unassembled reads. Roughly two-thirds of the reads had lengths between 100 and 400 bp. Four percent were longer than 1,000 bp, creating a long right-hand tail to the distribution. (C) The number of times a given nucleotide position is present in the reads used to assemble the contigs ranged from 1x to 8549.4x. Eighty-one percent were represented one to five times, while less than one percent had more than 100× coverage.

Automated Annotation

Reciprocal best BLAST (RBB) searches identified brittle star transcripts putatively corresponding to a total of 3,385 orthologous genes in other databases (Figure 5). The brittle star sequences were translated in all six reading frames, and blastp was used to query the SpBase sea urchin Glean3 protein models. There were 3,303 matches between brittle star and the sea urchin genome [7]. Of these, 1,863 also matched to the sea urchin combined UniGene transcriptome libraries [34]. The KEGG Orthology database [35] produced 1,368 matches. More than two-thirds (2,309 or 68%) of the identified brittle star genes had matches to more than one dataset. Almost a quarter (840 or 24.8%) matched to all three.

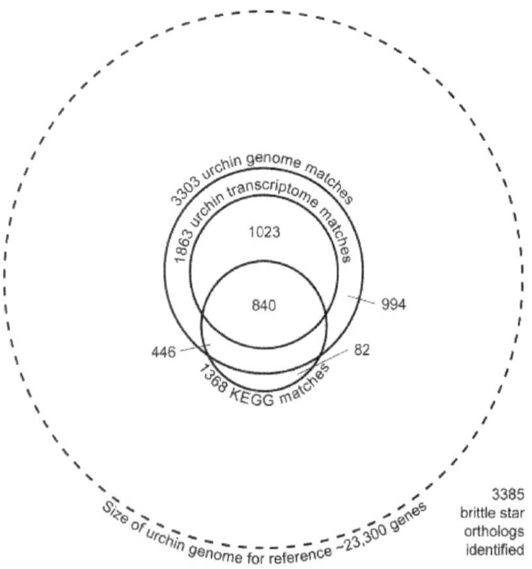

Figure 5: BLAST identification of brittle star genes. Automated BLAST was used to align*Ophiocoma wendtii* cDNA sequences to both the genome and transcriptome of the sea urchin *Strongylocentrotus purpuratus*, as well as to the KEGG Orthology database. The areas of the smaller circles represent the number of significant reciprocal best BLAST hits to the indicated datasets. Overlaps indicate matches of the same brittle star sequences to more than one dataset, and in nearly all such cases the matches from the different datasets are mutually consistent. For reference, the large dashed border represents the size of the*S. purpuratus* genome (~23,300 genes).

Note that the *O. wendtii* data were compared against data in each of the other datasets in Figure5 separately. Therefore, brittle star sequences with hits to multiple datasets do not necessarily represent RBB matches between every component in the annotation, but merely represent significant hits between brittle star and more than one of the other datasets independently. Examination of the results reveals that the individual hits are mutually consistent in terms of genes identified. The brittle star data set has many times more sequences than the sea urchin gastrula UniGene set available on the NCBI UniGene database, and the average length of the brittle star sequences is shorter. To assess whether we could make meaningful comparisons between these different data sets, we plotted the data as rarefaction curves. In ecology, rarefaction uses repeated random re-sampling of a large pool of samples to estimate the species richness as a function of the number of individuals sampled. Here we used it to estimate how thoroughly each data set represents the full transcriptome. In Figure 6, the curve for sea urchin has a much steeper initial slope, and therefore matches to a significant number of KEGG clusters even with many fewer sequences. This likely occurs because the sea urchin sequences are longer on average. The brittle star curve rises more gradually, but plateaus near the end, indicating that the sequencing captured most of the genes present in the transcriptome. If we assume that the two organisms express roughly the same number of genes at equivalent developmental stages, then the rarefaction curves indicate that this is indeed a meaningful comparison. Comparison of the brittle star gastrula transcriptome to the *S. purpuratus* genome as well as to the *S. purpuratus* gastrula transciptome and to the KEGG Orthology database, allowed us to identify 3,385 sequence matches.

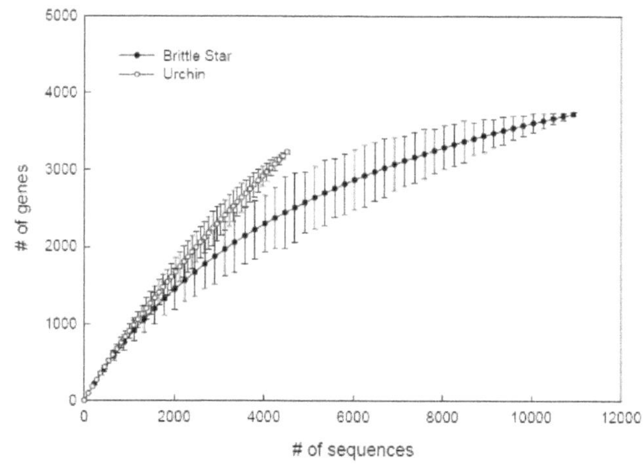

Figure 6: Rarefaction curves for sea urchin and brittle star. The steeper initial slope for the sea urchin curve indicates matches to a significant number of KEGG clusters even with many fewer sequences. The brittle star curve rises more gradually, but becomes asymptotic at the right, indicating that the sequencing captured most of the genes present in the gastrula transcriptome. If the two organisms express roughly the same number of genes at equivalent developmental stages, then the rarefaction curves indicate that comparison of these two data sets is indeed meaningful.

Gene Functional Classes

O. wendtii sequences were compared to the KEGG Orthology database [24] by reciprocal best BLAST (Figure 3A). The KEGG Orthology database contains clusters of genes orthologous among a large number of organisms. Of the 3,800 clusters relevant to animals, 1,368 (36%) had significant matches to brittle star. Similarly, 1,335 KEGG clusters (35%) had matches to sea urchin gastrula. These numbers include 840 KEGG clusters (22%) with matches to both organisms.

When sorted into functional classes (Figure 3B), an average of 43%, 39%, and 28% of the distinct KEGG clusters within each class had matches to brittle star, to sea urchin, and to both, respectively, with a range between 2% and 85%. Each KEGG functional class consists of a number of biochemical pathways. On average, 44%, 43%, and 30% of the KEGG clusters within each pathway had matches to brittle star, to sea urchin, and to both, respectively. Note that there is extensive overlap between the various KEGG functional classes and pathways, with many clusters falling into several different ones.

Overall, genes involved in metabolism and genetic information processing were the most highly conserved, as would be expected. The number of these 'housekeeping' genes found in sea urchins and brittle stars are similar, and the relationship between the number of genes observed in each group and the number shared between them is very consistent. There are fewer orthologs detected in the other KEGG orthology groups. Many pathways under 'Organ Systems' and 'Human Diseases' are vertebrate-specific and/or relate to functions which do not operate extensively until later stages of development or after metamorphosis and would not be expected to be expressed at the gastrula stage. This is true for both organisms. There is also more variation in the number of gene matches to sea urchins and brittle stars in these functional classes.

Genes involved with the cytoskeleton and cell junctions had considerably more matches to brittle star. Cell-adhesion genes are often large, with many exons, and with domains often repeated and shared between multiple genes [36]. These characteristics, along with the short lengths of the brittle star sequences, have the potential to produce an artificially high number of BLAST hits. However, this pattern was the exception, not the rule, across the other functional classes.

The sea urchin had a far greater number of matches to genes involved in endocytosis, lysosome and RNA degradation. Many of these genes again overlap with several other pathways, but there is no clear pattern to account for the disparity.

Comparison to Sea Urchin Developmental Gene Regulatory Network

The gene regulatory networks that underlie the differentiation of the basic tissue types in sea urchin embryos have been fairly well-characterized. The temporal and spatial expression of these genes has been determined and many of the regulatory interactions between the various genes have been determined, either directly or indirectly by interference with gene expression. The majority of these genes are expressed concurrently at the gastrula stage, which makes this stage an excellent point to identify a global set of genes important to the process of early cell differentiation. Here we use the sea urchin *S. purpuratus* gastrula GRNs at 21 to 30 h of development [3–6] as a reference to look for conservation of genes expressed in the brittle star gastrula at 40 h of development, which is equivalent morphologically. At this point the skeletal spicules have just begun forming, the archenteron is one third to halfway across the blastocoel cavity and the equivalent of secondary mesenchyme has formed. The gut is not yet partitioned and no mouth has formed. The presence of the same genes expressed at the same stage in these two organisms would suggest a conservation of GRNs and a shared gastrula 'toolkit' of proteins. The absence of genes expressed in either organism would indicate that there is either a temporal change in expression or that the gene is not expressed at all in the embryo of one group. Either is an indication of a change in a GRN. It is possible we could fail to detect some of those transcripts that are expressed at very low levels and recalcitrant to this RNA profiling. However, reciprocal BLAST searches using the brittle star gastrula transcriptome data and the *S. purpuratus* genome found homologs for a majority of genes involved in the sea urchin developmental gene regulatory network, including transcripts expressed at very low levels in *S. purpuratus*.

In sea urchins, a gradient of β-catenin initiated at the vegetal pole of the egg sets up and is soon reinforced by a circuit in the early embryo involving β–catenin/*lef1, wnt8, blimp1,* and *otx* in an intricate shifting relationship, creating a ring of gene expression which moves outward from the vegetal pole to specify endomesoderm [37]. *Hox11/13b* is also soon involved in this circuit [5,38]. Comparisons between sea urchins and starfish have revealed that just downstream from these early endomesoderm genes in the endoderm lies an extremely well-conserved kernel involving *blimp1/krox, otx, gatae, foxa,* and *brachyury*[39]. In starfish, *tbr (t-brain)* is also part of this kernel, a role which is likely deeply ancestral, as it is also expressed in vegetal pole endoderm precursors in both sea cucumbers and hemichordates [40, 41]. However, in sea urchins *tbr* has lost this role in endoderm and has instead been co-opted into skeletogenesis [42]. In sand dollars it appears to play both these roles [43].

In Table 1 we compare some key endomesoderm and endoderm specific genes in the sea urchin to the transcripts present in the brittle star gastrula. Brittle stars express – *catenin, lef1, otx, blimp1, wnt, hox/11/13b and foxa* genes, suggesting that components of the endomesoderm and endoderm GRNs expressed early in development are conserved. *Gatae,* however, is not expressed. Many animal phyla employ *gata* genes in gut formation [44]. *Gatae* is a key component of the endoderm GRN in sea urchins and forms a feedback loop that maintains expression of these genes in the endoderm [45]. *Otx* and *blimp1* constitute another portion of that feedback loop [5], and this could be sufficient for endoderm differentiation in brittle stars. Two genes that are activated by *gatae* in *S. purpuratus brachyury (bra)* and *krüppel (krl)*, are not expressed at the gastrula stage in brittle stars. *Krüppel* expression in sea urchins is highest in the early blastula, and is mostly gone by the time of gastrulation in *S. purpuratus*[46]. Its absence from the brittle star data may therefore reflect a small shift in timing and/or low transcript abundance at the onset of gastrulation. *T-brain* is not expressed in the brittle star gastrula. This would seem to indicate that *tbr* expression is not required for skeletogenesis in brittle star embryos as it is in sea urchins, or for endoderm formation as in starfish.

Table 1: Comparison of *Ophiocoma wendtii* gastrula transcripts to the*Strongylocentrotus purpuratus* endodermal and endomesodermal gene regulatory networks

Gene	Found in*O.w.* gastrula	RBB to*S.p.*Genome [SpBase:]	RBB to NCBI RefSeq Proteins [NCBI:]	Role in*S.p.*
β-Catenin	Y	β-Catenin	*S.p.* β-Catenin	Endoderm
		[SPU_004319]	[XP_786059.2]	
Otx	Y	Otx	*S.p.* Otx	Endoderm

		[SPU_010424]	[NP_999753.2]	
Wnt	Y	Wnt5	*S.k.* Wnt2	Endoderm
		[SPU_026277]	[NP_001158455.1]	
Blimp1	Y	Blimp1/Krox	*B.f.* Zn-finger	Endoderm
		[SPU_027235]	[XP_002587482.1]	
Hox11/13b	Y	Hox11/13b	*S.p.* Hox11/13b	Endoderm
		[SPU_002631]	[NP_999774.1]	
Bra	N			Endoderm
Krl	N			Endoderm
Myc	Y	Myc	*S.p.* Myc	Endoderm
		[SPU_003166]	[NP_999744.1]	
SoxB1	Y	SoxB1	*S.p.* SoxB1	Endoderm
		[SPU_022820]	[NP_999639.1]	
Brn1-2-4	Y	Brn1-2-4	*S.p.* Brn1-2-4	Endoderm
		[SPU_016443]	[XP_782909.2]	
Tgif	Y	Tgif	*I.s.* Tgif	Endoderm
		[SPU_018126]	[XP_002433653.1]	
Hnfl	N			Endoderm
Eve	N			Endoderm
Hh	N			Endoderm
VEGF	Y	VEGF	*H.p.* VEGF	Endoderm
		[SPU_030148]	BAI67115.1]	
Dac	Y	Dac	*I.s.* Dachsund	Endoderm
		[SPU_028061]	[XP_002407755.1]	
Endo16	N			Endoderm
FoxA	Y	FoxA	*S.p.* FoxA	Endo+SMC
		[SPU_006676]	[NP_001073010.1]	
GataE	N			Endo+SMC
Kakapo	Y	Syne1	*S.p.* Similar to CG33715-PD	Endo+SMC
		[SPU_013237]	[XP_784190.2]	
Apobec	Y	Hnrpr	*S.p.* Hnrpr	Endo+SMC
		[SPU_019557]	[XP_793277.1]	
Gelsolin	Y	Gelsolin	*S.p.* Gelsolin	Endo+SMC
		[SPU_003985]	[XP_788777.1]	

B.f. = Branchiostoma floridae, H.p. = Heliocentrotus pulcherrimus.

I.s. = Ixodes scapularis, O.w. = Ophiocoma wendtii.

S.k. = Saccoglossus kowalevskii, S.p. = Strongylocentrotus purpuratus.

The endoderm in *S. purpuratus* is derived from two tiers of blastomeres formed during cleavage from the macromeres: Veg2, closest to the vegetal pole, and Veg1 above that. The Veg2 derived endoderm in *S. purpuratus* expresses *myc, brn1/2/4, tgif and dac* genes at the gastrula stage [47]. All of these are expressed in the brittle star gastrula (Table 1). In contrast, the Veg1 gene*seve* and *hnf1* are not expressed in brittle stars. Together this suggests that a central early kernel of the endoderm GRN is conserved, although the expression of *gatae* and some genes it regulates are not. The expression of genes found in Veg2 endoderm is also largely conserved. The most likely explanation of our results is that the equivalent of Veg1 endoderm has not formed in the brittle star gastrula at the stage we examined. This suggests a heterochronic shift in the formation of the second tier of endoderm. This could also explain the absence of*brachyury*. It is a key player in gut formation in both protostomes and deuterostomes, though the details differ between taxa [48–51]. A shift in the timing of Veg1 endoderm formation could delay expression of *brachyury* in the brittle star. A less likely explanation is that a loss of this layer of endoderm has occurred in brittle stars, and that the gut is formed entirely by the equivalent of Veg2 endoderm. *Endo16*, one of the major differentiation gene products in endoderm is not expressed in brittle star gastrula.

Following endomesoderm specification, mesenchyme precursors all express *ets1/2, erg, and hex*in *S. purpuratus.* All three of these genes are expressed in the brittle star gastrula (Table 2). The sea urchin skeletogenic primary mesenchyme derived from the micromeres, the homologous vegetal plate mesoderm in starfish, and the larval structures that produce the adult skeletons in both animals all express many of the same genes as sea urchin micromeres [19, 52, 53], and a majority of these genes were found in the brittle star gastrula transcriptome as well (Table 2). In all cases of echinoderm skeleton formation studied, including brittle star embryos, *alx1* is expressed. This is consistent with reports that ectopic expression of *alx1* in sea urchin non-skeletogenic mesenchyme (NSM) induces skeleton formation [52]. *Ets1* is expressed both maternally and zygotically, and is involved in all the above cases, activating a great number of downstream genes. *Ets1* and *alx1* were both found in the brittle star gastrula transcriptome. Just downstream from these in both sea urchin micromere and starfish vegetal plate mesoderm are a group of three genes, *erg, hex,* and *tgif,* which form a "lockdown" mechanism, stabilizing the specification state by feeding back to each other and to *tbr* and *ets1*, and feeding forward into tissue- specific differentiation genes [53]. All three were present in brittle star, as was*deadringer (dri)*, which appears to play a similar role in all the skeletogenic cases. *Tbr* has not been found in brittle stars. In starfish, *tbr* is seen in both endoderm (discussed above) and mesoderm [53]. It does not appear to be involved in adult skeletogenesis in either starfish or sea urchins. Its absence in brittle stars reinforces the idea that it was not part of the ancestral skeletal GRN and that its role in sea urchin embryonic skeleton formation is derived.

Table 2: Comparison of *Ophiocoma wendtii* gastrula transcripts to the*Strongylocentrotus purpuratus* mesenchymal gene regulatory networks

Gene	Found in*O.w.*gastrula	RBB to *S.p.*Genome [SpBase:]	RBB to NCBI RefSeq Proteins [NCBI:]	Role in *S.p.*
HesC	Y	HesC	*S.p.* HesC	Mesenchyme
		[SPU_021608]	[XP_796692.1]	
Erg	Y	Erg	*S.p.* Erg	Mesenchyme
		[SPU_018483]	[NP_999833.1]	
Hex	Y	Hex	*S.p.* Hex	Mesenchyme
		[SPU_027215]	[XP_001197103.1]	
Ets1/2	Y	Ets1/2	*S.p.* Ets1/2	Mesenchyme
		[SPU_002874]	[NP_999698.1]	
Alx1	Y	Alx1	*S.p.* Alx1	PMC
		[SPU_025302]	[NP_999809.1]	
Tbr	N			PMC
Tgif	Y	Tgif	*I.s.* Tgif	PMC
		[SPU_18126]	[XP_002433653.1]	
FoxN2/3	N			PMC
Dri	Y	Dri	*S.p.* Dri	PMC
		[SPU_017106]	[NP_999799.1]	
FoxB	Y	FoxB	*S.p.* FoxB	PMC
		[SPU_004551]	[NP_999797.1]	
FoxO	Y	FoxO	*S.p.* FoxO	PMC
		[SPU_009178]	[XP_001183650.1]	
VEGFR	N			PMC
Delta	N			PMC
Spicule matrix genes	Possible	C-lectin [SPU_007882]	*S.p.* C-lectin [NP_999805.1]	Skeletal differentiation
MSP130	N			Skeletal differentiation
G-Cadherin	Y	G-Cadherin	*S.k.* G-Cadherin	Skeletal differentiation
		[SPU_015960]	[XP_002741140.1]	
Ficolin	Y	Fic	*B.f.* Ficolin	Skeletal differentiation
		[SPU_023548]	[XP_002594892.1]	

Cyclophilin	Y	CypL7	*D.m.* Cyclophilin 1	Skeletal differentiation
		[SPU_008305]	[NP_523366.2]	
Gcm	Y	Gcm	*S.k.* Gcm	SMC
		[SPU_006462]	[XP_002733441.1]	
Notch	N			SMC
Six1/2	N			SMC
Hnf6	Y	Hnf6	*S.p.* Hnf6	SMC
		[SPU_016449]	[NP_999824.1]	
GataC	N			SMC
Scl	N			SMC
Pks	Y	Pks	*S.p.* Pks	SMC
		[SPU_028395]	[NP_001239013.1]	
FoxF	Y	FoxF	*S.p.* FoxF	Small micromeres
		[SPU_000975]	[XP_794135.1]	
SoxE	N			Small micromeres
FoxY	N			Small micromeres

B.f. = Branchiostoma floridae, D.m. = Drosophila melanogaster, I.s. = Ixodes scapularis.

O.w. = Ophiocoma wendtii, S.k. = Saccoglossus kowalevskii.

S.p. = Strongylocentrotus purpuratus.

The downstream differentiation genes found in *S. purpuratus* skeletogenic cells at the gastrula stage are also found in the brittle star gastrula (Table 2). The spicule matrix proteins of the sea urchin endoskeleton contain a single C-type lectin domain and repetitious stretches rich in proline and glycine [54–56]. The apparently loose constraints on primary structure in these proteins, and the resulting low sequence conservation make identification of brittle star homologs difficult. However, the brittle star gastrula transcriptome contains several transcripts encoding C-type lectin domains and repetitive regions. Several other proteins, including Cyclophilin and Ficolin, are all expressed in sea urchin PMC cells and associated with the skeleton, though their exact functions remain unclear. The brittle star gastrula transcriptome contains matches for *Cyclophilin* and *Ficolin*, but not for *MSP130*, a major cell surface protein in sea urchin PMCs. Overall there is a remarkable conservation of the GRN leading to formation of mineralized tissue in the embryos of sea urchins and brittle stars.

In sea urchins, Delta-Notch signaling from the micromeres activates *gcm* in the adjacent NSM to form pigment cells [57, 58]. Brittle star embryos do not form embryonic pigment cells. Neither do starfish, but they express *gcm* in ectoderm rather than mesoderm, and it does not depend on Delta signaling [18]. Neither *notch* nor *delta* is expressed in the brittle star gastrula (Table 2). *Gcm* is expressed in brittle star gastrula, but *gatac, gatae. six1/2,* and *scl* are not. This suggests that the GRN leading to pigment cells, not surprisingly, is not conserved in brittle stars. Likewise, most of the genes that are expressed in the *S. purpuratus* small micromeres (i.e.*soxe, foxy*), which are not formed outside of euechinoids, are not expressed in the brittle star gastrula (Table 2).

In sea urchin ectoderm, Nodal patterns both the ventro-dorsal (oral-aboral) and left-right axes [59], but was not found to be transcribed in brittle star gastrula (Table 3); nor was its antagonist Lefty, which limits Nodal to the ventral side during sea urchin development [60]. On the other hand, a number of genes downstream from Nodal and key to specification of different ectodermal regions [61] were found in brittle star (Table 3). Most of the genes expressed in the*S. purpuratus* oral ectoderm are found in the brittle star gastrula transcriptome, including*chordin* and *BMP2/4*. Sea urchin BMP2/4 is expressed in the oral ectoderm, then diffuses to and specifies the aboral ectoderm by inhibiting Nodal [62], while Chordin helps pattern neural tissue in the ciliary band at the oral/aboral border by excluding BMP2/4 activity from the oral side [63]. Genes that, in the sea urchin, are activated by Nodal-independent early oral ectoderm input are found to be expressed in brittle star gastrula. These include *otxb1/2* and *hnf6.* Of the sea urchin genes that are activated at the boundary of ectoderm and endoderm,*foxj* is expressed in brittle star gastrula, but *lim1* and *nk1* are not.

Table 3: Comparison of *Ophiocoma wendtii* gastrula transcripts to the*Strongylocentrotus purpuratus* ectodermal gene regulatory network

Gene	Found in *O.w.* gastrula	RBB to *S.p.* Genome [SpBase:]	RBB to NCBI RefSeq Proteins [NCBI:]	Role in *S.p.*
Nodal	N			Oral ectoderm
Lefty	N			Oral ectoderm

Chordin	Y	Chordin	*S.k.* Chordin	Oral ectoderm
		[SPU_004983]	[NP_001158390.1]	
Sip1	N			Oral ectoderm
FoxG	N			Oral ectoderm
BMP2/4	Y	BMP2/4	*S.p.* BMP2/4	Oral ectoderm
		[SPU_000669]	[NP_001116977.1]	
FoxA	Y	FoxA	*S.p.* FoxA	Oral ectoderm
		[SPU_006676]	[NP_001073010.1]	
Bra	N			Oral ectoderm
Dri	Y	Dri	*S.p.* Dri	Oral ectoderm
		[SPU_017106]	[NP_999799.1]	
Hes	Y	Hes	*S.k.* Hes1	Oral ectoderm
		[SPU_006814]	[NP_001158466.1]	
Hnf6	Y	Hnf6	*S.p.* Hnf6	Oral ectoderm
		[SPU_016449]	[NP_999824.1]	
FoxJ1	Y	FoxJ1	*S.p.* FoxJ1	Ecto/Endo border
		[SPU_027969]	[NP_001073013.1]	
Nk1	N			Ecto/Endo border
Lim1	N			Ecto/Endo border
Tbx2/3	Y	Tbx2/3	*S.p.* Tbx2/3	Aboral ectoderm
		[SPU_023386]	[NP_001123280.1]	
Lhx2 (Lim2)	Y	Lhx2	*M.m.* Lhx2	Aboral ectoderm
		[SPU_021313]	[NP_034840.1]	
Dlx	N			Aboral ectoderm
Nk2.2	Y	Nk2.2	*S.p.* Nk2.2	Aboral ectoderm
		[SPU_000756]	[NP_001123283.1]	
Hox7	N			Aboral ectoderm
Msx	N			Aboral ectoderm
Klf7	Y	Klf2/4	*S.k.* Klf2	Aboral ectoderm
		[SPU_020311]	[NP_001161575.1]	
IrxA	N			Aboral ectoderm
Hmx	N			Aboral ectoderm

M.m. = *Mus musculus, O.w.* = *Ophiocoma wendtii, S.k.* = *Saccoglossus kowalevskii, S.p.* = *Strongylocentrotus purpuratus.*

Genes that are expressed in the sea urchin aboral ectoderm are not as uniformly expressed in brittle star gastrula. Genes expressed by 12 h of sea urchin development such as *sim* and *nk2.2*are expressed in brittle star gastrula, but not genes expressed later in sea urchin aboral ectoderm such as *hox7* and *msx*. The differentiation genes *spec1* and *spec2a* are also not found to be expressed in brittle star gastrula at the time examined. *Tbx2/3* is expressed in brittle star gastrula, but not *irxa* and *dlx*, which are activated by Tbx2/3 in sea urchins. Taken together this would suggest two heterochronic shifts in ectoderm determination between sea urchins and brittle stars. In sea urchins, all of the genes examined are expressed at the gastrula stage. In brittle stars, patterning by Nodal and Lefty is apparently complete by gastrulation and these genes are no longer expressed. Oral ectoderm is determined and specification of the aboral ectoderm is underway, but it appears that this process is not complete in the 40 h brittle star gastrula.

CONCLUSIONS

The brittle star *O. wendtii* exhibits radial holoblastic cleavages that are equal throughout, giving rise to uniform-sized blastomeres without the formation of the micromeres characteristic to sea urchins. Despite this, mesenchymal cells ingress and give rise to an embryonic skeleton, a developmental structure unique to echinoids and ophiuroids among the echinoderms. Mesenchymal cells also give rise to the coelomic pouches, but no pigment cells are formed in the embryo. Archenteron formation occurs much the same as in sea urchins, although there is a delay in gut elongation following invagination as well as in growth of the skeletal spicules initiated in the ventrolateral clusters. The resulting pluteus larva closely resembles that of sea urchins, albeit without pigment cells. The *O. wendtii* gastrula expresses genes from all functional classes at the gastrula stage. Brittle stars and sea urchins have comparable numbers of genes in most functional classes expressed at the gastrula stage.

A majority of the genes involved in the sea urchin gene regulatory network were also found in the brittle star gastrula transcriptome (Table 4). The brittle star pyrosequencing data are completely consistent with our earlier results using a PCR-based candidate gene approach (not shown). For example, transcripts of *alx1, dri, gabp, ets1*, and *erg* were found by both methods, whereas *tbr, gatac,* and *gatae* were not. While this does not completely rule out these genes being expressed but undetected by our methods, their absence is striking given the overall conservation of expression between the two groups. The percentage of genes involved in gene regulatory networks expressed in *S. purpuratus* gastrula that are also expressed in *O. wendtii*gastrula exceeds the percentage of transcripts conserved overall (Table 4). However, this conservation is not uniform across the different tissue types found in echinoderm gastrulae. Some of these differences can be explained by heterochronic shifts in gene expression, although loss of gene expression is also a possibility. Some of the endomesoderm genes that are expressed in sea urchin gastrula at declining levels could be undetectable by the brittle star gastrula stage. Examination of the aboral ectoderm genes expressed in *O. wendtii* relative to *S. purpuratus*indicates that specification of aboral ectoderm has begun but is delayed in the brittle star. The same could be true for the Veg1 endoderm. Other differences in gene expression correlate with differences in embryonic development. Brittle star embryos do not possess micromeres or pigment cells. The second lowest percentage of GRN genes conserved (33%) is seen in the genes expressed in *S. purpuratus* small micromeres and pigment cells (Table 4).

Table 4: Conservation of genes between *Strongylocentrotus purpuratus* and*Ophiocoma wendtii*

		% conserved in *O. wendtii*
S. purpuratus transcriptome		55
Early gastrula GRN		65
Endoderm		53
	Veg2 endoderm	70
	Veg1 endoderm	20
Primary mesenchyme		86
Non-skeletogenic mesenchyme		57
	Secondary mesenchyme	64
	Small micromeres	33
Oral ectoderm		67
Aboral ectoderm		58

The highest percentage of GRN conservation is seen in the skeletogenic mesenchyme cells (PMCs in sea urchins). This is not surprising, since all adult echinoderms form mineralized structures. The GRN and differentiation genes that lead to mineralized structures must be conserved in order for the adult skeleton to form. In sea urchins this GRN is activated in the embryo largely intact. The conservation of these genes in the *O. wendtii* gastrula suggests that is the case in brittle stars as well. An extensive analysis of spatial expression of the genes involved in these GRNs is the next step in confirmation of homology.

In sea urchins, Hesc is a transcriptional repressor ubiquitously expressed in the embryo, where its role is to repress the skeleton program. In the sea urchin micromeres, *hesc* is itself repressed by Pmar1 in response to nuclearized β-catenin, thereby de-repressing the skeleton circuits [64]. This double-negative *pmar1/hesc* gate appears unique to sea urchins as the mechanism that coupled the pre-existing programs of skeletogenesis and maternal β-catenin-mediated vegetal specification to produce the novelty of the embryonic skeleton, as it is not involved in adult sea urchin skeletogenesis [19, 53]. Recent evidence suggests that other, as yet unknown, mechanisms related to the unequal cleavage that produces the micromeres are also involved [65]. Starfish, which do not build an embryonic skeleton, also express *hesc* throughout most of the embryo, but this expression appears to have no effect on mesodermal genes shared with sea urchin skeletogenesis, and *pmar1* has never been found in starfish [54].

Sea urchins express *pmar1* from fourth cleavage through mid-blastula, so we would not expect to see it expressed in the *O. wendtii* gastrula transcriptome. Using PCR, we have searched for, but never found, *pmar1* transcripts from any stage of brittle star development. We have, however, successfully amplified the *pmar1* homolog from brittle star genomic DNA, identified as such by the presence of a conserved intron [unpublished]. This suggests that activation of the adult skeletal GRN in embryos occurred differently in brittle stars than in sea urchins. Overall, our data suggest that embryonic skeleton formation in sea urchins and brittle stars represents convergent evolution by independent co-optation of a shared pathway utilized in adult skeleton formation.

ACKNOWLEDGEMENTS

Mitch Ruzek assisted in collecting animals and culturing embryos. Darren Bauer prepared the RNA for sequencing. Pyrosequencing was performed at The Center for Genomics and Bioinformatics, Indiana University, Bloomington. Dan

Bergeron provided technical advice on the initial processing of the data. Tiehang Wu performed the rarefaction test. This work was supported by NSF grant 0909797 to BTL.

AUTHORS' CONTRIBUTIONS

BTL conceived of and oversaw the project, collected animals, cultured embryos, extracted RNA, and conducted BLAST comparisons against the sea urchin GRN. NG performed, and WKT oversaw and advised on, initial processing of the data and automated BLAST searches. RV assembled the annotation data, carried out the analyses and interpretation, and wrote most of the manuscript. JRG oversaw and advised on the analyses. All authors contributed to the final version of the manuscript.

REFERENCES

1. Davidson EH, Rast JP, Oliveri P, Ransick A, Calestani C, Yuh CH, Minokawa T, Amore G, Hinman V, Arenas-Mena C, Otim O, Brown CT, Livi CB, Lee PY, Revilla R, Rust AG, Pan Z, Schilstra MJ, Clarke PJ, Arnone MI, Rowen L, Cameron RA, McClay DR, Hood L, Bolouri H: A genomic regulatory network for development. Science. 2002, 295: 1669-1678. 10.1126/science.1069883.
2. Davidson EH, Rast JP, Oliveri P, Ransick A, Calestani C, Yuh CH, Minokawa T, Amore G, Hinman V, Arenas-Mena C, Otim O, Brown CT, Livi CB, Lee PY, Revilla R, Schilstra MJ, Clarke PJ, Rust AG, Pan Z, Arnone MI, Rowen L, Cameron RA, McClay DR, Hood L, Bolouri H: A provisional regulatory gene network for specification of endomesoderm in the sea urchin embryo. Dev Biol. 2002, 246: 162-190. 10.1006/dbio.2002.0635.
3. Oliveri P, Davidson EH: Gene regulatory network controlling embryonic specification in the sea urchin. Curr Opin Genet Dev. 2004, 204: 351-380.
4. Su YH: Gene regulatory networks for ectoderm specification in sea urchin embryos. Biochim Biophys Acta. 2009, 1789: 261-267. 10.1016/j.bbagrm.2009.02.002.
5. Peter IS, Davidson EH: The endoderm gene regulatory network in sea urchin embryos up to mid-blastula stage. Dev Biol. 2010, 340: 188-199. 10.1016/j.ydbio.2009.10.037.
6. Cameron RA, Samanta M, Yuan A, He D, Davidson E: SpBase: the sea urchin genome database and web site. Nucleic Acids Res. 2009, 37: D750-D754. 10.1093/nar/gkn887.http://spbase.org,
7. Sea Urchin Genome Sequencing Consortium: The genome of the sea urchin Strongylocentrotus purpuratus. Science. 2006, 314: 941-952.
8. Poustka AJ, Herwig R, Krause A, Hennig S, Meier-Ewert S, Lehrach H: Toward the gene catalogue of sea urchin development: the construction and analysis of an unfertilized egg cDNA library highly normalized by oligonucleotide fingerprinting. Genomics. 1999, 59: 122-133. 10.1006/geno.1999.5852.
9. Lee YH, Huang GM, Cameron RA, Graham G, Davidson EH, Hood L, Britten RJ: EST analysis of gene expression in early cleavage-stage sea urchin embryos. Development. 1999, 126: 3857-3867.
10. Zhu X, Mahairas G, Illies M, Cameron RA, Davidson EH, Ettensohn CA: A large-scale analysis of mRNAs expressed by primary mesenchyme cells of the sea urchin embryo. Development. 2001, 128: 2615-2627.
11. Poustka AJ, Groth D, Hennig S, Thamm S, Cameron A, Beck A, Reinhardt R, Herwig R, Panopoulou G, Lehrach H: Generation, annotation, evolutionary analysis, and database integration of 20,000 unique sea urchin EST clusters. Genome Res. 2003, 13: 2736-2746. 10.1101/gr.1674103.
12. Wei Z, Angerer RC, Angerer LM: A database of mRNA expression patterns for the sea urchin embryo. Dev Biol. 2006, 300: 476-484. 10.1016/j.ydbio.2006.08.034.
13. Materna SC, Nam J, Davidson EH: High accuracy, high-resolution prevalence measurement for the majority of locally expressed regulatory genes in early sea urchin development. Gene Expr Patterns. 2010, 10: 177-184. 10.1016/j.gep.2010.04.002.
14. Paul CRC, Smith AB: The early radiation and phylogeny of echinoderms. Biol Rev. 1984, 59: 443-481. 10.1111/j.1469-185X.1984.tb00411.x.
15. Littlewood DTJ, Smith AB, Clough KA, Emson RH: The interrelationships of the echinoderm classes: morphological and molecular evidence. Biol J Linnean Soc. 1997, 61: 409-438. 10.1111/j.1095-8312.1997.tb01799.x.
16. Harmon MC: The position of the ophiuroidea within the phylum echinodermata. 2005, University of South Florida, Biology Department
17. Davidson EH, Erwin DH: Gene regulatory networks and the evolution of animal body plans. Science. 2006, 311: 796-800. 10.1126/science.1113832.
18. Hinman VF, Davidson EH: Evolutionary plasticity of developmental gene regulatory network architecture. Proc Natl Acad Sci USA. 2007, 104: 19404-19409. 10.1073/pnas.0709994104.
19. Gao F, Davidson EH: Transfer of a large gene regulatory apparatus to a new developmental address in echinoid evolution. Proc Natl Acad Sci USA. 2008, 105: 6091-6096. 10.1073/pnas.0801201105.
20. Wilt FH, Ettensohn CA: Handbook of Biomineralization. The morphogenesis and biomineralization of the sea urchin larval skeleton. Edited by: Bauerlin E. 2007, Weinheim Germany: Wiley-VCH, 183-210.
21. Ettensohn CA: Lessons from a gene regulatory network: echinoderm skeletogenesis provides insights into evolution, plasticity and morphogenesis. Development. 2009, 136: 11-21. 10.1242/dev.023564.
22. Wray GA, McClay DR: The origin of spicule-forming cells in a "primitive" sea urchin (Eucidares tribloides) which appears to lack primary mesenchyme cells. Development. 1988, 103: 305-315.
23. Meyer E, Aglyamova GV, Wang S, Buchanan-Carter J, Abrego D, Colbourne JK, Willis BL, Matz MV: Sequencing and de novo analysis of a coral larval transcriptome using 454 GSFlx. BMC Genomics. 2009, 10: 219-10.1186/1471-2164-10-219.

24. Altschul SF, Gish W, Miller W, Myers EW, Lipman DJ: Basic local alignment search tool. J Mol Biol. 1990, 215: 403-410.

25. Lerat E, Daubin V, Ochman H, Moran NA: Evolutionary origins of genomic repertoires in bacteria. PLoS Biol. 2005, 3: e130-10.1371/journal.pbio.0030130.

26. Cooper VS, Vohr SH, Wrocklage SC, Hatcher PJ: Why genes evolve faster on secondary chromosomes in bacteria. PLoS Comput Biol. 2010, 6: e1000732-10.1371/journal.pcbi.1000732.

27. Flynn KM, Vohr SH, Hatcher PJ, Cooper VS: Evolutionary rates and gene dispensability associate with replication timing in the archaeon Sulfolobus islandicus. Genome Biol Evol. 2010, 2: 859-869. 10.1093/gbe/evq068.

28. Garnhart N, Bergeron RD: Homology Inspector (HomIn): A Tool for Exploring Homology.http://www.cs.unh.edu/~rdb/reports/homin.pdf,

29. Kanehisam K, Goto S: KEGG: Kyoto encyclopedia of genes and genomes. Nucleic Acids Res. 2000, 28: 27-30. 10.1093/nar/28.1.27.

30. Tatusov RL, Koonin EV, Lipman DJ: A genomic perspective on protein families. Science. 1997, 278: 631-637. 10.1126/science.278.5338.631.

31. Tatusov RL, Fedorova ND, Jackson JD, Jacobs AR, Kiryutin B, Koonin EV, Krylov DM, Mazumder R, Mekhedov SL, Nikolskaya AN, Rao BS, Smirnov S, Sverdlov AV, Vasudevan S, Wolf YI, Yin JJ, Natale DA: The COG database: an updated version includes eukaryotes. BMC Bioinforma. 2003, 4: 41-10.1186/1471-2105-4-41.

32. Ashburner M, Ball CA, Blake JA, Botstein D, Butler H, Cherry JM, Davis AP, Dolinski K, Dwight SS, Eppig JT, Harris MA, Hill DP, Issel-Tarver L, Kasarskis A, Lewis S, Matese JC, Richardson JE, Ringwald M, Rubin GM, Sherlock G: Gene ontology: tool for the unification of biology. The gene ontology consortium. Nat Genet. 2000, 25: 25-29. 10.1038/75556.

33. EcoSim: null models software for ecology: Version 7. 2011,http://garyentsminger.com/ecosim.htm

34. UniGene sea urchin embryo cDNA libraries. http://www.ncbi.nlm.nih.gov/UniGene/lbrowse2.cgi?TAXID=76680026;CUTOFF=1000,

35. KEGG Orthology Database.http://www.genome.jp/kegg/ko.html,

36. Whittaker CA, Bergeron KF, Whittle J, Brandhorst BP, Burke RD, Hynes RO: The echinoderm adhesome. Dev Biol. 2006, 300: 252-266. 10.1016/j.ydbio.2006.07.044.

37. Smith J, Theodoris C, Davidson EH: A gene regulatory network subcircuit drives a dynamic pattern of gene expression. Science. 2007, 318: 794-797. 10.1126/science.1146524.

38. Smith J, Kraemer E, Liu H, Theodoris C, Davidson E: A spatially dynamic cohort of regulatory genes in the endomesodermal gene network of the sea urchin embryo. Dev Biol. 2008, 313: 863-875. 10.1016/j.ydbio.2007.10.042.

39. Hinman VF, Nguyen AT, Cameron A, Davidson EH: Developmental gene regulatory network architecture across 500 million years of echinoderm evolution. Proc Natl Acad Sci USA. 2003, 100: 13356-13361. 10.1073/pnas.2235868100.

40. Maruyama YK: A sea cucumber homolog of the mouse T-Brain-1 is expressed in the invaginated cells of the early gastrula in Holothuria leucospilota. Zoolog Sci. 2000, 17: 383-387.

41. Tagawa K, Humphreys T, Satoh N: T-Brain expression in the apical organ of hemichordate tornaria larvae suggests its evolutionary link to the vertebrate forebrain. J Exp Zool B Mol Dev Evol. 2000, 288: 23-31. 10.1002/(SICI)1097-010X(20000415)288:1<23::AID-JEZ3>3.0.CO;2-H.

42. Hinman VF, Nguyen A, Davidson EH: Caught in the evolutionary act: precise cis-regulatory basis of difference in the organization of gene networks of sea stars and sea urchins. Dev Biol. 2007, 312: 584-595. 10.1016/j.ydbio.2007.09.006.

43. Minemura K, Yamaguchi M, Minokawa T: Evolutionary modification of T- brain (tbr) expression patterns in sand dollar. Gene Expr Patterns. 2009, 9: 468-474. 10.1016/j.gep.2009.07.008.

44. Patient RK, McGhee JD: The GATA family (vertebrates and invertebrates). Curr Opin Genet Dev. 2002, 12: 416-422. 10.1016/S0959-437X(02)00319-2.

45. Yuh CW, Dorman ER, Howard ML, Davidson EH: An otx cis-regulatory module: a key node in the sea urchin endomesoderm gene regulatory network. Dev Biol. 2004, 269: 536-551. 10.1016/j.ydbio.2004.02.025.

46. Howard EW, Newman LA, Oleksyn DW, Angerer RC, Angerer LM: SpKrl: a direct target of β-catenin regulation required for endoderm differentiation in sea urchin embryos. Development. 2001, 128: 365-375.

47. Peter IS, Davidson EH: A gene regulatory network controlling the embryonic specification of endoderm. Nature. 2011, 474: 635-639. 10.1038/nature10100.

48. Peterson KJ, Harada Y, Cameron RA, Davidson EH: Expression Pattern of Brachyury and Not in the Sea Urchin: Comparative Implications for the Origins of Mesoderm in the Basal Deuterostomes. Dev Biol. 1999, 207: 419-431. 10.1006/dbio.1998.9177.

49. Shoguchi E, Satoh N, Maruyama YK: Pattern of Brachyury gene expression in starfish embryos resembles that of hemichordate embryos but not of sea urchin embryos. Mech Dev. 1999, 82: 185-189. 10.1016/S0925-4773(99)00008-8.

50. Mitsunaga-Nakatsubo K, Harada Y, Satoh N, Shimada H, Akasaka K: Brachyury homolog (HpTa) is involved in the formation of archenteron and secondary mesenchyme cell differentiation in the sea urchin embryo. Zoology. 2001, 104: 99-102. 10.1078/0944-2006-2410018.

51. Gross JM, McClay DR: The role of Brachyury (T) during gastrulation movements in the sea urchin Lytechinus variegatus. Dev Biol. 2001, 239: 132-147. 10.1006/dbio.2001.0426.

52. Ettensohn CA, Kitazawa C, Cheers MS, Leonard JD, Sharma T: Gene regulatory networks and developmental plasticity in the early sea urchin embryo: alternative deployment of the skeletogenic gene regulatory network. Development. 2007, 134: 3077-3087. 10.1242/dev.009092.

53. McCauley BS, Weideman EP, Hinman VF: A conserved gene regulatory network subcircuit drives different developmental fates in the vegetal pole of highly divergent echinoderm embryos. Dev Biol. 2010, 340: 200-208. 10.1016/j.ydbio.2009.11.020.

54. Livingston BT, Killian CE, Wilt F, Cameron A, Landrum MJ, Ermolaeva O, Sapojnikov V, Maglott DR, Buchanan AM, Ettensohn CA: A genome-wide analysis of biomineralization-related proteins in the sea urchin Strongylocentrotus purpuratus. Dev Biol. 2006, 300: 335-348. 10.1016/j.ydbio.2006.07.047.

55. Mann K, Poustka AJ, Mann M: The sea urchin (Strongylocentrotus purpuratus) test and spine proteomes. Proteome Sci. 2008, 6: 22-10.1186/1477-5956-6-22.

56. Mann K, Wilt FH, Poustka AJ: Proteomic analysis of sea urchin (Strongylocentrotus purpuratus) spicule matrix. Proteome Sci. 2010, 8: 33-10.1186/1477-5956-8-33.

57. Ransick A, Davidson EH: cis-regulatory processing of Notch signaling input to the sea urchin glial cells missing gene during mesoderm specification. Dev Biol. 2006, 297: 587-602. 10.1016/j.ydbio.2006.05.037.

58. Croce JC, McClay DR: Dynamics of Delta/Notch signaling on endomesoderm segregation in the sea urchin embryo. Development. 2010, 137: 83-91. 10.1242/dev.044149.

59. Duboc V, Lepage T: A conserved role for the nodal signaling pathway in the establishment of dorso-ventral and left–right axes in deuterostomes. J Exp Zool B Mol Dev Evol. 2008, 310B: 41-53. 10.1002/jez.b.21121.

60. Duboc V, Lapraz F, Besnardeau L, Lepage T: Lefty acts as an essential modulator of Nodal activity during sea urchin oral-aboral axis formation. Dev Biol. 2008, 320: 49-59. 10.1016/j.ydbio.2008.04.012.

61. Saudemont A, Haillot E, Mekpoh F, Bessodes N, Quirin M, Lapraz F, Duboc V, Röttinger E, Range R, Oisel A, Besnardeau L, Wincker P, Lepage T: Ancestral regulatory circuits governing ectoderm patterning downstream of nodal and BMP2/4 revealed by gene regulatory network analysis in an Echinoderm. PLoS Genet. 2010, 6 (12): 1-31.

62. Lapraz F, Besnardeau L, Lepage T: Patterning of the dorsal-ventral axis in echinoderms: insights into the evolution of the BMP-chordin signaling network. PLoS Biol. 2009, 7: e1000248-10.1371/journal.pbio.1000248.

63. Bradham CA, Oikonomou C, Kühn A, Core AB, Modell JW, McClay DR, Poustka AJ: Chordin is required for neural but not axial development in sea urchin embryos. Dev Biol. 2009, 328: 221-233. 10.1016/j.ydbio.2009.01.027.

64. Revilla-i-Domingo R, Oliveri P, Davidson EH: A missing link in the sea urchin embryo gene regulatory network: hesC and the double-negative specification of micromeres. Proc Natl Acad Sci USA. 2007, 104: 12383-12388. 10.1073/pnas.0705324104.

65. Sharma T, Ettensohn CA: Activation of the skeletogenic gene regulatory network in the early sea urchin embryo. Development. 2010, 137: 1149-1157. 10.1242/dev.048652.

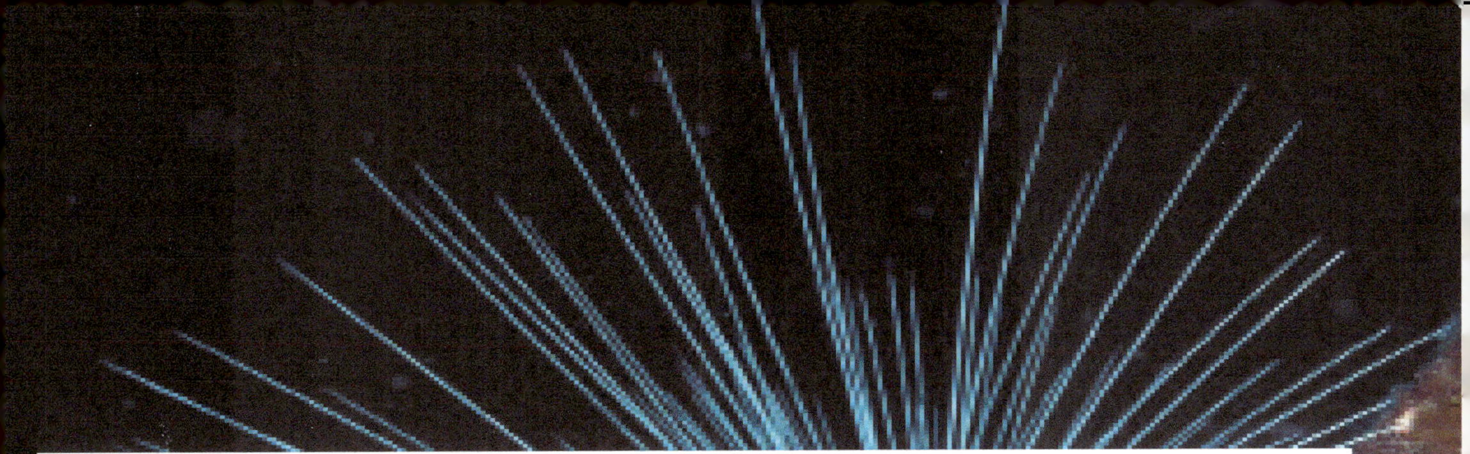

CHAPTER 19 SKELETAL REGENERATION IN THE BRITTLE STAR AMPHIURA FILIFORMIS

Anna Czarkwiani, Cinzia Ferrario, David Viktor Dylus, Michela Sugni and Paola Oliveri

Keywords: Brittle star, Echinoderms, Skeleton, Regeneration, Proliferation, C-lectin, p58b, p19

BACKGROUND

Regeneration, the ability to regrow missing body parts after self-induced or traumatic amputation, has fascinated scientists for a long time. In the context of adult organisms, this developmental process varies extensively among animals (reviewed in [1]) and also among different parts of the same organism. Some animals, such as cnidarians and flatworms, can regenerate the entire body from small fragments [2, 3, 4, 5], while others can restore only some cell tissues or particular body parts, as in vertebrates (reviewed in [6]). Echinoderms are marine deuterostomes and constitute an invertebrate phylum containing five extant classes, all possessing extensive regenerative abilities in both adult and larval forms (reviewed in [7]). Animals belonging to this phylum are characterized by 1) penta-radial symmetry, 2) large coelomic cavities, 3) a complex system of fluid-filled canals called the water vascular system and used for various aspects of animal life, 4) a well-developed nervous system, and 5) a calcareous endoskeleton [8]. Understanding how this group of animals regenerate entire body parts formed by different tissue types can provide valuable insight into our understanding of different mechanisms of animal regeneration and their evolutionary origin, and might help to explain why not all animals possess this postembryonic developmental mode.

The most prominent characteristic of all adult echinoderms is the mesoderm-derived skeletal system composed of epidermis-covered ossicles (dermaskeleton) and some internalized skeletal components (e.g.,vertebrae of ophiuroids) [8]. The degree of development of the skeleton varies among the different classes, from small isolated ossicles of holothurians to the pervasive skeleton of the sea urchin formed by several different calcified elements such as teeth, body armor and spines [8]. Ossicles have a three-dimensional porous structure formed by crystals of calcite and associated proteins, also referred to as stereom, which provides light but sturdy endoskeletal support [9]. The molecular and cellular mechanisms of skeleton development have been extensively studied in sea urchin embryos [10, 11, 12] and they provide the basis for investigations on adult skeleton development, regeneration, and comparison with other classes. In postembryonic skeletogenesis, the formation of spicules and the participation of skeletogenic cells in echinoderm juveniles has been previously described, and major morphological similarities between embryonic and juvenile skeletogenic cells have been observed [13]. Indeed, at both life stages the skeleton is formed by round-shaped mesenchymal cells with filopodia, capable of migrating to the location where new

Citation: Czarkwiani, Cinzia Ferrario, David Viktor Dylus, Michela Sugni and Paola Oliveri, Skeletal regeneration in the brittle star Amphiura filiformis, DOI: 10.1186/s12983-016-0149-x.

skeleton is deposited [13]. These morphological studies have been complemented by: 1) gene expression analyses, carried out in both sea urchin and starfish, which show that many of the genes involved in sea urchin embryonic skeletogenesis are also expressed in juvenile skeletogenic centers [14, 15]; and 2) proteomic studies that revealed an extended similarity of the molecular make-up of embryonic and adult isolated skeletal elements [16, 17, 18]. Adult sea urchin spine regeneration was described over 80 years ago [19]. This process has been studied only in terms of ultrastructure and growth process of the tip of the spine [20, 21]. Recently, a single study investigated the potential role of the Notch signaling pathway in sea urchin regeneration [22], but without providing any insight into the cellular or molecular aspects of this regenerative process. In other echinoderms regeneration is better understood at the level of morphogenesis, including the well-characterized nervous system regeneration in holothurians [23, 24], and arm regeneration in asteroids [25], crinoids [26] and ophiuroids [27]. Nevertheless, the cellular and molecular aspects of specifically skeletal regeneration have not been investigated in detail in those echinoderm clades.

Brittle stars (class: Ophiuroidea) are the most diverse group of echinoderms comprising over 2000 species with a global distribution [28, 29]. Adult brittle stars are able to regenerate their entire arms, making them an appealing system for studying regeneration of adult structures. Brittle star arms are complex structures composed of various tissue types organized in repetitive segments, here referred to as metameric units. Each such unit contains five different skeletal elements-the dermal oral, aboral and lateral shields (or plates), spines, and the internal vertebral ossicles-in addition to a set of two pairs of intervertebral muscles, intervertebral ligaments, and a pair of podia on each side. The radial water canal, the different specialized coelomic cavities (i.e., aboral coelomic canal, neural sinuses) and a radial nerve cord and peripheral sensory neurons are also present throughout the arm segments [30]. *Amphiura filiformis (Afi)*, a burrowing brittle star from the North Sea, has been used to understand various aspects of regeneration in this class [27, 31, 32, 33,34]. Some studies have shown significant variability in regeneration rates of these animals. This is likely due to differences in animal size, traumatic versus self-amputation, length of the lost arm, or the most pertinent function required at the time (i.e., the differentiation of sensory structures or the growth of arm for locomotion and feeding) [31]. This variability highlights the plasticity of the regenerative process of these animals. On the other hand, histological studies, which best describe the early repair and regeneration phases in two ophiuroid species [27], emphasize the consistency of the regenerative process in this class of echinoderms. Recent molecular studies identified genes expressed during regeneration [33, 34]. Importantly, it has been shown that several mesodermal genes including *alx1*, a transcription factor known to play a key role in the sea urchin and sea cucumber skeletogenic gene regulatory networks [10, 35], are expressed during both embryonic skeletogenesis [36] and arm regeneration in *A. filiformis* [37]. Importantly, members of the Cart/Alx3/Alx4 group of transcription factors are also involved in skeletal development in vertebrates [38, 39,40].

Despite these studies, little is known about the cellular process underlying skeletogenesis in adult *A. filiformis*, from the initial appearance of simple mineralized elements (spicules) to the formation of complex skeletal ossicles. Here, we address this lack of basic knowledge by using a combination of approaches to characterize skeletal development at the cellular and molecular level in the regenerating arm of *A. filiformis*. To address the problem of variability, we first divide the regenerative process into five stages using clear morphological landmarks, as done in other regeneration models like salamanders [41] and sea cucumbers [23]; and embryonic developmental models such as chicken embryos [42]. We thus use our new staging system to complete the histological survey presented in [27]. We then employ light microscopy and calcein staining to observe at what stage the primordial skeleton forms and in which domain of the regenerating arm. Additionally, we use histological staining and confocal imaging to further understand the position and morphology of the potential skeletogenic cells. We then identify the cells expressing known skeletogenic marker genes *Afi-c-lectin, Afi-p58b and Afi-p19* using in situ hybridization. Finally, we apply 5-ethynyl-2'-deoxyuridine (EdU) labeling to study cell proliferation in the early and late regenerating arms and to determine if differentiated skeletogenic cells are proliferative. Our work will provide the basis for understanding skeleton regeneration in the brittle star on a cellular and molecular level.

RESULTS

Staging of Early Phases of Regeneration Based on Morphological Features

The development of a regenerating structure is a dynamic process that proceeds in a step-wise fashion to reconstitute a fully functional structure. To facilitate experimental investigation of arm regeneration in *A. filiformis* we identified five stages, which rely on observable morphological changes. Based on observations of >100 individuals we propose a refined staging system (Fig. 1), which is relevant to the phases of regeneration when cell specification, differentiation and initial arm patterning begin to occur. The presence of clear morphological landmarks is supported by histological analysis carried out on paraffin sections stained with Milligan's trichrome technique [43], which detects collagen (cyan) and individual cells (pink/magenta) (Figs. 2 and 3). The average duration and standard deviation (S.D.) of time it takes for an arm to regenerate up to a given morphological stage is reported in Fig. 1A. The timing of regeneration shows a certain degree of variation in agreement to what has previously been reported [31].

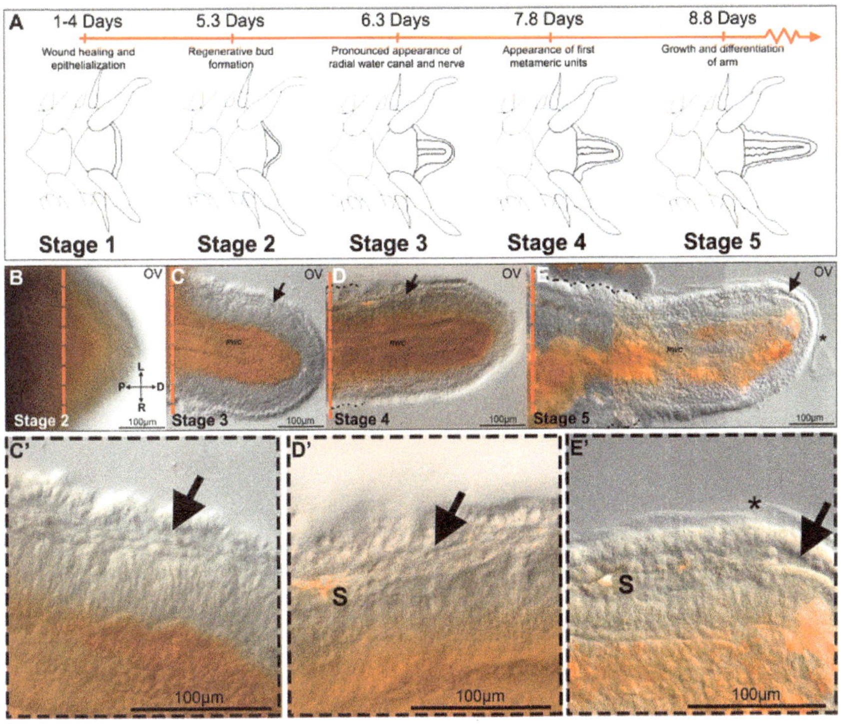

Figure 1: Early stages of arm regeneration in the brittle star *Amphiura filiformis*. A) Schematic diagram of early arm regeneration stages and average timing to those stages, based on morphological landmarks. Stage 1-wound healing and re-epithelialization. Stage 2-regenerative bud formation. Stage 3-appearance of the radial water canal, coelomic cavities (aboral coelomic canal, ectoneural and hyponeural sinuses) and radial nerve in the regenerative bud. Stage 4-appearance of first metameric units (arm segments). Stage 5-advanced extension of arm, formation of several metameric units at proximal end. B–E) Oral view of fixed arms at stages 2 (B), 3 (C), 4 (D) and 5 (E) shown using differential interference contrast (DIC) microscopy. C') Detail of C focusing on dermal layer. D') Detail of D. E') Detail of E. Arrows - the dermal layer, asterisk - cuticle, D - distal, L - left, OV - oral view, P - proximal, R - right, RWC - radial water canal, S – spicule.

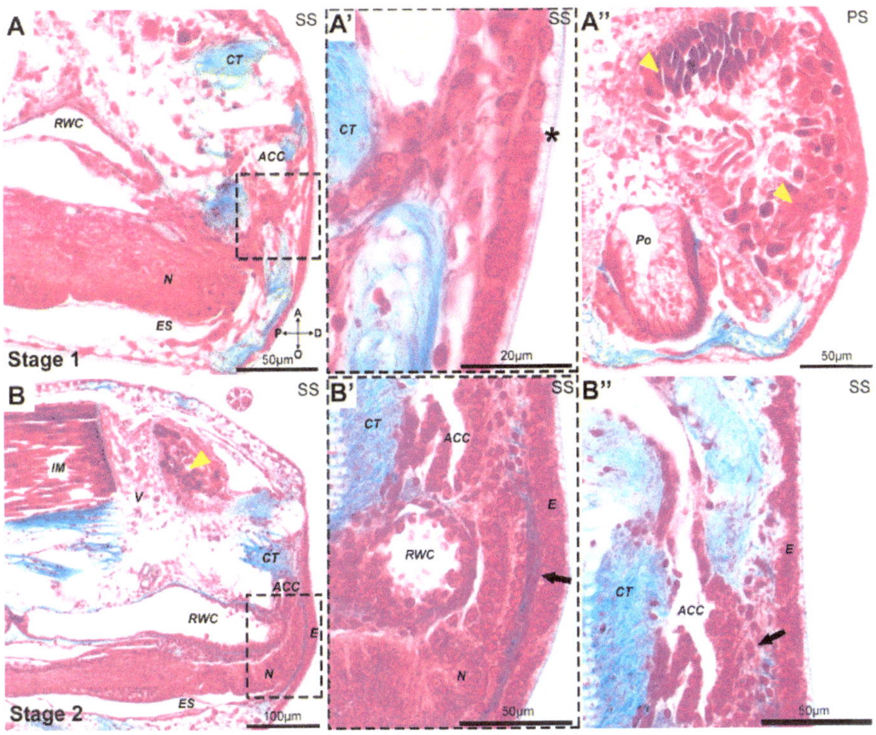

Figure 2: Histological sections of the earliest stages of arm regeneration in the brittle star *Amphiura filiformis*. Sagittal (A and A') and parasagittal (A") paraffin sections at stage 1 and sagittal paraffin sections at stage 2 (B) stained with Milligan's trichrome technique. Collagen stained cyan, all cells labeled pink/magenta. A) The wound is completely healed and re-epithelialization occurs. A') Detail on the new thin epithelium with a recognizable cuticle (asterisk). A") Histolysis (arrowheads) of intervertebral muscle bundles proximal to the amputation plane. B) The radial nerve, the radial water canal and the coelomic cavities start regenerating beneath the new epidermis. B') Detail of the mesenchymal cells (arrow). B") Detail of the mesenchymal cells at the level of the aboral dermal layer (arrow). A - aboral, ACC - aboral coelomic cavity, CT - connective tissue, D - distal, E - epidermis, ES - ectoneural sinus, intervertebral muscles - IM, N - radial nerve, O - oral, P - proximal, Po - podia, PS - parasagittal section, RWC - radial water canal, SS - sagittal section, V – vertebra.

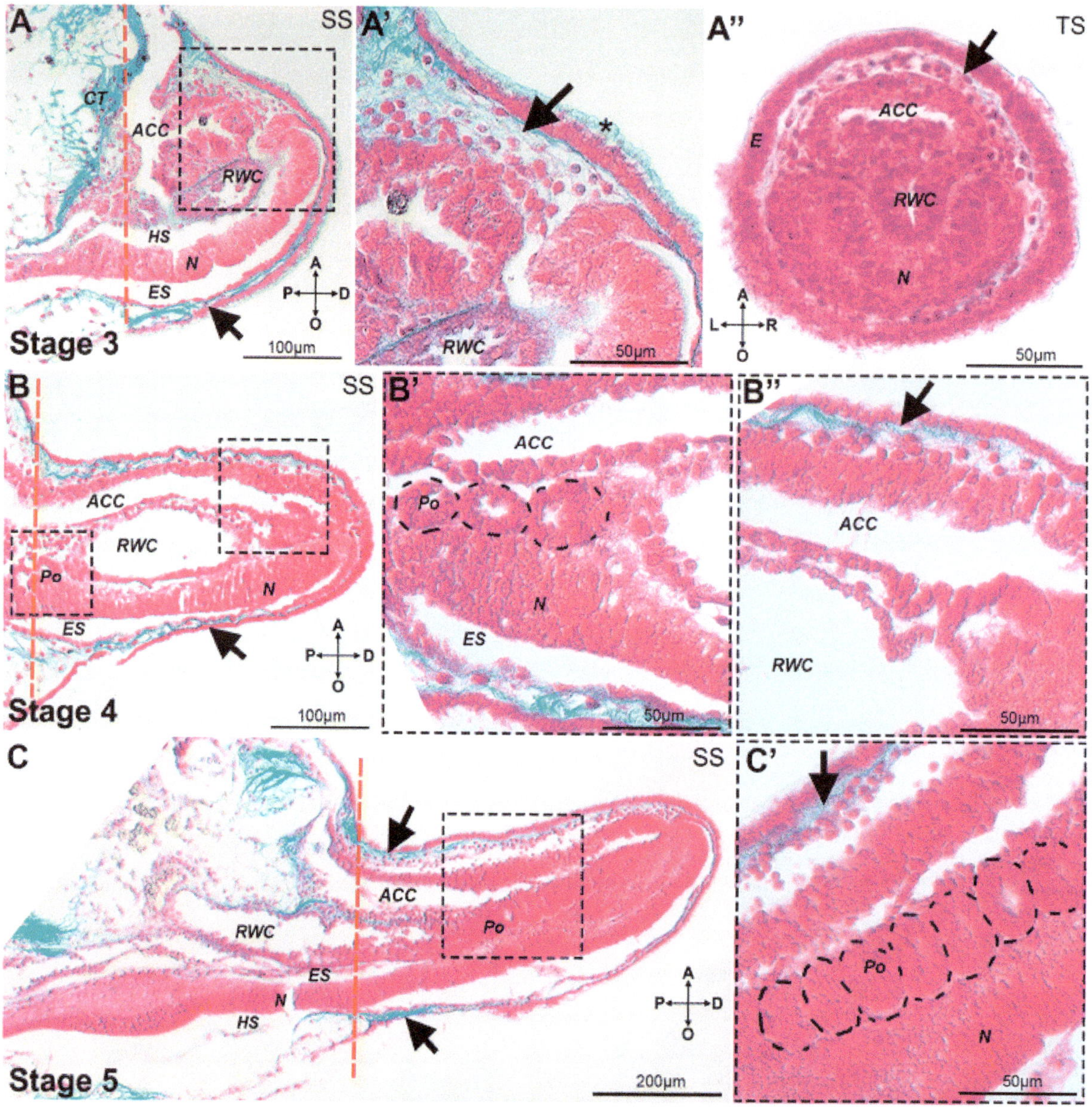

Figure 3: Histological sections of regenerating arms at stages 3, 4 and 5. Sagittal and transverse paraffin sections at stage 3 (A and A"), and sagittal sections of stages 4 (B), and 5 (C) stained with Milligan's trichrome technique. Collagen stained cyan, all cells labeled pink/magenta. Red dashed line indicates amputation plane. A) The three regenerating axial structures are enveloped by the dermal layer (arrow) and the new epidermis. A') Detail of A showing mesenchymal cells at the level of the aboral dermal layer (arrow) covered by the new epidermis with its cuticle (asterisk). A") Transverse section of a stage 3 regenerate showing the dermal layer (arrow), developing radial nerve, radial water canal and aboral coelomic cavity. B) The stage 4 regenerate is longer and the radial water canal shows the first signs of podia regeneration. B') Detail of B showing the developing podia (dotted line). B") Detail of B showing the scattered mesenchymal cells in the aboral dermal layer (arrow). C) Further development and differentiation of the three axial structures at stage 5. C') Detail of developing podia (dotted line). A - aboral, ACC - aboral coelomic cavity, CT - connective tissue, D - distal, E - epidermis, ES - ectoneural sinus, HS - hyponeural sinus, L - left, N - radial nerve, O - oral, P - proximal, Po - developing podia, RWC - radial water canal, R - right, SS - sagittal section, TS - transverse section.

All of the stages and corresponding average timings have been compiled based on arms amputated 1cm from the central disc. We categorize stage 1 of regeneration as wound healing and re-epithelialization. This phase occurs between 1 and 4 days post-amputation (dpa) (Fig. 1A) and involves changes which mediate the closure of the wound and re-epithelialization and remodeling of the existing tissue; however, from whole mount DIC (differential-interference contrast) observations little or no changes are evident at the amputation plane. On the contrary, at this stage histological sections show (Fig. 2) the aboral coelomic cavity (ACC), the ectoneural sinus (ES) and the radial water canal (RWC) are sealed off and the wound is completely re-epithelialized by epidermal cells, already covered by a thin and faint cuticle indicated by the asterisk in Fig. 2A'. The intervertebral muscles adjacent to the amputation site acquire a disorganized pattern and show morphological signs of histolysis: myocytes lose their elongated shape (Fig. 2A", B). The duration of time between amputation and stage 2 shows the highest variability. To assess whether the type of traumatic amputation might have an effect on the rate of the wound healing stage and the appearance of a regenerative bud, we amputated two arms from the same animal: one clean, blunt amputation similar to natural autotomy, and the other skewed. We used 5 animals of the same size to minimize the individual variability and we documented the process of regeneration of each arm for 10 days. These data reported reveal a general inter-individual variability in time to reach stage 2, which is not related to the type of amputation.

Once re-epithelialization is accomplished, a small regenerative bud protrudes from the distal end of the stump, in the oral region (Fig. 1A, B). We characterize this as stage 2, which occurs on average after 5.3 dpa ±0.91 S.D.. From whole mount DIC observations the regenerate appears optically homogeneous (Fig. 1B), however histological sections indicate already a certain degree of organization, in which the first outgrowths of the regenerating radial nerve, radial water canal and the aboral coelomic canal are visible (Fig. 2B). A thin layer of connective tissue is present below the wound epidermis where mesenchymal cells are embedded (Fig. 2B', B").

Stage 3, which occurs on average 6.3 dpa ±0.48 S.D. (Fig. 1A), differs from stage 2 in that the regenerate acquires a more organized and complex inner architecture with loss of the external optical homogeneity; at this stage the radial water canal and the sub-epidermal mesenchymal cells become clearly visible in whole mount DIC images (Fig. 1C, C'). Histological sections show the regenerate mainly contains projections of the radial nerve, radial water canal and aboral coelom, which protrude from the amputation plane (Fig. 3A, A"). Mesenchymal cells embedded in collagenous tissues can be detected throughout the dermal layer (Fig. 3A', A").

We define stage 4 (on average 7.8 dpa ±0.63 S.D., Fig. 1A) by the appearance of the first metameric unit, which is formed at the most proximal region of the developing regenerate. A small bulging of the epidermis reveals the early segregation of the regenerating segments (dotted lines, Fig. 1D). Outpocketing of the radial water canal system, which will eventually form the podia, can start to be distinguished in histological sections (Fig. 3B, B').

Finally, stage 5 occurs on average after 8.8 dpa ±0.63 S.D. (Fig. 1A). This stage is characterized by several small repetitive units, which can be observed bulging at the proximal side (dotted line, Fig. 1E) with several podia precursors beginning to form along the new regenerate (Fig. 3C, C').

As the regeneration process progresses metameric units form following a proximal-distal gradient. The late regeneration stages that we use in this work corresponds to the 50 % (2–3 weeks post-amputation) or 95 % (4–5 weeks post-amputation) regeneration stages described before [31, 37]. These arms have a 50 or 95 % differentiation index (DI), which is the ratio between the length of the arm that contains differentiated structures (like spines and podia) and the total regenerate length. The duration of regeneration time is highly variable between stage 5 and the 50 % stage depending on many aspects including length of arm lost, functional requirements, animal size or environmental factors [31].

Skeletogenesis during Early and Late Regeneration Stages

To define when and where skeletogenesis occurs during early stages of regeneration in the brittle star arm we combined light transmission microscopy observations of whole regenerates with calcein staining to detect the newly forming mineralized skeleton. Calcein is commonly used to visualize calcium carbonate deposition, which is shown as green fluorescence [44]. Single spots of green fluorescence, corresponding to the forming skeletal primordia, are consistently first observed at stage 3 indicating the earliest stage at which differentiated skeletogenic cells are present (Fig. 4a, $n=10$). No fluorescence can be detected at stages 1 and 2. At stage 4 more elaborate tri-radiated or tetra-radiated skeletal elements called spicules can be observed in the regenerate (Fig. 4a, $n=10$). At stage 5, the spicules present in the distal end of the regenerate have no obvious pattern of distribution. In contrast, skeletal elements in the newly forming metameric units, at the proximal end of the regenerate, are formed with a typical bilateral distribution that corresponds to where the future lateral shields will be (Fig. 4a , $n=10$). High magnification of the regenerates at different stages reveals that the single spicules form just below the well-developed epidermis in the dermal layer (Fig. 4b), corresponding to the position of the mesenchymal cells embedded in the collagenous matrix observed in histological sections (Figs. 2 and 3).

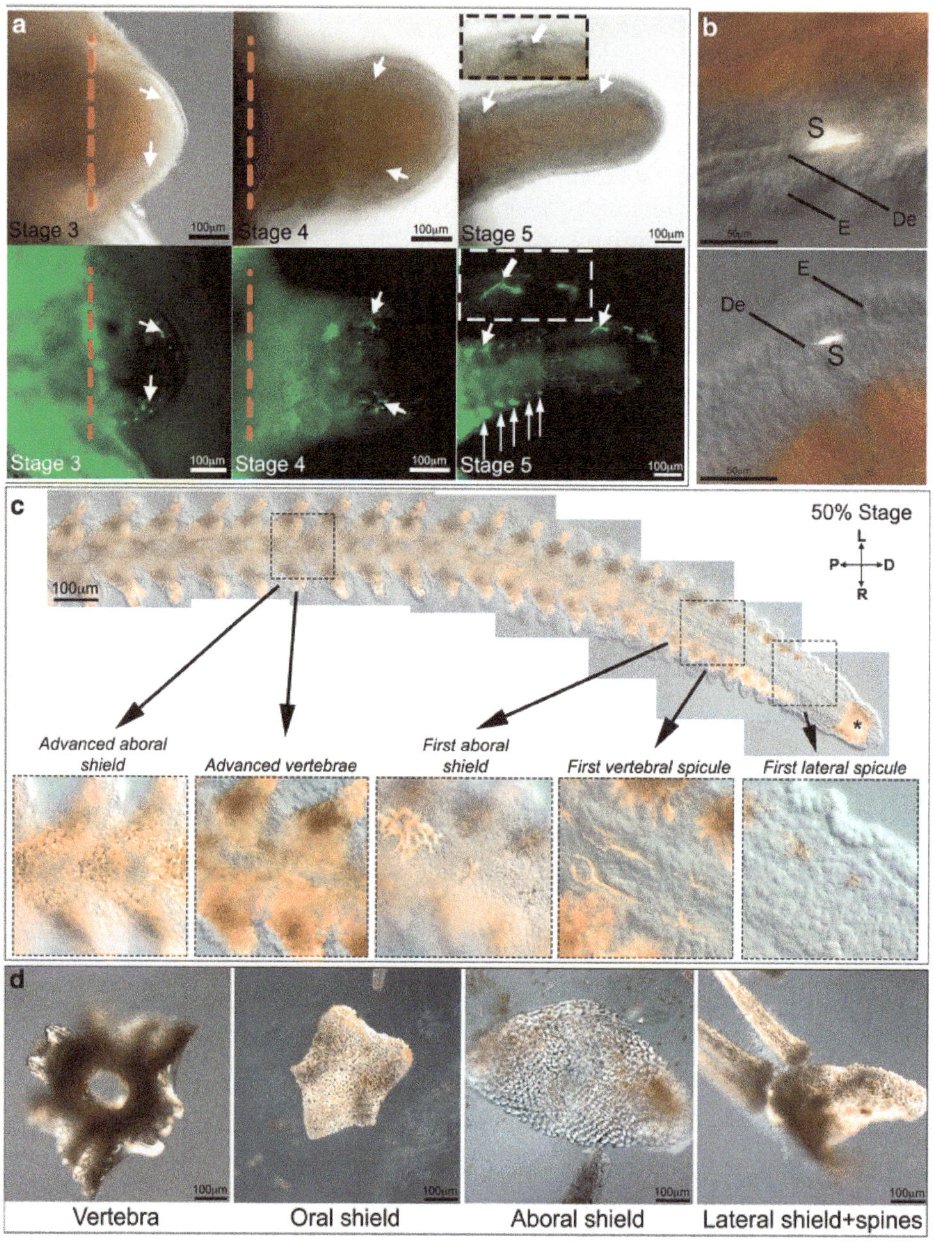

Figure 4: Skeletogenesis during early and late regeneration stages. a) DIC (top panels) and fluorescent images (bottom panels) of calcein-labeled spicules at stages 3, 4 and 5. White arrows show individual spicules, which can be compared in both images. Insets show detail of single spicules. b) High magnification DIC images showing single spicules localized in the dermal layer. c) 50 % differentiated arm showing the formation of the skeletal elements (reflective structures). The highly calcified distal cap (terminal ossicle) forms at the distal-most end of the regenerating arm. The spicules, which will form the lateral shields and spines, appear in the first metameric unit (see inset for detail). Proximally, eight metameric units later, first vertebral spicules can be observed (see inset for detail). At the proximal end of the regenerate skeletogenesis is already very advanced and forms the individual stereomic skeletal elements including the oral, aboral and lateral shields, spines and vertebrae (see insets for detail). d) Differentiated skeletal elements in an adult non-regenerating arm. Arrows - calcein-labeled spicules, asterisk - terminal ossicle, E - epidermis, De - dermis. L - left, R - right, P - proximal, D - distal. Red dashed line - amputation plane.

Later during regeneration, when the arm reaches the 50 % regeneration stage (50 % DI), the whole maturation of developing skeleton can be observed (Fig. 4c) simply by analysis of subsequent metameric units from proximal (most mature skeleton) to distal (new elements) and using light microscopy. This is due to the birefringent properties of calcified structures. At the distalmost end the distal cap becomes highly calcified (terminal ossicle). The first observable distal metameric unit already contains a tiny tri-radiated lateral spicule (Fig. 4c, first lateral spicule inset). Several metameric units (eight in Fig. 4c) separate the appearance of the first lateral spicules from the appearance of the first vertebral spicules, forming more proximally in the inner layers of the regenerating arm. Contrary to skeletal elements developing during early regeneration and at lateral positions during late regeneration, which always form multi-branched spicules, the vertebrae first appear as two long, non-radiated skeletal rods (Fig. 4c, first vertebral spicule inset). Later, as the skeletal elements mature, they also begin to branch out and the two parallel vertebral elements fuse together at the midline between distal and proximal ends of the metameric unit (Fig. 4c, advanced vertebrae inset). Oral and aboral shields begin to form approximately at the same level as the developing vertebrae (Fig. 4c, first aboral shield inset). In a distal to proximal gradient the segments contain increasingly complex spicule structures that will eventually form the stereom of skeletal ossicles, as is typical for echinoderms

(Fig. 4c, advanced shield formation inset). The fully formed structures found in an adult non-regenerating arm are shown in Fig. 4d. A schematic diagram of the organization of individual skeletal shields in the adult non-regenerating arm.

Expression of the Skeletogenic Marker Genes in the Early and Late Stages of Regeneration

To better characterize the cells involved in skeleton formation during regeneration in the brittle star we assessed the spatial expression pattern of the skeletogenic cell differentiation genes. We first analyzed the closest matching gene, identified in an *A. filiformis* embryonic transcriptome, to the sea urchin C-lectin. This spicule matrix protein containing a C-type lectin domain is expressed specifically in the skeletogenic mesodermal cells [45]. Proteomic studies identified the C-lectin protein as an integral part of biomineralized structures both in sea urchin larval spicules and adult test plates and spines [16, 17], and in ophiuroid adult arm skeletal elements of the species *Ophiocoma wendtii* [46]. The expression of the *A. filiformis c-lectin* gene has been first examined in the larval stages with extended skeletal elements (i.e., ophiopluteus). Indeed, only a group of mesenchymal cells arranged in a pattern coinciding with the skeletal elements express *Afi-c-lectin* . We have therefore used it as a marker for skeletogenic cells in adult regenerating arms (Fig. 5). It is first detected by *in situ* hybridization (ISH) at stage 2 of early regeneration in a broad sub-epidermal domain (Fig. 5b). It then becomes much more restricted to the dermal layer only by stage 4 (Fig. 5c, d). This is consistent with the developmental changes that occur between these two stages, when the regenerate acquires a more heterogeneous structure containing morphologically differentiated tissue types. Strong colorimetric staining has the tendency to diffuse in the tissue, rendering the precise boundary of expression not clearly distinguishable. To discriminate whether the expression of *Afi-c-lectin* is restricted specifically to the sub-epidermal mesenchymal cell layer, where spicules form, or extends to other domains of the regenerate, we additionally performed a fluorescent ISH on regenerating arm sections. We found that the expression of this gene is indeed specifically localized to a single cell layer just underneath the epidermis (Fig. 5e, f).

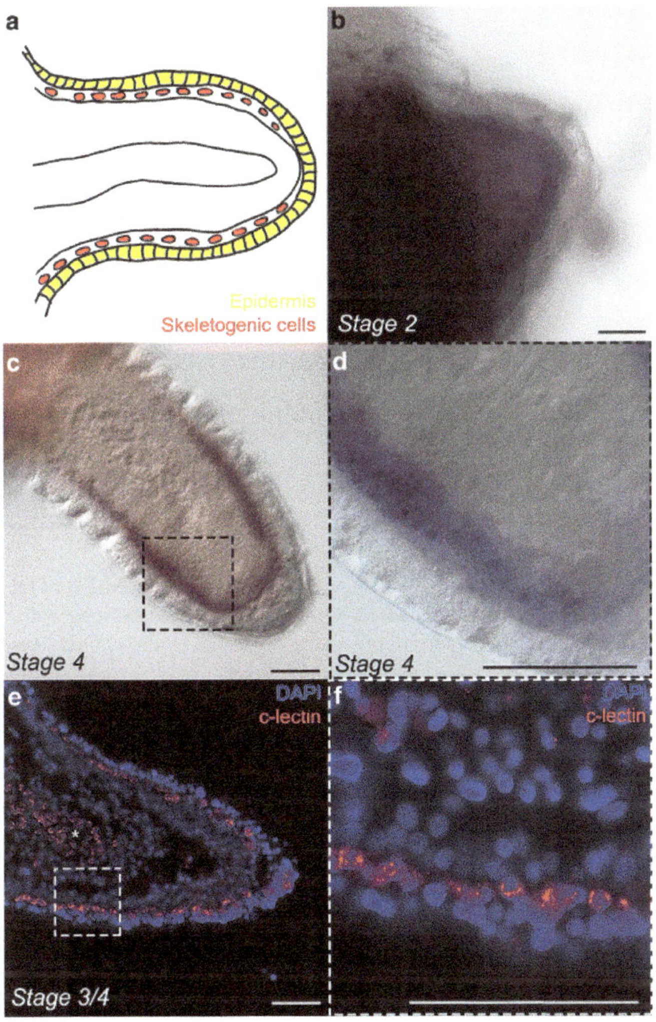

Figure 5: *Afi-c-lectin* expression during early arm regeneration stages. a) Schematic representation of a 3/4 stage regenerate.b–d) Chromogenic whole mount *in situ* hybridization (WMISH) showing expression of gene at stages 2 and 4. b) *Afi-c-lectin* is expressed at stage 2 in a broad, sub-epidermal domain. c) At stage 4 the expression of *Afi-c-lectin*becomes much more restricted to the sub-epidermal domain. d) Higher magnification of arm in c. e, f Fluorescent *in situ* hybridization of sections through stage 3/4 arm counterstained with the nuclear marker DAPI. e) Frontal paraffin section showing fluorescent ISH of *Afi-c-lectin* counterstained with DAPI. *Afi-c-lectin* is clearly localized to single cells just beneath the epidermis. Asterisk marks cells in sub-epidermal layer seen from oral view, due to the slanted plane of sectioning along the oral-aboral axis. f) Higher magnification of arm in **e**. Scale bars - 50µm.

Next, we examined the expression of *Afi-c-lectin* in the 50 % DI stage of the regenerating arm (Fig. 6). In the distalmost-undifferentiated segments of the regenerate, *Afi-c-lectin* is expressed in the dermal layer similar to the early stages of regeneration (Fig. 6A). Towards the proximal side of the regenerating arm, its expression becomes more complex, corresponding to the formation and patterning of the different skeletal elements (Fig. 6A). The scattered mesenchymal cells expressing the marker gene are arranged in regular and repetitive patterns, which mirror the pattern of the stereom structure of the oral, aboral, lateral shields with spines and vertebrae (Fig. 6A') in each of the differentiating segments of the arm. The expression pattern of *Afi-c-lectin*thus coincides specifically with the location and time of appearance of the different skeletal elements; therefore, confirming it as a reliable marker of skeletogenesis during regeneration as well as embryonic development. This is further supported by the expression pattern of *Afi-c-lectin* in the most advanced proximal segments of the 95 % DI regenerating arms (Fig. 6B–D). Two additional genes, identified in a recent embryonic study of *A. filiformis* larval skeleton development [36], were examined to confirm the molecular signature of the mesenchymal cells in the dermis and the cells in the differentiating skeletal elements of late stages of regeneration. Both *Afi-p19* and *Afi-p58b* have a similar expression pattern to *Afi-c-lectin* at the early stages of regeneration and at the distal end of the 50 % DI stage arms (expression in the dermis). The genes are more restrictively localized to individual skeletal elements in differentiating proximal segments of the arm. *Afi-p58b* is expressed in spines and vertebrae, whereas *Afi-p19* is localized to the oral and lateral shields, and vertebrae.

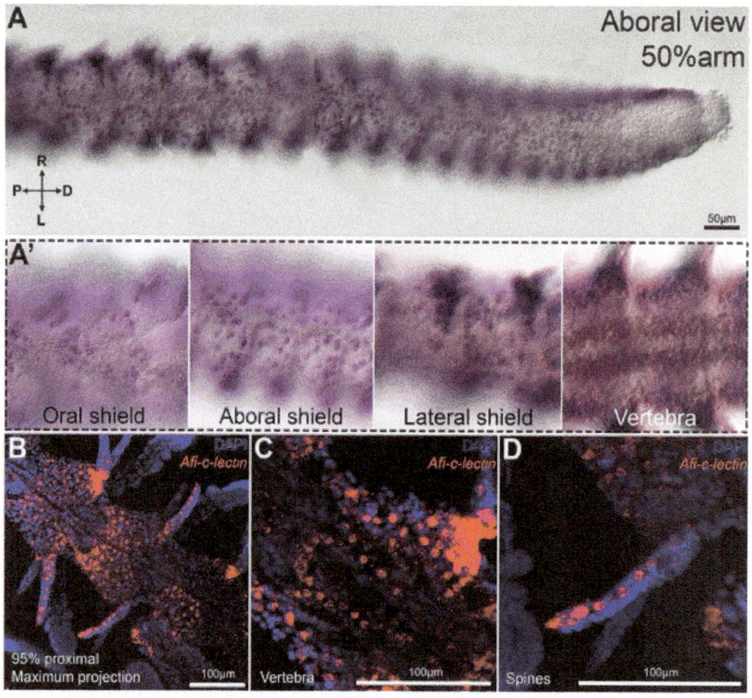

Figure 6: *Afi-c-lectin* expression at 50 and 95 % differentiation stages. A) Chromogenic WMISH of the whole 50 % regenerating arm from proximal differentiated segments to distal undifferentiated segments and terminal cap. At distal-most end staining is localized to sub-epidermal cells. In differentiating metameric units the expression expands to scattered mesenchymal cells covering the areas of future formation of oral, aboral and lateral shields as well as spines. A') Detail of gene expression in oral shields, aboral shields, lateral shields and vertebrae. B-D) Confocal images of fluorescent WMISH of *Afi-c-lectin* (red) in proximal segments of 95 % differentiated arms counterstained with nuclear stain DAPI (blue). B) Maximum projection showing *Afi-c-lectin* expression in three proximal segments of the arm is localized to cells in the aboral shields and spines. C) Single z-plane projection through one segment in B showing detail of *Afi-c-lectin* expression corresponding to the shape of the growing vertebra. D) Maximum projection through one segment in B showing detail of *Afi-c-lectin* expression along the spine. R - right, L - left, P - proximal, D – distal.

Differentiated Skeletogenic Cells are not Proliferative

To test when active cell proliferation is initiated post amputation and if mesenchymal cells - located in the dermal layer where the skeleton is formed - have the ability to proliferate during regeneration, we used the EdU assay to label cells in, or having gone through S-phase during the early stages of regeneration. In normal non-regenerating arms some proliferating cells, labeled by EdU, are identifiable in different tissue types including the epidermis, podia, radial water canal and in cells surrounding the vertebrae. On the contrary, no cells are labeled during the first hours (8–24) post-amputation in the plane of injury. Only at the end of stage 1 (between two and three dpa) is a marked increase in EdU labeling visible at the wound site prior to bulging of the regenerative bud, mainly in correspondence with the position of the radial nerve cord and the radial water canal in the oral half of the metameric unit. Stage 2, which marks the appearance of the regenerative bud, shows extensive cell proliferation in both the epidermis and the inner tissues, containing the outgrowths of the above-mentioned structures (Fig. 7A; $n=3$).

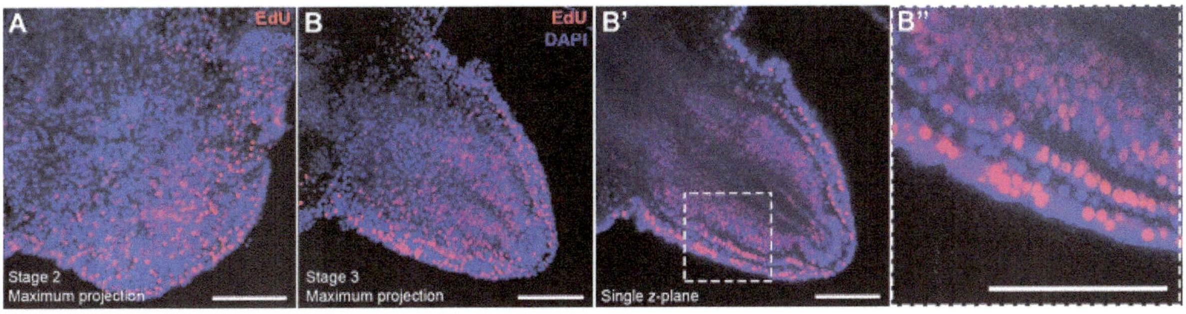

Figure 7: Confocal images showing EdU labeling (red) of early stage regenerating arms counterstained with nuclear stain DAPI (blue). A) Stage 2 maximum projection of confocal z-stack showing widespread cell proliferation in regenerative bud. B) Maximum projection of confocal z-stack of stage 3 arm showing continuing cell proliferation in early regenerate. B') Single z-plane of B. B") Detail of B' showing lack of EdU-labeled cells in dermal layer. Scale bars - 100μm.

Five individual arms were used for EdU labeling of stage 3 regenerates. Confocal maximum projections of whole z-stacks also show a large amount of proliferating cells at this stage (Fig. 7B). Notably, closer inspection of individual z-planes reveals that the dermal layer contains the only cells, that are not labeled with EdU, implying that they did not proliferate during the time course of labeling (Fig. 7B', B").

As the number of cells that express *Afi-c-lectin* clearly increases during the regeneration process, we investigated whether the cells have the ability to divide at these late stages. To answer this question we performed a fluorescent ISH using *Afi-c-lectin* on arms previously labeled with EdU (n=2). If skeletogenic cells marked by the expression of *Afi-c-lectin* were proliferating, cells with a red nucleus (EdU DNA incorporation) and surrounding green cytoplasm (*Afi-c-lectin* RNA expression) would be observed. Extensive cell proliferation can be observed throughout the regenerating arm at this stage (Fig. 8A). However, closer inspection of individual confocal z-planes shows that none of the green-labeled *Afi-c-lectin* cells overlap with the red EdU$^+$cells in a manner indicating that skeletogenic cells have nuclei in the S-phase of mitosis (Fig. 8). Specifically, none of the EdU labeled red nuclei are surrounded by green *Afi-c-lectin* labeled cytoplasm (Fig. 8B, D). This trend is also apparent in the whole 50 % stage arm as shown in the proximal segments (Fig. 8A, B) and the distalmost tip (Fig. 8C, D). Therefore, the *Afi-c-lectin* expressing cells do not proliferate throughout the whole arm regeneration process, in agreement with the role of *c-lectin* as a final differentiation gene in skeletogenesis. Notably, the distalmost part (terminal ossicle and podium) (Fig. 8C) of the regenerate contains very few proliferating cells. However, the area just proximal to it is a domain of major accumulation of EdU$^+$ cells. This is consistent with previous observations [37], and the expression patterns of genes here reported, and thus we conclude that the regenerating arm proliferative growth zone is located just underneath (proximally) to the terminal ossicle. New metameric units are then added proximal to the growing tip of the regenerate.

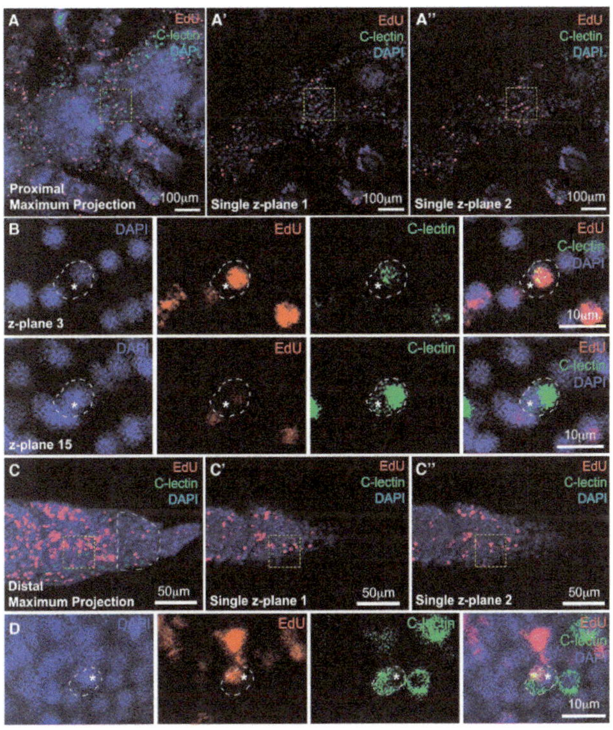

Figure 8: *Afi-c-lectin* expression (green) combined with EdU labeling (red) and counterstained with nuclear stain DAPI (blue) showing that skeletogenic cells do not proliferate. A) Maximum projection of proximal segments of 50 % regenerating arm. A' and

A") Single z-planes of A showing *Afi-c-lectin* expressing cells do not overlap with EdU-labeled cells. Yellow dashed box shows one case of yellow signal suggesting potential overlap of red and green signals. B) Magnified image in yellow dashed box from A showing cells in z-plane number 3 and z-plane number 15. The EdU labeled red nucleus is seen clearly in z-plane 3 where *Afi-c-lectin* expression is very faint, whereas on z-plane 15 where the whole cytoplasm of the green-labeled *Afi-c-lectin* cell is clearly visible, the nucleus of this cell is labeled with DAPI but not EdU (asterisk). C) Maximum projection of distal-most end of 50 % regenerating arm. White dashed half-circle indicates position of terminal ossicle. C' and C") Single z-planes of C showing *Afi-c-lectin*expressing cells do not overlap with EdU-labeled cells. Yellow dashed box shows one case of yellow signal suggesting potential overlap of red and green signals. D) Magnified image in yellow dashed box from C showing cells in single z-plane. Again the nucleus of the *Afi-c-lectin* expressing cell is labeled with DAPI (asterisk) and is in close proximity to the EdU-labeled nucleus causing the yellow overlap signal.

DISCUSSION

In this study we describe the major developmental events occurring during early phases of arm regeneration in the brittle star *Amphiura filiformis*, with a focus on the formation of the calcitic skeleton. The high regenerative capacity of this species, likely evolved to escape predation [31], is ideal for the investigation of the developmental process underlying the regrowth of a completely functional arm constituted by many different cell types and tissues.

The regeneration process, like embryonic development, is characterized by transient stages. At the base of any molecular and cellular investigation of dynamic developmental processes lays the clear understanding of the sequence of events taking place during these stages. Thus, in this study, we first identify five major early regenerative stages (Fig. 1) easily recognizable by external morphology in living animals. These stages subdivide the regenerate starting at the wound healing and repair phase (stage 1), to initial growth (stage 2; also described as blastema in other studies [27, 31]), into more and more complex cellular layers and structures (during stages 3 to 5) that foresee the organization of the future rebuilt arm. There are two previously published staging systems for arm regeneration in *Amphiura filiformis.* The division described by Dupont and Thorndyke [31] only distinguishes an early, uncharacterized phase (called blastema), from two rather advanced stages of regeneration (50 and 95 % DI), in which several metameric units of the regenerate are already present and contain differentiated structures. This staging system, although useful for studying late stages of regeneration and quantifying growth rate, completely bypasses early developmental and morphological events. According to this classification, all our early stages would be classified generally as 0 % DI; therefore, this system is not relevant for our aim of understanding cell specification and morphogenesis. On the other hand, the staging system devised by Biressi et al [27] is focused largely on the very early phases post amputation. This system first subdivides early regeneration into a repair phase (immediately after amputation, 1 dpa) and an early regenerative phase (1–3 dpa), both of which correspond to our stage 1. Our EdU labeling experiments show that the events taking place during this stage do not involve cell proliferation. The two remaining phases described, namely an intermediate regenerative phase (3–12 dpa) and an advanced regenerative phase (11–24 dpa), do not discriminate at high enough resolution the events taking place within this time frame. This is essential to understanding how morphology is generated, and how specification and differentiation of skeletogenic cells occur. For instance, we show that our stage 3 regenerate (average 6.3 dpa) is already characterized by a clear organization of different tissues (Fig. 3A), but is still quite different from stage 4 (average 7.8 dpa) when the metameric units that will form the arm begin to appear. Our revised staging system is more relevant for the precise study of the developmental processes taking place during *A. filiformis* regeneration. It also allows more homogeneous sampling by identifying regenerating arms at specific stages, which are directly comparable, despite coming from animals with different regeneration rates.

An important observation arising from this study concerns the mode of brittle star arm regeneration. By looking at the development of skeletal elements in the 50 % DI arms we can observe the whole gradient of regenerative developmental stages from the oldest differentiated proximal metameric units to the newest un-differentiated distal metameric unit with skeletal primordia. However, the distalmost tip containing the terminal ossicle and podium is differentiated, and the growth zone adding new metameric units is located just proximally to it. This suggests that after a transient stage of formation of the regenerative bud, the brittle star regenerates its arm following a distalization-intercalation model similar to what has been described in sea star [25] and planarian regeneration [4]. According to this model regenerating organisms first form the distalmost part of the regenerate, which acts as a re-organization centre, and then add new structures by sequential intercalation of newly generated tissues between the distal part and the stump following a distal-proximal gradient. The following observations are consistent with this model of regeneration: 1) the small amount of EdU^+ cells in the distalmost tip of late regenerates compared with the underlying growth zone (Fig. 8C) supports the differentiation status of the distalmost element; 2) the early appearance of skeletal elements in the distalmost tip, which suggests early differentiation of the terminal ossicle (stage 5, Fig. 4a); 3) the appearance of forming segments in an intermediate position between the terminal tip and the stump (Fig. 4). This is consistent with gene expression studies of transcription factors known to be involved in early specification of mesodermal cells (e.g., *alx1, ets1/2, gataC*) and reported to be expressed in the growth zone, but not in the distalmost tip [37]. A similar mode of regeneration has been recently reported for sea stars, in which the distalmost element is represented by the terminal tube foot and associated terminal ossicle and the growth zone is located just at the base of this structure [32]. Whether the terminal structures of both brittle stars and sea stars have 'simply' a protective function over the delicate growth zone or act as true signalling centres to the patterning of regenerating tissues remains to be elucidated.

Our observations of the formation of the skeleton during early regenerating stages suggest that cells undergo specification and differentiation events very early during the regeneration process. The spicule primordia observed at stage 3 resemble the granule-like skeletal rudiment of *A. filiformis* embryos [36], and both sea urchin embryos [11, 47] and juveniles [13] at the earliest step in the development of the skeleton, which then extend into tri-radiated and tetra-radiated spicules. The 50 % differentiated arm shows the developmental progress of the skeleton from single spicules up to the formation of complex mesh-like structures of the dermal plates (lateral, oral and aboral shields), spines and vertebrae. Notably, the vertebral spicules, which are internal skeletal elements, appear much later than those involved in the formation of the lateral shields and spines. As seen in SEM images, the complete vertebrae in adult non-regenerating arms of ophiuroids are composed of two conjoined ambulacral plates [28, 48, 49], which could explain why during regeneration the vertebrae appear to form by a fusion of spicule complexes from bilateral halves of each segment. The same SEM studies also show that they are clearly the most complex and dense skeletal elements in the ophiuroid arms [28, 48, 49]. Taken together, these data suggest molecular differences and possibly a separate developmental program might be involved in the formation of those internal-most skeletal structures compared to the sparser stereom constituting the lateral shields and spines. This is supported by differences in expression of *Afi-c-lectin*, present in all skeletal territories (Fig. 6), and *Afi-p19* and *Afi-p58b*, which are localized preferentially in the vertebrae in late regenerates. Certainly different positional cues must be required for different skeletal elements. While it is conceivable that the epidermis acts as a signaling center for the underlying dermal layer of skeletogenic cells, as the ectoderm provides essential positional cues in the sea urchin embryos [50], this is unlikely to be the case for the skeletogenic cells forming the vertebrae. It would be interesting to test what potential signaling pathways might be involved in the formation of individual skeletal ossicles during brittle star arm regeneration. The VEGF and FGF signaling pathways have both been shown to play important roles in the development of embryonic skeleton in sea urchins [50, 51, 52], as well as in vertebrates [53, 54], such as directing skeletogenic mesoderm cell migration. The expression of the ligands and receptors of these pathways have additionally been characterized in brittle star embryos and sea star juveniles, and are consistent with the potential role of these pathway components in skeletogenesis also in other echinoderm phyla [55]. It would thus be compelling to investigate their role in the developing skeletal elements in the context of *A. filiformis* adult arm regeneration, as it could potentially reveal similar molecular components underlying skeletogenesis among deuterostomes despite the fact that skeleton of vertebrates and ambulacrarians is considered to have evolved independently.

Skeletal regeneration is observed in other deuterostome groups: for example in cirri regeneration of amphioxus [56], and in appendage regeneration of different vertebrates (reviewed in [6]). It has even been suggested that adult bone repair and regeneration may recapitulate embryonic bone development at a molecular level [57]. Although the skeleton of echinoderms is composed of calcium carbonate, instead of calcium phosphate, similarities of its ontogeny can be observed when compared to vertebrates. For example, in both groups of animals the trunk skeletal precursor cells are mesoderm-derived, motile mesenchymal cells [44, 47, 58] suggesting conserved features of skeletogenesis in deuterostomes. In sea urchins and brittle stars the larval skeleton is formed by a group of mesenchymal cells (skeletogenic mesoderm, for *A. filiformis*), which ingress into the blastocoel prior to gastrulation, then migrate and secrete the molecules (proteins, inorganic matrix, etc.), which form the biomineralized spicules [36, 44, 47]. In regenerating adult arms of *A. filiformis*, we observed that the first skeletal elements are deposited within the dermal layer of the early regenerate, coinciding with the appearance of a pool of mesenchymal cells localized in this domain (Figs. 1, 2 and 3). Cell tracing experiments would be required to conclude that these mesenchymal cells are directly responsible for the deposition of the biomineralized skeleton. On the other hand, the molecular signature of the mesenchymal cells in the dermis (determined by examining the expression of *Afi-c-lectin, Afi-p19* and *Afi-p58b)* provides additional evidence to support their role in skeleton development in light of the extensive knowledge of echinoderm skeletogenesis [11, 12, 14, 16, 25, 46]. All three genes are homologs of well-characterized biomineralization genes found in sea urchin larvae [45] and adults [17, 18] and expression of the two latter genes in the skeletogenic mesoderm of *A. filiformis* embryos was recently shown [36]. Their highly restricted expression patterns differ significantly from the expression of previously described mesodermal transcription factors such as *Afi-alx1, Afi-ets1/2* or *Afi-gataC* that, although also absent from the epidermis, occupy a broader domain extending from the dermal layer into the remaining sub-epidermal tissue layers [37]. Late in regeneration, the *Afi-c-lectin* expression pattern highly resembles the expression of *Afi- coll* [37], a homolog of the vertebrate *col2A1*, a widely conserved skeletogenic collagen gene [59, 60].

We show that the regeneration of the brittle star arm is associated with a high degree of cell proliferation. Importantly, we show that *Afi-c-lectin* expressing cells are not labeled with EdU at any time during regeneration; however, their number increases during regeneration, as does the number of skeletal structures. This suggests that there must be a constant supply of *Afi-c-lectin* expressing cells, likely constituting the skeletogenic cell population, since they themselves have no proliferative capacity. For example, a small pool of local progenitor cells that proliferate and give rise to daughter cells, which then molecularly and morphologically differentiate, could maintain the cell population. The newly formed cells could also be supplied from the adjacent coelomic epithelium, which shows high levels of cell proliferation (Fig. 7) and also expresses genes like *Afi-alx1* and *Afi-ets1/2* [37]. Alternatively, the radial water canal has also been implicated as a major source of cells in the regenerative bud [27] and, as mesenchymal cells, skeletogenic cells could migrate from there into their final dermal location. It is possible that epidermal signals could be involved in guiding the cells into the appropriate position and cause them to differentiate and stop proliferating, similar to what has been shown in the sea urchin embryo [50].

Many questions remain to be answered concerning the origin of the skeletogenic cells in the regenerating arm and the signaling cues required for the correct positioning and shaping of the spicules, their extension, and formation of highly complex and divergent skeletal elements. Our work provides the basis for the further study of this crucial developmental process.

CONCLUSIONS

Here, we describe for the first time the developmental process of skeletal regeneration in a brittle star. Using different approaches we found that the cells expressing skeletal markers are localized in the dermal layer of the regenerating arm, where the deposition of the biomineralized skeletal spicules occurs during the regenerative process. During late regeneration stages these cells are arranged in a tightly controlled pattern that follows the shape of five skeletal elements of the adult arm, namely the oral, aboral and lateral shields, spines and vertebrae. Furthermore, no cell proliferation has been detected in the cell layer where the skeleton is synthesized. We thus conclude that the skeletogenic cells are likely to come from another source, which constantly supplies new cells to build the complex skeletal elements. Elucidating the cellular and molecular mechanisms of skeleton regeneration in brittle stars, including gene regulation involved in skeletogenic cell specification, and the differentiation, patterning, and origin of these cells in the regenerate, will greatly enhance our understanding of the mechanisms responsible for the extensive regenerative potential of echinoderms.

METHODS

Animal Handling, Sample Preparation and Fixation

Adult *A. filiformis* were collected at the Sven Lovén Centre for Marine Sciences in Kristineberg, Sweden, and transported to London where they were kept in flow-through tanks with filtered artificial seawater (ASW; Instant Ocean, Aquarium Systems, 30‰ salinity) at 14 °C. After a period of acclimatization of at least one week the animals (disc sizes between 0.5 and 1 cm and arm lengths up to ten times the length of the body) were anaesthetized in 3.5 % $MgCl_2(6H_2O)$ solution. The arms were then amputated (maximum two per animal) 1 cm from the disc and left to regenerate until they reached the desired stage (1–8 days for early stages and approximately 2 weeks for the 50 % DI stage). The regenerating arms were collected leaving 2–3 non-regenerating segments and then fixed in Bouin's fixative at 4 °C for at least 2 weeks for histological sections or in 4 % PFA in 1x PBT (Phosphate buffer saline, 0.1 % Tween-20) overnight at 4 °C for *in situ* hybridization experiments. To show the individual adult skeletal ossicles the non-regenerating arms of the brittle star were amputated as described before and placed in 1 % bleach for approximately 20 min to degrade the tissue prior to imaging.

Calcein Staining

Live animals were incubated with calcein (Sigma) at a dilution of 1:50 of a stock solution (1.25 mg/ml) in artificial seawater to label the newly deposited calcium carbonate and thus visualize the biomineralized structures during the early stages of arm regeneration. Calcein was replenished every day together with adding fresh ASW throughout the duration of the observation period. The animals were washed in filtered seawater several times and anaesthetized in the magnesium chloride solution prior to imaging.

Sectioning of Paraffin-Embedded Tissue

PFA-fixed samples for thick sections were first de-calcified in 0.5 M EDTA for 1-3 days at 4 °C and then washed several times with PBT. The Bouin's (VWR) fixed samples were washed several times in deionized water. Both types of samples were dehydrated in a series of increasing ethanol concentrated washes (30, 50, 70 %, 2× 100 %), washed twice (once at room temperature and once at 60 °C) with Histo-Clear (Fischer-Scientific) and then 3 times in paraffin wax at 60 °C (Thermo Scientific) before finally orientating them during the embedding. The wax was allowed to cool down and solidify overnight at room temperature. Samples were then mounted onto wooden blocks and sectioned at a thickness of 5–10 μm using a Leica RM2155 microtome, floated on slides with distilled water at 42°C and then left to dry overnight at 37 °C before further processing. The sections were de-waxed using Histo-Clear and processed for ISH or histology.

Histology

Milligan's trichrome technique was employed for the histological staining of Bouin's fixed sections with modifications in the timing of staining of acid fuchsin (3 min) and fast green (8 min) [43]. All sections were de-waxed, rehydrated, stained and mounted with coverslip using Histomount (National Diagnostics) before imaging. Each stage.

In situ Hybridization

Probe preparation and whole mount *in situ* hybridization were performed as described before [22]. The *Afi-c-lectin* sequence can be found in GenBank under the accession number KT936152. At least three arms from different animals were used per experiment. Fluorescent *in situ* hybridization was performed on sections on microscopy slides using the protocol described in [36] with the modification of hybridization temperature (50°C). DAPI (Sigma) was added at a dilution of 1:5000 (of stock solution 5mg/ml) for 15 min and then washed out once with buffer just before microscopy.

Cell Proliferation Assay

The cell proliferation assay was carried out using the Click-iT® EdU Alexa Fluor® 555HCS Assay (Life Technologies), which uses a modified thymidine analogue EdU that gets incorporated into newly-synthesized DNA and can be detected by fluorescent dyes using the highly selective copper-catalyzed azide-alkyne cycloaddition chemistry (click chemistry; [61]), without the need for antibody detection or harsh permeabilization treatments. Animals with non-regenerating arms and with arms at early regeneration stages were incubated in 5μM EdU in ASW for 2h, then the regenerating arms were amputated and fixed for 1h in 4 % PFA in PBT at room temperature. The arms were then washed with 1x PBT several times and permeabilized for 1h in 1x PBS with 0.1 % Triton X-100. After two more PBT washes 100μl of the reaction cocktail were added to the samples for 30 min (made according to manufacturer's instructions and using kit reagents). The solution was then removed and the samples were washed for 30 min in Click-IT reaction rinse buffer. The buffer was then also removed followed by two PBT washes and finally DAPI was added as described above.

Light and Confocal Microscopy

For differential interference contrast (DIC) images as well as epi-fluorescent images the Zeiss AxioImager M1 microscope was used together with a Zeiss AxioCamHRc camera. For confocal images of fluorescently labeled samples the Leica TCS SP2 confocal laser scanning microscope was used and the LAS-AF software implemented to capture the image stacks. Between 30 (for stage 2) and 109 (stage 3) confocal z-planes were taken at 1µm thickness for each confocal stack at early stages, and 90 z-planes at 1µm thickness were used for the 50 % regenerated arm samples. Images were processed using Fiji and Adobe Photoshop CS4.

ACKNOWLEDGEMENTS

We would like to thank the staff at the Sven Lovén Centre for Marine Sciences in Kristineberg, especially Olga Ortega-Martinez and Sam Dupont, for assistance during animal and sample collection. We thank Anoop Kumar for assistance with sectioning and histology. The authors also want to thank the two anonymous reviewers as well as Helen Robertson, Johannes Girstmair and Patrick Toolan-Kerr for constructive comments on the manuscript. This work was partly funded by the KVA fund SL2015-0048 from the Royal Swedish Academy of Science and the EU FP7 Research Infrastructure Initiative ASSEMBLE (ref. 227799). AC was funded by a Wellcome Trust PhD studentship. DVD was funded by a Systems Biology UCL studentship. CF was funded by an Erasmus Placement fellowship. MS was funded by a Young Researcher Grant of the University of Milan.

AUTHORS' CONTRIBUTIONS

AC and PO conceived the study and wrote the manuscript. AC carried out the experiments. DVD prepared the *Afi-c-lectin, Afi-p19 and Afi-p58b* probes, analyzed the embryonic expression of *Afi-c-lectin*, and optimized the fluorescent *in situ* hybridization protocol. AC and CF performed the histological sections and Milligan's trichrome staining. AC, CF, MS and PO analyzed the data. All authors read and approved the final manuscript.

REFERENCES

1. Brockes JP, Kumar A. Comparative aspects of animal regeneration. Annu Rev Cell Dev Biol. 2008;24:525–49.
2. Technau U, Cramer von Laue C, Rentzsch F, Luft S, Hobmayer B, Bode HR, Holstein TW. Parameters of self-organization in *Hydra* aggregates. Proc Natl Acad Sci U S A. 2000;97:12127–31.
3. Gierer A, Berking S, Bode H, David CN, Flick K, Hansmann G, Schaller H, Trenkner E. Regeneration of *Hydra* from reaggregated cells. Nat New Biol. 1972;239:98–101.
4. Agata K, Tanaka T, Kobayashi C, Kato K, Saitoh Y. Intercalary regeneration in planarians. Dev Dyn. 2003;226:308–16.
5. Montgomery JR, Coward SJ. On the minimal size of a planarian capable of regeneration. Trans Am Microsc Soc. 1974;93:386–91.
6. Ferretti P, Health C. Regeneration of vertebrate appendages, eLS. Chichester: Wiley; 2013. p. 1–7.
7. Candia Carnevali MD. Regeneration in Echinoderms: repair, regrowth, cloning. Invertebr Surviv J. 2006;3:64–76.
8. Brusca RC, Moore W, Shuster SM. Invertebrates. 3rd ed. Sunderland: Sinauer Associates Inc.; 2016.
9. Wilt FH, Killian CE, Livingston BT. Development of calcareous skeletal elements in invertebrates. Differentiation. 2003;71:237–50.
10. Oliveri P, Tu Q, Davidson EH. Global regulatory logic for specification of an embryonic cell lineage. Proc Natl Acad Sci U S A. 2008;105:5955–62.
11. Wilt FH, Ettensohn CA. The Morphogenesis and Biomineralization of the Sea Urchin Larval Skeleton. Handb Biominer Biol Asp Struct Form. 2008;1:182–210.
12. Rafiq K, Shashikant T, McManus CJ, Ettensohn CA. Genome-wide analysis of the skeletogenic gene regulatory network of sea urchins. Development. 2014;141:950–61.
13. Yajima M, Kiyomoto M. Study of larval and adult skeletogenic cells in developing sea urchin larvae. Biol Bull. 2006;211:183–92.
14. Gao F, Davidson EH. Transfer of a large gene regulatory apparatus to a new developmental address in echinoid evolution. Proc Natl Acad Sci U S A. 2008;105:6091–6.
15. Gao F, Thompson JR, Petsios E, Erkenbrack E, Moats RA, Bottjer DJ, Davidson EH. Juvenile skeletogenesis in anciently diverged sea urchin clades. Dev Biol. 2015;400:148–58.
16. Mann K, Wilt FH, Poustka AJ. Proteomic analysis of sea urchin (*Strongylocentrotus purpuratus*) spicule matrix. Proteome Sci. 2010;8:33.
17. Mann K, Poustka AJ, Mann M. In-depth, high-accuracy proteomics of sea urchin tooth organic matrix. Proteome Sci. 2008;6:33.
18. Mann K, Poustka AJ, Mann M. The sea urchin (*Strongylocentrotus purpuratus*) test and spine proteomes. Proteome Sci. 2008;6:22.
19. Chadwick HC. Regeneration of spines in *Echinus esculentus*. Nature. 1929;124:760–1.
20. Gorzelak P, Stolarski J, Dubois P, Kopp C, Meibom A. 26Mg labeling of the sea urchin regenerating spine: Insights into echinoderm biomineralization process. J Struct Biol. 2011;176:119–26.
21. Heatfield BM, Travis DF. Ultrastructural studies of regenerating spines of the sea urchin *Strongylocentrotus purpuratus*. II. Cell types with spherules. J Morphol. 1975;145:51–71.
22. Reinardy HC, Emerson CE, Manley JM, Bodnar AG. Tissue regeneration and biomineralization in Sea Urchins: role of notch signaling and presence of stem cell markers. PLoS One. 2015;10, e0133860.
23. Mashanov VS, Zueva OR, Heinzeller T. Regeneration of the radial nerve cord in a holothurian: a promising new model system for studying post-traumatic recovery in the adult nervous system. Tissue Cell. 2008;40:351–72.

24. Mashanov VS, Zueva OR, Garcia-Arraras JE. Organization of glial cells in the adult sea cucumber central nervous system. Glia. 2010;58:1581–93.

25. Ben Khadra Y, Ferrario C, Di Benedetto C, Said K, Bonasoro F, Candia Carnevali MD, Sugni M. Re-growth, morphogenesis, and differentiation during starfish arm regeneration. Wound Repair Regen. 2015;23:623–34.

26. Kondo M, Akasaka K. Regeneration in crinoids. Dev Growth Differ. 2010;52:57–68.

27. Biressi ACM, Zou T, Dupont S, Dahlberg C, Di Benedetto C, Bonasoro F, Thorndyke M, Carnevali MDC. Wound healing and arm regeneration in *Ophioderma longicaudum* and *Amphiura filiformis* (Ophiuroidea, Echinodermata): comparative morphogenesis and histogenesis. Zoomorphology. 2010;129:1–19.

28. Stöhr S, O'Hara TD, Thuy B. Global diversity of brittle stars (Echinodermata: Ophiuroidea). PLoS One. 2012. doi:10.1371/journal.pone.0031940.

29. O'Hara T, Hugall A, Thuy B, Moussalli A. Phylogenomic resolution of the class Ophiuroidea unlocks a global microfossil record. Curr Biol. 2014;24:1874–9.

30. Byrne M. Ophiuroidea, Microscopic anatomy of invertebrates. 1994. p. 247–343.

31. Dupont S, Thorndyke MC. Growth or differentiation? Adaptive regeneration in the brittlestar *Amphiura filiformis*. J Exp Biol. 2006;209:3873–81.

32. Hu MY, Casties I, Stumpp M, Ortega-Martinez O, Dupont S. Energy metabolism and regeneration are impaired by seawater acidification in the infaunal brittlestar *Amphiura filiformis*. J Exp Biol. 2014;217:2411–21.

33. Burns G, Ortega-Martinez O, Thorndyke MC, Peck LS, Dupont S, Clark MS. Dynamic gene expression profiles during arm regeneration in the brittle star *Amphiura filiformis*. J Exp Mar Bio Ecol. 2011;407:315–22.

34. Purushothaman S, Saxena S, Meghah V, Swamy CVB, Ortega-Martinez O, Dupont S, et al. Transcriptomic and proteomic analyses of *Amphiura filiformis* arm tissue-undergoing regeneration. J Proteomics. 2014;112:1–12.

35. McCauley BS, Wright EP, Exner C, Kitazawa C, Hinman VF. Development of an embryonic skeletogenic mesenchyme lineage in a sea cucumber reveals the trajectory of change for the evolution of novel structures in echinoderms. Evodevo. 2012;3:17.

36. Dylus DV, Czarkwiani A, Stångberg J, Ortega-Martinez O, Dupont S, Oliveri P. Large-scale gene expression study in the ophiuroid *Amphiura filiformis* provides insights into evolution of gene regulatory networks. Evodevo. 2016;7:2.

37. Czarkwiani A, Dylus DV, Oliveri P. Expression of skeletogenic genes during arm regeneration in the brittle star *Amphiura filiformis*. Gene Expr Patterns. 2013;13:464–72.

38. ten Berge D, Brouwer A, el Bahi S, Guénet JL, Robert B, Meijlink F. Mouse Alx3: an aristaless-like homeobox gene expressed during embryogenesis in ectomesenchyme and lateral plate mesoderm. Dev Biol. 1998;199:11–25.

39. Brouwer A, ten Berge D, Wiegerinck R, Meijlink F. The OAR/aristaless domain of the homeodomain protein Cart1 has an attentuating role in vivo. Mech Dev. 2003;120:241–52.

40. Zhao Q, Behringer RR, de Crombrugghe B. Prenatal folic acid treatment suppresses acrania and meroanencephaly in mice mutant for Cart1 homeobox gene. Nat Genet. 1996;13:275–83.

41. Young HE, Bailey CF, Dalley BK. Gross morphological analysis of limb regeneration in postmetamorphic adult*Ambystoma*. Anat Rec. 1983;206:295–306.

42. Hamburger V, Hamilton H. A series of normal stages in the development of the chick embryo. Dev Dyn. 1992;88:231–72.

43. Milligan M. Trichrome stain for formalin-fixed tissue. Am J Clin Pathol. 1946;10:184.

44. Guss KA, Ettensohn CA. Skeletal morphogenesis in the sea urchin embryo: regulation of primary mesenchyme gene expression and skeletal rod growth by ectoderm-derived cues. Development. 1997;124:1899–908.

45. Illies MR, Peeler MT, Dechtiaruk AM, Ettensohn CA. Identification and developmental expression of new biomineralization proteins in the sea urchin *Strongylocentrotus purpuratus*. Dev Genes Evol. 2002;212:419–31.

46. Seaver RW, Livingston BT. Examination of the skeletal proteome of the brittle star *Ophiocoma wendtii* reveals overall conservation of proteins but variation in spicule matrix proteins. Proteome Sci. 2015;13:1–12.

47. Okazaki K. Spicule formation by isolated micromeres of the sea urchin embryo. Integr Comp Biol. 1975;15:567–81.

48. Irimura S, Fujita T. Interspecific variation of vertebral ossicle morphology in the Ophiuroidea, Echinoderm Research. Banyuls-sur-mer: Proceedings 6th European Conference Echinoderm Research; 2003. p. 161–7.

49. Gage JD. Skeletal growth bands in brittle stars: microstructure and significance as age markers. J Mar biol Ass UK. 1990;70:209–24.

50. Duloquin L, Lhomond G, Gache C. Localized VEGF signaling from ectoderm to mesenchyme cells controls morphogenesis of the sea urchin embryo skeleton. Development. 2007;134:2293–302.

51. Röttinger E, Saudemont A, Duboc V, Besnardeau L, McClay D, Lepage T. FGF signals guide migration of mesenchymal cells, control skeletal morphogenesis and regulate gastrulation during sea urchin development. Development. 2008;135:353–65.

52. Adomako-Ankomah A, Ettensohn CA. Growth factor-mediated mesodermal cell guidance and skeletogenesis during sea urchin gastrulation. Development. 2013;140:4214–25.

53. Yu K, Ornitz DM. FGF signaling regulates mesenchymal differentiation and skeletal patterning along the limb bud proximodistal axis. Development. 2008;135:483–91.

54. Deckers MM, Karperien M, van der Bent C, Yamashita T, Papapoulos SE, Löwik CW. Expression of vascular endothelial growth factors and their receptors

during osteoblast differentiation. Endocrinology. 2000;141:1667–74.

55. Morino Y, Koga H, Tachibana K, Shoguchi E, Kiyomoto M, Wada H. Heterochronic activation of VEGF signaling and the evolution of the skeleton in echinoderm pluteus larvae. Evol Dev. 2012;14:428–36.

56. Kaneto S, Wada H. Regeneration of amphioxus oral cirri and its skeletal rods: implications for the origin of the vertebrate skeleton. J Exp Zool Part B Mol Dev Evol. 2011;316B:409–17.

57. Ferguson C, Alpern E, Miclau T, Helms JA. Does adult fracture repair recapitulate embryonic skeletal formation? Mech Dev. 1999;87:57–66.

58. Yang Y. Skeletal morphogenesis and embryonic development. In: Rosen CJ, editor. Primer on the metabolic bone diseases and disorders of mineral metabolism. Eighthth ed. Ames: Wiley; 2013. p. 1–14.

59. Ng LJ, Wheatley S, Muscat GE, Conway-Campbell J, Bowles J, Wright E, Bell DM, Tam PP, Cheah KS, Koopman P. SOX9 binds DNA, activates transcription, and coexpresses with type II collagen during chondrogenesis in the mouse. Dev Biol. 1997;183:108–21.

60. Metsaranta M, Toman D, De CB, Vuorios E. Mouse type 1 collagen gene. J Biol Chem. 1991;266(25):16862–9.

61. Meldal M, Tornoe CW. Cu-catalyzed azide - Alkyne cycloaddition. Chem Rev. 2008;108:2952–3015.

INDEX